BIM in the Construction Industry

BIM in the Construction Industry

Special Issue Editors

Hee Sung Cha
Shaohua Jiang

MDPI • Basel • Beijing • Wuhan • Barcelona • Belgrade • Manchester • Tokyo • Cluj • Tianjin

Special Issue Editors

Hee Sung Cha
Department of Architectural
Engieering, Ajou University
Korea

Shaohua Jiang
Department of Construction
Management, Faculty of
Infrastructure Engineering,
Dalian University of Technology
China

Editorial Office
MDPI
St. Alban-Anlage 66
4052 Basel, Switzerland

This is a reprint of articles from the Special Issue published online in the open access journal *Applied Sciences* (ISSN 2076-3417) (available at: https://www.mdpi.com/journal/applsci/special_issues/BIM_in_Construction_Industry).

For citation purposes, cite each article independently as indicated on the article page online and as indicated below:

LastName, A.A.; LastName, B.B.; LastName, C.C. Article Title. *Journal Name* **Year**, *Article Number*, Page Range.

ISBN 978-3-03936-573-9 (Hbk)
ISBN 978-3-03936-574-6 (PDF)

Contents

About the Special Issue Editors

Hee Sung Cha (Ph.D.) is a full professor at the Department of Architectural Engineering, Ajou University, Suwon, South Korea. He received his M.S. from Seoul National University and his Ph.D. in civil engineering from the University of Texas at Austin. He is currently the Program Director of Architectural Engineering at Ajou University. He is a scientific reviewer for many different associations and more than 10 international journals indexed in Journal Citation Reports (JCR); he has authored more than 100 journal and conference papers, including more than 18 published in SCI-indexed journals. He has been supervisor of more than 20 M.S. and Ph.D. students, a member of editorial boards of scientific JCR journals, and a member of the scientific committees in international conferences. His research activities include innovative building technology, project management in construction, project productivity/performance improvement, construction economics and project financing, building information modeling (BIM) applied to construction, and safety control in facility management.

Shaohua Jiang is an associate professor with the Department of Construction Management, Faculty of Infrastructure Engineering at Dalian University of Technology, Dalian, China. He received his M.S. and Ph.D. in civil engineering and in management science and engineering, respectively, from Dalian University of Technology. He is currently the head of the Department of Construction Management at Dalian University of Technology. He serves as a scientific reviewer for several of the world's top journals related to construction management and has authored more than 70 journal and conference papers. He has been the supervisor of more than 40 M.S. and Ph.D. students. He is an editorial advisory board member of the *Journal of Construction Innovation - Information, Process, Management*. His main research interests include digital construction and management, construction project information and knowledge management, building information modeling (BIM) applied to construction, and infrastructure sustainability.

Preface to "BIM in the Construction Industry"

In recent decades, the term building information modeling (BIM) has been mentioned in a wide range of construction research endeavors. BIM is a new solution for unprecedented recession in the construction industry, i.e., productivity loss, labor shortage, cost overrun, and severe competitiveness. BIM technology provides many benefits: prompt design clash detection, automatic deign regulatory check algorithm, augmented/virtual reality visualization, and collaboration work environment. BIM experts as well as industry practitioners are stressing the importance of BIM applications in the field of construction. Given the rapid development and adoption of BIM in the architecture, engineering, and construction (AEC) industry, new trends relevant to the research of BIM are emerging, being exceedingly helpful not only for academics but also for practitioners. These new trends include open BIM research to address the issue of information interoperability between different types of BIM software, and BIM-supported project lifecycle management, such as safety management and quality management, which are receiving increasing attention.

Hee Sung Cha, Shaohua Jiang
Special Issue Editors

Editorial

Special Issue on BIM in the Construction Industry

Hee Sung Cha [1],* and Shaohua Jiang [2]

[1] Department of Architectural Engineering, Ajou University, Suwon 16499, Korea
[2] Department of Construction Management, Dalian University of Technology, Dalian 116024, China; shjiang@dlut.edu.cn
* Correspondence: hscha@ajou.ac.kr

Received: 25 May 2020; Accepted: 10 June 2020; Published: 23 June 2020

1. Introduction

In recent decades, the term Building Information Modeling (BIM) has been mentioned in a wide range of construction research endeavors. BIM is considered to be a new solution for ever-challenging problems in the construction industry, i.e., productivity loss, labor shortage, cost overrun, and severe competitiveness.

Clearly, BIM technology brings quite a few benefits: prompt design clash detection, automatic design regulatory check algorithm, augmented/virtual reality visualization, and collaboration work environment are good examples. Not only the experts of BIM, but also industry practitioners, are stressing the importance of BIM applications in the field of construction.

On account of the rapid development and adoption of BIM in the AEC (architecture, engineering, and construction) industry, new trends relevant to the research of BIM are emerging, which are exceedingly helpful not only for academics but also for practitioners. These new trends include "open BIM" to address the issue of information interoperability between different types of BIM software, and BIM-supported project lifecycle management such as safety management and quality management, which are receiving increasing attention.

2. BIM in the Construction Industry

In view of the above, this Special Issue aimed to cover recent advances in the development of BIM technologies and applications of BIM technologies in the fields of architecture, engineering, construction, and facility management, as well as relevant theoretical research endeavors in this area. There were 60 papers submitted to this Special Issue, and 19 papers were accepted (i.e., 32% acceptance rate). The paper, authored by C. P. Schimanski, C. Marcher, G. P. Monizza and D. T. Matt, presents a review to investigate the relevant literature regarding integrations of Building Information Modeling technology and the Lean Construction's Last Planner System (LPS) in the construction execution phase. It seeks to help scientists and practitioners to obtain an overview of the state-of-the-art in terms of synergies between both approaches, and devises a conceptual model for integration based on the review's findings [1]. There is another review of openBIM, authored by S. Jiang, L. Jiang, Y. Han, Z. Wu and N. Wang. In the paper, the openBIM related standards, software platforms, and tools enabling information interoperability are introduced and analyzed comprehensively based on related websites and literature. Moreover, the research on engineering information interoperability supported by openBIM was comprehensively analyzed [2].

When reviewing innovation issues, many papers combine BIM with computer technology to improve work efficiency in the construction industry. The first paper by X. Y. Deng, H. H. Lai, J. Y. Xu and Y. F. Zhao uses XML schema to develop a generic language, by which the partial model can be extracted from an Industry Foundation Classes (IFC) model based on the proposed selection set [3]. The second paper proposes a method by combining a rule-based reasoning technique and a supervised machine learning technique, which can automatically screen for irrelevant clashes and distinguish them from lots of design clashes discovered by BIM software [4]. The third one adopts BIM and the particle swarm optimization algorithm to build an intelligent optimal design search system, which can save large amounts of life cycle energy and costs [5]. Y. Q. Xiao, S. W. Li and Z. Z. Hu introduce an approach to generate the logic chains of Mechanical/Electrical/Plumbing (MEP) systems using building information models semi-automatically and substantiate its accuracy with a real-world project [6]. The final paper presents an innovative approach, which combines mvdXML and semantic technologies to generate green construction knowledge. It can be used to improve the efficiency of green construction code checking [7].

There are four papers focused on BIM-based fire safety in the construction or operation stages. The first one, by K. Kim and Y.C. Lee, proposes a framework to automatically analyze, generate, and visualize the evacuation paths of multiple crews in 4D BIM, considering construction activities and site conditions at the specific project schedule [8]. The second one, authored by D. Zhang, J. Zhang, H. Xiong, Z. Cui and D. Lu, proposes a crowdsourcing application, iInspect, which can be used to recruit members of the general public to carry out fire safety inspection tasks with the assistance of BIM + VR and an indoor real-time location system [9]. The above two articles are applied in the construction stage, and the last two articles are focused on fire safety management in the operation stage. The first one, authored by S. Jung, H. S. Cha and S. Jiang, presents a prototype system for a building's fire information management using three-dimensional visualization, by deriving the relevant information required for mitigating building fire disasters. This system automatically provides reliable fire-related information through a computerized and systematic approach in conjunction with a BIM tool [10]. The other paper, authored by N. Khan, A. K. Ali, S. V. T. Tran, D. Lee and C. Park, focuses on a visual language approach for rule translation and a multi-agent-based construction fire safety planning simulation in BIM [11].

As one of the most important parts of project management, quality management has always been the focus of research. In this Special Issue, three papers put the focus on this field. The first one, authored by M. Mirshokraei, C. I. D. Gaetani and F. Migliaccio, investigates quality management during the execution phase of structural elements by proposing, developing, and testing a complete framework by integrating BIM and augmented reality technology [12]. The second paper, by M. Hamooni, M. Maghrebi, J. M. Sardroud and S. Kim presents a novel method of monitoring the maturity of concrete and providing reduced formwork removal time with the strength ensured in real-time [13]. The last one, authored by Y. Zhao, X. Deng, and H. Lai proposes a deep learning-based method [14].

There are three papers focused on BIM application. With the evolution of Industry 4.0, the construction industry has introduced a lot of digital technologies. The first paper, authored by R. Maskuriy, A. Selamat, K. N. Ali, P. Maresova and O. Krejcar, conducts a bibliometric mapping study to discuss the current trend of Industry 4.0, to identify its key areas and to provide suggestions for future research directions in the construction industry [15]. The second one analyzes the behavior and performance of BIM users with different specialties during BIM collaborative work, based on the data generated by a BIM software in its log files [16]. Previous studies on the effects of BIM are conducted for large-scale projects but ignore the small and medium projects. So, the last paper about BIM application authored by M. Yoo, J. Kim and C. Choi, proposes a BIM-based construction of a prefabricated steel framework from the perspective of SMEs, which verifies the major functions that will be installed in the system for the potential SME users and their perceived performance [17].

Moreover, the critical limiting factors to building information modeling implementation have attracted the attention of researchers in the case of the vigorous promotion of BIM applications in many regions. L. H. Liao, E. A. L. Teo and R. D. Chang identify critical factors hindering BIM implementation in Singapore's construction industry, analyze their interrelationships, and propose strategies for reducing these barriers through a survey of 87 experts and five post-survey interviews [18]. S. H. Hong, S. K. Lee, I. H. Kim and J. H. Yu analyze the factors that affect the acceptance of mobile BIM by construction practitioners and present the association of factors as a model to activate mobile BIM use [19].

3. Future Development of BIM Technology

BIM technologies in the AEC industry are still developing. Although it is taken for granted that computer and information technology has played a crucial role in the advent of BIM technology, the fruits of these technologies cannot be exploited without the proper applications. Making technology more adaptable to the development of industry will be the focus of research for a long time in the future.

Funding: This research received no external funding.

Conflicts of Interest: The authors declare no conflict of interest.

References

1. Schimanski, C.P.; Marcher, C.; Monizza, G.P.; Matt, D.T. The Last Planner® System and Building Information Modeling in Construction Execution: From an Integrative Review to a Conceptual Model for Integration. *Appl. Sci.* **2020**, *10*, 821. [CrossRef]
2. Jiang, S.; Jiang, L.; Han, Y.; Wu, Z.; Wang, N. OpenBIM: An Enabling Solution for Information Interoperability. *Appl. Sci.* **2019**, *9*, 5358. [CrossRef]
3. Deng, X.; Lai, H.; Xu, J.; Zhao, Y. Generic Language for Partial Model Extraction from an IFC Model Based on Selection Set. *Appl. Sci.* **2020**, *10*, 1968. [CrossRef]
4. Lin, Y.W.; Huang, Y. Filtering of Irrelevant Clashes Detected by BIM Software Using a Hybrid Method of Rule-Based Reasoning and Supervised Machine Learning. *Appl. Sci.* **2019**, *9*, 5324. [CrossRef]
5. Liu, S.; Ning, X. A Two-Stage Building Information Modeling Based Building Design Method to Improve Lighting Environment and Increase Energy Efficiency. *Appl. Sci.* **2019**, *9*, 4076. [CrossRef]
6. Xiao, Y.; Li, S.; Hu, Z. Automatically Generating a MEP Logic Chain from Building Information Models with Identification Rules. *Appl. Sci.* **2019**, *9*, 2204. [CrossRef]
7. Jiang, S.; Wu, Z.; Zhang, B.; Cha, S.H. Combined MvdXML and Semantic Technologies for Green Construction Code Checking. *Appl. Sci.* **2019**, *9*, 1463. [CrossRef]
8. Kim, K.; Lee, Y. Automated Generation of Daily Evacuation Paths in 4D BIM. *Appl. Sci.* **2019**, *9*, 1789. [CrossRef]
9. Zhang, D.; Zhang, J.; Xiong, H.; Cui, Z.; Lu, D. Taking Advantage of Collective Intelligence and BIM-Based Virtual Reality in Fire Safety Inspection for Commercial and Public Buildings. *Appl. Sci.* **2019**, *9*, 5068. [CrossRef]
10. Jung, S.; Cha, H.S.; Jiang, S. Developing a Building Fire Information Management System Based on 3D Object Visualization. *Appl. Sci.* **2020**, *10*, 772. [CrossRef]
11. Khan, N.; Ali, K.A.; Van-Tien Tran, S.; Lee, D.; Park, C. Visual Language-Aided Construction Fire Safety Planning Approach in Building Information Modeling. *Appl. Sci.* **2020**, *10*, 1704. [CrossRef]
12. Mirshokraei, M.; De Gaetani, I.C.; Migliaccio, F. A Web-Based BIM–AR Quality Management System for Structural Elements. *Appl. Sci.* **2020**, *10*, 3984. [CrossRef]
13. Hamooni, M.; Maghrebi, M.; Majrouhi Sardroud, J.; Kim, S. Extending BIM Interoperability for Real-Time Concrete Formwork Process Monitoring. *Appl. Sci.* **2020**, *10*, 1085. [CrossRef]
14. Zhao, Y.; Deng, X.; Lai, H. A Deep Learning-Based Method to Detect Components from Scanned Structural Drawings for Reconstructing 3D Models. *Appl. Sci.* **2020**, *10*, 2066. [CrossRef]
15. Maskuriy, R.; Selamat, A.; Ali, N.K.; Maresova, P.; Krejcar, O. Industry 4.0 for the Construction Industry—How Ready Is the Industry? *Appl. Sci.* **2019**, *9*, 2819. [CrossRef]
16. Forcael, E.; Martínez-Rocamora, A.; Sepúlveda-Morales, J.; García-Alvarado, R.; Nope-Bernal, A.; Leighton, F. Behavior and Performance of BIM Users in a Collaborative Work Environment. *Appl. Sci.* **2020**, *10*, 2199. [CrossRef]
17. Yoo, M.; Kim, J.; Choi, C. Effects of BIM-Based Construction of Prefabricated Steel Framework from the Perspective of SMEs. *Appl. Sci.* **2019**, *9*, 1732. [CrossRef]
18. Liao, L.; Teo, A.E.; Chang, R. Reducing Critical Hindrances to Building Information Modeling Implementation: The Case of the Singapore Construction Industry. *Appl. Sci.* **2019**, *9*, 3833. [CrossRef]
19. Hong, S.; Lee, S.; Kim, I.; Yu, J. Acceptance Model for Mobile Building Information Modeling (BIM). *Appl. Sci.* **2019**, *9*, 3668. [CrossRef]

Review

The Last Planner® System and Building Information Modeling in Construction Execution: From an Integrative Review to a Conceptual Model for Integration

Christoph Paul Schimanski [1,2,*], Carmen Marcher [1,2], Gabriele Pasetti Monizza [2] and Dominik T. Matt [1,2,*]

[1] Free University of Bozen-Bolzano, Faculty of Science and Technology, Piazza Università 5, 39100 Bolzano, Italy; carmen.marcher@natec.unibz.it (C.M.); gabriele.pasettimonizza@fraunhofer.it (G.P.M.)

[2] Fraunhofer Italia Research, Via A.-Volta 13A, 39100 Bolzano, Italy

[*] Correspondence: christophpaul.schimanski@natec.unibz.it (C.P.S.); dominik.matt@unibz.it (D.T.M.)

Received: 31 December 2019; Accepted: 21 January 2020; Published: 23 January 2020

Featured Application: This review aims to investigate the relevant literature regarding integrations of Building Information Modeling technology and the Last Planner® System in the construction execution phase. It is supposed to help scientists and practitioners to overview the state-of-the-art in terms of synergies between both approaches and devises a conceptual model for integration based on the review's findings.

Abstract: Many researchers have stated that lean and building information modeling (BIM) have positive synergies. This integrative literature review aims at exploring this body of knowledge within the scope of combinations of BIM and the Last Planner® System, as an important Lean construction method, in the phase of construction execution. The research motivation is to find out whether a comprehensive understanding of how to take advantage of these synergies exists. Eventually, the question arises of how to condense this understanding—if existing—into a robust conceptual model for integration. As a theoretical backbone, we will make use of the original BIM-Lean interaction matrix. The hypothesis is that new BIM functionalities have been evolved since the first formulation of this interaction matrix almost 10 years ago. These new BIM functionalities cause new interactions with existing lean principles. We will focus on interactions that refer directly or indirectly to production planning and control and use them to find the most relevant literature for this review. Within the content analysis, as a part of this review, we focus on existing conceptual models and frameworks for integration of BIM and the Last Planner® System and reveal their shortcomings. Eventually, we will propose a new conceptual model.

Keywords: BIM; lean construction; production planning and control; data-driven construction

---

1. Introduction

1.1. Starting Point

Except for a few scientific publications in the field of Lean Construction and digitization, both movements seem to be treated independently in the scientific context [1]. In recent years, the architectural, engineering, and construction (AEC) industry has been experiencing an ever-increasing wave of digitization. This is visible in this sector thanks to the adoption of building information modeling (BIM) [2]. BIM was originally conceived as the last generation of object-oriented CAD systems in which intelligent objects collectively represent the design of a building, coexisting in a single virtual structure.

Over time, connotation has evolved with the term and is now also used to represent digital innovation in the construction sector [3].

At the same time, the movement of lean construction management (LCM) has established itself on the level of construction process management as an adaptation of lean production from the automotive industry and is gaining more and more influence in today's construction management practice [4]. This resulted among others in new methods for design and/or site management such e.g., the Last Planner® System (LPS) but also in holistic project management and delivery approaches, such as e.g., integrated project delivery (IPD) concepts [5]. The principles of lean philosophy are primarily aimed at maximizing value for the customer by eliminating non-value-adding activities referred to as waste [6,7]. In lean jargon, basically, everything that is not needed is considered as being waste [8]. BIM can help to define what is needed and thus help to identify and remove waste, since it can precisely—digitally—describe what, how much and in which quality is needed on-site. The latter refers already to an often-cited positive interaction of BIM and Lean [9–11]. These positive interactions—or in other words, synergies—are receiving more and more attention in both academia and practice [11]. This tendency of the growing importance of both movements can also be seen in the increasing number of scientific studies over the last 16 years in both fields BIM and Lean Construction as depicted in Figure 1.

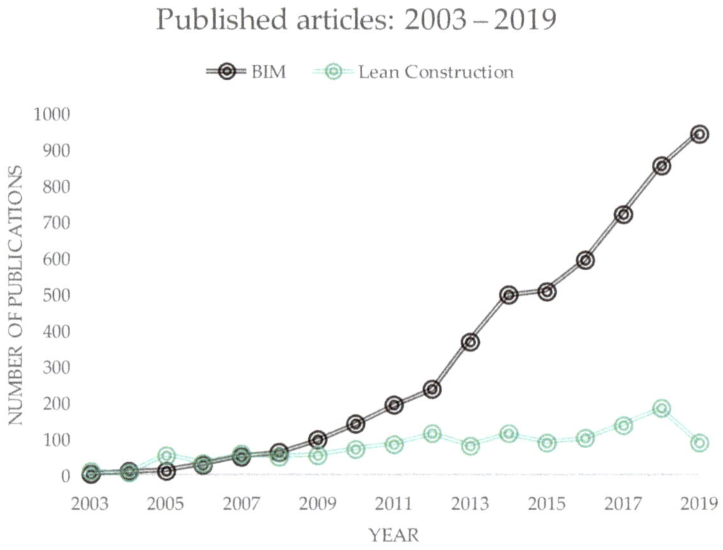

Figure 1. Building information modeling (BIM)/Lean construction papers published: 2003–2019; (data retrieved from the Scopus Database; search terms TITLE-ABS-KEY (("building information modeling" or "building information modelling"))" and ("lean construction"))).

The indicator of scientific publications clearly shows the proactive role of the AEC-industry in the adoption of both BIM and LCM [12]. It can be also seen that especially the number of BIM publications is growing steadily and rapidly. On the one hand, this is due to the general trend towards digitization in our society today. On the other hand, it is also and particularly due to the pressure triggered by legal and social requirements to increase productivity and efficiency. This is why the discrepancy with other industry sectors is still very large, but the construction industry, however, accounts for a very important share of the economic performance of industrialized countries [12,13]. BIM is seen here as a key technology [14]. This is also made clear by the fact that in many countries BIM will be mandatory in the near future for the planning of public-sector projects [15].

In addition to the growing popularity of both approaches, it is important to emphasize that BIM and Lean are in principle independent of each other and could be used concurrently [16]. Nevertheless, the scientific community and also practitioners are certain that (new) BIM functionalities can generate numerous interaction effects with the existing principles of the lean philosophy. In most cases, these can be considered as positive synergies and thus add value to the construction industry [10].

Considering this potential added value of an integrated BIM-lean adaption, interest in combining both approaches has arisen in the scientific world in recent years. The resulting synergies have been brought to the surface, both through scientific-theoretical treatises [10,17–19] and by proposing BIM-Lean supporting IT systems [20–23].

On the theoretical side two works deserve special mention in this context.: the original work of Sacks et al. [10], who describe the combination of BIM functionalities and Lean principles in an interaction matrix, and its update from Oskouie et al. [9], which together form the basis of numerous scientific publications in the tension between BIM and Lean, we will make use of the original interaction matrix as a guideline for this literature review. More specifically, it will serve as an a-priori filter for the identification of relevant literature in the scoped area of BIM-LPS combinations in the construction phase. It will also help to analyze only papers dealing with on-site production that combine both approaches. The rationale for limiting this review to the Last Planner® System rather than considering lean construction, in general, is explained in more detail in Section 1.2.

On the one hand, this review aims at finding out to what extent the interactions that were attributed as "cannot be considered mature technology" in 2010, when the original interaction matrix was first published, have evolved over the last decade. In addition, it will be also analyzed where still existing gaps in terms of technology maturity could be closed with new BIM-Lean IT systems to be developed in the future.

On the other hand, we claim that within the last 10 years the BIM functionalities have expanded and that the possibilities of using the information available through BIM are no longer reserved for IT specialists only, but have also opened up for architects and engineers through the introduction of e.g., visual programming languages (VPL). Consequently, these new functionalities also result in new interactions with existing lean principles, which are formulated in this study and serve as a "lens" for this literature review, too.

In summary, and as a starting point for this literature review, it can be said that numerous studies have been published in the field of BIM–Lean interaction. The present inadequacy, however, is that there is much literature on synergies but—to our best knowledge—no summarizing review for the specific case of BIM-LPS integration in the phase of construction execution.

This study inserts in this niche and aims at disclosing the relevant body of knowledge within this scope. This goal is pursued by means of the research method "integrative review", which will be guided by the original BIM–Lean interaction matrix as theoretical backing.

The body of this review article is organized as follows. Section 2 explains the applied research strategy for conducting this review. Section 3 provides the details of the followed review protocol allowing for replication of the procedure. Section 4 contains the actual results of the review in terms of a description of the final literature sample, a qualitative content analysis, and implications. Section 5 provides a final discussion on the findings and their potential meaning for practice. Section 6 concludes this study.

1.2. Research Question and Scope

This integrative literature review deals exclusively with the integration of the Last Planner® System and BIM in the construction execution phase.

This review, therefore, differs from other reviews about BIM and Lean [12,24] in its integrative nature, which resulted in the inference of a new conceptual model for integration and its focus on the Last Planner® System instead of dealing with the field of lean construction in general. The scope is limited to the execution phase, because Lean has started in the construction phase and is now moving towards lean design management [25], while BIM is widely accepted as a design method, hence first standards and norms exist for this phase [26]. However, there are still few guidelines for the beneficial use of BIM in construction execution processes, with a few exceptions like e.g., studies dealing with the design of BIM models for the sake of constructability [27,28], or proposals for so-called BIM-stations [29,30]. The motivation is, therefore, to make use of the complementary "maturity levels" within the two project phases and to combine BIM and lean—using the concrete example of the LPS—during construction execution.

This raises the following research questions, which will be addressed with the help of this integrative literature review: What is the current situation of conceptual models for the integration of BIM and LPS in the construction execution phase? Are there any models or frameworks that allow BIM data to be brought systematically, continuously and software vendor-neutral to the site enabling the application of LPS? If not, how do available models have to be adapted to achieve this goal? For this purpose, short introductions to BIM, LPS as well as BIM–lean interactions identified in the literature are given in the following sections.

1.3. Building Information Modeling

Building Information Modeling is a process for creating and managing the information of a building, infrastructure or facility through its life cycle [31]. One of the fundamental results of this process is the Building Information Model, the digital representation of all relevant aspects of the built edifice. This model is developed on the basis of the information that is inserted and updated collaboratively during the course of the project. This definition, given by the British NBS [32], highlights the three fundamental concepts that characterize BIM:

- BIM is a digital model, functioning as a container of data and information, which must be read, enriched, modified throughout the life cycle of a building, infrastructure or facility
- BIM is a process, or—in other words—a network of activities to manage the information contained within the models in order to use them beneficially
- BIM imposes collaboration, i.e., so that the information models are always up to date and usable, all operators must collaborate at appropriate times during the process and according to certain rules

BIM differs from the traditional CAD approach in that there is only one single place of editing—the BIM model itself—and no longer all plan deductions such as views, sections, and floor plans have to be adapted individually when changes are requested [33]. Azhar and Nadeem [33] point also out that BIM models are composed of intelligent and semantically structured objects such as walls or slabs whilst CAD drawings only represent geometry by means of interconnected lines or arcs.

Well-known benefits, such as automated clash detection, code checking, visualization by means of the help of augmented/virtual reality technology and enhanced collaboration among stakeholders becoming more and more exploited in the regular business of many practitioners [34]. These positive effects of BIM technologies have overcome the status of constituting exclusively research topics. However, since BIM technology is evolving and access to BIM data is becoming increasingly convenient—even for non-IT specialists—also new research potential arises continuously.

1.4. Lean Construction Management and the Last Planner® System

The term lean construction management (LCM) is derived from lean principles being applied in the production industry and refers to the original concepts of the Toyota Production System (TPS). One of the core elements of lean management, in general, is the aim for maximization of value from the customer's perspective by eliminating waste. Waste in the construction industry can e.g., be referred to as rework or waiting time [35].

The Last Planner® System has been developed for the construction industry as a production planning and controlling system. It aims at both increasing schedule reliability and smoothening flow of work [36], and is been considered as one of the most import lean methods in construction execution [10]. The LPS strives for the achievement of its purposes through collaboration, transparency, continuous improvement as well as reliable commitments from the responsible persons for actual work completion, the Last Planners [37].

According to Ballard [36], every possibly occurring task, needed for the completion of a construction project, can be technically categorized into four groups during its execution phase: (1) SHOULD-tasks include in principal all tasks needed for achieving global milestones defined by an existing master schedule or client's requirements. (2) CAN-tasks for a preliminarily defined project look-ahead window consist out of SHOULD-tasks, which are free of constraints and thus ready for immediate execution. (3) WILL-tasks are considered as CAN-tasks whose execution by a defined due date has been assured by consensus of all involved and explicitly by the responsible last planner. (4) DID-tasks in turn, are all accordingly completed WILL-tasks of the prevenient LPS-cycle. The latter ratio in LPS applications can be expressed as a percent planned complete value (PPC).

The SHOULD-CAN-WILL-DID (SCWD) logic is supposed to be applied in five consecutive steps which are characterized by corresponding plan phases with increasing level of detail: (1) master scheduling, (2) phase scheduling, (3) make-ready-planning, (4) commitment-planning and (5) learning and control [38]. This procedure is shown in Figure 2.

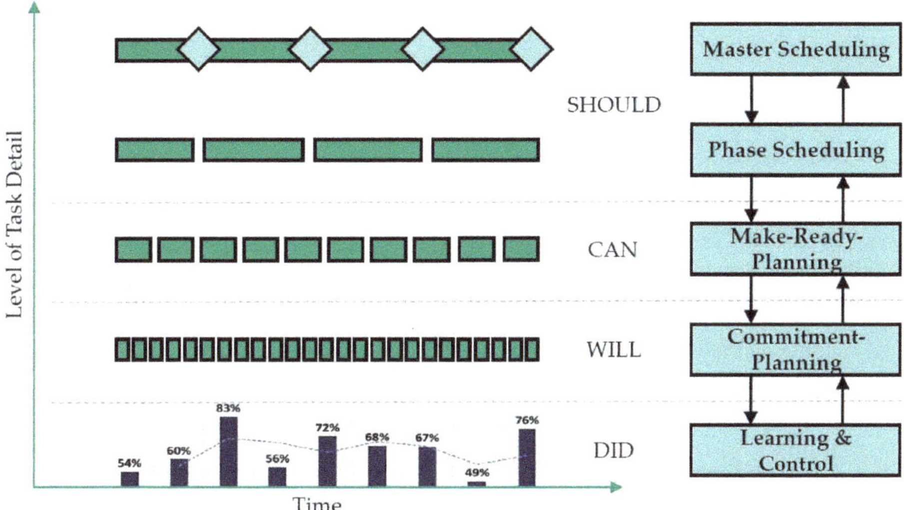

Figure 2. Last Planner® System (LPS), adapted from Hamzeh and Bergstrom (2010).

The least mentioned step learning and control accounts for the analysis of completed tasks with respect to their commitment. Crucial at this point is a profound analysis of reasons for non-completion (RNC) of tasks that have not been completed as committed in terms of both quality and due date in order to avoid those mistakes to re-occur, as part of a continuous improvement process (CIP) [39].

Of equal importance is the social process going along with the implementation and application of the LPS in construction projects. These interpersonal matters are especially characterized by open and transparent discussions between involved trades represented by their last planners, increasing trust among them as well as collectively emerging learning cycle effects. The latter can be seen as an inherent consequence of the analysis of RNC [40,41].

According to Ballard [36], the main advantages of the LPS include the following points:

- Stabilization of project-based production systems
- Making control proactive

- Reduction of waiting times and project durations
- Decentralization of decision-making processes
- Early communication of insufficient project status

As with many lean-based process changes, the necessary commitment of all those involved and thus resistance to change can be stated as disadvantages. At the process level, it should be noted that the LPS does not offer any functionalities with regard to cost control and is therefore not sufficient as the sole method for holistic project management with regard to the optimization of the three target variables quality, time and budget.

1.5. BIM—Lean interaction matrix

As indicated above, we want to make use of the BIM–Lean interaction matrix by Sacks et al. [10] as a filter to find the most relevant literature within the scope of this review. This interaction matrix contains synergies between the Lean philosophy and the BIM approach as known as of the year 2010. Synergies in this context are referred to as interaction effects that could result when certain Lean principles are combined with BIM functionalities. Sacks et al. [10] elaborated originally a total of 56 interactions.

To have an even finer a-priori filter, we reduced this original interaction matrix to those Lean principles and BIM functionalities that refer directly or indirectly to production planning and control and the Last Planner® system. Additionally, and according to our hypothesis of newly appeared BIM functionalities over the last decade, new interactions with Lean principles have evolved too, which we also added to the reduced interaction matrix. It is true that the original interaction matrix has been already extended by Oskouie et al. [9]. However, most of the extensions mentioned deal with BIM functionalities for the operation and maintenance phase which is out of the scope of this paper. Nonetheless, we will consider the BIM functionalities mentioned in that work, which are dealing with production planning and control and have not been covered by the original interaction matrix yet: "Facilitating real-time construction tracking and reporting" [22]. All these elements together result in a specialized BIM-LPS interaction matrix which is given Table 1.

The resulting interactions of the new BIM-LPS interaction matrix are given in Appendix A.

We will use this list of interactions for eliminating those papers from the literature sample that should not be part of the final content analysis. More in detail, the topics of the here listed interactions will be condensed and abstracted to a high-level. This allows for deciding which of the found sub-clusters in the second literature screening round should not be considered in the qualitative content analysis. Furthermore, we are particularly interested in how the interactions that were considered as being negative back then (e.g., called "cannot be considered mature technology") have developed over time.

Table 1. New BIM-LPS interaction matrix.

| Column key | Last Planner® principles | Reuse of model data for predictive analysis | Automated generation of drawings and contents | Rapid generation and evaluation of multiple construction plan alternatives | | | Online/electronic object-based communication | | | Customized data retrieval & manipulation | | Facilitating real-time construction tracking and reporting | |
| | Sacks et al. [10] | | | | | | | | | Authorship of this study | | Oskouie et al. [9] | |
		Automated cost estimation (4)	Automated generation of drawings and contents (8)	Automated generation of construc-tion tasks (11)	Construction process simulation (12)	4D simulation of con-struction schedules (13)	Visual-ization of process status (14)	Online commu-nication of product and process information (15)	Provision of context for status data collection on site/ off site (18)	IFC toolkits & Visual programming allow for developing customized applications (19*)	BIM objects can be highlighted/ visualized for application-specific purposes (20*)	Monitoring construc-tion process [21*]	Real-time evaluation of productivity [22*]
	Reduce cycle times												
C	Reduce production cycle duration	12	22	25	25	25	26	26				61	61
E	**Reduce batch size**		53				30	30	30				
	Select an appropriate production control approach												
H	Use pull systems						34	34	34	58	58, 59	62	62
I	Level the production									57	58, 59, 60	62	62
	Use visual management												
L	Visualize production methods					40		38			60		
M	Visualize production process					40	34	34			60, 61. 62		
	Design the production system for flow and value												
N	Simplify			−41	−41								
P	Use only reliable technology		54				−42	−42	−42	57			
Q	Ensure the capability of the production system		54			58,59					60		
R	**Ensure comprehensive requirement capture**										59, 60		

* refer to added BIM functionalities to the original BIM-Lean interaction matrix.

2. Research Strategy

We have decided to conduct an integrative literature review on the integration of BIM and the Last Planner® System in construction execution. The rationale for this decision is that we would like to cover all the relevant literature in this field involving both quantitative and qualitative studies, which is the essence of an "integrative" review [42]. However, this review remains systematic in the sense that it follows a strict review protocol allowing for replication of the chosen procedure in line with the specifications for performing systematic reviews by Kitchenham [43].

In addition to that, integrative literature reviews analyze, criticize and integrate the considered literature sample in a way that new models or frameworks on the topic can be devised [44]. We will make use of this technique to formulate a new conceptual model for the integration of BIM and the Last Planner® System as a basis for improved production management systems in construction.

3. Review Protocol and Implementation

This review is divided into the parts of (i) data collection; (ii) initial screening and (iii) an in-depth analysis which is mainly characterized by a qualitative content analysis of the final sample literature. The review protocol is shown in Figure 3 and described in the following subsections.

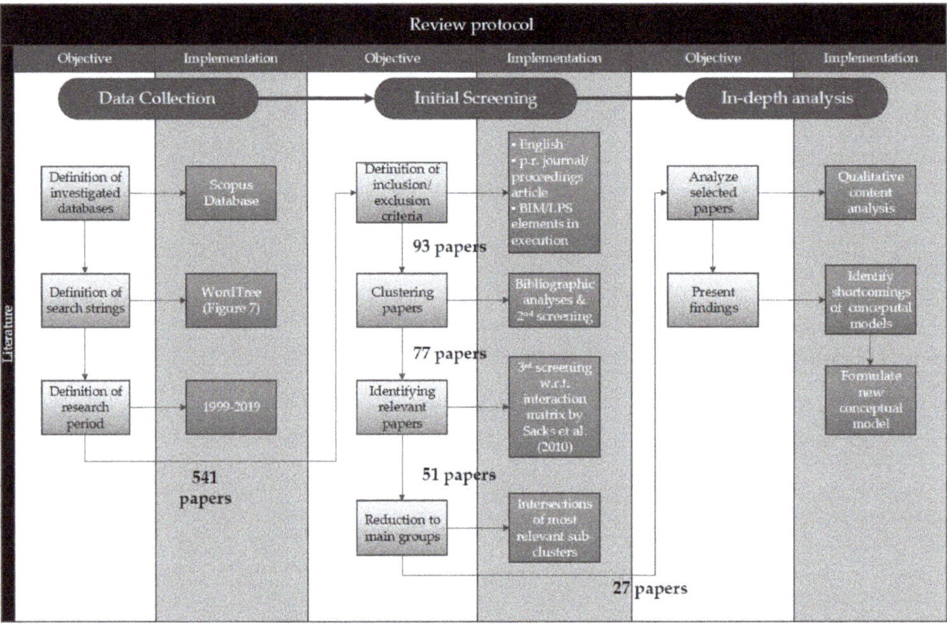

Figure 3. Review protocol.

3.1. Data collection

In line with Saunders et al. [45], we considered as reliable sources only peer-reviewed journal articles and conference proceedings for this literature review. We chose the Scopus database as a search engine and used as search string any combinations of the terms "BIM" and "Last Planner System", their synonyms, as well as possible subsystems such as e.g., "Look-ahead planning" in the case of the LPS. These different terms are shown in the word tree in Figure 4, whilst Table 2 contains the resulting and full search string that has been entered into the database. Only title and keywords were considered in the search engine, since the search string contains many general terms and would output too many hits if the abstract field was also enabled. The resulting sample size would not be manageable with the procedure according to this review protocol. The data collection part comprises all entries in the research period from 1999 to 2019 in the Scopus database that were available until October 2019, when the database search was performed. The database search was performed by the first author of this review article. This first step of data collection resulted in a total of 541 papers.

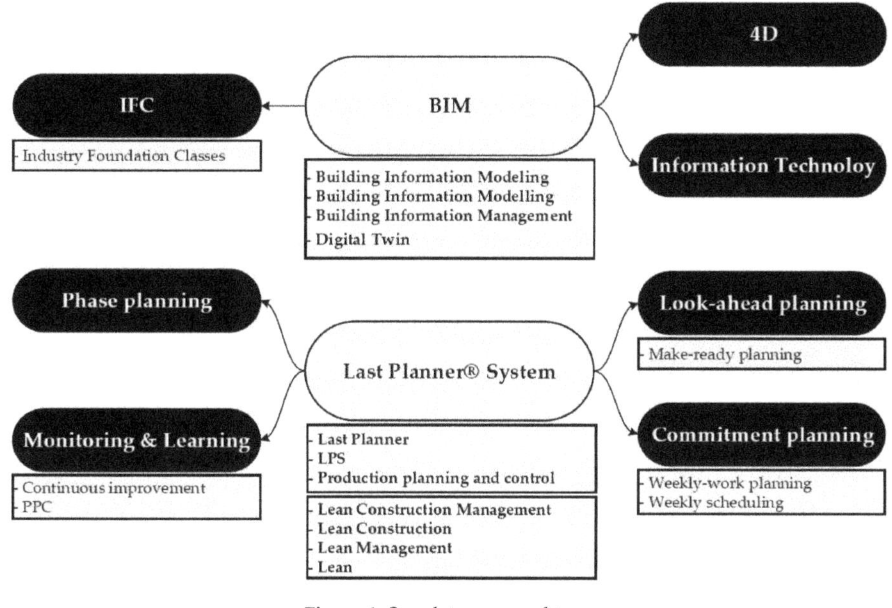

Figure 4. Search terms word tree.

Table 2. Search string.

String
(TITLE ("BIM" OR "Building Information Modeling" OR "Building Information Modelling" OR "Building Information Management" OR "Digital Twin" OR "IFC" OR "Industry Foundation Classes" OR "4D" OR "Information Technology") OR KEY("BIM" OR "Building Information Modeling" OR "Building Information Modelling" OR "Building Information Management" OR "Digital Twin" OR "IFC" OR "Industry Foundation Classes" OR "4D" OR "Information Technology")) AND (TITLE("last planner*" OR "LPS" OR "Lean" OR "Phase planning" OR "Look-ahead" OR "Make-ready" OR "Commitment planning" OR "Weekly Work Plan*" OR " Continuous Improvement*" OR "PPC") OR KEY("last planner*" OR "LPS" OR "Lean" OR "Phase planning" OR "Look-ahead" OR "Make-ready" OR "Commitment planning" OR "Weekly Work Plan*" OR "Continuous Improvement*" OR "PPC")) AND DOCTYPE(ar OR cp) AND PUBYEAR > 1998

3.2. Initial Screening

3.2.1. Inclusion/Exclusion Criteria

Within the initial screening phase of 541 papers, which resulted from data collection, we first defined several inclusion and exclusion criteria to reduce the sample size and to narrow the considered literature down to the focus area of joint BIM-LPS applications in execution. The criteria for this purpose were defined as (i) written in English; (ii) peer-reviewed journal/conference papers; (iii); related to the phase of construction execution; (iv) the studies have to deal with at least partial elements of BIM and the LPS, such as e.g., look-ahead planning in the case of LPS (Table 3). The latter has been verified by reading the title and abstract (where necessary) of the papers within the initial sample. Applying these criteria reduced the sample size to 93 papers.

Table 3. Inclusion and exclusion criteria.

Inclusion	Exclusion
• Written in English	• Written in a language other than English
• Peer-reviewed journal/conference papers	• Not peer-reviewed; contribution constitutes book, book section, technical report and similar
• Related to the construction phase	• Related only to the design phase
• Deals with at least one element of each concept (BIM **AND** LPS)	• Deals only with one of the concepts (BIM **OR** LPS)

3.2.2. Bibliometric Analyses and Paper Clustering

Bibliometric analyses of these remaining 93 papers comprise at this stage statistics about

- Documents by author (Figure 5a)
- Documents by country (Figure 5b)
- Documents by subject (Figure 6a)
- Documents by type (Figure 6b)

The bibliometric analysis shows in Figure 5b that most of the relevant documents were published by American researchers (25). Also, the United Kingdom (11), Brazil (10), Israel (10) and Germany (9) appear to be very active in the field of joint BIM-LPS applications.

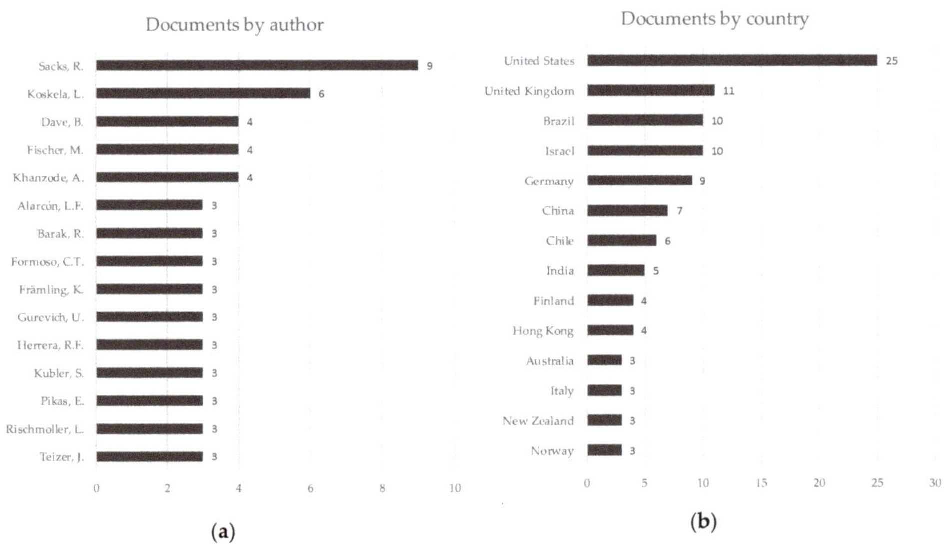

(a) (b)

Figure 5. Documents by author (**a**); documents by country (**b**).

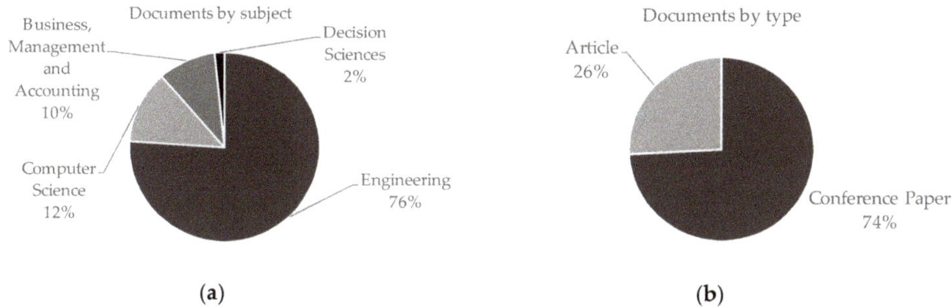

(**a**) (**b**)

Figure 6. Documents by subject (**a**); documents by type (**b**).

Figure 5a presents the document distribution by authors. Here, Rafael Sacks is the author of the most publications. This underlines his significance in this field of research, who is already playing an important role in this review as the first author of the original BIM–Lean interaction matrix.

On the other hand, Figure 6 shows the document distribution by subject (a) and type (b). Especially the type distribution is noticeable as it reveals that the majority of publications are presented at research conferences. This could be an indication that many observations and considerations in this research area have not yet been consolidated and that, accordingly, there are not yet enough reliable findings for extensive publication in journals.

In addition to that, science mapping visualizations were conducted with *VOSViewer* software (Figure 7). *VOSViewer* is a freely usable software system based on text-mining algorithms and network clustering techniques which has been developed by van Eck and Waltman [46]. It allows for distance-based cluster visualization of bibliometric networks [47], where the size of the nodes in our settings renders absolute occurrences of analyzed terms in title and abstract fields of the literature sample. The distance between the nodes, on the other hand, visualizes how much related to each other the terms are, meaning how often they occur in the same paper. The clustering in different colors is done automatically by the software. However, the user can decide how many clusters the system should generate and by doing so, fine-tune clustering results. The detailed settings that we chose are summarized in Table 4.

The rationale for this clustering approach was to identify the most relevant macro topics that BIM-LPS co-applications in scientific studies deal with. The results of the cluster visualization are shown in Figure 7.

Figure 7 shows the most occurring terms in the considered literature sample of 93 papers, which have been grouped into three different clusters by the *VOSViewer* clustering algorithm:

1. The red cluster contains characteristic terms such as e.g., site, construction process, production, control, operation, worker, workflow, so that we grouped these terms to the cluster of production planning and control.
2. The green cluster contains characteristic terms such as e.g., information modeling, information technology, implementation, analysis, communication, so that we grouped these terms to the cluster of information systems and implementations.
3. The blue cluster contains characteristic terms such as e.g., coordination, team, project participant, project team, so that we grouped these terms into the cluster of collaboration.

These found clusters are also indicated in Table 4 and represent the macro topics of the considered literature sample. Furthermore, the visual science mapping approach with VOSViewer software indicates also that the majority of the papers dealing with IT-systems and digitization advancements in construction (green cluster) are more often concerned with supporting production planning and control (red cluster) rather than supporting collaboration among the persons on site (blue cluster). This is evident in Figure 7 by the spatial proximity between the red and green cluster, while the blue cluster is more distant.

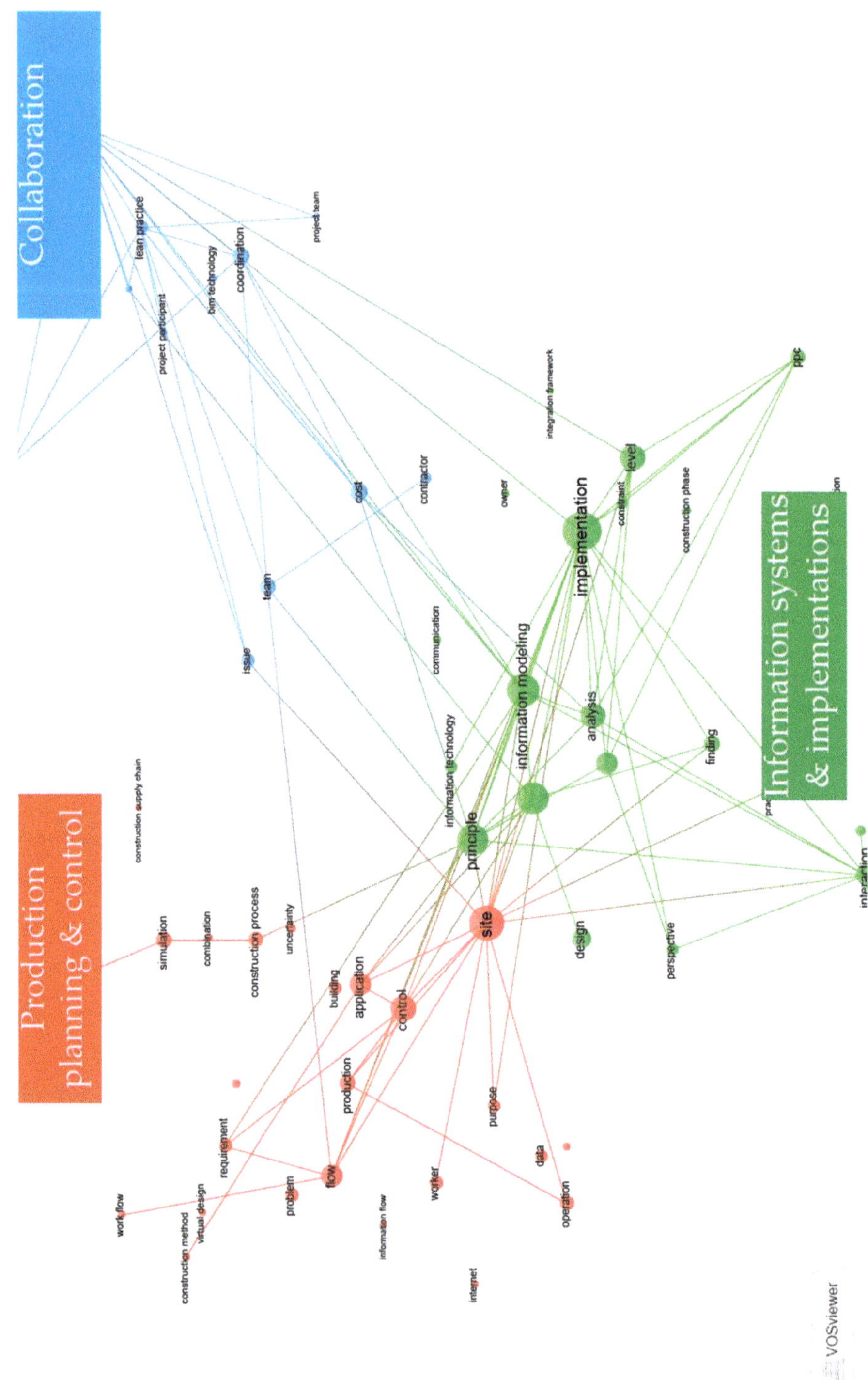

Figure 7. Science mapping using VOSViewer.

Table 4. VOSViewer settings and found clusters.

General Settings	Excluded Terms	Found Clusters *
• Analyze only title and abstract fields	lpds php	• Production Planning and Control
• Do not ignore structure abstract labels • Full counting	idm addition component	• Collaboration • Information Systems and Implementations
• Minimum number of occurrences = 6	thing part	
• 60% relevance level	bim station	
• Manual exclusion of less interesting or too general terms	researcher lps year goal type method need contribution objective term bim model lack last planner system industry model	

* Naming of the clusters has been defined by the authors based on the terms within the clusters.

At this stage of the review procedure and following the review protocol (Figure 3), a second round of screening was conducted. This time also the main text body of the papers was considered where necessary. The aim was to filter out papers, that could not be grouped into one of the three identified main topics (clusters), since they were considered as dealing with less popular topics. This further reduction of the literature sample resulted in a total of 77 papers. We defined a total of 9 sub-clusters for these 77 papers based on their content as another result of the second literature screening round. These sub-clusters are presented in Table 5.

Table 5. Sub-clusters after second screening.

#	Sub-Clusters
1	BIM–Lean interactions
2	Augmented Reality (AR)
3	Supply Chain Management and Logistics
4	Production
5	4D Simulations
6	Conceptual Models and Frameworks
7	Internet of Things (IoT)
8	Metrics
9	Agent-based Simulations/Optimizations

3.2.3. Identifying Relevant Papers for Content Analysis

The remaining 77 papers were considered for identifying the really relevant papers for the content analysis within the scope of this review. At this point, the reduced BIM-LPS interaction matrix comes back into play for helping to define which of the sub-clusters are relevant for this review. Since the interactions of the BIM-LPS interaction matrix are mainly dealing with arguments regarding production planning and control supported by BIM, it has been decided to take over the sub-clusters of "BIM–lean interactions" and "Production" to the next step of detailed qualitative content analysis. In addition to that, we took also the cluster "4D simulation" into consideration since it helps practitioners to test different production strategies by means of visualization in the BIM model. For this reason, it is also in line with the introduced BIM-LPS interaction matrix filter. Lastly, we also considered the sub-cluster of "conceptual models and frameworks" within the next round to analyze the foundations in terms of theory in the area of BIM-LPS co-applications.

All the papers that are not part of one of these sub-clusters were removed from the literature sample which further reduced the sample size to 51 papers.

3.3. Final Sample Overview

The remaining sample contains 51 papers being published from 2009 to 2019. This sample is listed in Table 6.

Table 6. Articles included in the content analysis.

No.	Article
1	Assessing the Impacts of an IT LPS Support System on Schedule Accomplishment in Construction Projects [48]
2	BIM-based Last Planner System tool for improving construction project management [49]
3	Site logistics planning and control for engineer-to-order prefabricated building systems using BIM 4D modeling [50]
4	BIM-based takt-time planning and takt control: Requirements for digital construction process management [51]
5	Mobile BIM implementation and lean interaction on construction site: A case study of a complex airport project [52]
6	Exploring the BIM and lean synergies in the Istanbul Grand Airport construction project [53]
7	Using Building Information Modelling to achieve Lean principles by improving efficiency of work teams [54]
8	BIM-enhanced collaborative smart technologies for LEAN construction processes [55]
9	Guidelines to develop a BIM model focused on construction planning and control [28]
10	Using technology to achieve lean objectives [56]
11	Development of an integrated BIM and lean maturity model [57]
12	CAN BIM furnish lean benefits—An Indian case study [58]
13	Envision of an integrated information system for project-driven production in construction [59]
14	Evaluation of a case study to design a BIM-based cycle planning concept [60]
15	An exploration of BIM and lean interaction in optimizing demolition projects [61]
16	Using BIM-Based sheets as a visual management tool for on-site instructions: A case study [62]
17	Modelling and simulating time use of site workers with 4d BIM [63]
18	Integration of Building Information Modeling (BIM) and Prefabrication for Lean Construction [64]
19	From conventional to it based visual management: A conceptual discussion for lean construction [1]
20	Virtual design and construction: Aligning BIM and lean in practice [65]
21	Contributions of information technologies to last planner system implementation [66]
22	Lean production controlling and tracking using digital methods [67]
23	Building information modeling: A report from the field [26]
24	Experiences from the use of BIM-stations [30].
25	Exploration of a lean-BIM planning framework: A last planner system and BIM-based case study [68]
26	BIM and sequence simulation in structural work—Development of a procedure for automation [69]
27	Last Planner & Bim Integration: Lessons From a Continuous Improvement Effort [70]

Table 6. *Cont.*

No.	Article
28	Bim—Stations: What It Is and How It Can Be Used To Implement Lean [29]
29	Simulating and vizualising emergent production in construction (Epic)using agents and BIM [71]
30	Intelligent products: Shifting the production control logic in construction (with lean and Bim) [72]
31	Constructible bim elements -a root cause analysis of work plan failures [27]
32	Bim and lean in the design-production interface of eto components in complex projects [73]
33	An empowered collaborative planning method in a swedish construction company-a case study [74]
34	Using BIM for Last Planner System: Case Studies in Brazil [75]
35	Addressing information flow in lean production management and control in construction [22]
36	Bim based conceptual framework for lean and green integration [76]
37	Using 4D models for tracking project progress and visualizing the owner's constraints in fast-track retail renovation projects [77]
38	Adapted use of andon in a horizontal residential construction project [78]
39	Examination of the effects of a KanBIM production control system on subcontractors' task selections in interior works [21]
40	Complementarity between the Building Information Modeling (BIM) and Product Life Cycle Management (PLM) through the Lean Construction (LC) [79]
41	BIM-lean synergies in the management on MEP works in public facilities of intensive use—A case study [80]
42	KanBIM workflow management system: Prototype implementation and field testing [81]
43	BIM and Lean interactions from the bim capability maturity model perspective: A case study [82]
44	Integration framework of bim with the last planner system [83]
45	Extending the interaction of building information modeling and lean construction [9]
46	Visilean: Designing a production management system with lean and BIM [23]
47	Field tests of the KanBIM™ lean production management system [84]
48	Interaction of lean and building information modeling in construction [10]
49	Requirements for building information modeling based lean production management systems for construction [20]
50	Analysis framework for the interaction between lean construction and Building Information Modelling [85]
51	Visualization of work flow to support lean construction [86]

Most of the contributions (ca. 65%) were made through the annual conference of the International Group for Lean Construction (IGLC), which provides for a separate BIM and Lean track even though this conference is focusing on Lean Construction. None of the other mediums account for more than 8% of the papers in the final literature sample, which can be seen in the Pareto chart in Figure 8. The top journal to be mentioned here is *Automation in Construction*, which ranked second right after the IGLC proceedings.

A total of 37 of the papers of this sample deals with interactions of BIM and Lean (ca. 73%), whilst 19 papers of this subset focus in particular on BIM-supported production planning and control systems which accounts for ca. 37% of the entire sample. Another 16 papers of the sub-cluster BIM–Lean interactions have a strong focus on theory in terms of conceptual models or frameworks for BIM–Lean integration or new production (management) systems, which are highly relevant to this review with respect to the defined scope. The same is true for the earlier mentioned sub-set of "production planning and and control supported by BIM", so that we decided to perform the qualitative content analysis on their elements. The sub-cluster of 4D simulations with only 10 papers was not considered there.

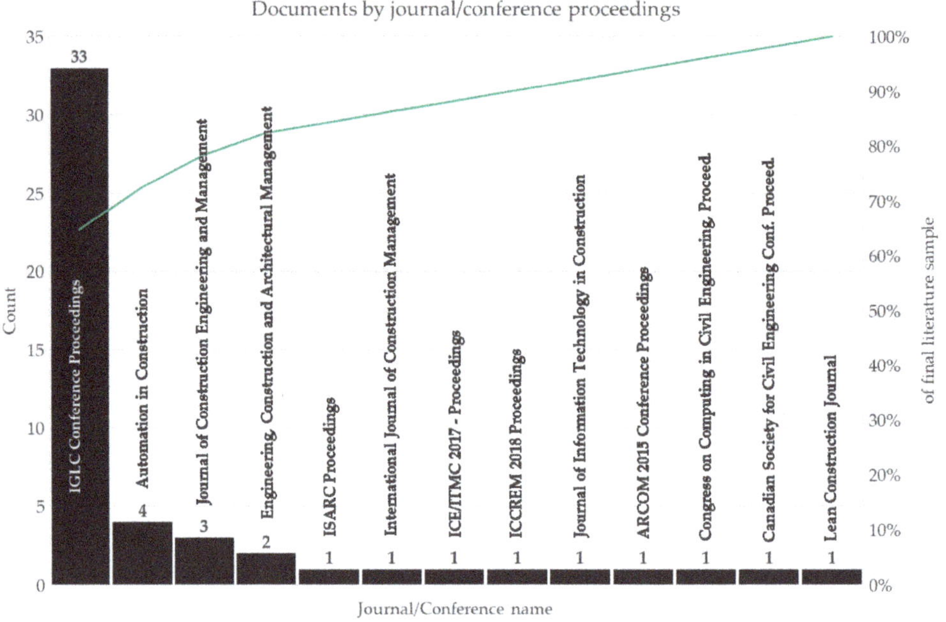

Figure 8. Documents by journal/conference proceedings.

The remaining sub-clusters of this 51-papers-sample are presented in Table 7. These were further abstracted to the two groups of "production planning and control supported by BIM" and "theory of BIM-LPS integration", as indicated above.

Table 7. Relevant sub-clusters and abstract groups for content analysis.

Sub-Clusters	Count/(%) *	Groups for Content Analysis	Count/(%)	Total Sample Size for Content Analysis ****
❶ BIM-Lean interactions	37/(73%)	① Production planning and control supported by BIM**	19 (37%)	27
❷ Production	30/(59%)			
❸ 4D Simulations	10/(20%)			
❹ Conceptual Models and Frameworks	20/(39%)	② Theory of BIM-LPS integration***	16 (31%)	

* The sum of the count does not sum up to the sample size of 51 (100%), since several papers deal with multiple sub-clusters. ** Intersection of sub-cluster ❶ and ❷. *** Intersection of sub-cluster ❶ and ❹. **** A total of 8 papers is part of both groups ① and ②.

The papers within these two groups—a total of 27 papers—became then the subject of qualitative content analysis as part of the in-depth analysis according to the review protocol.

4. In-Depth Analysis and Results

In this phase, we qualitatively analyze the remaining papers to provide the reader with a comprehensive summary of their main focus (Section 4.1) and resulting implications (Section 4.2), before we proceed with the discussion of the findings.

4.1. Qualitative Content Analysis

The in-depth analysis of the final literature sample was carried out by means of the research technique referred to as qualitative content analysis. In the scientific context, content analysis, in general, is an approach for identifying the most important characteristics of written research studies allowing for valid inferences [87]. It is a very useful technique to gather and organize information for obtaining insights into current trends and patterns in the considered literature sample [88]. The qualitative aspect, in particular, refers to sorting the content into groups and summarizing the key messages [87]. The results of the qualitative content analysis for the groups of (i) production planning and control supported by BIM and (ii) theory of BIM-LPS integration are presented in the next sub-sections.

4.1.1. Group 1: Production Planning and Control Supported by BIM

Group 1 refers to studies concerned with the management of construction processes according to Lean principles that are also supported by BIM technology. Thanks to the a-priori BIM-LPS interaction matrix filter and the rigorous focalization of this review procedure, most of such comprise BIM-LPS co-applications in construction execution. Several authors developed so-called "Lean IT-tools" which support one or many aspects of the LPS and provide a link to BIM models [22,23,49,55,84,86,89]. However, none of them supports the Last Planner® System to its full extent, since the tools either support only collaborative phase planning [55], even neglect the collaborative aspects such as hand-offs discussion among last planners [22,23,49] or systematic task-breakdown [84,86,90] as stipulated in the current LPS process benchmark [91].

Besides methodological insufficiencies with respect to LPS, this literature sample reveals that the BIM integrations are also not fully developed up to the data processing level: Either, only 3D-representations of BIM models are considered [84,86,90] in case of some integrated IT-systems, LPS elements were included in a non-digital fashion in regular BIM coordination meetings [83], or lean production philosophy was applied in parallel with BIM models but not truly integrated [27,50,70].

This literature sample is also characterized by case studies in the construction execution phase of different project types to find and demonstrate evidence of the described BIM–Lean synergies [54,70,73,74]. Examples here are engineer-to-order (EtO) projects [50,73] or areas of mechanical, electrical, and plumbing (MEP) works [70,80,83].

Another aspect of this literature sample deals with BIM models made available at the construction site to support the implementation of lean principles, which are referred to as BIM-stations [29].

4.1.2. Group 2: Theory of BIM-LPS Integration

Besides the fundamental theoretical work of BIM–Lean interactions of Sacks and his co-authors [10,85], the second group of the final literature sample regarding the theory of BIM-LPS integrations contains two main aspects (i) conceptual models and frameworks for integrations [1,52,64,68,76], and (ii) findings of applications in case studies of such [1,53,75,92]. Some of the analyzed papers cover both aspects [1,52,82].

Regarding the first aspect, we found examples for BIM–lean frameworks [68,83], that propose to "better use BIM models together with LPS" [68]. These kinds of frameworks are basically instructions on where and when to take a look at the BIM model during the Last Planner® process to improve the process itself. In this very example, an Autodesk Revit® BIM model is used, which does not constitute an integrated BIM-LPS IT-system, but requires the use of external, purchasable software.

Another piece of research of this aspect deals with the digital transformation on construction sites by means of mobile applications that enable simultaneously BIM practices and lean principles [52]. Here, the focus is on the delivery of BIM data on-site through tablets according to lean production principles. Also, here the framework is leaning towards commercial BIM software products. Other frameworks focus on prefabrication [64], the integration of BIM, Lean and Green concepts providing for the identification of operational and tactical relations [76] or the maturity of BIM adaption from stakeholder's perspective and its role on BIM–Lean interactions [82].

Findings of BIM-LPS co-applications, on the other hand, indicate among other things that both Lean and BIM approaches are inextricably linked. However, this link is not fully explored yet [92]. BIM can support LPS decision-making processes thanks to the availability of the right information at the right time [75] and the level of BIM maturity plays a significant role in enabling Lean principles on construction sites [82].

4.2. Findings and Implications

In the following, the central claims from the qualitative content analysis are evaluated and the subsequent implications are presented. The main points listed below both represent shortcomings of current BIM-LPS applications in practice and theory and the resulting motivation to propose a new conceptual model for integration.

4.2.1. Shortcomings and Gaps

As it has become evident in the previous qualitative content analysis, insufficiencies in current attempts of linking the Last Planner® System and the BIM approach conceptually consist mainly of three points:

- Parallel applications rather than true integration which would need provision for detailed insights on how to capitalize BIM data systematically and in a software-vendor independent way and vice versa: how to retrieve BIM data for the LPS process and how to store and potentially visualize LPS process information within a BIM model. In addition to that, a sound conceptual model for integration should also cover BIM functionalities other than 3D representations and automated clash detections, especially when striving for new, BIM-based production management systems in construction.
- There is a lot of scientific evidence focusing on traditional, "stand-alone" LPS that describes its implementation as less successful when not implemented to its full extent in terms of methodology [93–95]. Current attempts of implementing the LPS together with BIM show the same methodological inappropriateness which might lead to the same non-satisfactory results. This could be avoided if an accepted reference framework that guides future users through correct methodological implementations in a BIM environment was established.
- Combinations of BIM and LPS in both frameworks and practical implementations in case studies are infrequently connected to solid theory, other than the nowadays known BIM–lean interactions themselves. As claimed by Kasanen et al. [96], this connection is especially needed in the case of frameworks or conceptual models, which often derive from constructive research based on a design science research (DSR) approach.

As for gaps and the combination of digital technology such as BIM with the Last Planner® System, also digital representations of the LPS entities should be considered in a conceptual model for making a link of these entities to BIM objects possible. Thus, LPS being conceptually a Kanban system [97], intuitively brings up the suggestion of digital Kanban systems, where traditional sticky notes could become digital on a touch-whiteboard.

We further think that clear roles with distinct responsibilities could add value to a framework for BIM-LPS integrations in order to improve process compliance in terms of methodology and to avoid the omission of important LPS steps. We will consider this when devising a new conceptual model for integration in the next section.

4.2.2. Devising a New Conceptual Model

Given the results of the integrative literature review, we take both the lack of a complete conceptual model that is accepted by the scientific and practitioner community and the shortcomings mentioned above as an opportunity to propose a new conceptual model for BIM-LPS integration.

This new conceptual model aims to account for the BIM-LPS interactions presented in Table 1 and the above-mentioned needs of adding new functionalities. These new functionalities are:

- Establishing clear roles (e.g., similar to the Scrum framework)
- Deploying on (digital) Kanban boards → Enhancing visualization, emphasizing LPS' pull-system characteristics, adhering to digitization [97–99]

This addition is intended to give the model on a conceptual level the properties that make it possible to derive holistic, IT-oriented production management systems based on BIM and LPS. These could become the source of new information systems. For this purpose, the proposed model also highlights a software-neutral information flow from BIM to the LPS process steps and vice versa.

In order to establish a connection to existing theory, and thus meeting the requirements of the DSR approach, our model (Figure 9) is based on the original LPS model by Ballard [36]. Our model refers to the definition of Greca and Moreira [100], who stated that a conceptual model comprises an "external representation created by researchers, teachers, engineers, etc., that facilitates the comprehension or the teaching of systems or states of affairs in the world".

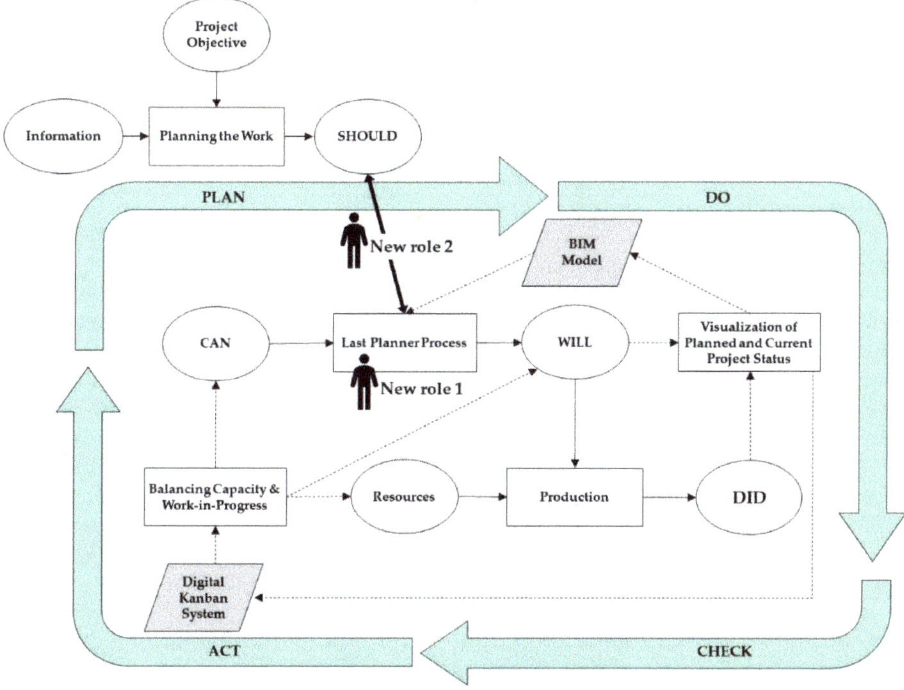

Figure 9. Conceptual model for BIM-LPS integration.

On the one hand, this original model is extended by the BIM part which serves both as input and output/visualization instrument. On the other hand, a (digital) Kanban system should help to balance load to capacity [36], to optimize flow through limited Work-in-Progress (WIP) [101], and to improve visual management [98,102]. What is more, Kanban, being a system to provide information as pull signals along value-adding-chains, applied to the LPS, it could help to release new work, when tasks are completed as demanded by Ballard [103].

Lastly, we introduce new roles for enhanced process compliance and customer satisfaction in the style of Scrum. We need a role ("new role 1") that precisely implements the LPS methodology in conjunction with BIM to prevent partial implementations as we have seen in pure LPS implementations. Besides that, we want to introduce another role ("new role 2") that ensures that process planning and construction execution according to this model is in line with the customer requirements. Of course, both roles could also be filled by the same person, for example, the site manager in a general contractor environment or an external project manager.

As a consequence, using LPS terminology, the new role 2 is responsible for aligning what SHOULD be done to what WILL be and to what was done (DID) from the customer perspective. The new role 1 has been introduced to define a person who moderates the combined BIM–Last Planner® process according to this conceptual model. The BIM model serves as an information input source for the planning process according to the LPS and for status visualization. The latter can also be visualized by means of a digital Kanban system. Furthermore, the Kanban system can be capitalized for defining what CAN be done when load is visually balanced to capacity (will). Comprising a circular process, this model can be interpreted as being embedded in the plan–do–check–act (PDCA) cycle by Deming [104].

5. Discussion

Drawing a line to the initially stated research questions, it can be said, that—to our best knowledge—there is currently no sufficient conceptual model for the integration of BIM and the Last Planner® System in construction execution. The findings of this integrative literature review demonstrated this and led eventually to the formulation of a new conceptual model for integration as a response to the discovered shortcomings of current integration attempts.

This new conceptual model is intended to serve as an inspiration for future joint BIM and LPS implementations in construction processes, and also—but not exclusively—to develop new integrated information systems that make these implementations possible. These implementations shall be independent of commercial software solutions, such as the selected BIM modeling and scheduling software. A special role is played here by the open BIM exchange format Industry Foundation Classes (IFC):

Even though this conceptual model foresees the use of BIM data as input information for the LPS process planning and the subsequent visualization of the process parameters, it is inevitable for actual practical implementations—as also claimed by Toledo et al. [68]—that this becomes actually feasible in practical, real-world tools. In other words, it has to be made sure, that process parameters can be actually represented in the IFC data structure or that IFC entities and/or attributes can be linked to digital LPS entities, such as e.g., quantity take-off information. With the emergence of new, open BIM toolkits and binding libraries based on higher programming languages [105], and new IFC entities—from IFC version 4 onwards—such as e.g., *IfcTaskTime*, both becomes possible at the present time. In addition, it is possible to extend the IFC data structure in accordance with official standards according to buildingSMART [106]. Also, so-called model view definitions (MVD), as defined by buildingSMART could be used to create valid IFC sub-sets [107], which allow for BIM-LPS data exchange. All this was not possible at the time of the formulation of the original BIM-Lean interaction matrix. Thus, these new possibilities can be interpreted as new BIM functionalities in line with the hypothesis presented at the beginning of this paper.

Logically, these functionalities were not listed in the original interaction matrix and their absence may have contributed to the lack of robust conceptual models for integration that we have encountered in this review. The same applies to the negative interactions that have been entitled e.g., "cannot be considered mature technology" back then. Now and with the help of new software solutions following this conceptual model, they could be potentially transformed into actual positive synergies. Eventually, this justifies our formulation of a new conceptual model and, at the same time, links our work to the previous body of knowledge in this field of research.

6. Conclusions

This integrative literature review provides for a comprehensive overview of the joint application of BIM and the Last Planner® System in the area of both practical case studies and theoretical contributions. Furthermore, this study reveals gaps at the conceptual level of BIM-LPS integrations and makes a proposal on how to fill these gaps. Concrete results of this study and their conclusions are:

- A specialized BIM-LPS interaction matrix as an update of the general BIM-Lean interaction matrices by Sacks et al. [10] and Oskouie et al. [9] has been formulated.
- On overview of the existing literature in the area of BIM-LPS co-applications in the phase of construction execution from the perspective of both theory and practice has been presented.
- The content of this literature has been condensed and key findings have been elaborated. Among them:

  ○ When BIM and the LPS are implemented together, they are often not entirely implemented (e.g., only parts of the LPS such as look-ahead planning). We know from past, pure LPS implementations, that this way of implementation can become less successful.

  ○ BIM and LPS are mainly used in parallel, but not truly integrated as e.g., in the form of an integrated information system.

  ○ Lack of a comprehensive and holistic framework or conceptual model for BIM-LPS integration in construction execution.

  ○ Software support is not treated thoroughly.

- A new conceptual model for BIM-LPS integration has been devised based on the findings of the integrative literature review.

The latter could be a reference within the existing body of knowledge to point out new "ways-of-doing" related to the fruition of BIM technology in the practice of construction execution. Future steps could foresee to actually take this theoretical conceptual model to develop new IT-tools that are based on BIM data and support the production on-site in line with LPS principles. These tools should be compliant to current BIM–lean theory and thoroughly tested under real conditions on pilot construction sites to evaluate robustness, applicability and especially utility in practice, since many important concepts—e.g., in lean production—have been erected based on findings in practical analyses [108]. In addition, the conceptual model itself does not claim to be complete or perfectly mature. Improvements based on real-implementation experiences should be continuously incorporated. Also, other aspects of production management might be interesting for assessment and potential integration into this conceptual model. For example, the role of cost control could be examined given that the LPS itself does not support this function.

Author Contributions: Conceptualization, C.P.S., C.M.; methodology, C.P.S.; investigation, C.P.S.; writing—original draft preparation, C.P.S.; supervision, C.M., G.P.M., D.T.M. All authors have read and agreed to the published version of the manuscript.

Funding: This work was supported by the Open Access Publishing Fund of the Free University of Bozen-Bolzano.

Acknowledgments: This work is part of BIM Simulation Lab—FESR 1086, a research project financed by the European Development Fund (ERDF) Südtirol/Alto Adige. Special thanks are directed to Stefania Benedicti for proofreading.

Conflicts of Interest: The authors declare no conflict of interest.

Appendix A

Table A1. BIM-LPS interactions according to the matrix presented in Table 1.

Authors	Index *	Explanation
Sacks et al. [10]	12	"Use of software capable of model integration (such as Solibri/Navisworks/Tekla) to merge models, identify clashes, and resolve them through iterative refinement of the different discipline specefic model results in almost error free installation on site."
	22	"Quick turnaround of structural, thermal, and acoustic performance analyses; of cost estimation; and of evaluation of conformance to client program, all enable collaborative design, collapsing cycle times for building design, and detailing."
	25	"All three functions serve to reduce cycle time during construction itself because they result in optimized operational schedules, with fewer conflicts."
	26	"Where process status is visualized through a BIM model, such as in the KanBIM system, series of consecutive activities required to complete a building space can be performed one after the other with little delay between them. This shortens cycle time for any given space or assembly."
	30	"Online visualization and management of process can help implement production strategies designed to reduce work-in-process inventories and production batch sizes (number of spaces in process by a specific trade at any given time), as in the KanBIM approach."
	34	"Process visualization and online communication of process status are key elements in allowing production teams to prioritize their subsequent work locations in terms of their potential contribution to ensuring a continuous subsequent flow of work that completes spaces, thus implementing a pull flow. This is central to the KanBIM approach, which extends the last planner system."
	38	"Online access to production standards, product data, and company protocols helps institutionalize standard work practices by making them readily available and, within context, to work teams at the work face. This relies, however, on provision of practical means for workers to access online information."
	40	"BIM provides an ideal visualization environment for the project throughout the design and construction stage and enables simulation of production methods, temporary equipment, and processes. Modeling and animation of construction sequences in "4D" tools provide a unique opportunity to visualize construction processes for identifying resource conflicts in time and space and resolving constructability issues. This enables process optimization improving efficiency and safety and can help identify bottlenecks and improve flow."
	−41	"Detailed planning and generation of multiple fine-grained alternatives can be said to increase complexity rather than simplify management."
	−42	"These applications cannot be considered mature technology."
	54	"Automated drawing generation improves engineering capacity when compared with 2D drafting, and it is a more reliable technology because it produces properly coordinated drawing sets."

Table A1. *Cont.*

Authors	Index *	Explanation
Authorship of this study	57	Having reliable information for construction components in terms of quantities and materials immediately available, (through a BIM viewer connected to Last Planner® process design) allows for more accurate estimation of duration and resource demand as well as better leveling of production. When correctly modelled, these information are more reliable than manual estimates from 2D plans.
	58	Quantity/material take offs and the visualization of information origin in terms of BIM objects, can help to establish pull systems and understand intuitively downstream demand
	59	First Run Studies for operation design according to the Last Planner® System could be done by making use of local 4D simulations operations that are linked to BIM objects
	60	Visualization of completed BIM components, can support the pull planning requisite of task-completion releasing new work Thanks to the BIM-based information visualization, task specifications and sequencing are clearly visible to workers or respectively to Last Planners.
Oskouie et al. [9]	61	"With BIM interfaces, real- time productivity can be measured, so that any defects in terms of production in workstations can be rapidly detected and resolved in early stages of work. This will result in learning from experience and consequently improves future methods of implementation"
	62	"Comparing generated As- Built model with As- Planned, discrepancies between these two can be easier detected. In addition, quality assurance and control can be accelerated through this procedure resulting in a better product to the owner."

* We have taken the indices from the original BIM-Lean interaction matrix by Sacks et al. [10] and continued the counter accordingly.

References

1. Tezel, A.; Aziz, Z. From conventional to it based visual management: A conceptual discussion for lean construction. *J. Inf. Technol. Constr.* **2017**, *22*, 220–246.
2. Ratajczak, J.; Riedl, M.; Matt, D.T. BIM-based and AR Application Combined with Location-Based Management System for the Improvement of the Construction Performance. *Buildings* **2019**, *9*, 118. [CrossRef]
3. Maskuriy, R.; Selamat, A.; Ali, K.N.; Maresova, P.; Krejcar, O. Industry 4.0 for the Construction Industry—How Ready Is the Industry? *Appl. Sci.* **2019**, *9*, 2819. [CrossRef]
4. Dallasega, P.; Rauch, E.; Matt, D.T.; Fronk, A. Increasing productivity in ETO construction projects through a lean methodology for demand predictability. In Proceedings of the IEOM 2015—5th International Conference on Industrial Engineering and Operations Management, Dubai, UAE, 3–5 March 2015.
5. Matthews, O.; Howell, G.A. Integrated Project Delivery An Example Of Relational Contracting. *Lean Constr. J.* **2005**, *2*, 46–61.
6. Koskela, L.; Howell, G.; Ballard, G.; Tommelein, I.D. The foundations of lean construction. In *Design and Construction: Building in Value*; Butterworth-Heinemann: Oxford, UK, 2002; pp. 211–226.
7. Matt, D.T. Template based production system design. *J. Manuf. Technol. Manag.* **2008**, *19*, 783–797. [CrossRef]
8. Womack, J.P.; Jones, D.T. *Lean Thinking: Banish Waste and Create Wealth in Your Corporation*, 1st ed.; Free Press: New York, NY, USA, 2003.
9. Oskouie, P.; Gerber, D.J.; Alves, T.; Becerik-Gerber, B. Extending the interaction of building information modeling and lean construction. In Proceedings of the IGLC 2012—20th Conference of the International Group for Lean Construction, San Diego, CA, USA, 18 July 2012.
10. Sacks, R.; Koskela, L.; Dave, B.A.; Owen, R. Interaction of lean and building information modeling in construction. *J. Constr. Eng. Manag.* **2010**, *136*, 968–980. [CrossRef]

11. Kröger, S. *BIM und Lean Construction—Synergien zweier Arbeitsmethodiken*; Beuth: Berlin, Germany, 2018; ISBN 9783410267416.
12. Saieg, P.; Sotelino, E.D.; Nascimento, D.; Caiado, R.G.G. Interactions of Building Information Modeling, Lean and Sustainability on the Architectural, Engineering and Construction industry: A systematic review. *J. Clean. Prod.* **2018**, *174*, 788–806. [CrossRef]
13. Li, J.; Hou, L.; Wang, X.; Wang, J.; Guo, J.; Zhang, S.; Jiao, Y. A project-based quantification of BIM benefits. *Int. J. Adv. Robot. Syst.* **2014**, *11*, 1–13. [CrossRef]
14. Schimanski, C.P.; Monizza, G.P.; Marcher, C.; Matt, D.T. Pushing digital automation of configure-to-order services in small and medium enterprises of the construction equipment industry: A design science research approach. *Appl. Sci.* **2019**, *9*, 3780. [CrossRef]
15. Bolpagni, M.; Burdi, L.; Luigi, A.; Ciribini, C. The implementation of building information modelling and lean construction in design firms in Massachusetts. In Proceedings of the 25th Ann. Conf. of the Int'l. Group for Lean Construction, At Haifa, Israel, 9 July 2017; Volume II, pp. 235–242.
16. Arokiaprakash, A.; Kannan, S.; Manikanda Prabhu, S. Formulize construction and operation management by integrating BIM and lean. *Int. J. Civ. Eng. Technol.* **2017**, *8*, 991–1001.
17. Mollasalehi, S.; Fleming, A.; Talebi, S.; Underwood, J. Development of an Experimental Waste Framework Based on Bim / Lean Concept in Construction Design. In Proceedings of the 24th Annual Conference of the International Group for Lean Construction, Boston, MA, USA, 18 July 2016; pp. 193–202.
18. Junior, J.S.; Baldauf, J.P.; Formoso, C.T.; Tzortzopoulos, P. Using BIM and Lean for Modelling Requirements in the Design of Healthcare Projects. In Proceedings of the 26th Annual Conference of the International Group for Lean Construction (IGLC), Chennai, India, 16–22 July 2018; pp. 571–581.
19. Al Hattab, M.; Hamzeh, F. Using social network theory and simulation to compare traditional versus BIM-lean practice for design error management. *Autom. Constr.* **2015**, *52*, 59–69. [CrossRef]
20. Sacks, R.; Radosavljevic, M.; Barak, R. Requirements for building information modeling based lean production management systems for construction. *Autom. Constr.* **2010**, *19*, 641–655. [CrossRef]
21. Gurevich, U.; Sacks, R. Examination of the effects of a KanBIM production control system on subcontractors' task selections in interior works. *Autom. Constr.* **2014**, *37*, 81–87. [CrossRef]
22. Dave, B.; Kubler, S.; Främling, K.; Koskela, L. Addressing information flow in lean production management and control in construction. In Proceedings of the 22nd Annual Conference International Group for Lean Construction IGLC, Oslo, Norway, 4 June 2014; pp. 581–592.
23. Dave, B.; Boddy, S.; Koskela, L. Visilean: Designing a production management system with lean and BIM. In Proceedings of the 19th Annual Conference of the International Group for Lean Construction, Lima, Peru, 1 January 2011; pp. 514–524.
24. Tezel, A.; Taggart, M.; Koskela, L.; Tzortzopoulos, P.; Hanahoe, J.; Kelly, M. Lean Construction and BIM in Small and Medium-Sized Enterprises (SMEs) in Construction: A Systematic Literature Review. *Can. J. Civ. Eng.* **2019**. [CrossRef]
25. Franco, J.V.; Picchi, F.A. Lean design in building projects: Guiding principles and exploratory collection of good practices. In Proceedings of the IGLC 2016—24th Annual Conference of the International Group for Lean Construction, Boston, MA, USA, 13 August 2016; pp. 113–122.
26. Harris, B.N.; Alves, T.D.C.L. Building information modeling: A report from the field. In Proceedings of the IGLC 2016—24th Annual Conference of the International Group for Lean Construction, Boston, MA, USA, 18–24 July 2016; pp. 13–22.
27. Spitler, L.; Feliz, T.; Wood, N.; Sacks, R. Constructible Bim Elements: A Root Cause Analysis Of Work Plan Failures. In Proceedings of the International Group for Lean Construction, Perth, Norway, 29–31 July 2015; pp. 351–360.
28. De Vargas, F.B.; Bataglin, F.S.; Formoso, C.T. Guidelines to develop a BIM model focused on construction planning and control. In Proceedings of the 26th Annual Conference of the International Group for Lean Construction: Evolving Lean Construction Towards Mature Production Management Across Cultures and Frontiers, Chennai, India, 16–22 July 2018; Volume 2, pp. 744–753.
29. Vestermo, A.; Murvold, V.; Svalestuen, F.; Lohne, J.; Lædre, O. BIM-stations: What it is and how it can be used to implement lean principles. In Proceedings of the IGLC 2016-24th Annual Conference of the International Group for Lean Construction, Boston, MA, USA, 18–24 July 2016; pp. 33–42.

30. Murvold, V.; Vestermo, A.; Svalestuen, F.; Lohne, J.; Lædre, O. Experiences from the use of BIM-stations. In Proceedings of the IGLC 2016-24th Annual Conference of the International Group for Lean Construction, Boston, MA, USA, 18–24 July 2016; pp. 23–32.
31. Eastman, C.M.; Teicholz, P.; Sacks, R.; Liston, K. *BIM Handbook: A Guide to Building Information Modeling for Owners, Managers, Designers, Engineers and Contractors*, 1st ed.; Wiley: Hoboken, NJ, USA, 2008; ISBN 9780470541371.
32. NBS National BIM Report, Building Information Modelling. Available online: https://www.thenbs.com/about-nbs/ (accessed on 20 October 2019).
33. Azhar, S.; Nadeem, A.; Mok, J.Y.N.; Leung, B.H.Y. Building Information Modeling (BIM): A New Paradigm for Visual Interactive Modeling and Simulation for Construction Projects. In Proceedings of the First International Conference on Construction in Developing Countries (ICCIDC-I), Karachi, Pakistan, 4 August 2008; pp. 435–446.
34. Mirshokraei, M.; De Gaetani, C.I.; Migliaccio, F. A web-based BIM-AR quality management system for structural elements. *Appl. Sci.* **2019**, *9*, 3984. [CrossRef]
35. Kalsaas, B.T. Measuring waste and workflow in construction. In Proceedings of the Annual Conference of the International Group for Lean Construction, Fortaleza, Brazil, 29 July–2 August 2013; pp. 33–42.
36. Ballard, G. *The Last Planner System of Production Control*; University of Birmingham: Birmingham, UK, 2000.
37. Schimanski, C.P.; Monizza, G.P.; Marcher, C.; Matt, D.T. Conceptual Foundations for a New Lean BIM-Based Production System in Construction. In Proceedings of the 27th Annual Conference of the International Group for Lean Construction, Dublin, Ireland, 1–7 July 2019; pp. 877–888.
38. Hamzeh, F.; Bergstrom, E. The Lean Transformation: A Framework for Successful Implementation of the Last Planner TM System in Construction. In Proceedings of the International 46th Annual Conference Associated Schools of Construction, Boston, MA, USA, 7–10 April 2010; p. 8.
39. Daniel, E.I.; Pasquire, C.; Dickens, G. Exploring the implementation of the last planner® system through iglc community: Twenty one years of experience. In Proceedings of the 23rd Annual Conference of the International Group for Lean Construction: Global Knowledge-Global Solutions, Perth, Australia, 28 July 2015; Volume 23, pp. 153–162.
40. Hamzeh, F.R. The Lean Journey: Implementing The Last Planner® System in Construction. In Proceedings of the 19th Annual Conference of the International Group for Lean Construction, Lima, Peru, 13–15 July 2011; pp. 379–390.
41. Gao, S.; Low, S.P. The Last Planner System in China's construction industry—A SWOT analysis on implementation. *Int. J. Proj. Manag.* **2014**, *32*, 1260–1272. [CrossRef]
42. Whitehead, D. Searching and reviewing the research literature. In *Nursing & Midwifery Research: Methods and Appraisal for Evidence-Based Practice*; Schneider, Z., Whitehead, D., LoBiondo-Wood, G., Habe, R.J., Eds.; Elsevier: Sidney, Australia, 2013; pp. 35–56.
43. Kitchenham, B. *Procedures for Performing Systematic Reviews*; Joint Technical Report TR/SE-0401; Computer Science Department, Keele University: Keele, UK, 2004; Volume 33.
44. Torraco, R.J. Writing Integrative Literature Reviews: Guidelines and Examples. *Hum. Resour. Dev. Rev.* **2005**, *4*, 356–367. [CrossRef]
45. Saunders, A.; Lewis, P.; Thornhill, A. *Research Methods for Business Students*; Pearson Education Limited: Harlow, UK, 2012.
46. Van Eck, J.N.; Waltman, L. Software survey: VOSviewer, a computer program for bibliometric mapping. *Scientometrics* **2010**, *84*, 523–538. [CrossRef]
47. Jin, R.; Yuan, H.; Chen, Q. Science mapping approach to assisting the review of construction and demolition waste management research published between 2009 and 2018. *Resour. Conserv. Recycl.* **2019**, *140*, 175–188. [CrossRef]
48. Lagos, C.I.; Herrera, R.F.; Alarcón, L.F. Assessing the Impacts of an IT LPS Support System on Schedule Accomplishment in Construction Projects. *J. Constr. Eng. Manag.* **2019**, *145*, 04019055. [CrossRef]
49. Heigermoser, D.; de Soto, B.; Abbott, E.L.S.; Chua, D.K.H. BIM-based Last Planner System tool for improving construction project management. *Autom. Constr.* **2019**, *104*, 246–254. [CrossRef]
50. Bortolini, R.; Formoso, C.T.; Viana, D.D. Site logistics planning and control for engineer-to-order prefabricated building systems using BIM 4D modeling. *Autom. Constr.* **2019**, *98*, 248–264. [CrossRef]

51. Melzner, J. BIM-based takt-time planning and takt control: Requirements for digital construction process management. In Proceedings of the 36th International Symposium on Automation and Robotics in Construction, Banff, AB, Canada, 21–24 May 2019; pp. 50–56.
52. Koseoglu, O.; Nurtan-Gunes, E.T. Mobile BIM implementation and lean interaction on construction site: A case study of a complex airport project. *Eng. Constr. Arch. Manag.* **2018**, *25*, 1298–1321. [CrossRef]
53. Koseoglu, O.; Sakin, M.; Arayici, Y. Exploring the BIM and lean synergies in the Istanbul Grand Airport construction project. *Eng. Constr. Arch. Manag.* **2018**, *25*, 1339–1354. [CrossRef]
54. Zhang, X.; Azhar, S.; Nadeem, A.; Khalfan, M. Using Building Information Modelling to achieve Lean principles by improving efficiency of work teams. *Int. J. Constr. Manag.* **2017**, *18*, 293–300. [CrossRef]
55. Guerriero, A.; Kubicki, S.; Berroir, F.; Lemaire, C. BIM-enhanced collaborative smart technologies for LEAN construction processes. In Proceedings of the 2017 International Conference on Engineering, Technology and Innovation: Engineering, Technology and Innovation Management Beyond 2020: New Challenges, New Approaches, ICE/ITMC 2017, Piscataway, NJ, USA, 27–29 June 2017; pp. 1023–1030.
56. Surendhra Babu, P.R.; Hayath Babu, N. Using technology to achieve lean objectives. In Proceedings of the 26th Annual Conference of the International Group for Lean Construction: Evolving Lean Construction Towards Mature Production Management Across Cultures and Frontiers, Chennai, India, 16–22 July 2018; pp. 1069–1078.
57. Mollasalehi, S.; Aboumoemen, A.A.; Rathnayake, A.; Fleming, A.; Underwood, J. Development of an integrated BIM and lean maturity model. In Proceedings of the 26th Annual Conference of the International Group for Lean Construction: Evolving Lean Construction Towards Mature Production Management Across Cultures and Frontiers, Chennai, India, 16–22 July 2018; pp. 1217–1228.
58. Singhal, N.; Ahuja, R. CAN BIM furnish lean benefits—An Indian case study. In Proceedings of the 26th Annual Conference of the International Group for Lean Construction: Evolving Lean Construction Towards Mature Production Management Across Cultures and Frontiers, Chennai, India, 16–22 July 2018; pp. 90–100.
59. Antunes, R.; Poshdar, M. Envision of an integrated information system for project-driven production in construction. In Proceedings of the 26th Annual Conference of the International Group for Lean Construction: Evolving Lean Construction Towards Mature Production Management Across Cultures and Frontiers, Chennai, India, 16–22 July 2018; pp. 134–143.
60. Häringer, P.; Borrmann, A. Evaluation of a case study to design a BIM-based cycle planning concept. In Proceedings of the 26th Annual Conference of the International Group for Lean Construction: Evolving Lean Construction Towards Mature Production Management Across Cultures and Frontiers, Chennai, India, 16–22 July 2018; pp. 58–67.
61. Elmaraghy, A.; Voordijk, H.; Marzouk, M. An exploration of BIM and lean interaction in optimizing demolition projects. In Proceedings of the 26th Annual Conference of the International Group for Lean Construction: Evolving Lean Construction Towards Mature Production Management Across Cultures and Frontiers, Chennai, India, 16–22 July 2018; pp. 112–122.
62. Matta, G.; Herrera, R.F.; Baladrón, C.; Giménez, Z.; Alarcón, L.F. Using BIM-Based sheets as a visual management tool for on-site instructions: A case study. In Proceedings of the 26th Annual Conference of the International Group for Lean Construction: Evolving Lean Construction Towards Mature Production Management Across Cultures and Frontiers, Chennai, India, 16–22 July 2018; pp. 144–154.
63. Vrijhoef, R.; Dijkstra, J.T.; Koutamanis, A. Modelling and simulating time use of site workers with 4d BIM. In Proceedings of the 26th Annual Conference of the International Group for Lean Construction: Evolving Lean Construction Towards Mature Production Management Across Cultures and Frontiers, Chennai, India, 16–22 July 2018; pp. 155–165.
64. Goyal, M.; Gao, Z. Integration of Building Information Modeling (BIM) and Prefabrication for Lean Construction. In Proceedings of the ICCREM 2018: Innovative Technology and Intelligent Construction-Proceedings of the International Conference on Construction and Real Estate Management 2018, Charleston, SC, USA, 9–10 August 2018; pp. 78–84.
65. Fosse, R.; Ballard, G.; Fischer, M. Virtual design and construction: Aligning BIM and lean in practice. In Proceedings of the 25th Annual Conference of the International Group for Lean Construction, Heraklion, Greece, 9–12 July 2017; pp. 499–506.

66. Lagos, C.I.; Herrera, R.F.; Alarcón, L.F. Contributions of information technologies to last planner system implementation. In Proceedings of the 25th Annual Conference of the International Group for Lean Construction, Heraklion, Greece, 9–12 July 2017; pp. 87–94.

67. Von Heyl, J.; Teizer, J. Lean production controlling and tracking using digital methods. In Proceedings of the 25th Annual Conference of the International Group for Lean Construction, Heraklion, Greece, 9–12 July 2017; pp. 127–134.

68. Toledo, M.; Olivares, K.; González, V. Exploration of a lean-BIM planning framework: A last planner system and BIM-based case study. In Proceedings of the IGLC 2016-24th Annual Conference of the International Group for Lean Construction, Boston, MA, USA, 18–24, July 2016; pp. 3–12.

69. Haiati, O.; von Heyl, J.; Schmalz, S. Bim and Sequence Simulation in Structural Work—Development of a Procedure for Automation. In Proceedings of the 24th Annual Conference of the International for Lean Construction, Boston, MA, USA, 18–24, July 2016; pp. 73–82.

70. Tillmann, P.; Sargent, Z. Last Planner & Bim Integration: Lessons From a Continuous Improvement Effort. In Proceedings of the International Group for Lean Construction, Boston, MA, USA, 18–24, July 2016; pp. 113–122.

71. Ben-Alon, L.; Sacks, R. Simulating and vizualising emergent production in construction (Epic)using agents and BIM. In Proceedings of the 23rd Annual Conference of the International Group for Lean Construction: Global Knowledge—Global Solutions, Perth, Australia, 29–31 July 2015; pp. 371–380.

72. Dave, B.; Kubler, S.; Pikas, E.; Holmström, J.; Singh, V.; Främling, K.; Koskela, L.; Peltokorp, A. Intelligent Products: Shifting the Production Control Logic in Construction (With Lean and BIM). In Proceedings of the 23rd Annual Conference of the International Group for Lean Construction, Perth, Australia, 29–31 July 2015; pp. 341–350.

73. Tillmann, P.; Viana, D.; Sargent, Z.; Tommelein, I.; Formoso, C. Bim and lean in the design-production interface of eto components in complex projects. In Proceedings of the 23rd Annual Conference of the International Group for Lean Construction: Global Knowledge-Global Solutions, Perth, Australia, 29–31 July 2015; pp. 331–340.

74. Tallgren, M.V.; Roupé, M.; Johansson, M. An empowered collaborative planning method in a swedish construction company-a case study. In Proceedings of the 31st Annual Association of Researchers in Construction Management Conference, ARCOM 2015, Lincoln, UK, 7–9 September 2015; pp. 793–802.

75. Garrido, M.C.; Mendes, R.; Scheer, S.; Campestrini, T.F. Using BIM for last planner system: Case studies in Brazil. In Proceedings of the Congress on Computing in Civil Engineering, Austin, TX, USA, 21–23 June 2015; pp. 604–611.

76. Ahuja, R.; Sawhney, A.; Arif, M. Bim based conceptual framework for lean and green integration. In Proceedings of the 22nd Annual Conference of the International Group for Lean Construction: Understanding and Improving Project Based Production, Oslo, Norway, 25–27 June 2014; pp. 123–132.

77. Toledo, M.; González, V.A.; Villegas, A.; Mourgues, C. Using 4D models for tracking project progress and visualizing the owner's constraints in fast-track retail renovation projects. In Proceedings of the 22nd Annual Conference of the International Group for Lean Construction: Understanding and Improving Project Based Production, Oslo, Norway, 25–27 June 2014; pp. 969–980.

78. Biotto, C.; Mota, B.; Araújo, L.; Barbosa, G.; Andrade, F. Adapted use of andon in a horizontal residential construction project. In Proceedings of the 22nd Annual Conference of the International Group for Lean Construction: Understanding and Improving Project Based Production, Oslo, Norway, 25–27 June 2014; pp. 1295–1305.

79. Bouguessa, A.; Forgues, D.; Doré, S. Complementarity between the Building Information Modeling (BIM) and Product Life Cycle Management (PLM) through the Lean Construction (LC). In Proceedings of the Proceedings, Annual Conference-Canadian Society for Civil Engineering, Montreal, QC, Canada, 29 May–1 June 2013; pp. 1139–1148.

80. Clemente, J.; Cachadinha, N. BIM-lean synergies in the management on MEP works in public facilities of intensive use-A case study. In Proceedings of the 21st Annual Conference of the International Group for Lean Construction 2013, Fortaleza, Brazil, 29 July–2 August 2013; pp. 70–79.

81. Sacks, R.; Barak, R.; Belaciano, B.; Gurevich, U.; Pikas, E. KanBIM workflow management system: Prototype implementation and field testing. *Lean Constr. J.* **2013**, *2013*, 19–35.

82. Hamdi, O.; Leite, F. BIM and Lean interactions from the bim capability maturity model perspective: A case study. In Proceedings of the IGLC 2012-20th Conference of the International Group for Lean Construction, San Diego, CA, USA, 18–20 July 2012.

83. Bhatla, A.; Leite, F. Integration Framework of BIM with the Last Planner System. In Proceedings of the 20th Annual Conference of the International Group for Lean Construction, San Diego, CA, USA, 18–20 July 2012; p. 10.

84. Sacks, R.; Barak, R.; Belaciano, B.; Gurevich, U.; Pikas, E. Field tests of the KanBIM lean production management system. In Proceedings of the 19th Annual Conference of the International Group for Lean Construction 2011, Lima, Peru, 13–15 July 2011; pp. 465–476.

85. Sacks, R.; Dave, B.A.; Koskela, L.; Owen, R. Analysis framework for the interaction between lean construction and Building Information Modelling. In Proceedings of the 17th Annual Conference of the International Group for Lean Construction, Taipei, Taiwan, 15–17 July 2019; pp. 221–234.

86. Sacks, R.; Treckmann, M.; Rozenfeld, O. Visualization of work flow to support lean construction. *J. Constr. Eng. Manag.* **2009**, *135*, 1307–1315. [CrossRef]

87. Siraj, N.B.; Fayek, A.R. Risk Identification and Common Risks in Construction: Literature Review and Content Analysis. *J. Constr. Eng. Manag.* **2019**, *145*, 03119004. [CrossRef]

88. Shelley, M.; Krippendorff, K. Content Analysis: An Introduction to its Methodology. *J. Am. Stat. Assoc.* **1984**, *79*, 240. [CrossRef]

89. Sacks, R.; Radosavljevic, M.; Barak, R. A Building Information Modelling Based Production Control System for Construction. In Proceedings of the W078-Special Track 18th CIB World Building Congress May 2010, Salford, UK, 10–13 May 2010; p. 1.

90. Khan, S.; Tzortzopoulos, P. Effects of the interactions between lps and bim on workflow in two building design projects. In Proceedings of the 22nd Annual Conference of the International Group for Lean Construction, Oslo, Norway, 25–27 June 2014; pp. 933–944.

91. Ballard, G.; Tommelein, I. *Current Process Benchmark for the Last Planner System*; Project Production Systems Laboratory, UC Berkeley: Berkeley, CA, USA, 2016.

92. Gerber, D.J.; Becerik-Gerber, B. Building Information Modeling and Lean Construction: Technology, Methodology and Advances From Practice. In Proceedings of the 18th Annual Conference, International Group for Lean Construction, Haifa, Israel, 14–16 July 2010.

93. Bortolazza, R.; Formoso, C.T. A quantitative analysis of data collected from the last planner system in brazil. In Proceedings of the 14th International Group for Lean Construction, Santiago de Chile, Chile, 25–27 July 2006; pp. 625–635.

94. Alarcón, L.F.; Diethelm, S.; Rojo, O.; Calderon, R. Assessing the impacts of implementing lean construction. In Proceedings of the 13th Annual Conference of the International Group for Lean Construction, Sydney, Australia, 19–21 July 2005; pp. 387–393.

95. Friblick, F.; Olsson, V.; Reslow, J. Prospects for implementing Last Planner in the construction industry. In Proceedings of the 17th Annual Conference of the International Group for Lean Construction, Taipei, Taiwan, 15–17 July 2019; pp. 197–206.

96. Kasanen, E.; Lukha, K.; Siitonen, A. The constructive approach in management accounting research. *J. Manag. Account. Res.* **1993**, *5*, 243–264.

97. Rybkowski, Z.K. Last planner and its role as conceptual kanban. In Proceedings of the Iglc 2010, Haifa, Israel, 14–16 July 2010; 2010; pp. 63–72.

98. Mossman, A. *Last Planner®: 5 + 1 Crucial & Collaborative Conversations for Predictable Design & Construction Delivery*; The Change Business Ltd.: Stroud, Gloucestershire, UK, 2013.

99. Matt, D.T.; Rauch, E. Implementing Lean in Engineer-to-Order Manufacturing. In *Handbook of Research on Design and Management of Lean Production Systems*; Modrák, V., Semančo, P., Eds.; IGI Global: Hershey, PA, USA, 2014; pp. 148–172. ISBN 97814666503981.

100. Greca, I.M.; Moreira, M.A. Mental models, conceptual models, and modelling. *Int. J. Sci. Educ.* **2000**, *22*, 11. [CrossRef]

101. Anderson, D.J. *Kanban-Successful Evolutionary Change for Your Technology Business*; Blue Hole Press: Sequim, WA, USA, 2010; Volume 209, ISBN 978-0-9845214-0-1.

102. Modrich, R.-U.; Cousins, B.C. Digital kanban boards used in desing and 3D coordination. In Proceedings of the 25th Ann. Conf. of the Int'l. Group for Lean Construction, Heraklion, Greece, 9–12 July 2017; pp. 663–670.

103. Ballard, G. Phase Scheduling. *LCI White Pap.* **2000**, *7*, 1–3.
104. Deming, W.E. *Out of the Crisis*, 1st ed.; The MIT Press: Cambridge, UK, 1982; ISBN 0911379010.
105. Jiang, S.; Jiang, L.; Han, Y.; Wu, Z.; Wang, N. OpenBIM: An Enabling Solution for Information Interoperability. *Appl. Sci.* **2019**, *9*, 5358. [CrossRef]
106. Scherer, R.J.; Schapke, S.-E. *Informationssysteme im Bauwesen 1: Modelle, Methoden und Prozesse*; Springer: Berlin/Heidelberg, Germany, 2014; ISBN 978-3-642-40882-3.
107. Jiang, S.; Wu, Z.; Zhang, B.; Cha, H.S. Combined MvdXML and semantic technologies for green construction code checking. *Appl. Sci.* **2019**, *9*, 1463. [CrossRef]
108. Matt, D.T.; Rauch, E. Design of a scalable modular production system for a two-stage food service franchise system. *Int. J. Eng. Bus. Manag.* **2012**, *4*, 32. [CrossRef]

Review

OpenBIM: An Enabling Solution for Information Interoperability

Shaohua Jiang *, Liping Jiang, Yunwei Han, Zheng Wu and Na Wang

Department of Construction Management, Dalian University of Technology, Dalian 116024, China; pome16@163.com (L.J.); hanyunwei0627@sina.com (Y.H.); wuzheng0709@gmail.com (Z.W.); sweetwzy@126.com (N.W.)
* Correspondence: shjiang@dlut.edu.cn; Tel.: +86-411-8470-7482

Received: 31 October 2019; Accepted: 4 December 2019; Published: 8 December 2019

Abstract: The expansion of scale and the increase of complexity of construction projects puts higher requirements on the level of collaboration among different stakeholders. How to realize better information interoperability among multiple disciplines and different software platforms becomes a key problem in the collaborative process. openBIM (building information model), as a common approach of information exchange, can meet the needs of information interaction among different software well and improve the efficiency and accuracy of collaboration. To the best of our knowledge, there is currently no comprehensive survey of openBIM approach in the context of the AEC (Architecture, Engineering & Construction) industry, this paper fills the gap and presents a literature review of openBIM. In this paper, the openBIM related standards, software platforms, and tools enabling information interoperability are introduced and analyzed comprehensively based on related websites and literature. Furthermore, engineering information interoperability research supported by openBIM is analyzed from the perspectives of information representation, information query, information exchange, information extension, and information integration. Finally, research gaps and future directions are presented based on the analysis of existing research. The systematic analysis of the theory and practice of openBIM in this paper can provide support for its further research and application.

Keywords: openBIM; information interoperability; standards; software

---

1. Introduction

The construction industry, as one of the traditional industries in the world, boosts the economic development and accelerates the realization of urbanization. However, the development of more complicated construction projects has also created a series of collaboration issues among different stakeholders and software platforms, which has placed higher requirements on the collaboration over the whole lifecycle, including design, construction, operation, and maintenance. In most cases, data in the construction field is fragmented and distributed in different sources, which is not convenient for information collection and representation. How to better solve information collaboration issues is closely related to the success of projects.

Building information model (BIM) software is widely used in design, construction, operation, and maintenance stages, providing great convenience for construction industry. However, data heterogeneity among different software leads to a series of problems, such as interoperability issues. As a universal approach for collaborative design, construction, and operation of buildings based on open standards and workflows [1], openBIM provides solutions for these problems by reducing collaborative errors, improving the interoperability between software, and ensuring accuracy of multi-party collaboration, thus improving the efficiency of the whole project.

At present, studies related to openBIM have received extensive attention, but there is still a lack of a comprehensive overview of openBIM. Therefore, openBIM related standards, software platforms and tools are summarized in this paper by analyzing openBIM related websites and literature from Scopus, Engineering Village, and so on. Furthermore, engineering information interoperability research supported by openBIM is concluded to provide a systematic review of openBIM related research and practice. Moreover, the limitations of existing research and open research challenges as well as possible suggestions are outlined to assist both the engineers and researchers in giving a guide for future research.

The whole paper is organized as follows. The research methodology of the paper is analyzed in Section 2; systematic analysis of openBIM standards, software platforms, and tools are given in Section 3; engineering information interoperability research supported by openBIM is summarized in Section 4; discussion about research gaps and future directions is given in Section 5; and the conclusion is made in Section 6.

2. Literature Search

To explore the current research status of openBIM among different countries, the paper refers to related literature from several main and authoritative databases in the world and summarizes the distribution of openBIM related publications from the perspectives of databases, standards, time, publishers, and countries, respectively.

2.1. Distribution of Publications by Databases

Papers related to openBIM were collected from five databases, namely Web of Science, Scopus, Engineering Village, Science Direct, and CNKI, which is one of the most authoritative academic databases in China. In order to comprehensively consider the development process of openBIM related research in recent years, the search period was set as from January 2000 to October 2019, giving the consideration for 20 years. The number of openBIM related papers in different databases were counted respectively by taking the combination of the abbreviation of the standard and BIM as search field, such as "IFC and BIM" when searching Industry Foundation Classes (IFC)-related papers. The statistical data according to the standards and databases are listed in Table 1. It can be seen from the table that the most widely adopted openBIM standard is IFC, while Information Delivery Manual (IDM) and Model View Definition (MVD) rank second and third respectively.

Table 1. Distribution of publications by databases.

		Scopus	Web of Science (Core Collection)	Engineering Village	CNKI	Science Direct	Total
1	IFC	738	392	620	534	147	2431
2	IDM	56	31	49	38	9	183
3	MVD	43	28	46	21	16	154
4	COBie	47	19	32	9	4	111
5	LandXML	26	15	20	13	7	81
6	ifcOWL	22	22	21	7	7	79
7	IFD	4	8	1	22	0	35
8	BCF	8	9	5	1	2	25
9	bsDD	2	3	2	0	0	7
	Total	946	527	796	645	192	3106

2.2. Distribution of Publications by Standards

It can be seen from Figure 1, the openBIM standards are widely used. As the core of openBIM, IFC, MVD, and IDM are adopted by most of the papers while IFC takes a lead with a number of 2431 which takes up around 78% of the total papers.

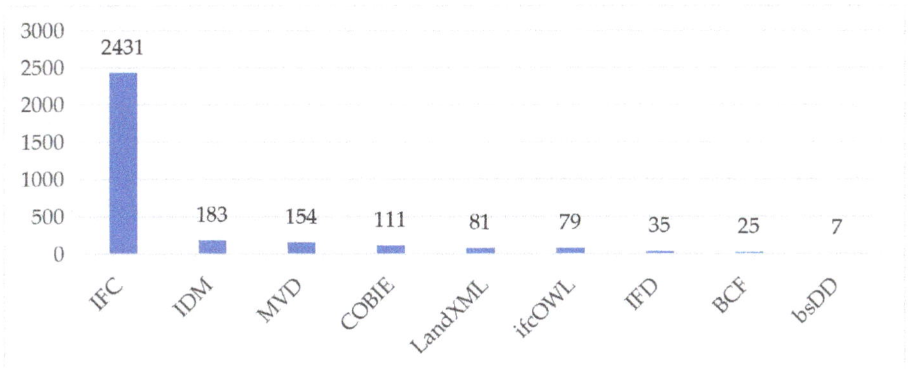

Figure 1. Distribution of publications by standards.

2.3. Distribution of Publications Over Time

It can be seen from Figure 2, the published papers related to openBIM are from 2000 to 2018, most of them are from 2011 to 2018. The number of these published papers has increased rapidly in recent years, which reflects how openBIM related research has been paid more and more attention.

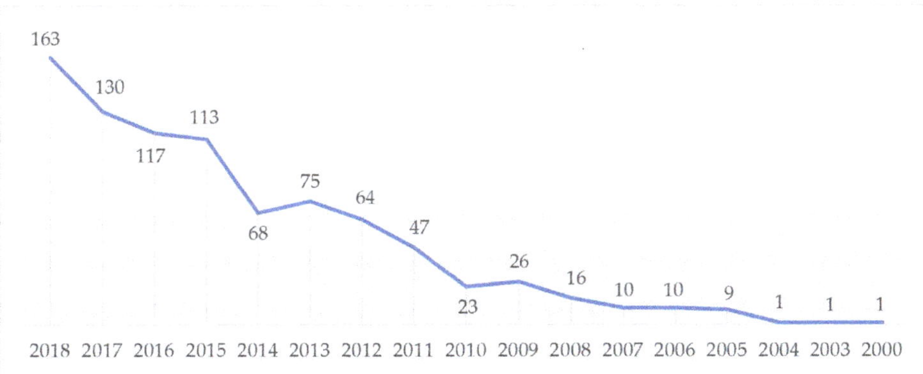

Figure 2. Distribution of publications over time.

2.4. Distribution of Publications by Publishers

It can be seen from Figure 3, most of the research related openBIM was published in journals, such as Automation in Construction, Advanced Engineering Informatics, and Journal of Computing in Civil Engineering. In particularly, papers in the journal of Automation in Construction take around 50% of the total papers.

2.5. Distribution of Publications by Countries

The statistical data of distribution of publications by countries is based on the authors of the published paper. As for papers with several different authors, the country of each author is recorded and the same country is just recorded for once. For example, if A (USA), B (Israel), C (China), and D (China) are authors for a specific paper, then the country of USA, Israel, and China is recorded once, respectively. It can be seen from Figure 4, the research related to openBIM is mainly from USA, Germany, China, and some other European countries.

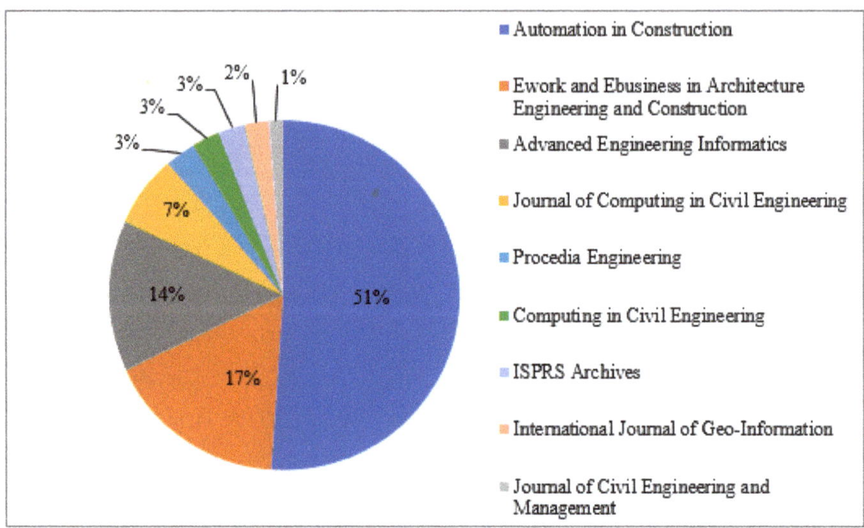

Figure 3. Distribution of publications by publishers.

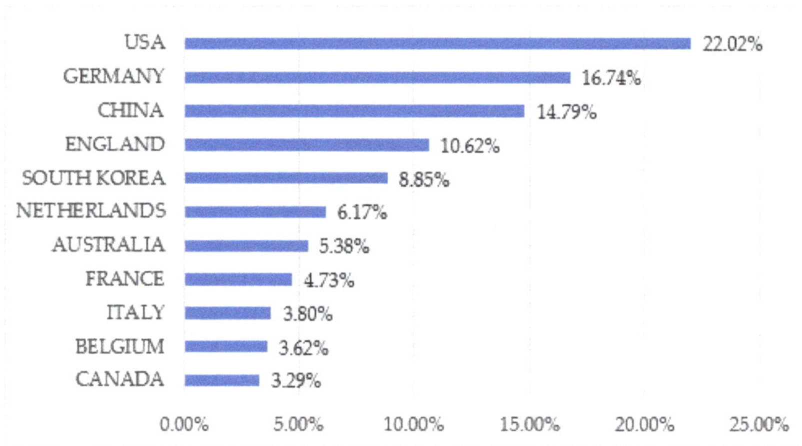

Figure 4. Distribution of publications by countries.

In conclusion, according to the above analysis result, it can be seen that: (1) among all the openBIM standards, IFC has been widely adopted by most of the fields, ranking in the top of the total application. (2) the development of openBIM varies among countries, while USA ranks first. (3) the application of openBIM standards are continuously growing over time, especially after the year of 2011.

3. Overview of OpenBIM

3.1. OpenBIM Standards

buildingSMART is an open, neutral, and international not-for-profit organization which is committed to creating and adopting open, international standards and solutions for infrastructure and buildings so as to drive the digital transformation of the built asset industry [2]. Based on the analysis of openBIM standards from the official website of buildingSMART, openBIM standards are divided into three main categories in this paper: buildingSMART standards, buildingSMART candidate standards, and buildingSMART related other standards. buildingSMART standards are the final standards that have been voted by the Standards Committee. buildingSMART candidate standards are the activities that are waiting for acquiring international consensus before being submitted to the Standards Committee for the final vote [3]. buildingSMART related other standards are formed to supplement the openBIM standards.

3.1.1. buildingSMART Standards

buildingSMART standards, as the most important part of openBIM standards, mainly includes five basic standards, the names and functions of these five basic standards are briefly summarized one by one in Table 2 [4].

Table 2. Brief summary of five basic standards.

Name	What It Does
IFC Industry Foundation Classes	Transports information or data
IDM Information Delivery Manual	Describes processes
MVD Model View Definitions	Translates processes into technical requirements
BCF BIM Collaboration Format	Change coordination
bsDD buildingSMART Data Dictionary	Defines BIM objects and their attributes

1. IFC—Data Standard

IFC is a common data schema for describing building information model data, which aims to establish a standard method of data representation and storage, so that all kinds of software can import and export building data in this format, thus promoting data sharing among different specialties and different software throughout the whole life cycle. IFC standards can be used to unify the format of information generated by different types of software so as to realize the free conversion of building information [2]. Up to now, the maturity level of each IFC version can be vividly represented in Figure 5.

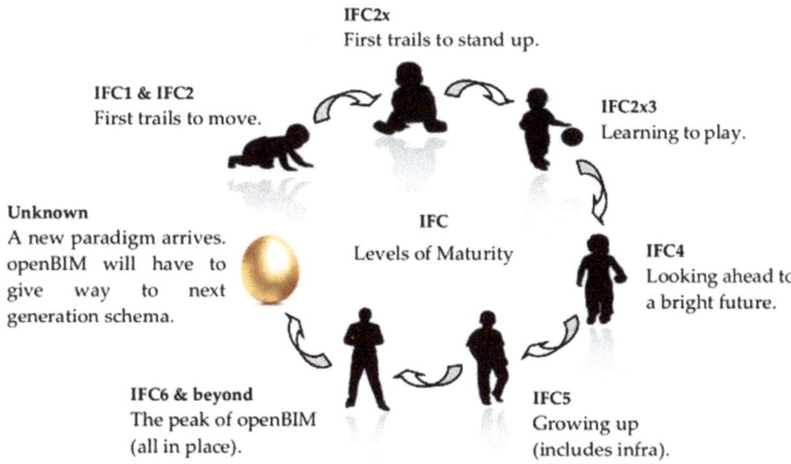

Figure 5. The maturity levels of Industry Foundation Classes (IFC).

2. IDM—Process Standard

IDM is a method for capturing and specifying the whole process and information flow during building life cycle, which aims to ensure relevant data is exchanged in a way that can be interpreted by the recipient software. The construction and maintenance of buildings involves many different participants, knowing what information needs to be communicated among them and when it is communicated is important. IDM can facilitate this information communication process by making full use of business process modelling notation (BPMN) and templates for Exchange Requirements [2].

3. MVD—Process Translation

The Model View Definition defines a subset of IFC schema that needs to meet one or more exchange requirements in the AEC industry. Together with this subset of IFC schema, a set of implementation instructions and validation rules are issued, and the method of publishing concepts and related rules is mvdXML [5]. The mvdXML is an encoding format, through which MVD can be encoded and allowable values in specific attributes of a specific data type can be defined. In practice, software applications can use mvdXML statically to support specific model views or use it dynamically to support any model view.

4. BIM Collaboration Format (BCF)—Change Coordination

The BIM collaboration format (BCF) is a simplified and open standard XML format used to encode information which enables workflow communication between different BIM software tools [3]. The development of BCF mainly includes two parts, namely a XML file format and a RESTful web service, in which the open file XML format called bcfXML can support workflow communication in BIM processes; besides, the RESTful web service called bcfAPI can enable software applications to seamlessly exchange BCF data in BIM workflows [6].

5. buildingSMART Data Dictionary (bsDD)—Standard Library

buildingSMART Data Dictionary (bsDD), as a reference library based on the IFD standard (ISO 12006-3), is one of a set of buildingSMART standards that enables collaborative work. To understand bsDD standard better, it is necessary to understand the concept of IFD standard. IFD is a terminology library-oriented or ontology-oriented standard, which ensures the accuracy of BIM information exchange and sharing by establishing the mapping relationship between IFC and information representations of different languages and vocabularies, as well as specifying the related conceptual semantics of objects [2]. Based on the IFD standard, the bsDD provides a shared system for identifying and validating the names of objects and their attributes used in building information models, which help to better understand the meaning of general terms, thereby improving the information interoperability during the construction industry [7].

In summary, each of buildingSMART's standards plays a significant role for sharing and exchanging structured building information openly during the whole life cycle, the process of using these five basic standards for collaborative work is shown in the Figure 6.

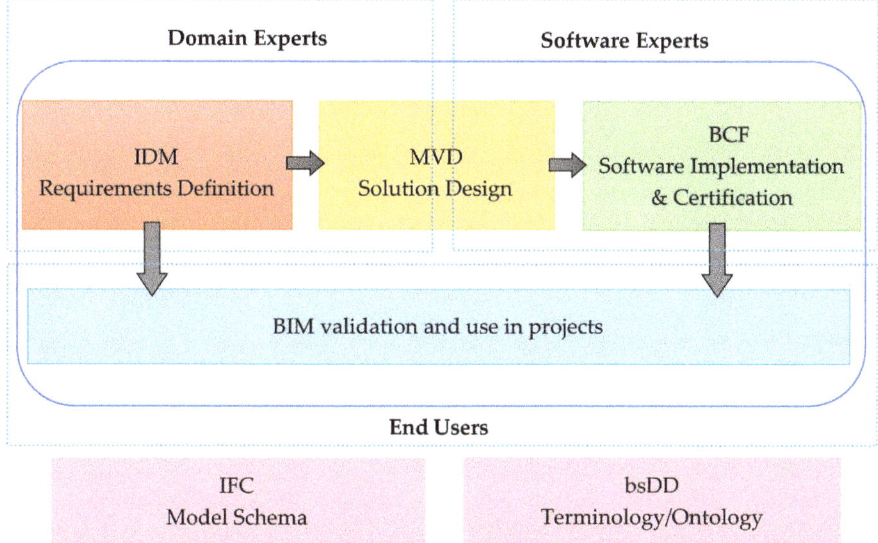

Figure 6. Process of using five basic standards for collaborative work.

3.1.2. BuildingSMART Candidate Standards

IfcOWL, as the only buildingSMART candidate standard, is used for providing a Web Ontology Language (OWL) representation for the IFC standard. To represent IFC data model as a schema, some schema modelling languages, such as EXPRESS and XML Schema Definition (XSD), are used to achieve this representation. By using OWL to provide corresponding representation of IFC schema, ifcOWL ontology is developed which has the same status as the EXPRESS and XSD schemas of IFC [8].

3.1.3. BuildingSMART Related Other Standards

1. Construction operations building information exchange (COBie)

The COBie is an information exchange specification for collecting and delivering information required by facility managers during the life cycle of the buildings. COBie can be viewed in design, construction, maintenance software, and simple spreadsheets, and this versatility enables COBie to be used in all projects regardless of project size and technical complexity [9].

2. LandXML

LandXML, as an electronic data transfer standard, was developed by the Open Geospatial Consortium in 1994. LandXML schema includes surveying, road design, and digital terrain model (DTM), which is capable of capturing, validating and displaying civil engineering and survey measurement data. Furthermore, infrastructure spatial objects based on the Geography Markup Language (GML) systems have been developed which help to realize the integration and transformation with the LandXML system [10].

3.2. OpenBIM Software Platforms and Tools

Based on the analysis about the existing research about openBIM, the mainly used openBIM software platforms and tools are listed as below (see Figure 7).

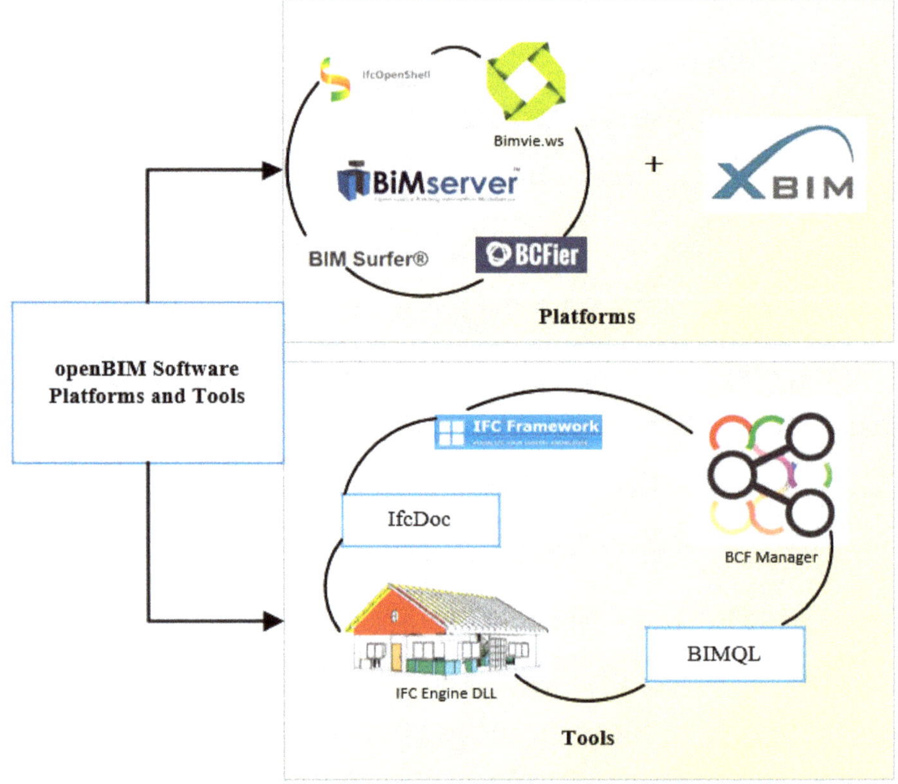

Figure 7. openBIM (building information model) software platforms and tools.

3.2.1. Platforms

Participants and stakeholders of construction projects can use different BIM software to realize the timely sharing of information [11]. At present, BIM software mainly includes Autodesk, Bentley, and Trimble series. There is a high compatibility between the same company's software, but the interoperability between different companies' software is relatively low, which leads to the difficulty of information sharing between software. Although some of the software has plugin development interface, but still cannot fully meet the construction industry's BIM software requirements about customization and modularization. This section introduces two typical open source software platforms.

1. BIMserver

BIMserver is an open and stable software, which can easily build reliable BIM software tools for supporting dynamic collaboration processes of AEC domain users. BIMserver (BIMserver.org) platform enables users to create their own "BIM operating system". The core of the software is based on the open IFC standard, so it can handle IFC data well. In BIMserver, the intelligent core interprets IFC data and stores it in the underlying database as an object. The main advantage of this method is that it can query, merge and filter BIM data. In addition to the core database functions of merging, model checking, authorization, authentication, and comparison, there are many other functions that can lower the threshold for developers [12]. BIMserver's common plugins are as follows:

*   IfcOpenShell is an open source software library that helps users and software developers to parse IFC models [13].

- bimvie.ws developed an online viewing software tool of BIM built in HTML and Javascript. It is compatible with the BIM Service interface exchange (BIMSie) API interface and works with BIM on the cloud platform. BIMSie is a standard API for BIM Web Services to get BIM in the cloud. Users can use bimvie.ws with any BIMSie-compatible online BIM service [14].
- BIMsurfer is an open source IFC model viewer developed by SceneJS and WebGL to parse BIM data from IFC and glTF formats [15].
- BCFier is an application that can handle BCF files, and it can be integrated directly with BIM software as a plugin. BCFier allows users to create and open BCF files in Autodesk Revit, add multiple views and comments, and easily share them with other team members [16].

2. Extensible Building Information Modeling (xBIM)

Extensible building information modeling (xBIM) is a free open source software development platform that allows developers to create custom BIM middleware for IFC-based applications [17]. xBIM provides a rich API for IFC data standards. It allows developers to read, write and update IFC files in a few lines of codes.

xBIM has a complete geometric information engine that converts IFC geometric data objects (such as IfcSweptAreaSolid) into fully functional boundary representation (Brep) geometric models. These models support all Boolean operations, intersection, union, and calculation behaviors such as volume, area, and length. Geometric information engine not only provides optimized 3D triangulation and meshing for visualization, but also provides overall model optimization for repeated recognition and transformation to maps.

3.2.2. Tools

There are many supporting tools for openBIM standard. Among them, five kinds of common tools are introduced as follows.

- IFC Document Generator (IfcDoc) is a software tool for generating IFC documents (starting with IFC4) and developing MVD [18]. This tool is based on the mvdXML specification and can be applied to all IFC versions.
- BCF Manager enables users to create, filter, and find issues in BIM model. Users can save, load issues, or synchronize them from BIMcollab to share issues with project members who use the same or different BIM tools shown in Figure 8, thereby improving the reliability of issue management and narrowing the gap between BIM tools [19].
- Building information model query language (BIMQL) is an open domain specific query language for building information model. The query language can select and update data stored in IFC. It is currently used in BIMserver.org platform [20].
- The apstex IFC Framework is a Java-based object-oriented toolkit that provides full access to IFC-based BIM models for reading, writing, modifying, and creating IFC models. The IFC Framework provides tools for accessing and visualizing IFC-based BIM to facilitate software developers to integrate information libraries into products. End-users can also perform visual operations and model checking [21].
- IFC Engine DLL is a STEP toolbox, which can generate 3D models for the latest version of IFC. This component can load, edit, and create Step Physical Files and its framework through its own object database [22].

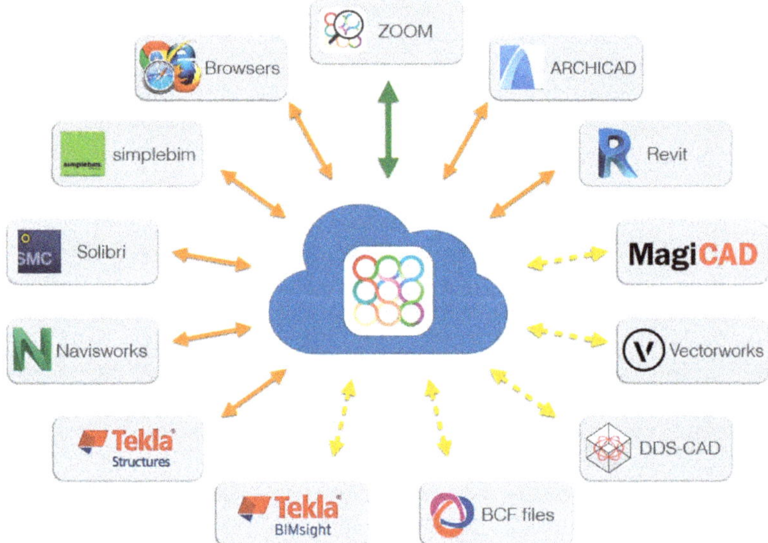

Figure 8. Building information model collaboration format (BCF) Managers are plugins for building information model (BIM) tools.

4. Engineering Information Interoperability Research Supported by OpenBIM

As a common approach of information exchange, openBIM has been applied widely in many research fields. In this paper, some related research of engineering information interoperability supported by openBIM is analyzed and summarized in five main aspects as shown in Figure 9, i.e., information representation, information query, information exchange, information extension, and information integration. The list of literatures by systematic analysis is summarized in Table 3.

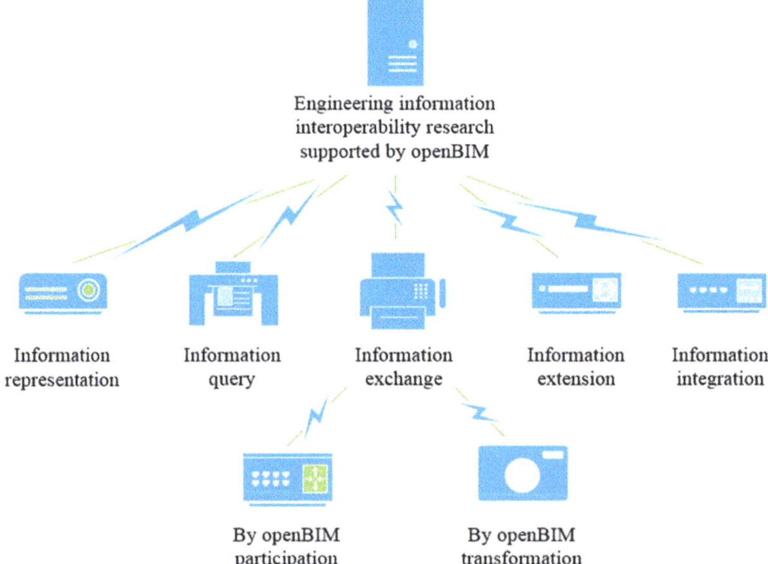

Figure 9. Engineering information interoperability research supported by openBIM.

Table 3. List of literatures summarized by systematic analysis.

Research Direction	Specific Classification	Reference
Information representation	Domain information representation	Ma et al. [23]; Sacks et al. [24]
	Semantic information representation	Sacks et al. [25]; Pauwels et al. [26]
Information query	Use openBIM only	Mazairac et al. [20]; Khalili et al. [27]; Kang [10]; Nepal et al. [28]
	Combine openBIM with other technologies	Ghang et al. [29]; Lin et al. [30]; Kuo et al. [31]
Information exchange	By openBIM participation	Hu et al. [32]; Lee et al. [33]; Solihin et al. [34]; Choi et al. [35]; Park et al. [36]; Caldas et al. [37]; Babič et al. [38]; Caffi et al. [39]; Ding et al. [40]; Lu et al. [41]; Hu et al. [42]; Edmondsona et al. [43]; Vanlande et al. [44]; Lin et al. [45]; Jeong et al. [46]; Wang et al. [47]; Venugopal et al. [48]; Lee et al. [49]; Lee et al. [50]; Lee et al. [51]; Pärn et al. [52]; Alreshidi et al. [53]; Lee et al. [54]; Du [55]; Beetz et al. [56]; Isikdag [57]; Ma et al. [58]; van Berlo et al. [59]; Solihin et al. [60]
	By openBIM transformation	Solihin et al. [61]; Oldfield et al. [62]; Terkaj et al. [63]; Le et al. [64]; Lee et al. [65]; Pauwels et al. [66]
Information extension	Domain information extension	Lee et al. [67]; Ji et al. [68]
	Attribute information extension	Patacas et al. [69]
	Semantic information extension	Venugopal et al. [70]; Belsky et al. [71]; Fahad et al. [72]
Information integration	BIM-based workflow	Andriamamonjya et al. [73]
	Integration of BIM and GIS	Laat et al. [74]; Jusuf et al. [75]; Liu et al. [76]; El-Mekawy et al. [77]; Zhu et al. [78]; Kang et al. [79]; Amirebrahimi et al. [80]; Teo et al. [81]

4.1. Information Representation

Information representation is the basis of information interoperability. openBIM standards, such as IFC and IDM, provide convenience for information representation in many domains. For example, in building domain, Ma et al. developed an BIM application with reasoning support in which BIM data was assumed to be stored in files that conformed to the Industry Foundation Classes standards (IFC files) for better extensibility and reusability [23]. In the field of bridge, Sacks et al. proposed a Semantic Enrichment Engine for Bridges (SeeBridge) system for bridge inspection, in which IDM was responsible for compiling specific information while MVD was mainly used for identifying required information based on IFC. The proposed system can meet the requirements of automatic information inspection and can also be further applied to other infrastructure [24]. Overall, the application of openBIM facilitates the process of domain information representation, thus improving information management of construction industry.

In addition to domain information representation, openBIM can also be applied to semantic information representation. For example, Sacks et al. used IDM for identifying needed domain data and related elements for topological rule processing in the proposed procedure which was targeted to improve the shortage of Semantic Enrichment Engine for BIM (SeeBIM) tool in enriching building model [25]. Furthermore, Pauwels et al. optimized geometric data in ifcOWL semantic representation and proposed four alternative representations of ifcOWL geometric aspects. They also quantified the influence of these metrics on the size of ifcOWL ontology and instance models, eventually they suggested to use the well-known text (WKT) representation as an additional component for ifcOWL ontology, which greatly reduced the scale and complexity of the ifcOWL construction model [26].

The framework of information representation using openBIM is shown in Figure 10.

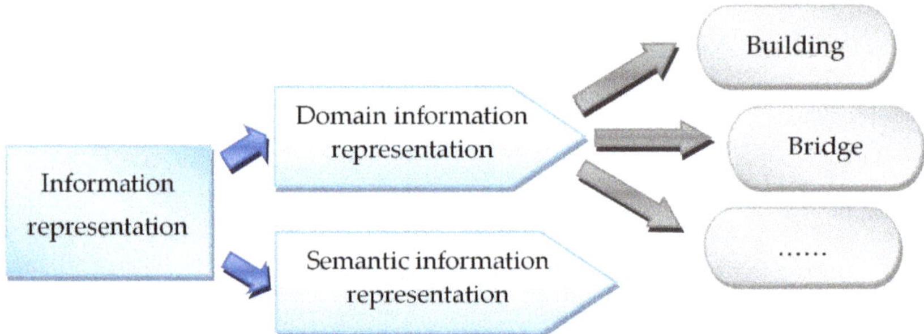

Figure 10. openBIM for information representation.

4.2. Information Query

In addition to using openBIM for information representation, more operations can be performed on the subsequent processing of engineering information with openBIM. The standards, such as IFC, IFD, and BCF, can be used in querying information to realize data retrieval and analysis. For example, Mazairac et al. presented a framework based on a domain specific and open query language which can be used to select, update and delete the data stored in IFC. An outline of existing approaches, conceptual sketches were provided, and the implementation status was recorded as a prototype plugin which was developed for bimserver.org [20]. To speed up topological queries among building elements, Khalili et al. used IFC as a new topology driving method, and proposed a new schema called the Graph Data Model (GDM), through which 3D objects such as building elements could be mapped into a set of nodes, and their relationships could be converted to a set of edges by using IFC standard. In addition, a new IFC-based algorithm was used to deduce the topological relationships among building elements [27]. Due to the difficulty of obtaining civil engineering and building model objects represented by LandXML schema and IFC respectively using linkage manner based on shape information, Kang proposed an effective method of BIM-integrated object query using LandXML and IFC through an object query method based on the BIM linkage model. In addition, the object type and simple query language were defined [10]. Moreover, Nepal et al. proposed an efficient method of extracting and querying information from BIM models by combing ifcXML and spatial XML data, which can save time and reduce errors of extracting information manually [28].

In addition to use openBIM only, some other advanced technologies, such as database technology, natural language processing, and machine learning, can also be combined with openBIM to solve information query issues. For example, combining the advantage of traditional relational databases, Ghang et al. developed an innovative object–relational IFC server which can improve query performance of traditional IFC servers [29]. Lin et al. proposed an approach to realize intelligent data retrieval and representation based on natural language processing. The proposed concepts "keywords" and "constraints" can be mapped in IFC entities or properties through IFD to enable data retrieval and analysis [30]. Kuo et al. proposed machine learning steps to extract and process BCF data. A conceptual framework of queryable knowledge discovery system was introduced to integrate knowledge for future problem prediction [31].

Intelligent information query can improve the efficiency and effectiveness of traditional information query. The data standards provided by openBIM play an indispensable role in intelligent information query process. The combination of intelligent technologies such as natural language processing and machine learning with openBIM standards is a very interesting research direction of information query. The framework of information query supported by openBIM and other advanced technologies is shown in Figure 11.

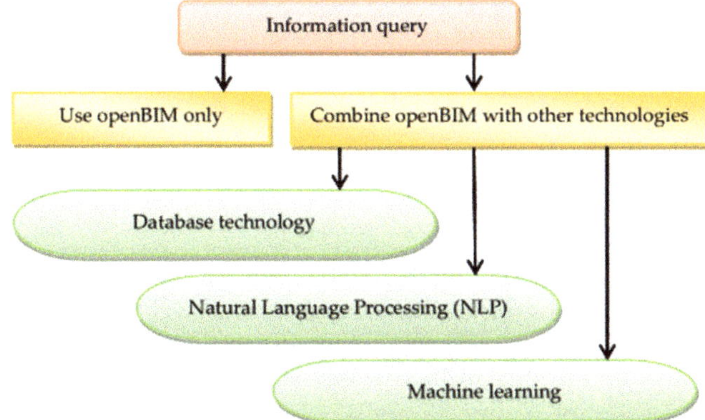

Figure 11. Information query supported by openBIM and other advanced technologies.

4.3. Information Exchange

The development of BIM software contributes to the improvement of work efficiency for different disciplines, which facilitates the process of informatization in construction industry. However, the interoperability among different BIM software becomes a key problem during the collaboration process of stakeholders and how to realize information exchange effectively matters the whole collaboration process. By taking the advantages of openBIM, information exchange can be smoother and easier. Based on the analysis and summary of the existing literature, the information exchange process supported by openBIM is divided into two ways: (1) openBIM standards or platforms participate in information exchange; (2) information exchange is realized through openBIM standard data transformation. The specific classification of these two ways is shown in Figure 12.

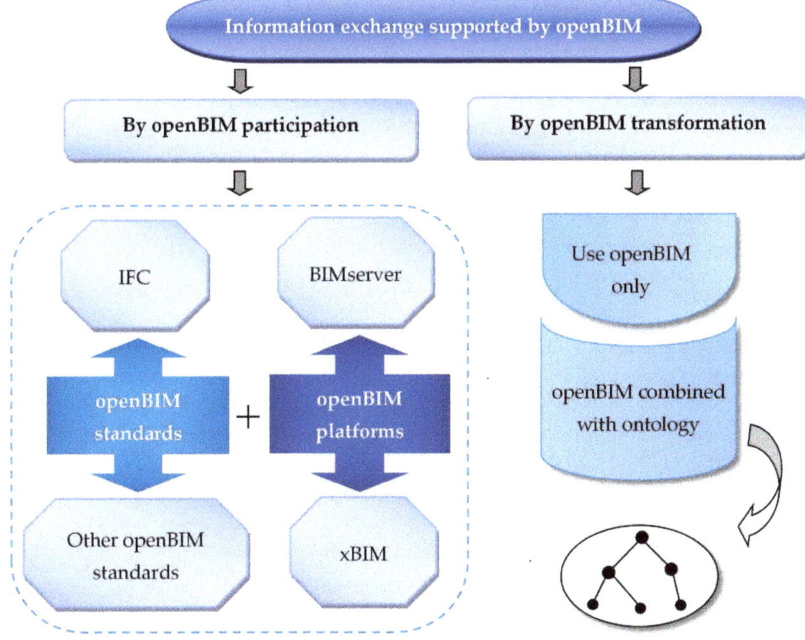

Figure 12. Information exchange supported by openBIM.

4.3.1. Information Exchange by OpenBIM Participation

1. Information exchange by openBIM standards participation

(1) Use IFC only

Some researchers applied openBIM in the process of information exchange. At present, the most widely adopted openBIM standard is IFC, which is applied in building life cycle, including design, construction, operation, and maintenance phases, as shown in Figure 13.

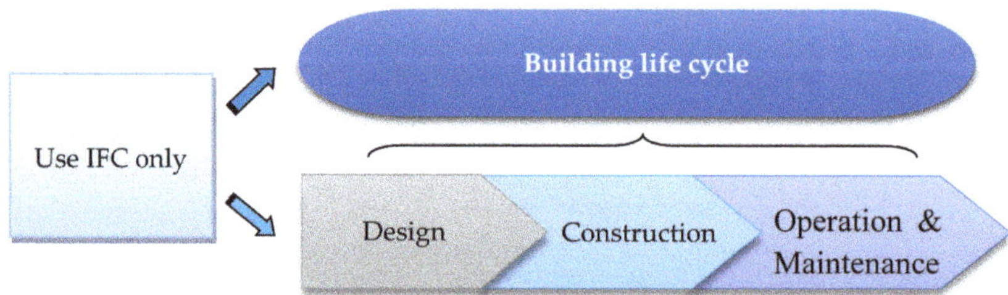

Figure 13. Information exchange by IFC participation.

- Design phase

To improve the interoperability of the design phase, Hu et al. proposed a new method of combining IFC based unified information model with various algorithms, in which IFC based unified information model was used to form the comprehensive central information layer of the model transformation. Moreover, through the proposed method, the entities, attributes, and relationships required for the conversion can be standardized to overcome the inconsistent representation of data and information in different structural analysis applications [32]. Lee et al. described the implementation process of the Building Environment Rule and Analysis (BERA) Language related to BIM, which could contribute to design rule checking and many other analysis purposes, in which IFC was used as a given building information model to transfer BIM data into the BERA Language framework [33]. Solihin et al. demonstrated multiple representations for BIM 3D geometry data using a traditional relational database model to support high-performance rule checking related queries, IFC acts an important role in this information exchange process [34]. Choi et al. proposed a Quantity Take-Off (QTO) process and QTO prototype system within the framework of openBIM to improve the low reliability of early design phase estimation. As long as the worker inputs the necessary task data, the concept of IFC model can progress throughout the entire task of the building lifecycle [35].

- Construction phase

Park et al. proposed a visualization method supported by web and database technology to realize real-time construction progress information sharing and visualization of daily 4D BIM, in which the 3D BIM model had been previously created and followed the IFC standard. By parsing the BIM models based on the IFC file format, the information of the BIM model entity could be recorded to the 4D BIM database [36]. In order to achieve integration of AEC/facility management (FM) information which stored in text documents, Caldas et al. proposed a method to automatically integrate AEC/FM project documents in model-based IFC compatible systems, in which IFC model was used as a benchmark to classify text documents and associate project objects, thereby improving the information interoperability during the construction phase [37]. Babič et al. proposed a system architecture for the interconnection of subsystems in construction company, in which the BIM was used to link the information from an enterprise resource planning (ERP) system with the building object related information. The system architecture is based on the IFC standard, therefore the level of interoperability among project participants is improved and the process of industrialized construction is promoted [38]. Caffi et al. declared the use of INNOVance database for construction process management, and with the adoption of the IFC protocol, interoperability at the end of the project is guaranteed [39]. Ding et al. developed an Industrial Foundation Classes-based Inspection Process Model (IFC-IPM). Within the IFC-IPM schema, a physical, schedule and quality management model exist to meet the requirement of real-time quality-related information exchange during the construction process [40].

- Operation and Maintenance phase

Lu et al. developed a semiautomatic image-driven system for building the original BIM objects in IFC from the images of existing buildings at the stage of operation and maintenance. Three subsystems were developed for this purpose, in which the IFC BIM object generation subsystem used ifcengine to create BIM objects in IFC, and selected IFC2 × 3 and IFC4 as its basic schema standard, so that the subsystem can automatically convert the identified objects into IFC BIM objects [41]. Hu et al. proposed an IFC-based practical multi-scale BIM which was described in detail with the information required by the mechanical, electrical and plumbing (MEP) components for both construction management (CM) and facility management, in which IFC2 × 4 was used to represent this information and exchange them among different BIM applications. Based on the proposed multi-scale BIM, BIM-based CM system and BIM-based FM system were presented to support the collaborative management with multi-scale functions among MEP project partners at the stage of operation and maintenance [42]. Some advanced technologies, such as Internet of Things (IOT), can also be combined with openBIM to solve the information interoperability problems. For example, Edmondsona et al. designed and developed a Smart Sewer Asset Information Model (SSAIM) of existing sewerage network using IFC4, which integrated distributed intelligent sensors to realize real-time monitoring and reporting for sewer asset. In the end an approach of sensor data analysis was proposed to promote the real-time prediction of flooding [43].

- Building life cycle

Vanlande et al. used IFC as a model to define the elements and relations of the construction projects, then constructed an information system which was exclusively used in the building life cycle and developed a platform called Active3D to facilitate information sharing and exchange process in the whole life cycle of the AEC projects [44]. Lin et al. proposed a new approach to deal with 3D indoor space path planning by using IFC file as input, in which IFC was used to restore both geometric information and rich semantic information of building components to support life cycle data sharing [45]. Jeong et al. carried out a structured set of detailed benchmark tests of data exchange between BIM tools. These tests illustrated that a mutually agreed standard which defines how building elements are modeled and mapped into IFC schema is needed to achieve totally effective interoperability [46]. To achieve the efficient transformation between architectural and structural models, Wang et al. developed an IFC-based transformation software which can facilitate transformation process [47]. The relevant literature, which only uses IFC standard to participate in information exchange, is summarized in Table 4 according to the classification of building life cycle.

Table 4. Literature relating to information exchange by Industry Foundation Classes (IFC) participation.

Building Life Cycle Classification	Literature
Design	Hu et al. [32]; Lee et al. [33]; Solihin et al. [34]; Choi et al. [35]
Construction	Park et al. [36]; Caldas et al. [37]; Babič et al. [38]; Caffi et al. [39]; Ding et al. [40]
Operation and Maintenance	Lu et al. [41]; Hu et al. [42]; Edmondsona et al. [43]
Life cycle	Vanlande et al. [44]; Lin et al. [45]; Jeong et al. [46]; Wang et al. [47]

(2)   Use other openBIM standards

Apart from IFC, other openBIM standards were also adopted for information exchange. For example, Venugopal et al. applied MVD for information exchange in sub-domain of building by embedding semantics into information exchange, to solve the insufficiency of IFC in defining entities, attributes, and relationships during the specific information exchange process [48]. Lee et al. proposed a method called extended process to product modeling (xPPM), which can implement the integration and seamless development of IDM and MVD. Based on this method, corresponding tools were developed to analyze the validity of xPPM by copying the existing IDM and MVD, thereby achieving efficient and seamless exchange of building data [49]. To validate the accuracy of BIM model information change, Lee et al. conducted a survey based on available software (such as IFC server ActiveX Component, the IfcDoc, and so on) on conformity to IFC, MVD, and design requirements, which was aimed to provide integrated approaches for improving interoperability [50]. Lee et al. suggested a robust MVD validation process using a modularized validation platform that evaluated an IFC instance file according to diverse types of rule sets of MVDs to address challenges existing in MVD validation process [51]. To realize the integration of BIM and facility management for promoting the process of information exchange in project operation and maintenance (O&M) phase and reducing the cost of the FM team to update and maintain the BIM, Pärn et al. iteratively developed a FinDD API plug-in which could manage the semantic FM data in BIM, in which COBie could provide relevant parameters that meet the requirements of O&M phase. The FinDD was a bespoke extension of COBie, and the data requirements and model structure of its API plug-in were mainly influenced by COBie standard [52]. Alreshidi et al. analyzed the requirements for BIM collaborative platform and pointed out that compatibility among standard-based software, such as IFC-based software, was a big setback for advance of BIM adoption and IDM can play an important role in the BIM collaborative process [53]. Lee et al. suggested a new approach for evaluating BIM data in accordance with diverse requirements of MVD, and examined their embedded checking rule types and categorized corresponding implementation scenarios [54]. Based on IFC and IDM, Du put forward an innovative cloud-based building information interaction framework based on cloud computing and open web environment [55].

2.   Information exchange by openBIM platforms participation

(1)   Use BIMserver

Except for openBIM standards, some researchers explored to realize information exchange and sharing progress with the aid of some software platforms related to openBIM, such as BIMserver and xBIM. As an open and stable software core, BIMserver can easily build reliable BIM software tools to support the dynamic collaboration process of AEC users. And the BIMserver.org platform enables users to create their own "BIM operating system". For example, Beetz et al. introduced the open source IFC server BIMsever.org and gave an outline of existing IFC toolkits and servers. An implement-independent model based on IFC STEP EXPRESS schema was proposed and used as a database persistency framework. The authors also introduced a model server application based on this framework which can realize interoperability about storage, maintenance, and query of BIM among different stakeholders [56]. Isikdag proposed three service-oriented architecture design modes, BIM AJAX, BIM SOAP Facade, and RESTful BIM, in which IFC model was used to promote information exchange and sharing progress. In the RESTful BIM mode, IFC object tree was used for information exchange, and the REST function of the BIMserver could further promote the collaborative use of web-based BIM [57]. In order to reduce the workload of inspectors, promote their cooperation and prevent inspection omissions in the construction quality inspection, Ma et al. proposed a comprehensive application system based on BIM technology and indoor positioning technology, in which IFC format was responsible for storing the BIM model and establishing the detection task generation algorithm, and the BIMserver.org platform was used to manage the BIM model, or was used as a IFC analyzer and a viewer [58]. Van Berlo et al. took advantages of online 3D viewer, BIMserver and the developed open source BCF server to streamline workflow, which can provide convenience for collaborative design [59].

(2) Use xBIM

As a free open source platform, xBIM allows developers to create bespoke BIM middleware for IFC-based applications. For example, Solihin et al. proposed a framework to integrate totally different models in a federated environment through the adoption of IFC schema and standardized procedures. In the validation and testing stage, a prototype implementation was developed by using a modified xBIM toolkit to parse IFC files, import and integrate IFC data [60].

4.3.2. Information Exchange by OpenBIM Transformation

1. Use openBIM only

As a common standard of information exchange, openBIM can not only participate information exchange, but also support information exchange through its standard data transformation. For example, to transform BIM data into a form which can be easily queried, Solihin et al. focused on the definition of a schema and transformation rules and proposed a methodology to transform IFC files to the BIMRL schema (an open and queryable database schema), which can realize flexible and efficient queries into the BIM data using standard SQL [61]. Oldfield et al. followed a standard BIM approach of first defining the requirements using IDM and then translating the process described in the IDM into technical requirements through the use of MVD. The modelling of an MVD or a subset of the IFC data model helped to create and exchange boundary representations of topological objects, which can be combined into a 3D legal space overview map [62].

2. openBIM combined with ontology

As an efficient way of structured information representation and the core of semantic technology, ontology has been paid much attention in recent years. It is important to realize information exchange by integrating ontology with openBIM. For example, Combining IFC with ontology, Terkaj et al. explored the conversion from IFC EXPRESS to OWL and proved the usability of ifcOWL, which provided a new idea for further application of ifcOWL ontology [63]. To realize the interconnection of life cycle data spaces and support the highway asset management, Le et al. proposed a life cycle data exchange framework based on ontology, in which LandXML was used for describing the various design data in the XML format of civil engineering, and some rules were proposed to transform LandXML design data into RDF graphs. The authors also tested the mechanism using a sample road project retrieved from Landxml.org [64]. Due to the lack of logical link in the data transformation process from IDM to MVD, redundant requirements and rules of data exchange may be caused. Under this circumstance, Lee et al. adopted ontology theories to generate IDM in the precast concrete field and link its MVD with formal information modellings to satisfy the requirement of identifying the intent of mapped MVDs and keeping track of the mapping issues. In addition, in order to integrate IDM and MVD, the authors parsed and transformed ontology-based IDM from OWL/XML to mvdXML, which automatically generated MVD documents in IfcDoc tool [65]. To improve the interoperability of construction projects, Pauwels et al. took the ifcOWL ontology as a connection between the semantic web technology and the IFC standard. By analyzing the corresponding application standards required by the key features of the recommend ifcOWL ontology, a conversion program from the EXPRESS mode to the OWL ontology of the IFC was proposed, and the transformation results of the program was used as the recommended ifcOWL ontology [66]. The literature that suggests how openBIM transformation support information exchange is summarized in Table 5.

Table 5. Literature related to information exchange by openBIM transformation.

Classification	Literature	OpenBIM Transformation
Use openBIM only	Solihin et al. [61]	IFC to BIMRL
	Oldfield et al. [62]	IDM to MVD
openBIM combined with ontology	Terkaj et al. [63]	IFC EXPRESS to OWL
	Le et al. [64]	LandXML to RDF
	Lee et al. [65]	OWL/XML to mvdXML
	Pauwels et al. [66]	IFC EXPRESS to OWL

4.4. Information Extension

OpenBIM covers a large range of industry information, such as building components information. But for some fields, such as road, bridge, and port field, openBIM is still not enough to support specific research. Therefore, it is necessary to carry out information extension to meet the requirements. For example, in road domain, based on the IFC framework, Lee et al. added the main building components of road structures to the IFC, and developed an IFC expansion model for road structures, thus extending the BIM technology of road structures [67]. In bridge domain, Ji et al. proposed a possible extension of the existing IFC-Bridge draft, in which an object-oriented data model was used to capture the geometric description of the parameters, and then it will be converted into an EXPRESS pattern and integrated with the current IFC-Bridge draft. The proposed neutral data structure lays the foundation for exchanging parameter bridge models among different software applications [68].

OpenBIM standard can also be used for attribute information extension. For example, Patacas et al. investigated whether IFC and COBie can provide facility manager with required assets data and information from a perspective of a whole life cycle by focusing on some specific use cases including asset registration and service life plan. The result showed that though IFC and COBie cannot meet all the information requirements of asset registration and service life plan by default, they permit users to add some unsupported information in the form of attribute sets by using Revit shared parameters [69].

In addition to domain information extension and attribute information extension, openBIM can also make great contributions in the field of semantic information extension. For example, the lack of semantic clarity in mapping entities, attributes, and relationships leads to multiple expressions of the same information when mapping between different federate modellings. Based on this consideration, Venugopal et al. examined IFC from a viewpoint of ontological framework to make the definition of IFC more accurate, consistent, and unambiguous. Ontology would build interoperability of BIM by providing formal and consistent classification and structure to extend IFC and define subsets as MVD [70]. To solve the problem in existing application of BIM tools, Belsky et al. proposed a new approach which can infer meaningful concepts from BIM model automatically. What is more, the proposed approach can also enrich IFC files and achieve MVD validation by checking spatial topology of elements [71]. In the aspect of achieving automatic verification, Fahad et al. extended ifcOWL ontology with bsDD vocabulary, and compared MVDXML and Semantic Web Rule Language (SWRL) technologies for conformance checking of IFC models [72].

The framework of using openBIM for information extension is shown in Figure 14.

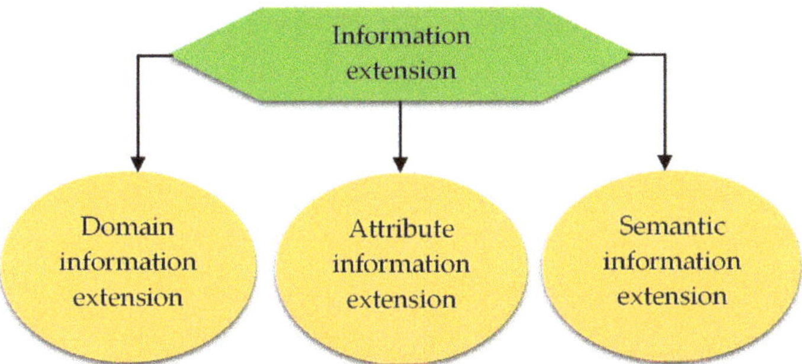

Figure 14. openBIM for information extension.

4.5. Information Integration

Information integration means that the information of the subsystems and users in the system uses unified standards, specifications and codes to realize the information sharing of the whole system, and then the interaction can be realized. openBIM provides a general standard of data exchange in this process.

- BIM-based workflow

It is evident that there is a pressing need for the standardization and uniformity of modelling. A well-defined workflow based on BIM can exploit and extend openBIM standard to improve information integration. Under this circumstance, Andriamamonjya et al. described the significant elements of an integrated BIM-based workflow, explained that openBIM comprised a standardized file format and stated what can be achieved with the help of the already available technology named IDM and MVD. Python language and open source IfcOpenShell were used in case study to illustrate the benefits of this workflow [73].

- Integration of BIM and geographic information system (GIS)

In recent years, the integrated application of BIM and GIS is gaining more and more attention. The method of realizing information interaction through simple model conversion in different fields makes only a small amount of semantic information retained, which leads to the dispersion and independence of the application. IFC and CityGML are the common data model standards of BIM and GIS respectively, and the geometric and semantic information sharing will lay the foundation for the integration of BIM and GIS.

For example, in order to import semantic IFC data into the GIS environment, Laat et al. described the development of CityGML extension called GeoBIM, which is mainly the development and the implementation of GeoBIM extension for IFC data on CityGML, thus facilitated the integration of BIM and GIS [74]. To achieve Neighborhood Scale Modelling, Jusuf et al. focused on ways of exchanging information and bringing together CityGML and IFC. The transformation system was developed using Feature Manipulation Engine (FME) by Safe Software. With the help of FME, data model (IFC) can be restructured and transformed to the destination data format (CityGML). The test results showed that CityGML format, as well as a Sketchup file, could be generated from detailed BIM models. These models can be imported to web visualization applications for urban energy modeling [75]. Liu et al. reviewed the development and difference of BIM and GIS, the existing integration methods, and discussed their potential in many applications. The author pointed out that semantic web technology provided a promising generalized integration solution and openness was the key factor to the integration of BIM and GIS [76]. El-Mekawy et al. described a new method of data integration based on the unified building model (UBM), which not only encapsulated CityGML and IFC models, but also avoided the translations between the models and information loss. The case scenario and four queries have verified that the developed UBM can integrate CityGML data and IFC data in a seamless manner [77]. Zhu et al. reviewed relevant research papers to (1) determine the most relevant data modellings used in BIM/GIS integration; (2) look for the possibility of other data modellings which can be used for data-level integration; and (3) provide guidance for future's BIM/GIS data integration [78]. In order to solve the problem of integration of BIM and GIS heterogeneous models, Kang et al. defined the integration process of BIM-GIS through the BIM-GIS conceptual mapping (B2GM) standard and proposed a mapping mechanism related to it. Based on IFC and CityGML, the schema structure of BIM and GIS were analyzed from the perspective of BIM-GIS integration. In addition, the IFC standard was also used for database integration and query process based on B2GM [79].

BIM and GIS information integration can effectively solve specific practical problems. Amirebrahimi et al. designed a new conceptual data model, which was used to integrate detailed assessment and three-dimensional visualization of flood damage of buildings. In this design process, investigation was conducted to figure out how concepts are represented in IFC format or CityGML format so that they can create a mapping between the concepts of this proposed model and IFC or CityGML [80]. Teo et al. proposed a multi-purpose geometric network model (MGNM) based on BIM and discussed the method of outdoor and indoor network connection to optimize emergency response and pedestrian route plan. To achieve these goals, the conversion between IFC and MGNM was discussed and validated by case study. The IFC-to-MGNM conversion included the following: (1) extraction of building information from IFC, (2) isolation of the MGNM information from building information, and (3) establishment of the topological relationship of MGNM to GIS geodatabase [81]. International Building Performance Simulation Association (IBPSA) Project 1 provides a BIM/GIS and Modelica framework for building and community energy system design and operation, in which Modelica is used for building and regional energy system performance modeling. It aims to create open-source software which lays the foundation for the next generation of computing tools for the design and operation of building and regional energy and control systems. All the work is open-source and based on three standards: IFC, CityGML, and Modelica [82].

In the past few years, the efforts of information integration especially the integration of BIM and GIS have been increasing from the semantic point of view. But the information loss and change are still normal in the process of data exchange. Openness and collaboration are the key to information integration. The sharing standards provided by openBIM should be fully utilized to make information integration more widely used.

The framework of using openBIM for information integration is shown in Figure 15.

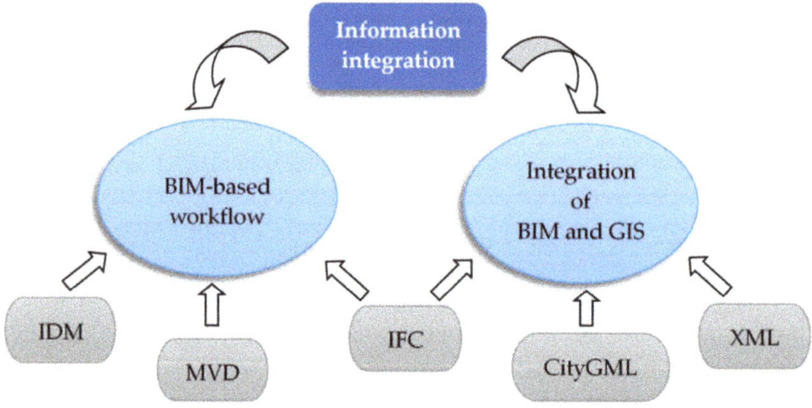

Figure 15. openBIM for information integration.

5. Research Gaps and Future Directions

This paper outlines the achievements of openBIM in various aspects of engineering information interoperability in recent years, such as information representation, query, exchange, extension, and integration. Although the application of openBIM provides favorable environment for the management and exchange of information and data, and to some extent overcomes the obstacles of cross professional cooperation in the project, more basic and applied research work are still needed. At present, the research gaps and future directions are mentioned below:

1. Rule checking

With the increase of project complexity and the number of project participants, unified BIM data exchange standards and robust validation frameworks are needed to ensure the seamless sharing and consistent maintenance of BIM data. At present, researchers have developed rule checking function based on MVD by using IfcDoc tool, and discussed the identified knowledge of BIM data validation using MVD, involving procedures, scope and complexity [83].

However, there are still some challenges in MVD-based rule checking: (1) although several versions of IFC schema provide the schematic basis of BIM data exchange standards, current rule checking cannot cover numerous types of modeling, representation, and interpretations scenarios that the IFC schema supports, thus sets back the process of validation of BIM data using MVD, which will lead to syntax problems, semantic errors, and unintended geometric transformations. (2) each field has different data and exchange processes which should be included in an MVD. However, domain experts and software vendors refer to multiple interpretations and different ways when defining their domain knowledge, which results in a broad scope of model view definitions and makes the formal rule checking process very complex and difficult.

In order to solve the limitations of current research, it is necessary to consistently examine various methods to identify an explicit link between conceptual rules and formal information models. In addition, to ensure the reliable validation of BIM data exchange and the consistent definitions of model view specifications, the existing MVDs and the related rules should be further investigated. In the future, industries, academia, and governments need to cooperate to develop formal processes for MVD development, and jointly establish generalized libraries containing knowledge in various domains and IFC schema to ensure data interoperability and quality.

2. Ontology development

Due to the lack of semantic clarity of IFC in mapping entities and relationships, it is too generic to use IFC exchange schema when exchanging information between BIM tools, which results in the inability to capture full semantic meaning required for the direct use of different construction project stakeholders' BIM tools. As a representation of concepts, relations, and rules in a specific domain, domain ontology can store, query, and share information in the domain. Therefore, checking IFC from the perspective of ontology framework can make the definition of IFC more formal, consistent, and clear. Applying ontology to engineering can support the following aspects of information interoperability: (1) realizing knowledge sharing among multiple human or software agents; (2) dealing with semantic interoperability between computer-based systems and data sources; and (3) using formal axioms accessible to machines to capitalize experts' knowledge [84].

At present, the research of data interoperability based on ontology has made some progress in some fields, such as construction, energy, and infrastructure. However, there are still technological and methodological gaps between academic researches and industrial practices. In order to take full advantage of the potential of ontology, further development and utilization of ontology should be carried out from the following aspects: (1) the development and (re-)use of inter-connected ontologies including upper- and domain-level ontologies; (2) a library of modelling alternatives to satisfy different representational needs; and (3) more effective software environments which can support different phases in the ontology-based software lifecycle.

3.   Integration of BIM and GIS

Based on the developed IFC and CityGML semantic models, users can flexibly choose the corresponding BIM-GIS integration methods according to the specific problems, including extracting data from one system to another, merging data using semantic models or integrating data of both models by a third-party platform [85].

However, the above integration technologies still have the following defects: (1) when converting IFC to CityGML, the model information mismatch between BIM and GIS will lead to geometric information mismatch. Because the huge and complex IFC data stored in BIM models has various attributes and geometric expressions, most of the information representing the attributes and associations between features in IFC may disturb the geometric mapping process between IFC and CityGML. (2) when transforming data between IFC and CityGML, semantic loss of features may occur. For instance, the outer and inner walls of a real building become the same object after conversion.

Therefore, the future research should focus on improving the accuracy of BIM-GIS data integration, and can be carried out from the following directions: (1) interoperability, transformation process and detailed information processing of these two kinds of data should be focused. First, the upgrade of IFC and CityGML semantic models is needed to improve the accuracy and matching degree of semantic information; second, in addition to consider one-way information transformation from IFC to CityGML, future work will focus on the two-way transformation of model information. (2) when using the third-party platform for data integration, it is necessary to ensure the consistency of these two semantic models and the accuracy of data mapping. Data can be classified based on data hierarchy, and the integrated data from both models using a third-party platform can be developed in the direction of customer-orientation, information sharing, and simplified operations. Besides, an open multi-party data sharing platform can be developed for future use. (3) in order to obtain optimal solutions and planning decisions using BIM and GIS data, it is necessary to ensure the accuracy and diversity of the original data. The process of data measurement, collection and statistical analysis should be comprehensive and careful, not only the macro geographic information, but also the detailed information of the internal space of the building is needed.

4.   Combination with IT technologies

In order to fully utilize the value of BIM and promote intelligent decision based on BIM data, it is necessary to combine openBIM with IT technologies such as cloud computing and big data. For example, the distributed architecture of cloud computing technology can solve the problem of limited capacity of a single server and facilitate information exchange; Big data technology can deeply analyze and mine massive data, discover hidden regularities in data, so as to realize intelligent decision and management; IOT technology can expand information source of projects, and realize dynamic and real-time management of projects.

The combination of cloud technology and BIM is only for storage and retrieval in the early stage, but the great advantage of cloud technology is providing BIM analysis with massive heterogeneous data spanning the whole life cycle of construction projects. The shorter the processing from original data to information representation, the less information loss will be. Therefore, the development of BIM combined with cloud technology should focus more on data-driven information exploration.

At present, some researches have discussed how to use big data analysis to extract effective information from the original BIM data that was previously considered unavailable or ignored. However, many of these researches are in the embryonic stage of concept. The possibility of using big data should be further discussed, especially the joint analysis of cloud technology and big data on the whole life cycle of the building. Accurate analysis can be made in many fields such as energy and structure by exploring the potential connection of data.

The summary of this study's findings is presented in Figure 16.

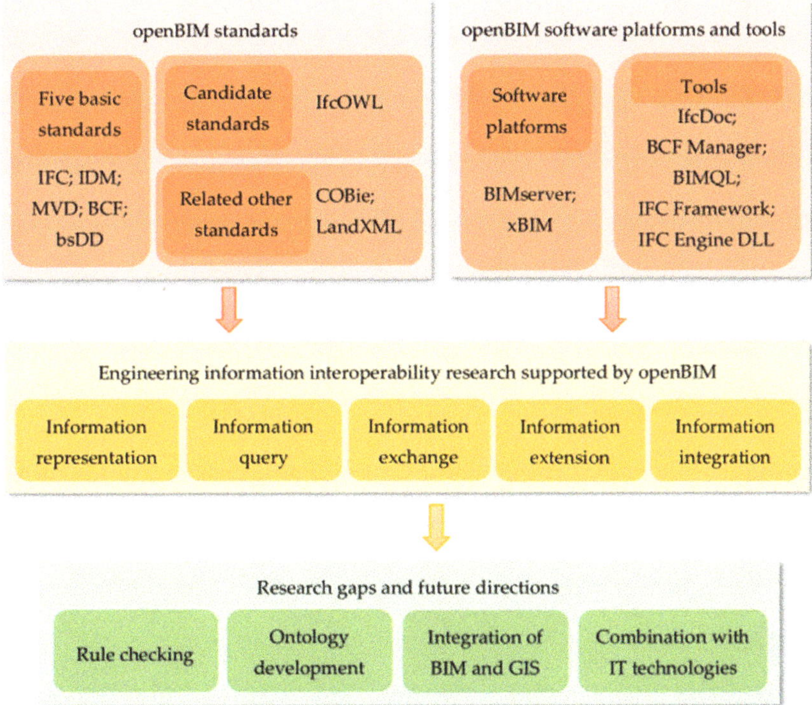

Figure 16. Summary of the findings.

6. Conclusions

The occurrence of openBIM lays a foundation for collaboration among users of the BIM platform and participants of the projects, and provides convenience for the management and exchange of information. In this paper, firstly, based on the literature from main databases like Scopus, the authors analyze the research status of openBIM in the world. And on the premise of elaborating the definitions, standards, and software of openBIM, the paper summarizes engineering information interoperability research supported by openBIM from five perspectives: Information representation, information query, information exchange, information extension, and information integration. This review provides an in-depth understanding of existing openBIM research. Moreover, the study makes a discussion of research gaps and future directions, which can provide valuable insights for academia and industry.

In conclusion, as an effective solution to the interoperability and sharing problem, openBIM has played a great role in the construction field and facilitated the collaboration process of projects. So with the continuous development of openBIM, it will provide more comprehensive solutions to the problems existing in the construction field.

Author Contributions: S.J. conceived the idea for the paper, oversaw its development and supervised all aspects of this research. L.J. and Y.H. wrote the overview of openBIM, engineering information interoperability research supported by openBIM and research gaps and future directions. Z.W. and N.W. designed the methodology. All authors have reviewed and approved the final manuscript.

Funding: This research was funded by the National Key R&D Program of China grant number 2016YFC0702107.

Acknowledgments: The work described in this paper was supported by National Key R&D Program of China (grant no. 2016YFC0702107). The authors also would like to greatly thank the anonymous reviewers who gave invaluable suggestions on improving the quality of the manuscript.

Conflicts of Interest: The authors declare no conflict of interest.

References

1. Official Definition of Open BIM—Open BIM. Available online: http://open.bimreal.com/bim/index.php/2012/03/20/what-is-the-official-definition-of-open-bim/ (accessed on 16 July 2019).
2. buildingSMART—The Home of BIM. Available online: https://www.buildingsmart.org/ (accessed on 16 March 2019).
3. Standards Library—buildingSMART. Available online: https://www.buildingsmart.org/standards/bsi-standards/standards-library/ (accessed on 16 July 2019).
4. Home—Welcome to buildingSMART-Tech.org. Available online: https://technical.buildingsmart.org/ (accessed on 16 July 2019).
5. MvdXML—Welcome to buildingSMART-Tech.org. Available online: https://technical.buildingsmart.org/standards/mvd/mvdxml/ (accessed on 16 July 2019).
6. BIM Collaboration Format (BCF)—Welcome to buildingSMART-Tech.org. Available online: https://technical.buildingsmart.org/standards/bcf/ (accessed on 16 July 2019).
7. Cloud-based Data Dictionary Launched—buildingSMART. Available online: https://www.buildingsmart.org/cloud-based-data-dictionary-launched/ (accessed on 16 March 2019).
8. ifcOWL—Welcome to buildingSMART-Tech.org. Available online: https://technical.buildingsmart.org/standards/ifc/ifc-formats/ifcowl/ (accessed on 16 July 2019).
9. buildingSMART Alliance Construction Operations Building Information Exchange (COBie) Project—National Institute of Building Sciences. Available online: https://www.nibs.org/page/bsa_cobie (accessed on 16 March 2019).
10. Kang, T.W. Object composite query method using IFC and LandXML based on BIM linkage model. *Autom. Constr.* **2017**, *76*, 14–23. [CrossRef]
11. National Institute of Building Sciences. *National Building Information Modeling Standard Version 1.0—Part 1 Overview, Principles, and Methodologies*; buildingSMART Initiative: St. Louis, MO, USA, 2007.
12. Open Source BIMserver—In the Heart of Your BIM! Available online: http://bimserver.org/ (accessed on 16 March 2019).
13. IfcOpenShell. Available online: http://ifcopenshell.org/ (accessed on 16 March 2019).
14. Giving Perspective to Your BIM Projects—Bimvie.ws. Available online: http://bimvie.ws/ (accessed on 16 March 2019).
15. BIM Surfer®. Available online: http://bimsurfer.org/ (accessed on 16 March 2019).
16. BCFier. Available online: http://bcfier.com/ (accessed on 16 March 2019).
17. OpenBIM. Available online: http://www.openbim.org/ (accessed on 16 March 2019).
18. ifcDoc Tool Summary—Welcome to buildingSMART-Tech.org. Available online: http://www.buildingsmart-tech.org/specifications/specification-tools/ifcdoc-tool/ (accessed on 16 March 2019).
19. BCF Manager—BIMcollab®. Available online: https://www.bimcollab.com/en/BCF-Manager/BCF-Manager (accessed on 16 March 2019).
20. Mazairac, W.; Beetz, J. BIMQL—An open query language for building information models. *Adv. Eng. Inform.* **2013**, *27*, 444–456. [CrossRef]
21. Apstex IFC Framework. Available online: http://www.apstex.com/ (accessed on 16 March 2019).
22. Weblet Importer. Available online: http://rdf.bg/product-list/ifc-engine/ (accessed on 16 March 2019).
23. Ma, Z.; Liu, Z. Ontology- and freeware-based platform for rapid development of BIM applications with reasoning support. *Autom. Constr.* **2018**, *90*, 1–8. [CrossRef]
24. Sacks, R.; Kedar, A.; Borrmann, A.; Ma, L.; Brilakis, I.; Hüthwohl, P.; Daum, S.; Kattel, U.; Yosef, R.; Liebich, T.; et al. SeeBridge as next generation bridge inspection: Overview, Information Delivery Manual and Model View Definition. *Autom. Constr.* **2018**, *90*, 134–145. [CrossRef]
25. Sacks, R.; Ma, L.; Yosef, R.; Borrmann, A.; Daum, S.; Kattel, U. Semantic Enrichment for Building Information Modeling: Procedure for Compiling Inference Rules and Operators for Complex Geometry. *J. Comput. Civ. Eng.* **2017**, *31*. [CrossRef]
26. Pauwels, P.; Krijnen, T.; Terkaj, W.; Beetz, J. Enhancing the ifcOWL ontology with an alternative representation for geometric data. *Autom. Constr.* **2017**, *80*, 77–94. [CrossRef]
27. Khalili, A.; Chua, D.K.H. IFC-Based Graph Data Model for Topological Queries on Building Elements. *J. Comput. Civ. Eng.* **2013**, *29*. [CrossRef]

28. Nepal, M.P.; Staub-French, S.; Pottinger, R.; Webster, A. Querying a building information model for construction-specific spatial information. *Adv. Eng. Inform.* **2012**, *26*, 904–923. [CrossRef]

29. Ghang, L.; Jiyong, J.; Jongsung, W.; Chiyon, C.; Seok-joon, Y.; Sungil, H.; Hoonsig, K. Query Performance of the IFC Model Server Using an Object-Relational Database Approach and a Traditional Relational Database Approach. *J. Comput. Civ. Eng.* **2014**, *28*, 210–222.

30. Lin, J.R.; Hu, Z.Z.; Zhang, J.P.; Yu, F.Q. A Natural-Language-Based Approach to Intelligent Data Retrieval and Representation for Cloud BIM. *Comput. Civ. Infrastruct. Eng.* **2016**, *31*, 18–33. [CrossRef]

31. Kuo, V.; Oraskari, J. A Predictive Semantic Inference System using BIM Collaboration Format (BCF) Cases and Machine Learning. *CIB World Build. Congr.* **2016**, *3*, 368–378.

32. Hu, Z.Z.; Zhang, X.Y.; Wang, H.W.; Kassem, M. Improving interoperability between architectural and structural design models: An industry foundation classes-based approach with web-based tools. *Autom. Constr.* **2016**, *66*, 29–42. [CrossRef]

33. Lee, J.K.; Eastman, C.M.; Lee, Y.C. Implementation of a BIM Domain-specific Language for the Building Environment Rule and Analysis. *J. Intell. Robot. Syst. Theory Appl.* **2015**, *79*, 507–522. [CrossRef]

34. Solihin, W.; Eastman, C.; Lee, Y.C. Multiple representation approach to achieve high-performance spatial queries of 3D BIM data using a relational database. *Autom. Constr.* **2017**, *81*, 369–388. [CrossRef]

35. Choi, J.; Kim, H.; Kim, I. Open BIM-based quantity take-off system for schematic estimation of building frame in early design stage. *J. Comput. Des. Eng.* **2015**, *2*, 16–25. [CrossRef]

36. Park, J.; Cai, H.; Dunston, P.S.; Ghasemkhani, H. Database-Supported and Web-Based Visualization for Daily 4D BIM. *J. Constr. Eng. Manag.* **2017**, *143*. [CrossRef]

37. Caldas, C.H.; Soibelman, L. *Integration of Construction Documents in IFC Project Models*; American Society of Civil Engineers (ASCE): Preston, WV, USA, 2004; pp. 1–8.

38. Babič, N.Č.; Podbreznik, P.; Rebolj, D. Integrating resource production and construction using BIM. *Autom. Constr.* **2010**, *19*, 539–543. [CrossRef]

39. Caffi, V.; Re Cecconi, F.; Pavan, A.; Pasini, D.; Maltese, S.; Daniotti, B.; Chiozzi, M.; Spagnolo, S.L. INNOVance: Italian BIM Database for Construction Process Management. In Proceedings of the applications of IT in the Architecture, Engineering and Construction Industry, Orlando, FL, USA, 23–25 June 2014; pp. 641–648.

40. Ding, L.; Li, K.; Zhou, Y.; Love, P.E.D. An IFC-inspection process model for infrastructure projects: Enabling real-time quality monitoring and control. *Autom. Constr.* **2017**, *84*, 96–110. [CrossRef]

41. Lu, Q.; Lee, S.; Chen, L. Image-driven fuzzy-based system to construct as-is IFC BIM objects. *Autom. Constr.* **2018**, *92*, 68–87. [CrossRef]

42. Hu, Z.Z.; Zhang, J.P.; Yu, F.Q.; Tian, P.L.; Xiang, X.S. Construction and facility management of large MEP projects using a multi-Scale building information model. *Adv. Eng. Softw.* **2016**, *100*, 215–230. [CrossRef]

43. Edmondson, V.; Cerny, M.; Lim, M.; Gledson, B.; Lockley, S.; Woodward, J. A smart sewer asset information model to enable an 'Internet of Things' for operational wastewater management. *Autom. Constr.* **2018**, *91*, 193–205. [CrossRef]

44. Vanlande, R.; Nicolle, C.; Cruz, C. IFC and building lifecycle management. *Autom. Constr.* **2008**, *18*, 70–78. [CrossRef]

45. Lin, Y.H.; Liu, Y.S.; Gao, G.; Han, X.G.; Lai, C.Y.; Gu, M. The IFC-based path planning for 3D indoor spaces. *Adv. Eng. Inform.* **2013**, *27*, 189–205. [CrossRef]

46. Jeong, Y.S.; Eastman, C.M.; Sacks, R.; Kaner, I. Benchmark tests for BIM data exchanges of precast concrete. *Autom. Constr.* **2009**, *18*, 469–484. [CrossRef]

47. Wang, X.; Yang, H.; Zhang, Q.L. Research of the IFC-based Transformation Methods of Geometry Information for Structural Elements. *J. Intell. Robot. Syst. Theory Appl.* **2015**, *79*, 465–473. [CrossRef]

48. Venugopal, M.; Eastman, C.M.; Sacks, R.; Teizer, J. Semantics of model views for information exchanges using the industry foundation class schema. *Adv. Eng. Inform.* **2012**, *26*, 411–428. [CrossRef]

49. Lee, G.; Park, Y.H.; Ham, S. Extended Process to Product Modeling (xPPM) for integrated and seamless IDM and MVD development. *Adv. Eng. Inform.* **2013**, *27*, 636–651. [CrossRef]

50. Lee, Y.C.; Eastman, C.M.; Lee, J.K. Validations for ensuring the interoperability of data exchange of a building information model. *Autom. Constr.* **2015**, *58*, 176–195. [CrossRef]

51. Lee, Y.C.; Eastman, C.M.; Solihin, W.; See, R. Modularized rule-based validation of a BIM model pertaining to model views. *Autom. Constr.* **2016**, *63*, 1–11. [CrossRef]

52. Pärn, E.A.; Edwards, D.J. Conceptualising the FinDD API plug-in: A study of BIM-FM integration. *Autom. Constr.* **2017**, *80*, 11–21. [CrossRef]

53. Alreshidi, E.; Mourshed, M.; Rezgui, Y. Requirements for cloud-based BIM governance solutions to facilitate team collaboration in construction projects. *Requir. Eng.* **2018**, *23*, 1–31. [CrossRef]

54. Lee, Y.C.; Eastman, C.M.; Solihin, W. Logic for ensuring the data exchange integrity of building information models. *Autom. Constr.* **2018**, *85*, 249–262. [CrossRef]

55. Juan, D. The research to open BIM-based building information interoperability framework. In Proceedings of the 2013 2nd International Symposium on Instrumentation and Measurement, Sensor Network and Automation, IMSNA 2013, Toronto, ON, Canada, 23–24 December 2013; pp. 440–443.

56. Beetz, J.; Berlo, L. Van BIMSERVER.ORG—An open source IFC model server. In Proceedings of the 27th International Conference on Information Technology in Construction CIB W78, Cairo, Egypt, 16–18 November 2010; pp. 16–18.

57. Isikdag, U. Design patterns for BIM-based service-oriented architectures. *Autom. Constr.* **2012**, *25*, 59–71. [CrossRef]

58. Ma, Z.; Cai, S.; Mao, N.; Yang, Q.; Feng, J.; Wang, P. Construction quality management based on a collaborative system using BIM and indoor positioning. *Autom. Constr.* **2018**, *92*, 35–45. [CrossRef]

59. van Berlo, L.; Krijnen, T. Using the BIM Collaboration Format in a Server Based Workflow. *Procedia Environ. Sci.* **2014**, *22*, 325–332. [CrossRef]

60. Solihin, W.; Eastman, C.; Lee, Y.C. A framework for fully integrated building information models in a federated environment. *Adv. Eng. Inform.* **2016**, *30*, 168–189. [CrossRef]

61. Solihin, W.; Eastman, C.; Lee, Y.C.; Yang, D.H. A simplified relational database schema for transformation of BIM data into a query-efficient and spatially enabled database. *Autom. Constr.* **2017**, *84*, 367–383. [CrossRef]

62. Oldfield, J.; van Oosterom, P.; Beetz, J.; Krijnen, T. Working with Open BIM Standards to Source Legal Spaces for a 3D Cadastre. *ISPRS Int. J. Geo-Inf.* **2017**, *6*, 351. [CrossRef]

63. Terkaj, W.; Šojić, A. Ontology-based representation of IFC EXPRESS rules: An enhancement of the ifcOWL ontology. *Autom. Constr.* **2015**, *57*, 188–201. [CrossRef]

64. Le, T.; David Jeong, H. Interlinking life-cycle data spaces to support decision making in highway asset management. *Autom. Constr.* **2016**, *64*, 54–64. [CrossRef]

65. Lee, Y.C.; Eastman, C.M.; Solihin, W. An ontology-based approach for developing data exchange requirements and model views of building information modeling. *Adv. Eng. Inform.* **2016**, *30*, 354–367. [CrossRef]

66. Pauwels, P.; Terkaj, W. EXPRESS to OWL for construction industry: Towards a recommendable and usable ifcOWL ontology. *Autom. Constr.* **2016**, *63*, 100–133. [CrossRef]

67. Lee, S.H.; Kim, B.G. IFC extension for road structures and digital modeling. *Procedia Eng.* **2011**, *14*, 1037–1042. [CrossRef]

68. Ji, Y.; Borrmann, A.; Beetz, J.; Obergrießer, M. Exchange of Parametric Bridge Models Using a Neutral Data Format. *J. Comput. Civ. Eng.* **2012**, *27*, 593–606. [CrossRef]

69. Patacas, J.; Dawood, N.; Vukovic, V.; Kassem, M. BIM for facilities management: Evaluating BIM standards in asset register creation and service life planning. *J. Inf. Technol. Constr.* **2015**, *20*, 313–331.

70. Venugopal, M.; Eastman, C.M.; Teizer, J. An ontology-based analysis of the industry foundation class schema for building information model exchanges. *Adv. Eng. Inform.* **2015**, *29*, 940–957. [CrossRef]

71. Belsky, M.; Sacks, R.; Brilakis, I. Semantic Enrichment for Building Information Modeling. *Comput. Civ. Infrastruct. Eng.* **2016**, *31*, 261–274. [CrossRef]

72. Fahad, M.; Bus, N.; Andrieux, F. Towards Mapping Certification Rules over BIM. In Proceedings of the 33rd CIB W78 Conference, Brisbane, Australia, 31 October–2 November 2016.

73. Andriamamonjy, A.; Saelens, D.; Klein, R. An automated IFC-based workflow for building energy performance simulation with Modelica. *Autom. Constr.* **2018**, *91*, 166–181. [CrossRef]

74. Laat, R.; Berlo, L. Integration of BIM and GIS: The Development of the CityGML GeoBIM Extension BT—Advances in 3D Geo-Information Sciences. In *Advances in 3D Geo-Information Sciences*; Springer: Berlin/Heidelberg, Germany, 2011; pp. 211–225. ISBN 978-3-642-12670-3.

75. Jusuf, S.; Mousseau, B.; Godfroid, G.; Soh, J. Path to an Integrated Modelling between IFC and CityGML for Neighborhood Scale Modelling. *Urban Sci.* **2017**, *1*, 25. [CrossRef]

76. Liu, X.; Liu, R.; Wright, G.; Cheng, J.; Wang, X.; Li, X. A State-of-the-Art Review on the Integration of Building Information Modeling (BIM) and Geographic Information System (GIS). *ISPRS Int. J. Geo-Inf.* **2017**, *6*, 53. [CrossRef]
77. El-Mekawy, M.; Östman, A.; Hijazi, I. A Unified Building Model for 3D Urban GIS. *ISPRS Int. J. Geo-Inf.* **2012**, *1*, 120–145. [CrossRef]
78. Zhu, J.; Wright, G.; Wang, J.; Wang, X. A Critical Review of the Integration of Geographic Information System and Building Information Modelling at the Data Level. *ISPRS Int. J. Geo-Inf.* **2018**, *7*, 66. [CrossRef]
79. Kang, T. Development of a Conceptual Mapping Standard to Link Building and Geospatial Information. *ISPRS Int. J. Geo-Inf.* **2018**, *7*, 162. [CrossRef]
80. Amirebrahimi, S.; Rajabifard, A.; Mendis, P.; Ngo, T. A BIM-GIS integration method in support of the assessment and 3D visualisation of flood damage to a building. *J. Spat. Sci.* **2016**, *61*, 317–350. [CrossRef]
81. Teo, T.A.; Cho, K.H. BIM-oriented indoor network model for indoor and outdoor combined route planning. *Adv. Eng. Inform.* **2016**, *30*, 268–282. [CrossRef]
82. IBPSA Project 1. Available online: https://ibpsa.github.io/project1/ (accessed on 27 November 2019).
83. Lee, Y.C.; Solihin, W.; Eastman, C.M. The Mechanism and Challenges of Validating a Building Information Model regarding data exchange standards. *Autom. Constr.* **2019**, *100*, 118–128. [CrossRef]
84. Sanfilippo, E.; Terkaj, W. Editorial: Formal Ontologies meet Industry. *Procedia Manuf.* **2019**, *28*, 174–176. [CrossRef]
85. Wang, H.; Pan, Y.; Luo, X. Integration of BIM and GIS in sustainable built environment: A review and bibliometric analysis. *Autom. Constr.* **2019**, *103*, 41–52. [CrossRef]

applied sciences

Article

Generic Language for Partial Model Extraction from an IFC Model Based on Selection Set

Xueyuan Deng [1], Huahui Lai [2,*], Jiayi Xu [1] and Yunfan Zhao [1]

[1] Department of Civil Engineering, School of Naval Architecture, Ocean and Civil Engineering, Shanghai Jiao Tong University, Shanghai 200240, China; dengxy@sjtu.edu.cn (X.D.); xiaoyi1213@sjtu.edu.cn (J.X.); zhaoyunfan@sjtu.edu.cn (Y.Z.)

[2] Shenzhen Municipal Design & Research Institute Co., Ltd., Shenzhen 518029, China

* Correspondence: laihuahui81665@alumni.sjtu.edu.cn

Received: 31 January 2020; Accepted: 11 March 2020; Published: 13 March 2020

Abstract: During data sharing and exchange of building projects, the particular business task generally requires a part of the complete model. This paper adopted XML schema to develop a generic language to extract the partial model from an Industry Foundation Classes (IFC) model based on the proposed Selection Set (called PMESS). In this method, the Selection Set was used to integrate users' requirements, which could be mapped into IFC data. To ensure the validity of the generated partial IFC models in syntax and semantics, seven rules—including three basic rules for a valid IFC file, three extraction rules based on the Selection Set, and a processing rule for redundant information—were defined. Through defining PMESS-based configuration files, the required data can be extracted and formed as a partial IFC model. Compared with the existing methods, the proposed PMESS method can flexibly extract the user-defined required information. In addition, these PMESS-based configuration files can be stored as templates and reused in other tasks, which prevents duplicated work for defining extraction requirements. Finally, a practical project was used to illustrate the utility of the proposed method.

Keywords: Building Information Modeling (BIM); Industry Foundation Classes (IFC); partial model extraction; query language; selection set

1. Introduction

In traditional computer-aided design (CAD) practices, most interdisciplinary data exchanges take place with two-dimensional (2D) drawings, documents, or reports. As unstructured forms, these data files are hardly used by other software tools, resulting in remodeling work for their own tasks. Due to the lack of semantic meanings of elements (e.g., points, lines, and planes) in electronic documents, designers have to manually interpret these elements based on their experience in order to identify and extract the required data [1]. The CAD-based process limits the reusability of project data through the whole building life cycle. Compared with the traditional CAD-based method, Building Information Modeling (BIM) technology is able to represent the geometry, properties, and relations of building objects based on the object-oriented method [2]. In the early stage of BIM technology, its purpose is to make a complete model available for every participant. Nowadays, the use of BIM technology in the architecture, engineering, construction, and facility management (AEC/FM) industry is becoming more widespread [3], and it results in numerous structured data in various domains, which can be interpreted by different software tools and used for different business tasks. The business task means that participants use some project data to carry out some activities for a professional application, and it can take place during the building lifecycle. It is useful for participants to use the complete model because of its rich and precise information.

However, as projects become larger and more complicated, project information increases dramatically [4], resulting in the huge file size of the BIM model. It is difficult for participants to use such a large model, and it inevitably takes much time to process the complete model for their required information. In general, the designer/engineer requires a part of model data for their own business tasks, rather than a complete one. For example, the structural engineer mainly focuses on the structural objects (such as columns, walls, and slabs) from the architectural model for structural design and analysis, rather than the overall architectural model. The method or system to automatically extract the required information from BIM models can improve the quality and productivity by preventing unnecessary work. Due to the lack of effective tools to extract the required data, the designer/engineer generally deals with the original model in manual way. Such a large body of information makes it difficult for designers/engineers to directly process, leading to the inefficiencies in data sharing and exchange between software tools. Extracting the required data from the original BIM model has become one of the problems that must be addressed in BIM uses [5].

In general, commercial software tools in the AEC/FM industry have the functions of querying or extracting objects. However, only specific partial model data which have been defined in software tools can be exported. Or designers/engineers can use embedded filtering in BIM authoring tools to select the components of interest (e.g., only make the structural components visible, only make the beams visible), and then save this model for use. In this way, it is different to extract other model data according to specific requirements or purposes (e.g., only extract the concrete components). Consequently, a number of plug-ins in designated software tools were developed to extract the required information. It is noted that their functions are unavailable for other software tools, only can be used in specific software tools. Although the objects can be selected one by one manually for extraction, this process is cumbersome and prone to error. In addition, the manual extraction method cannot be stored as templates for reuse. Hence, extracting partial models based on a public data schema is inevitable in order to meet requirements from diverse business tasks.

Industry Foundation Classes (IFC) was developed to support the full range of data exchange [6]. Many studies related to partial model extraction have been carried out, and more details about these studies will be presented in the following section. However, most methods were developed to extract specific model data for specific business tasks. The users hardly extract model data by their intents, and sometimes manually select the required objects. It is necessary to develop an innovative method to extract the required model data based on the users' requirements. In this paper, a generic language was designed to extract partial models from IFC models based on the eXtensible Markup Language (XML) format. By using the proposed language, users can design a configuration file to define extraction requirements, and then the required model data is automatically extracted and formed as a partial model. The proposed method supports diverse definitions of extraction requirements, including object types, attributes, and relations. To make the definitions of extraction requirements more rigorous and standardized, the Selection Set was proposed to represent extraction requirements. Furthermore, mathematical logic and set theory [7] were used to describe rules for partial model extraction, so IFC data with multiple representations could be processed to form valid partial IFC models.

The rest of the paper is organized as follows. A review of related work for partial model extraction from two aspects (that is, task-specific and user-defined methods) was first introduced. Second, according to comparative analysis, the concept of the Selection Set was proposed to integrate users' requirements. Third, seven rules in the extraction of syntactically and semantically valid partial IFC models were designed. Subsequently, with the adoption of XML schema, a generic language was developed for partial model extraction from an IFC model based on the Selection Set (PMESS), and then the proposed method was validated through a test case. The final section summarizes the most important conclusions.

2. Related Work

The methods of extracting partial models could be classified into different areas [8,9]. This section mainly presents these methods according to the task-specific and user-defined requirements.

2.1. Partial Model Extraction According to Specific Tasks

At present, many commercial software tools have developed data interfaces for specific tasks, such as Revit, ArchiCAD, and Tekla Structures. The required objects and attributes can be extracted from the original model by using these tools. However, these data interfaces could not be applied to other software tools. As commercial software tools, their algorithms or codes are not public, so it is difficult to modify these algorithms for extracting different model data by users' intention. Besides the native model filtering for export, some software products provide the functionality for exporting IFC files. For example, IFC Translator in ArchiCAD [10] can be used to export different IFC models according to options: (1) selected elements only; (2) visible elements, on either all stories or the current story; (3) entire project. The first and second options are used for partial model extraction, and the third option is for the complete one. Similarly, a functionality called 'IFC Export Setup Options' exists in Revit [11], which can export only elements visible in the view. The aforementioned functionality is mainly used to extract physical objects according to the type or view. However, it can hardly extract the partial model according to other requirements, such as relationships and the required attributes.

IFC is a de facto standard to support a full range of business tasks in the AEC/FM industry [12]. It is a rich schema for representing diverse information throughout the building life cycle. Rather than exchange requirements (ERs) for specific tasks, IFC schema focuses on the complete building information among all actors at every stage of a building. Consequently, Information Delivery Manual (IDM) and Model View Definition (MVD) were proposed by buildingSMART International (bSI). The main purpose of IDM is to define exchange requirements of specific tasks in a non-technical term, and as a subset of IFC schema, MVD is a technical standard that translates IDM-based exchange requirements to Model Views for software implementation. Hence, software tools are able to export required partial models by defining different MVDs. So far, bSI has published several MVDs, such as IFC4 Reference View and IFC4 Design Transfer View [13]. Based on the approach of the Georgia Tech Process to Product Modeling (GTPPM) [14], Lee et al. [15] developed an eXtended Process to Product Modeling (xPPM) tool to automatically generate process maps (PMs), ERs, functional parts (FPs), and MVDs. The xPPM promotes consistent and reliable implementations of IDM and MVD. The implementation of IDM and MVD standardizes exchanged information in domain-specific tasks. For example, the recent version of software Revit is capable of exporting IFC models according to a set of MVDs, such as Ifc2 × 2 Singapore BCA e-Plan Check, Ifc2 × 3 Coordination View 2.0, Ifc2 × 3 Basic FM Handover View, and IFC4 Reference View. However, except these mentioned MVDs, other published ones have not been widely supported by BIM software tools yet. There are several tools to define and document new MVDs, such as IfcDoc [16] and ViewEdit [17]. It is useful for users to generate their own MVDs according to the requirements of business tasks. However, it still needs additional work to develop corresponding software tools to realize these MVDs.

Currently, there are numerous research projects on how to efficiently extract geometric information from data models [18–20]. However, business tasks require different kinds of information, not just geometric information. Hence, different frameworks and methods were developed to extract the required information for specific business tasks. For example, in the structural domain, some information—such as the type, location, geometry, and material of objects—should be extracted from architectural models. Qin et al. [21] proposed the Structural General Format (SGF) based on XML and developed an algorithm to automatically extract structural information from IFC-based architectural models to generate SGF-based models. Besides IFC data format, the SGF-based model could be translated into different finite element models for structural analysis. Hu et al. [22] proposed an IFC-based Unified Information Model with conversion algorithms between the architectural and structural models, and among various structural analysis models. Besides the structural domain, other business tasks also need to extract information from upstream models, e.g., energy simulation [23], construction schedule [24], etc. The aforementioned studies mainly extract the required information for specific business tasks.

2.2. Partial Model Extraction According to User-Defined Requirements

Due to the multidisciplinary, multi-stage and multi-party nature of building projects, business tasks require different information. The studies related to partial model extraction can be divided into two types: the schema level-based and the instance level-based methods. The former method is to develop a definition format with various mappings for data exchange, and the latter directly deals with the data within the original model. The schema level-based method refers to extract model data according to a predefined model data structure. Consequently, it is necessary to define all possible transformations in advance. The instance level-based method focuses on the specific information of project objects, and this method enables the extraction of some designated model data according to the user-defined requirements. In the IFC file, the specific information of objects means corresponding IFC instances. This method gives users enough flexibility and well-understanding, but more complex querying algorithms are required than the schema level-based method.

2.2.1. Methods at the Schema Level

Partial Model Query Language (PMQL) [25] and Generalized Model Subset Definition schema (GMSD) [26] are two early partial model extraction methods based on the schema level. PMQL was developed based on XML, Structured Query Language (SQL), and Simple Object Access Protocol (SOAP). It aims to extract a partial instance model from the original IFC model through the select, update, and delete operations. Inspired by PMQL and SPARQL Protocol and RDF Query Language (SPARQL), Mazairac and Beetz [9] adopted the BIM Server to develop an open query language (BIMQL, Building Information Model Query Language). The purpose of this language is to select, set, create, and delete IFC model data for managing BIM models, and the functions 'select' and 'set' have been developed. However, IFC schema includes numerous logical relations. When extracting specific relations using the PMQL, it requires many iterative cycles of request and response [26]. Furthermore, the PMQL method has room for improvement, such as path expression, nested queries, and inheritance hierarchies. As a result, GMSD was developed to support the dynamic selection of object instances and the filtering of a building model through predefined model view definitions. GMSD was designed to support EXPRESS-based models for consistency with IFC. Users need to define or edit MVDs within the GMSD method, but the MVD definition is a challenge for users. To solve this problem, bSI developed an official standardization specification data format for capturing MVDs based on the XML format, called mvdXML [27]. The mvdXML is a machine-interpretable representation for information exchange in IFC schema, and can be easily processed by software tools. More and more software tools are expected to support mvdXML. Inspired by this method, the proposed method in this paper was also developed based on the XML.

The new model view definitions generated based on the schema level need to be validated in the syntax and semantics, so Yang and Eastman [28] defined a serial of rules for subset generation by using set theory, aiming at supporting specific exchanges through the generation of valid model views. Furthermore, Lee [29] proposed the 'minimal set', the smallest complete subset of a schema related to a concept. Several conditions for extracting valid subsets from EXPRESS schema were defined to match the concepts. A tool called 'IFC Model View Extractor' (alpha version) was developed for generating subschema from IFC schema. According to subset generation rules [28], Yu et al. [30] proposed a semi-automatic generation method for MVDs, which could extract partial models according to core concepts of specific tasks. However, these core concepts need to be accurately predefined by users. In this study, with reference to the rule-based subset generation method [28], the rules for Selection Set-based partial model extraction were designed based on mathematical logic and set theory.

Some other researchers attempted to convert IFC models into a generic data schema. Given the requirement on spatial analysis, Daum and Borrmann [31,32] carried out a topological analysis of BIM models and proposed Query Language for Building Information Models (QL4BIM). This method enables users to extract the partial model by defining boundary representation. Fuchs and Scherer [33] developed a language called Multi-Model Query Language (MMQL), which required homogeneous data access to link and filter multi-model information, such as the bill of quantities, building, and schedule. Nevertheless, the export results are documented in textual format, rather than IFC-based data format or other model formats. Zhang and El-Gohary [34,35] developed an automated BIM Information Extraction method to extract the required information from IFC models with semantic Natural Language Processing techniques and Java Data Access Interface. Some limitations still exist in this method. For example, IFC relations in the extracted model are not yet fully aligned with the proposed semantic logic-based representation. Pauwels and Terkaj [36] proposed a procedure to convert IFC EXPRESS schema to an IfcOWL ontology for construction industry.

2.2.2. Methods at the Instance Level

The extraction method at the schema level always needs to be defined in a formal language for the data schema [15]. It is a complex and difficult task for users in the AEC/FM industry. Methods at the instance level (e.g., the object, property, etc.) were proposed to meet user-defined exchange requirements. Katranuschkov et al. [8] adopted the semantic query method to extend the GMSD definition at the instance level and developed Multi-model View Generator to extract partial models from BIM and non-BIM data. Based on the GMSD work, Windisch et al. [37] proposed a generic framework for consistent generation of BIM-based model views, which aims to provide the filtering at class and object level, and the generation of ad-hoc and multi-model views. However, the relations between these levels need to be further studied and implemented. To avoid the definition of data schema, Won et al. [38] proposed a no-schema algorithm to extract a partial model from an IFC model depending on user-defined object types or predefined ERs. The current version could not extract the partial model under combinatorial conditional expressions.

Currently, there are some good open source software libraries that help users and software developers to work with BIM IFC files, such as IfcOpenShell [39], xBIM [40], and IfcPlusPlus [41]. The user can use one of these IFC libraries to read IFC files according to the requirement, and IfcPlusPlus was selected as the IFC library in this study.

2.3. Summary on Related Research Works

The partial model extraction methods in the first subsection are mainly used in domain-specific tasks. The second subsection presents two types of generic partial model extraction methods related to user-defined requirements. One method is to extract partial models through definitions at the schema level. Even though it is general enough to meet various requirements of business tasks, users are required to be familiar with definitions within these methods. The other provides the selection function to extract partial models at the instance level. Nevertheless, the current tools mainly extract some common physical elements, but not fully support the extraction under some restrictions.

To support the extraction with user-defined requirements, the Selection Set was proposed to integrate user-defined requirements, and the XML format was used to design a generic language to automatically extract partial models. By using the proposed method, the required objects and their attributes can be extracted from the original model according to the user-defined requirements, and other objects that are not required can be filtered out. The key characteristics of the proposed method compared to related studies are summarized as follows:

(1)   The mathematical logic and set theory were used to define Selection Set and extraction rules for partial model extraction. The mappings between IFC data and user-defined requirements were developed by using the mathematical method, ensuring the stability of the proposed algorithm. When the version of IFC schema is updated, only some definitions in the Selection Set or extraction

rules need to be updated or revised, rather than the entire extraction algorithm. The structure of the proposed language is stable and independent of IFC versions.

(2)   A generic language for partial model extraction was designed based on Selection Set (PMESS). In order to extract different model data, data extraction requirements were analyzed and classified according to IFC schema. Subsequently, the technical structure of the partial model extraction language was developed by using the software-independent XML schema. The purpose of Selection Set is to standardize and integrate the users' extraction requirements, and its elements can be used to map into different user-defined requirements. By using the proposed language, the partial model can be extracted according to the user-defined extraction requirements, including objects, properties, and relations.

3. Concept of the Selection Set

During the process of partial model extraction, the software tool firstly identifies the extraction requirements defined by users, then extracts information which meets the requirements, and finally forms a valid data model based on the extracted information. Therefore, the extraction requirements can be regarded as input parameters. In this study, user-defined extraction requirements are integrated into the Selection Set, which can be assumed as some basic sets with specific semantic to extract partial models.

The Selection Set is an information set that is formed based on the requirements, such as object types, attributes, relations, and mixed ones. Elements in the Selection Set are used as input parameters of the proposed method. According to referencing relations between IFC data, the proposed method queries IFC data based on input parameters and then exports the required IFC model data, that is, a partial model or sub-model.

Extraction requirements can be classified as different semantics and relationships, such as object types, properties, relations, and mixed cases. In terms of data representation in the IFC schema, the entities and rules can be used to describe these extraction requirements. Hence, the first condition for the Selection Set is defined as follows:

Condition 1: *A Selection Set includes a set of Entities and Rules, and has at least one Entity.*

$$\exists e \exists r \forall S [[e \in E \land r \in R \land S = \{E, R\}] \rightarrow |S| \geq 1] \tag{1}$$

where e is ENTITY; r is Rule; S is Selection Set; E is a non-null set of Entities; and R is a set of Rules.

In an IFC model, every IFC instance represents a specific meaning, and it is illegal to have an abstract IFC entity in the IFC model. The proposed method is to query and extract IFC instances from the IFC model according to the Selection Set, so the abstract IFC entities should not be included in the Selection Set. The Condition 2 for Selection Set is listed as follows. This condition is similar to the rule BR02 defined by Yang and Eastman [28].

Condition 2: *A Selection Set cannot include an abstract entity.*

$$\forall e [e \in S \rightarrow e \notin A_{abs}] \tag{2}$$

where A_{abs} is a set of Abstract entity data types.

Business tasks require diverse information, so a set of exchange requirements may need to be defined for partial model extraction. In some cases, the partial model may only need to meet one of many Selection Sets, while other cases require to meet many Selection Sets. In conclusion, the former relation among Selection Sets is 'union', and the latter one is 'intersection'. Hence, the theorem for forming new Selection Sets is defined as follows:

Theorem 1: *Forming new Selection Sets*

The union of many Selection Sets is still a Selection Set, and the intersection of many Selection Sets is also a Selection Set. Let s_i denote a Selection Set:

$$(1)\ \forall s_i[[s_i \subset S \wedge \bigcup_{i \geq 1} s_i = s] \rightarrow s \subset S];\ \text{and}$$

$$(2)\ \forall s_i[[s_i \subset S \wedge \bigcap_{i \geq 1} s_i = s] \rightarrow s \subset S].$$

Proof:

(1) Let a be an element of $s = \bigcup_{i \geq 1} s_i$, that is, $a \in s$. There is at least one s_i, so that a belongs to s_i.

Because $a \in s_i \subset S$ and $\forall a \in s, s \subset S$;

(2) Let b be an element of $s = \bigcap_{i \geq 1} s_i$, that is, $b \in s$. For any s_i, b belongs to s_i.

Because $b \in s_i \subset S$ and $\forall b \in s, s \subset S$. $\square$

4. Rules for Partial Model Extraction

The output of the partial model extraction method is the IFC file, so the file must comply with the IFC schema. During the extraction process, the proposed method is to process the original IFC model depending on rules for partial model extraction and then export the partial IFC model by integrating required model data. These rules for partial model extraction can be categorized into three types: basic rules for a valid IFC file, extraction rules based on Selection Set, and processing rule for redundant information, as shown in Figure 1.

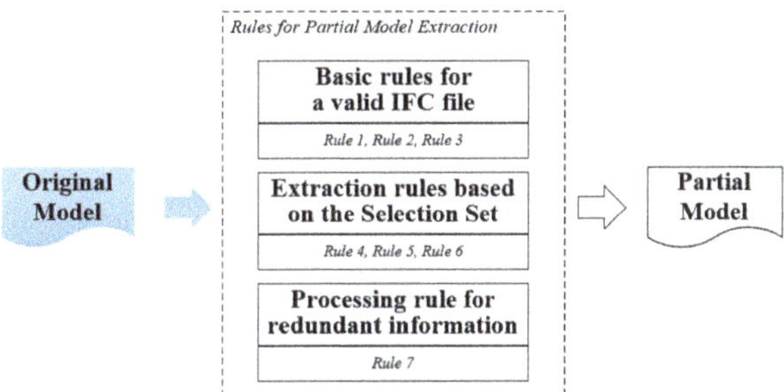

Figure 1. Rules for partial model extraction.

Basic rules for a valid IFC file: The basic rules refer to the fundamental requirements that should be complied with when an IFC file is formed. The proposed method is to export an IFC file, so the extracted partial model should also comply with these basic rules.

Extraction rules based on Selection Set: The extraction requirements were included in the Selection Set. In order to query and extract the matching IFC data, the mapping from elements in the Selection Set to IFC model data was developed.

Processing rule for redundant information: Required data could be identified according to the basic rules and extraction rules. In the final step to export the IFC file, the information that is not required by business tasks should be filtered to ensure the validity of the IFC file.

Traditionally, the existing extraction methods generally process BIM data based on one IFC version. When the IFC version is updated, it is clear that defining a mapping between one version and the updated one is a major undertaking. It may be needed to modify the corresponding algorithm. Through the proposed rules, the data processing of partial model extraction was divided into different steps. It can significantly reduce the modification work of the proposed algorithm because of the updated IFC version, ensuring the stability of this method. The data processing flow can be concluded as follows. The first step is to query corresponding IFC data according to the user-defined requirements. These requirements (such as object types, properties, and relations) can be obtained from the Selection Set. In Step 2, the target object will be located through the IFC reference relationship, and all of its corresponding attributes will be remained. Finally, all the target objects and their attributes will be extracted to form a new IFC file. The following subsection describes the proposed rules in detail.

4.1. Basic Rules for a Valid Industry Foundation Classes (IFC) File

The *IfcProject* is an important entity in an IFC file. It is not only a foundation of space structure entities (such as *IfcSite*, *IfcBuilding*, and *IfcBuildingStorey*), but also contains unit, owner history, geometric representation and other basic information of a building project. An IFC file has only one *IfcProject* entity [42]. Based on this entity, some basic project information can be queried from the IFC model data, such as site, building, and unit. Hence, a partial model should contain the *IfcProject* entity and related entities which are referenced by attributes of the *IfcProject*.

Rule 1: The partial model has only one *IfcProject* entity and an entity set which consists of other entities referenced by attributes of the *IfcProject*.

$$\forall M_o[[[e_{pro} \in M_o] \wedge |e_{pro}| = 1] \rightarrow R_{pro} \subset M_p] \tag{3}$$

where M_o is the original model; e_{pro} is the *IfcProject* entity; R_{pro} is a set of entities referenced by *IfcProject* entity's attributes; and M_p is the partial model.

In the IFC schema, the IFC entity mainly contains explicit and inverse attributes [43]. The explicit attributes are scalar values or the information computed from other attributes, while the inverse ones are identified relationally through other entities.

To ensure the completeness of the partial model, when extracting one designated entity (called 'target entity'), all entities referenced by the explicit attributes of the target entity should be extracted together. The entity set consisting of these referenced entities is assumed to define as the Essential Set (ES) of the target entity.

Rule 2: The Essential Set of the target entity is included in the partial model.

$$\forall e[e \in S \rightarrow R_e \subset M_p] \tag{4}$$

where R_e is the entity set referenced by the attributes of e.

Besides explicit attributes, some information of the target entity is represented by other IFC instances defined in inverse attributes. Similarly, the entities defined in inverse attributes of the target entity should be extracted. According to the referencing and inheritance structure of the IFC model, the entities in the ES also need to be queried to find out corresponding entities defined in inverse attributes. It is noteworthy that some particular IFC entities are used to represent basic information of a building project, which may be referenced by many IFC instances. A representative entity is the *IfcOwnerHistory*. If these entities were queried to search entities defined in inverse attributes, some entities that were not required would be extracted as well as some repeat entities. To avoid this situation, these entities comprising the Particular Set (PS) were designed as ending points of the query process. Besides the *IfcOwnerHistory*, the basic entities—such as *IfcDirection*, *IfcCartesianPoint*, *IfcAxis2Placement3D/IfcAxis2Placement2D*, *IfcLocalPlacement, and IfcGeometricRepresentationContext/IfcGeometricRepresentationSubContext*—were set as the particular entities in this study. When encountering these particular entities during the query of inverse attributes, the proposed method will stop the running and enter into the next query. After this step, the non-target IFC entities cannot be extracted. The entity set including entities in the ES and entities defined in inverse attributes is assumed to be the Individual Set (IS) of the target entity. The IS contains complete information of the target entity.

Rule 3: The partial model includes entities defined in inverse attributes of the target entity, which are not queried from the Particular Set.

$$\forall e[[\exists R_{inv}[e \in E_{es} \wedge e \notin E_{ps} \wedge R_{inv}(e, e_{inv})] \wedge e_{inv} \in E_{inv}] \\ \rightarrow E_{inv} \subset M_p] \tag{5}$$

where e_{inv} is an entity defined in inverse attributes of the target entity e; R_{inv} is the inverse relation from e to e_{inv}; E_{ps} is a set of particular entities in the Particular Set; and E_{inv} is a set of required entities e_{inv}.

4.2. Extraction Rules Based on Selection Set

According to Condition 1, the Selection Set has at least one Entity e. In the IFC model, all IFC instances matching the Entity e should be extracted. Furthermore, other IFC instances related to explicit and inverse attributes of the target entity should be extracted according to Rule 2 and Rule 3.

Rule 4: IFC instances matching Entity e in the Selection Set are contained in the partial model.

$$\forall e[e \in S \wedge \exists f_M[f_M(e) = E_e] \rightarrow E_e \subset M_p] \tag{6}$$

where E_e is the set of entities related to Entity e; and f_M is a function from e to E_e, working on original model M_o.

Numerous complex relations exist between various objects in a building project. For example, the binary relation can be divided into different types of relations, such as containment, parallel, and crosscutting relations. Consequently, relation entities should be set into Selection Set. All corresponding IFC instances within the user-defined relations in the Selection Set would be extracted to form the partial model. In this study, the object which contains other object(s) or is relied by other object(s) is set as the relating object, and other object(s) are called related object(s).

Rule 5: The corresponding relating entity and related entities are included in the partial model, when a relation entity is included in the Selection Set.

$$\forall e_{rel} \exists e_{relating} \exists E_{related} \\ [[e_{rel} \in S \wedge IFCREL(e_{rel}, e_{relating}, E_{related})] \\ \rightarrow [(e_{relating} \cup E_{related}) \subset M_p]] \tag{7}$$

where e_{rel} is a relation entity; $e_{relating}$ is a relating object entity; $E_{related}$ is the set of related object entities; and $IFCREL$ is a function to test if the relation entity e_{rel} exists between relating object entity $e_{relating}$ and related object entities $E_{related}$.

The attributes of IFC entity represent different essential characteristics from other entities. These attributes can form different rules. As a result, the according model data can be extracted by designing different rules.

Rule 6: The entities included in the partial model satisfy the rules in the Selection Set.

$$\forall r[r \in S \rightarrow r \cap M_p] \tag{8}$$

where r is a Rule.

4.3. Processing Rule for Redundant Information

According to elements in the Selection Set (Rule 4, 5, and 6), the matching IFC instances can be extracted from the original model, while the related necessary IFC instances are extracted based on Rule 1, 2, and 3. The IFC entities which are undefined in the Selection Set or cannot be inferred from the Selection Set are not supposed to be included in the partial model. These entities are called redundant information in this study. The redundant information, including IFC instances and related attributes, must be filtered before the export of the partial model.

Rule 7: The partial model cannot include entities which are not defined or inferred in the Selection Set.

$$\forall \neg e [[\neg e \in M_o \wedge (\neg e \notin S)] \rightarrow \neg e \notin M_p] \tag{9}$$

where $\neg e$ is an Entity undefined in the Selection Set or cannot be inferred from the Selection Set.

5. Generic Language for Partial Model Extraction Based on the Selection Set

To ensure the proposed method to be interpreted by software tools, the XML schema was adopted to define a generic language for partial model extraction based on the Selection Set (PMESS). The overall architecture of the PMESS is shown in Figure 2.

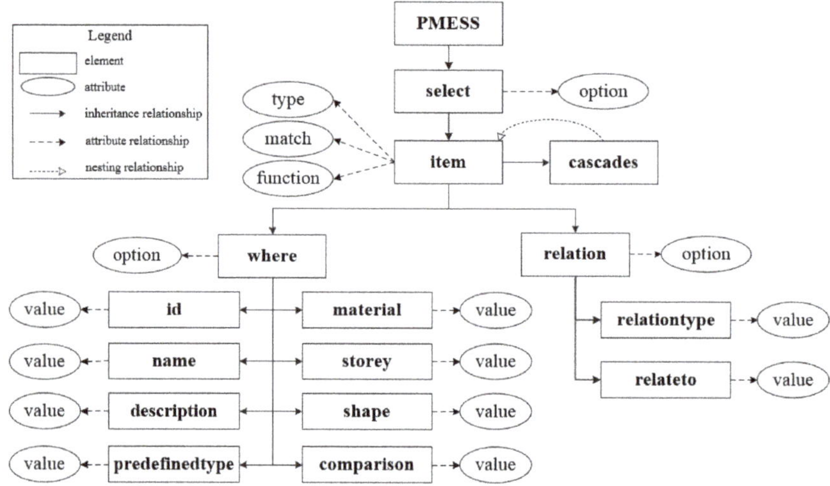

Figure 2. Architecture of the Partial Model Extraction based on the Selection Set (PMESS).

The 'PMESS' element is the root element at the first level of the overall architecture. The element is represented by the box symbol. The element at the second level is the 'select' element, which means the proposed method is to extract the partial model based on the Selection Set. The 'select' is a child element of the 'PMESS', which is connected by an arrow with a solid line. Condition 1 and 2 are mainly prescribed by elements 'item', 'relation', and 'where'. The 'item' element defines the entity type, including object entity and attribute entity, and the 'relation' for the relation entity. The 'item' element has three attributes: type, match, and function. The relationship between the element and the attribute is represented by the dotted arrow. The 'where' is used to represent the rules for extracting partial models. It is noted that the 'item' and the 'cascades' are connected by the hollow arrow with dotted line. This means that the structure of the 'cascades' is the same as the 'item'. More details will be discussed in the following section. An example of a PMESS-based configuration file used for extracting concrete columns is shown in Figure 3.

```
<?xml version="1.0" encoding="UTF-8"?>
<PMESS>
 <select>
  <item type="ELEMENT" match="column" function="extract">
   <where>
    <material value="concrete"/>
   </where>
  </item>
 </select>
<PMESS>
```

Figure 3. PMESS-based configuration file for extracting concrete columns.

5.1. 'Select'-Mechanism

The 'select' element has one unique attribute 'option', which is either 'AND' or 'OR'. The 'option' is designed to comply with the Theorem mentioned above. The use of 'AND' and 'OR' is defined as the intersection and union relations between several items, respectively. While the value of 'option' is 'AND', IFC instances will be extracted from the original model only when they match all the defined items. On the contrary, the 'OR' is required to match any one of the defined items. The default value of 'option' is 'OR'.

5.2. 'Item'-Classification

The type of entities in the Selection Set can be defined by the attribute 'type' of 'item'. Another two attributes of 'item' are 'match' and 'function'. The 'type' was designed to comply with Condition 1. Its value includes two types: ELEMENT and ATTRIBUTE, and it is required not to include abstract entity types (as mentioned in Condition 2). The 'match' enables users to describe the name of ELEMENT or ATTRIBUTE. The mapping between the value of 'match' and IFC entities/attributes has been established, which can automatically query IFC data according to the user-defined requirements.

When the 'type' is 'ELEMENT', the proposed method will extract IFC object entities which comply with the requirements defined in 'match' and 'where' (as mentioned in Rule 4). Particularly, if the value of 'match' is 'SET', the partial model extraction is required to comply with the rules defined in the 'relation'.

The matching attribute entities in the IFC model will be extracted, if the 'type' is 'ATTRIBUTE' (as mentioned in Rule 4). The proposed method queries and extracts object entities which have the matching attribute entities.

The third attribute of 'item' is 'function', which is currently limited to the 'extract' for partial model extraction. The 'filter', 'modify', and 'add' in the 'function' will be further studied in the next paper.

5.3. 'Relation'-Rule

According to representations of IFC relation entities, objects entities within a relationship can be divided into relating object entity and related object entity (entities), as shown in Figure 4a. In the IFC schema, there are many sub-entities within the *IfcRelationship* entity to represent diverse relations, such as *IfcRelContainedInSpatialStructure*, *IfcRelAggregates*, and *IfcRelAssignsToGroup*. Figure 4b illustrates an example of relating object entity and related object entities defined by the *IfcRelAssignsToGroup* entity.

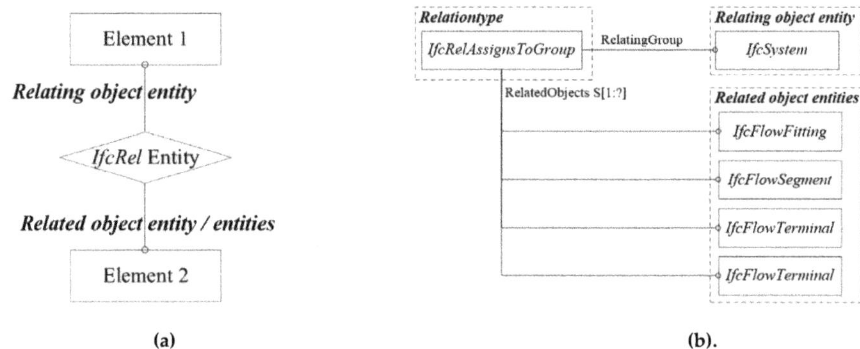

Figure 4. The relation between object entities: (**a**) Relation between relating object entity and related object entity/entities; (**b**) Relation using *IfcRelAssignsToGroup* entity.

The relating object entity and related object entity (entities) are defined in the sub-element 'relateto' of 'relation' (as shown in Figure 2), while the sub-element 'relationtype' is for the type of 'relation' (Rule 5). Currently, the proposed method supports the extraction of relations of building storey, group, and element assembly. As mentioned above, in the case of 'type=ELEMENT' and 'match=SET', the proposed method will query the matching relation entity according to the definition in 'relation', and extract IFC object entities referenced by the relation entity.

5.4. 'Where'-Rule

The elements mentioned above are mainly used to extract the objects with a certain type or relation, but not for the objects with some given characteristics. Hence, the 'where' element was designed to define rules for extracting specific objects according to the user-defined semantics (Rule 6). According to the characteristics of objects defined in the IFC schema, the object semantics could be classified as direct and indirect semantics. While direct semantics could be directly attained from IFC instances, indirect ones have to be inferred or computed from other IFC instances. The direct ones include the Identity Document (ID), name, description, and predefined type; and the material, storey, shape, and comparison for the indirect ones, as shown in Figure 2.

When these semantic meanings are defined in the PMESS document, the proposed method can query the target data to form a valid IFC model.

Figure 5 presents an example of all attributes in *IfcBeam* entity and the corresponding IFC entities for 'where' rules. The ID, name, and description are derived from the *IfcRoot* entity, a root entity in the IFC schema storing the most fundamental information. The predefined type, an extension of the IFC4 version to the attribute in the *IfcBeam* entity, is used to define different types of the object (*IfcBeam* in this example). The indirect semantics are required to query other IFC entities, for example, the material. In general, the *IfcMaterial* entity is associated with the IFC object entity through the *IfcRelAssociatesMaterial* entity, a subtype of the *IfcRelAssociates* entity.

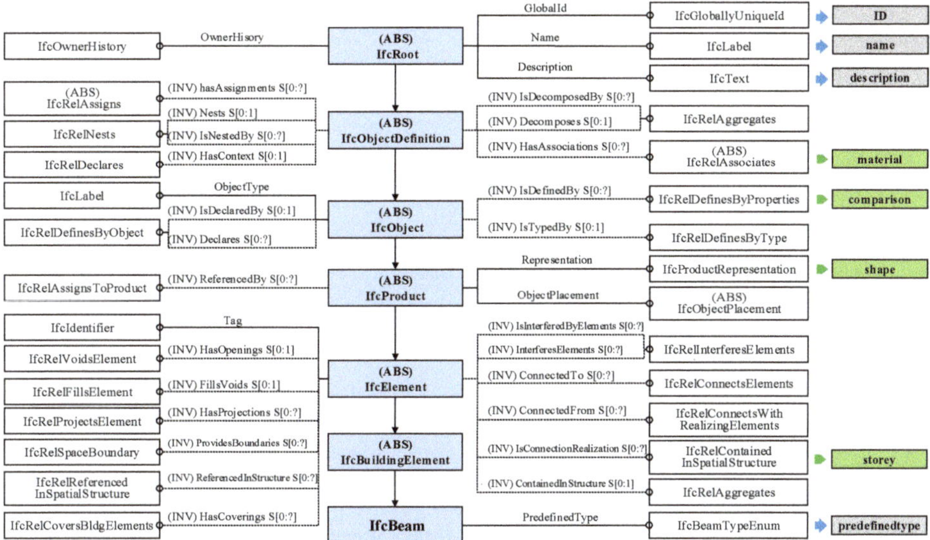

Figure 5. Mapping between *IfcBeam* entity's attributes and 'where' rules. ABS (Abstract): abstract entity of data types.

5.5. 'Cascades'-Rule

The definitions in 'item' are overall requirements for extracting the partial model, and the 'cascades' can be used to further prescribe the requirements. The structure of 'cascades' is the same as 'item' to ensure the uniform definition. Figure 6 shows an example of the PMESS-based configuration file for extracting beams under the 'where'-rule that the construction time is '2019-09-20'. In this case, the proposed method firstly queries all beams in the building project, and then queries the specified beams which match the 'cascades'.

```xml
<?xml version="1.0" encoding="UTF-8"?>
<PMESS>
<select>
<item type="ELEMENT" match="beam" function="extract">
<cascades>
<item type="ATTRIBUTE" match="property" function="extract">
<where>
<name value="construction time"/>
<comparison value="2019-09-20"/>
</where>
</item>
</cascades>
</item>
</select>
</PMESS>
```

Figure 6. PMESS-based configuration file for extracting beams (construction time is '2019-09-20').

The XML schema was adopted to define PMESS elements for complying with Rule 4-Rule 6. Rule 1-Rule 3 are the fundamental rules to form a valid IFC file, while Rule 7 is used to process redundant information within the extracted IFC instances. These four rules (Rule 1, 2, 3, and 7) have been embedded in the data process engine for implementation, not required to be defined by users.

6. Test Case

C++ programing language was used to develop two data interfaces for the implementation of the proposed method (PMESS). One is to read the PMESS-based configuration file, and the other is to extract and export the partial model. For further applications, these data interfaces were embedded into the proposed IFC-based platform. Different IFC models exported from many software tools (such as ArchiCAD, MagiCAD, Revit, and Tekla Structures) have been used to verify the feasibility of PMESS. In this section, a practical project model created by ArchiCAD was used to introduce the utility of the proposed method.

The test case was conducted using a building model of a shopping mall project. The building has eight floors with a construction area of 148,564 m^2, including six floors above ground and two underground floors. This model was built by ArchiCAD and exported as an IFC file by default settings. The file size of this IFC model was about 101 M, with 1,862,673 IFC instances. Figure 7 shows the visualization of this project in the proposed IFC-based platform. Due to the large IFC file size of this project, it is necessary to extract partial models for different business tasks. Through the proposed method, several partial models were extracted under the following extraction requirements.

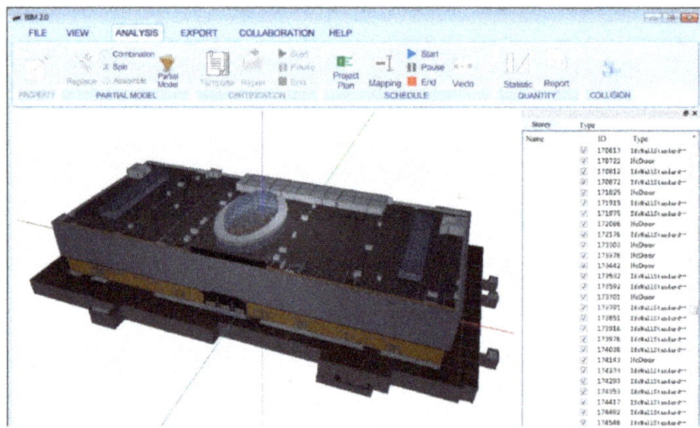

Figure 7. IFC model of the shopping mall project in the proposed IFC-based platform.

6.1. Partial Model Extraction for the User-Defined Extracted Objects

Through setting different 'item' elements, the required objects could be extracted. As examples for types of extracted objects, Table 1 shows four examples of partial models extracted from the original model. The first three partial models extract a certain type of object, while two types of objects are extracted in the fourth one. Figure 8 depicts the PMESS-based configuration file for the fourth partial model in the proposed platform. Through the PMESS, physical objects from different disciplines (architecture, structure, MEP, etc.) can be extracted from the original BIM model, such as the door and window for architecture, the beam and column for structure, the equipment and pipeline for MEP.

Table 1. Some examples of partial model extraction for the required objects by using PMESS.

Extracted Objects	Original Model			Partial Model		Visualization
	IFC Entity	Amount	Amount	File Size/M	Instances	
Slab	*IfcSlab*	5284	5284	19.9 (19.7%)	292,912 (15.7%)	
Door	*IfcDoor*	911	911	19.2 (19.0%)	309,332 (16.6%)	
Transport element	*IfcTransport Element*	64	64	15.7 (15.5%)	253,984 (13.6%)	
Beam and column	*IfcBeam* and *IfcColumn*	6886 (4002&2884)	6886 (4002&2884)	13.8 (13.7%)	173,763 (9.3%)	

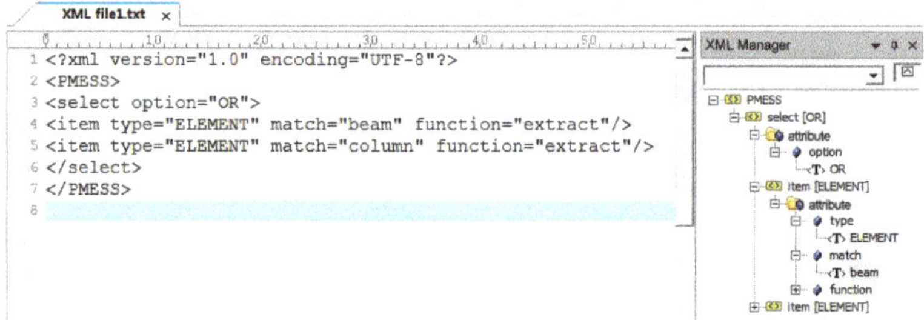

Figure 8. PMESS-based configuration file for extracting beams and columns.

By using the IFC File Analyzer [44], IFC model data can be analyzed in detail. As shown in Table 1, all IFC object entities which match the user-defined requirements were correctly extracted. Moreover, through filtering out other objects, the resulting models only contain the required objects and their attributes. For example, the number of IFC instances in the fourth partial model was 173,763, and 90.7% instances were filtered out. Accordingly, the file size decreased to 13.8 M (only 13.7% of the original model). The results show that the proposed method correctly identifies the user-defined requirements and extracts all the semantically required objects from the original model. On the other hand, the decreasing of the partial models in file size is apparent compared with the original model, which avoids filtering redundant information manually and facilitates the fulfillment of downstream business tasks based on useful building information.

Except the extraction according to the object type, other semantics could be used to extract the required BIM data, such as the relationship and the rules (as shown in the following subsections).

6.2. Partial Model Extraction Based on the User-Defined Relations

This project is composed of underground and overground structures, so it needs to be built by different designers. According to this requirement, the 'relation'-rule was used to extract all objects in the different parts of this building. An example of the PMESS-based configuration file for extracting the underground structure is illustrated in Figure 9, and the resulting partial model is presented in Figure 10. The file size of the extracted partial model is 32.1 M, including 942 columns, 1380 beams, 1253 walls, 351 doors, 69 slabs, and 148 stairs. These extracted objects are consistent with the original model. The proposed method is capable of querying and extracting the required information depending on the user-defined relation rule.

```
<?xml version="1.0" encoding="UTF-8"?>
<PMESS>
 <select>
   <item type="ELEMENT" match="SET" function="extract">
    <relation relationtype="building storey" relateto="related"/>
    <where>
     <name value="Floor B1"/>
    </where>
   </item>
   <item type="ELEMENT" match="SET" function="extract">
    <relation relationtype="building storey" relateto="related"/>
    <where>
     <name value="Floor B2"/>
    </where>
   </item>
 </select>
</PMESS>
```

Figure 9. PMESS-based configuration file for extracting building storeys named Floor B1 and B2.

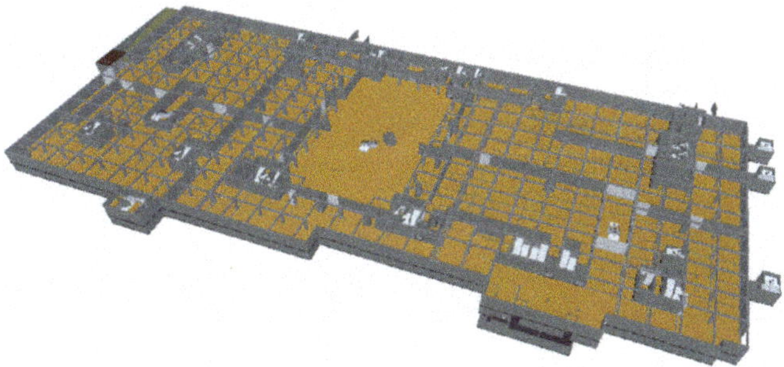

Figure 10. Partial model composed of the underground Floor B1 and B2.

6.3. Partial Model Extraction Based on the User-Defined Rules

Numerous curtain walls were contained in this building project, such as peripheral curtain walls, and the skylight on the sixth floor. Due to the complexity of curtain walls, the models of curtain walls were required to set as separate models, which would be designed by different curtain wall engineers. For this purpose, curtain walls in different placements were extracted from the complete architectural model, and imported back into the original software for further design and analysis, as shown in Figure 11. The main contents of the PMESS-based configuration files were presented in the middle part of Figure 11.

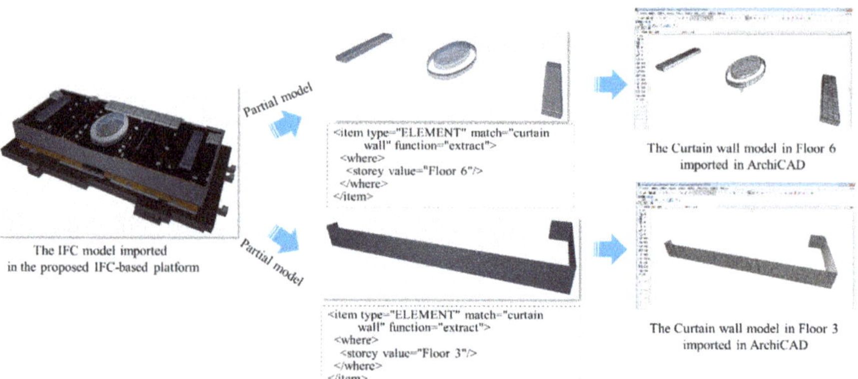

Figure 11. Partial models of curtain walls in different placements.

The file sizes of these two partial models were 2.37 M and 0.59 M, respectively, which were much smaller than the original model's (101 M). It is beneficial for engineers to make a detailed design based on the reduced models. These extracted partial models could be imported back to the original software ArchiCAD for detailed design. In addition, these extracted partial models were IFC compliant models, and could be used to import to other BIM software tools (such as Revit and Tekla) for professional design. It demonstrated that the extracted IFC files were syntactically valid.

7. Conclusions

A building project always consists of different types of information from multiple disciplines. However, business tasks require only a part of the complete building information model. Meanwhile, the required information varies depending on business tasks. A common method for partial model extraction which meets user-defined extraction requirements is necessary. For this purpose, a generic language for partial model extraction based on the Selection Set was proposed to extract a partial model from the IFC model.

The Selection Set was designed to represent extraction requirements. Elements in the Selection Set work as input parameters of the partial model extraction method. Due to the complexity of requirements for business tasks, several extraction requirements could be defined in the intersection or union form.

Furthermore, seven rules were defined to extract the partial model based on mathematical logic and set theory. These proposed rules ensure the syntactical and semantic validity of the partial IFC model during the extraction process. Firstly, the proposed method queries IFC data which matches the requirements defined in the Selection Set, such as IFC entities, properties, and relations. Subsequently, according to seven rules for partial model extraction, these extracted IFC data are defined as the nodes to query other related IFC data, and redundant information is filtered for forming a valid partial model.

Considering the processability of the computer and the readability of users, the XML schema was adopted to design the generic language for partial model extraction. Given the definitions of building information in the IFC schema, this study developed a mapping between IFC data and the elements defined in the PMESS method, which could meet diverse requirements defined by users. Through the PMESS method, users can extract the required information from the original model under different extraction requirements, such as objects, properties, and relations. In addition, the PMESS-based configuration file can be saved as a common template for reuse, which improves the efficiency of the definitions of extraction requirements.

To demonstrate the feasibility of the proposed method, a practical project was used to extract different partial models under three conditions: object types, object relations, and specific rules.

Compared with the original model, the required objects were correctly extracted, which showed the validity of partial models at the semantic level. Furthermore, the extracted partial models could be imported back into the original software tool, which demonstrated the syntactical validity of the extracted IFC file.

Currently, although some commercial software products can be used to extract the required objects, it needs users to manually select the required objects, and the partial model cannot be extracted according to the particular rules. Some researchers have carried out research projects for partial model extraction, and mainly focus on some specific information, such as geometric information. In this study, the proposed PMESS method makes users automatically extract the partial model by requirement definition. Furthermore, users can define different requirements to extract the required partial model based on the PMESS, which can accommodate more applications.

This study is an important step in data sharing and exchange of building projects, and it also has room for improvement. For example, bSI proposed IDM and MVD for exchange requirements, and defined several templates for practical tasks. To extend the applicability of the proposed method, the PMESS should be mapped to IDM and MVD. Given the fact that the PMESS was designed based on the XML schema, the mapping mechanism between the PMESS and mvdXML could be further studied.

Author Contributions: Conceptualization, X.D. and H.L.; Methodology, X.D. and H.L.; Validation, H.L., J.X., and Y.Z.; Writing—original draft preparation, X.D. and H.L.; Writing—review and editing, J.X., and Y.Z.; Visualization, H.L.; Supervision, H.L. and Y.Z.; Project administration, X.D. and J.X.; Funding acquisition, X.D. and H.L. All authors have read and agreed to the published version of the manuscript.

Funding: The research work was supported by the National Key Research and Development Program of China during the 13th Five-year Plan (no. 2016YFC0702001), the Industrial Internet Innovation and Development Project 2019 from the Ministry of Industry and Information Technology of China, and Project Funded by China Postdoctoral Science Foundation (no. 2019M663115).

Conflicts of Interest: The authors declare no conflict of interest.

References

1. Sanguinetti, P.; Abdelmohsen, S.; Lee, J.M.; Lee, J.K.; Sheward, H.; Eastman, C. General system architecture for BIM: An integrated approach for design and analysis. *Adv. Eng. Inform.* **2012**, *26*, 317–333. [CrossRef]
2. Olawumi, T.; Chan, D. Building information modeling and project information management framework for construction projects. *J. Civ. Eng. Manag.* **2019**, *25*, 53–75. [CrossRef]
3. Juan, Y.K.; Lai, W.Y.; Shih, S.G. Building information modeling acceptance and readiness assessment in Taiwanese architectural firms. *J. Comput. Civ. Eng.* **2016**, *23*, 356–367. [CrossRef]
4. Oh, M.; Lee, J.; Hong, S.W.; Jeong, Y. Integrated system for BIM-based collaborative design. *Automat. Constr.* **2015**, *58*, 196–206. [CrossRef]
5. Belsky, M.; Sacks, R.; Brilakis, I. Semantic Enrichment for Building Information Modeling. *Comput.-Aided Civ. Infrastruct. Eng.* **2016**, *31*, 261–274. [CrossRef]
6. Eastman, C.; Teicholz, P.; Sacks, R.; Liston, K. *BIM Handbook: A Guide to Building Information Modeling for Owners, Managers, Designers, Engineers and Contractors*, 2nd ed.; Wiley: New York, NY, USA, 2011.
7. Wilson, P.R. EXPRESS and Set Theory (Draft). Available online: http://citeseer.ist.psu.edu/viewdoc/summary?doi=10.1.1.52.6106&rank=1 (accessed on 3 August 2018).
8. Katranuschkov, P.; Weise, M.; Windisch, R.; Fuchs, S.; Scherer, R.J. BIM-based generation of multi-model views. In Proceedings of the CIB W78 2010: 27th International Conference—Applications of IT in the AEC Industry, Cairo, Egypt, 16–18 November 2010.
9. Mazairac, W.; Beetz, J. BIMQL—An open query language for building information models. *Adv. Eng. Inform.* **2013**, *27*, 444–456. [CrossRef]
10. Graphisoft. Filter Model at Export. Available online: https://helpcenter.graphisoft.com/user-guide/65795/ (accessed on 18 February 2020).
11. Autodesk. IFC Export Setup Options. Available online: http://help.autodesk.com/view/RVT/2020/ENU/?guid=GUID-E029E3AD-1639-4446-A935-C9796BC34C95 (accessed on 18 February 2020).

12. Zhang, C.; Beetz, J.; Vries, B.D. Towards model view definition on semantic level: A state of the art review. In Proceedings of the 20th International Workshop of the European Group for Intelligent Computing in Engineering, Vienna, Austria, 1–3 July 2013.

13. BuildingSMART International. Model View Definition Summary. Available online: http://www.buildingsmart-tech.org/specifications/ifc-view-definition (accessed on 12 January 2019).

14. Lee, G.; Sacks, R.; Eastman, C. Product data modeling using GTPPM—A case study. *Automat. Constr.* **2007**, *16*, 392–407. [CrossRef]

15. Lee, G.; Park, Y.H.; Ham, S. Extended process to product modeling (xPPM) for integrated and seamless IDM and MVD development. *Adv. Eng. Inform.* **2013**, *27*, 636–651. [CrossRef]

16. BuildingSMART International. IfcDoc Tool Summary. Available online: http://www.buildingsmart-tech.org/specifications/specification-tools/ifcdoc-tool (accessed on 10 February 2019).

17. Scherer, R.J.; Weise, M.; Katranuschkov, P. Adaptable views supporting long transactions in concurrent engineering. In Proceedings of the Joint International Conference on Computing and Decision Making in Civil and Building Engineering, Montreal, QC, Canada, 14–16 June 2006; pp. 3677–3686.

18. Nepal, M.P.; Staub-French, S.; Pottinger, R.; Webster, A. Querying a building information model for construction-specific spatial information. *Adv. Eng. Inform.* **2012**, *26*, 904–923. [CrossRef]

19. Teo, T.A.; Yu, S.C. The extraction of indoor building information from BIM to OGC IndoorGml. *Int. Arch. Photogramm. Remote Sens. Spat. Inf. Sci.* **2017**, 167–170. [CrossRef]

20. Zhou, Y.; Hu, Z.Z.; Zhang, W.Z. Development and application of an Industry Foundation Classes-based Metro protection information model. *Math. Probl. Eng.* **2018**, 1820631. [CrossRef]

21. Qin, L.; Deng, X.Y.; Liu, X.L. Industry foundation classes based integration of architectural design and structural analysis. *J. Shanghai Jiaotong Univ. (Sci.)* **2011**, *16*, 83–90. [CrossRef]

22. Hu, Z.Z.; Zhang, X.Y.; Wang, H.W.; Kassem, M. Improving interoperability between architectural and structural design models: An industry foundation classes-based approach with web-based tools. *Automat. Constr.* **2016**, *66*, 29–42. [CrossRef]

23. Kim, H.; Shen, Z.; Kim, I.; Kim, K.; Stumpf, A.; Yu, J. BIM IFC information mapping to building energy analysis (BEA) model with manually extended material information. *Automat. Constr.* **2016**, *68*, 183–193. [CrossRef]

24. Kim, H.; Anderson, K.; Lee, S.; Hildreth, J. Generating construction schedules through automatic data extraction using open BIM (building information modeling) technology. *Automat. Constr.* **2013**, *35*, 285–295. [CrossRef]

25. Adachi, Y. Overview of Partial Model Query Language. Available online: http://cic.vtt.fi/projects/ifcsvr/ (accessed on 3 December 2017).

26. Weise, M.; Katranuschkov, P.; Scherer, R.J. Generalised Model Subset Definition Schema. In Proceedings of the 20th CIB W78 International Conference on Information Technology in Construction, Waiheke Island, Auckland, New Zealand, 23–25 April 2003; pp. 440–447.

27. Chipman, T.; Liebich, T.; Weise, M. mvdXML Specification 1.1. Available online: https://standards.buildingsmart.org/MVD/RELEASE/mvdXML/v1-1/mvdXML_V1-1-Final.pdf (accessed on 12 March 2019).

28. Yang, D.; Eastman, C. A rule-based subset generation method for product data models. *Comput.-Aided Civ. Infrastruct. Eng.* **2007**, *22*, 133–148. [CrossRef]

29. Lee, G. Concept-based method for extracting valid subsets from an EXPRESS schema. *J. Comput. Civ. Eng.* **2009**, *23*, 128–135. [CrossRef]

30. Yu, F.Q.; Zhang, J.P.; Liu, Q. Semi-automatic generation of the BIM model view definition based on the IFC standard. *J. Tsinghua Univ. (Sci. Technol.)* **2014**, *54*, 987–992. [CrossRef]

31. Daum, S.; Borrmann, A. Boundary representation-based implementation of spatial BIM queries. In Proceedings of the EG-ICE Workshop on Intelligent Computing in Engineering, Vienna, Austria, 1–3 July 2013; pp. 1–10.

32. Daum, S.; Borrmann, A. Processing of topological BIM queries using boundary representation based methods. *Adv. Eng. Inform.* **2014**, *28*, 272–286. [CrossRef]

33. Fuchs, S.; Scherer, R.J. MMQL—A language for multi-model linking and filtering. In Proceedings of the 10th European Conference on Product and Process Modeling, Vienna, Austria, 17–19 September 2014; pp. 273–280.

34. Zhang, J.; El-Gohary, N.M. Automated information transformation for automated regulatory compliance checking in construction. *J. Comput. Civ. Eng.* **2015**, *29*, B4015001. [CrossRef]

35. Zhang, J.; El-Gohary, N.M. Automated extraction of information from building information models into a semantic logic-based representation. In Proceedings of the 2015 International Workshop on Computing in Civil Engineering, Austin, TX, USA, 21–23 June 2015; pp. 173–180.
36. Pauwels, P.; Terkaj, W. EXPRESS to OWL for construction industry: Towards a recommendable and usable IfcOWL ontology. *Automat. Constr.* **2016**, *63*, 100–133. [CrossRef]
37. Windisch, R.; Katranuschkov, P.; Scherer, R.J. A generic filter framework for consistent generation of BIM-based model views. In Proceedings of the 19th International Workshop of the European Group for Intelligent Computing in Engineering, Munich, Germany, 4–6 July 2012.
38. Won, J.; Lee, G.; Cho, C. No-schema algorithm for extracting a partial model from an IFC instance model. *J. Comput. Civ. Eng.* **2013**, *27*, 585–592. [CrossRef]
39. IfcOpenShell. Available online: http://www.ifcopenshell.org/ (accessed on 20 February 2020).
40. XBIM. Available online: https://github.com/xBimTeam/XbimWebUI (accessed on 20 February 2020).
41. IfcPlusPlus. Available online: https://github.com/ifcquery/ifcplusplus (accessed on 20 February 2020).
42. Liebich, T. IFC 2x Edition 3 Model Implementation Guide. Available online: https://standards.buildingsmart.org/documents/Implementation/IFC2x_Model_Implementation_Guide_V2-0b.pdf (accessed on 15 January 2018).
43. Lee, Y.C.; Eastman, C.M.; Solihin, W. An ontology-based approach for developing data exchange requirements and model views of building information modeling. *Adv. Eng. Inform.* **2016**, *30*, 354–367. [CrossRef]
44. National Institute of Standards and Technology. IFC File Analyzer. Available online: https://www.nist.gov/services-resources/software/ifc-file-analyzer (accessed on 8 March 2019).

Article

Filtering of Irrelevant Clashes Detected by BIM Software Using a Hybrid Method of Rule-Based Reasoning and Supervised Machine Learning

Will Y. Lin [1,*] and Ying-Hua Huang [2]

[1]   Department of Civil Engineering, Feng Chia University, Taichung 407, Taiwan
[2]   Department of Civil and Construction Engineering, National Yunlin University of Science and Technology, Yunlin 640, Taiwan; huangyh@yuntech.edu.tw
*   Correspondence: weiylin@mail.fcu.edu.tw; Tel.: +886-973-531-289

Received: 9 October 2019; Accepted: 2 December 2019; Published: 6 December 2019

Abstract: Construction projects are usually designed by different professional teams, where design clashes may inevitably occur. With the clash detection tools provided by Building Information Modeling (BIM) software, these clashes can be discovered at an early stage. However, the number of clashes detected by BIM software is often huge. The literature states that the majority of those clashes are found to be irrelevant, i.e., harmless to the building and its construction. How to filter out these irrelevant clashes from the detection report is one of the issues to be resolved urgently in the construction industry. This study develops a method that automatically screens for irrelevant clashes by combining the two techniques of rule-based reasoning and supervised machine learning. First, we acquire experts' knowledge through interviews to compile rules for the preliminary classification of clash types. Subsequently, the results of the initial classification inferred by the rules are added into the training dataset to improve the predictive performance of the classifiers implemented by supervised machine learning. The average predictive performance obtained by using the hybrid method is up to 0.96, which has been improved from the traditional machine learning process only using individual or ensemble learning classifiers by 6%–17%.

Keywords: clash detection; supervised machine learning; building information modeling (BIM)

1. Introduction

Design conflicts refer to the errors in which building components overlap with each other spatially when compiling various types of engineering drawings. Since engineering drawings are generally formed after compiling the designs by engineers of different professions, design conflicts between different system components are often common [1,2]. Minor design conflicts often result in rework and increase the project costs. In severe cases, design changes may be required, resulting in cost overruns, delay in progress, and compromising the structural safety. As pointed out by previous studies, unresolved design conflicts will hugely impact on the project success [3].

In recent years, the emergence of BIM software has enabled the easy detection of design conflicts; conflict checking has become one of the important functions of BIM software. Since resolving design conflicts is critical to the success of a project, many countries have mandated all public projects to execute clash detection. In the United Kingdom, for example, the design team must perform clash detection once every week or every two weeks to ensure that the engineering design receives complete coordination and is free of conflicts, thereby reducing the probability of change orders [4]. However, the clash detection algorithms of most BIM software are simple; as long as two building components are spatially overlapping, in contact, or within a given distance, they will be identified as a conflict/clash and listed in the detection report. Therefore, even for small projects, the number of clashes detected using BIM software can be enormous [5–7]. As many studies discovered, 50% or more of the clashes detected from BIM software are found to be "irrelevant clashes"; that is, these conflicts will not have a substantial impact on the projects, or they can be directly resolved by site engineers during the construction phase [7]. However, the clash detection report of BIM software does not identify these irrelevant clashes. Ideally, every single clash in a detection report should be evaluated by engineers to decide whether the resolution is needed. This is an extremely time-consuming job. According to our interviews with senior project managers, many projects in Taiwan cannot afford the high incurred costs. Thus, their BIM managers merely selectively review a few clashes or even neglect the entire report. It is why many studies have pointed out that unless filtering those irrelevant clashes is automated, the clash detection report with an overwhelming number of clashes will become trivial and meaningless [6–8].

Scholars have proposed methods to resolve this issue from three aspects, namely: clash avoidance, clash detection improvement, and clash filtering [6,7]. Clash avoidance begins with the modeling method with the emphasis on adopting collaboration or strengthened coordination to reduce the occurrence of clashes. Undoubtedly, this method will increase the burden on the design staff [7]. For those project participants without direct contractual relationships, collaboration is also difficult to implement [6,9]. By contrast, other scholars consider that the algorithm to detect clashes in BIM software can be improved by increasing the accuracy of its detection, thereby reducing the number of irrelevant clashes [5]. However, as some studies pointed out, the refinement of the clash detection algorithm cannot effectively prevent irrelevant clashes caused by human errors from happening [7,10]. Recent studies suggest that an alternative is to directly identify those irrelevant clashes and filter them out from the clash detection report generated by BIM software. This method is broadly divided into two approaches. One is to apply rules to identify the dependency relationship of conflicting/clashing components, thereby filtering out irrelevant clashes [7]. However, the constructing dependency relationships between components and query algorithms is often time-consuming and labor-intensive when acquiring and maintaining the rules. Jiang et al. proposed a rule-based knowledge system to automate the code-checking process for green construction [11]. They found that the domain knowledge is usually dispersed and fragmented, so rule acquisition requires human experts from many professional fields. Therefore, the process of knowledge representation and acquisition is often a complex and time-consuming task [12]. The other is the use of machine learning methods that train classifiers of machine learning through historical data to filter out irrelevant clashes [6]. However, researchers using machine learning on complex problems usually observe that a favorable classification performance often requires a larger training dataset that allows a more complex model with more features [13]. Nevertheless, identifying and labeling a large number of clashes requires tremendous and expensive manpower; therefore, the prediction accuracy of machine learning is often insufficient before a sufficiently large number of cases are collected [7]. The industry is in urgent need of more cost-effective solutions on this issue.

In the field of machine learning, many studies applied a combination of two or more sophisticated methods on specific domains and obtained better results than using individual methods. A hybrid method was developed for discretizing continuous attributes to enhance the accuracy of the naïve Bayesian classifier [14]. An algorithm based on support vector machine (SVM), 2D fast Fourier transform (FFT), and hybrid fuzzy c-mean techniques was proposed to recognize and visualize the cracking incurred in the structure [15]. The hybrid ML algorithm performs better to recognize cracks with higher accuracy than the traditional SVM. Another hybrid computational model based on genetic algorithm (GA) and support vector regression (SVR) was developed to predict bridge scour depth near piers and abutments [16]. The proposed hybrid model achieved 80% more accurate error rates than those obtained using other methods, such as regression tree, chi-squared automatic interaction detector, artificial neural network, and ensemble models. These studies provide examples that demonstrate the effect of using a hybrid method.

This study attempted to combine the techniques of rule-based reasoning and supervised machine learning to develop an algorithm that can automatically filter out irrelevant clashes from the BIM-generated clash detection reports. The main purpose of this study is to explore whether the hybrid method we proposed can enhance the predictive accuracy by machine learning algorithms without a large number of training cases as well as without increasing the development manpower. Unlike most rule-based systems that require an exhaustive knowledge acquisition process, the rule-based reasoning in this study is not intended to obtain accurate results, because this often requires a great amount of efforts regarding knowledge acquisition. Instead, we intend to first obtain a preliminary classification of clashes by applying a simple rule set acquired from the same experts of labeling for the subsequent machine learning process and incorporate these preliminary results in the machine learning process to see if they can help improve the prediction accuracy.

2. Related Work

In most construction projects, structural, mechanical, electrical, and plumbing (MEP) engineers develop their designs based on the architectural model. This base model is often constantly updated along with the progress of the design work. Without the adequate synchronization of all the updates among these design teams, there will be so-called "design clashes". It refers to a conflict of building components overlapping each other spatially when various types of engineering drawings are compiled. As pointed out by some scholars, if a collaboration environment exists between design teams, most clashes can be avoided [7]. However, in the participatory action research of the United Kingdom, researchers introduced a collaboration environment in a multi-floor large-scale construction project, where the engineers were assisted by software to avoid design clashes. However, there were still more than 400 clashes found between the structural model and the MEP model [4]. Their study pointed out that collaboration can indeed reduce design conflicts, but clash detection is still a necessary operation.

In the era of 2D drawings, design conflicts were not easily detected at the design phase, but remained until the construction, leading to reworks or even change orders. Clashes have been regarded as one of the major factors causing cost overruns and project delays. The emergence of BIM software enables easy design conflict/clash detection; conflict checking has been one of the essential functions of BIM software. However, Helm et al. [5] found that those clash detection algorithms in most BIM software are relatively simple: as long as two building components are spatially overlapping, touching, or within a given distance, they are recognized as conflicts and are listed in the detection report. Identifying and resolving those detected clashes is a time-consuming and laborious task [4,6,7]. The clashes detected by BIM software can be roughly divided into three categories: (1) errors, which are the clashes that will affect the project and must be resolved, such as structural components penetrated by pipes; (2) deliberate clashes, which includes intentional clashes originating from the designer, such as the pipes and conduits penetrating through the slabs; (3) pseudo clashes, which are permissible clashes appearing to be errors. As Wang and Leite [17] discovered, the proportions of deliberate and pseudo clashes, which are also known as "irrelevant clashes", in a particular project were up to 50% [10]. Among the cases considered in our study, this proportion was even higher. Scholars termed these conflicts that do not have substantial impacts on the project as "irrelevant clashes" [6,7]. These irrelevant clashes can be discovered in subsequent project stages and easily handled by the site engineers themselves; therefore, there is no need to resolve them during the clash detection. However, the clash detection report of BIM software does not disclose these irrelevant clashes, which means they must be manually identified by BIM managers instead. As pointed out by Hu et al. [7], in practice, many projects can have millions of clashes, so automating the filtering of irrelevant clashes is an important and urgently needed function [4,6,7].

Existing methods of reducing irrelevant clashes can be roughly divided into three aspects: clash avoidance, clash detection improvement, and clash filtering [7]. Clash avoidance begins with the modeling method during the design phase, emphasizing the collaboration and coordination among the design teams to avoid the occurrence of clashes from the beginning. Mehrbod et al. [18] established taxonomy to classify the causes of clashes into three categories, namely, process-based, model-based, and physical design [12]. They aimed to understand the causes of design conflicts and the consideration factors for conflict/clash resolution. With the aim of reducing the occurrence of clashes through automated coordination, Wang and Leite [17] constructed a sematic schema for MEP coordination that was used to present and acquire the experience and knowledge hidden behind the coordination issues. Both Hartmann [19] and Gijezen [20] re-examined the BIM model via a more organized work breakdown structure (WBS) to reduce the number of irrelevant clashes. However, scholars believe that this approach undoubtedly increases the burden on the design teams [7]. Collaboration would be difficult to implement in many projects because project participants may not have a mutual contractual relationship [6,9].

Meanwhile, some scholars consider that improving the clashes detection algorithms in BIM software can increase the accuracy of its detection, thereby reducing the number of irrelevant clashes [5]. These methods include the sphere-trees method [21], approximate polyhedra with spheres and bounding volume hierarchy [22,23], oriented bounding boxes or OBB-trees method [24], and ray–triangle intersection algorithm [5]. These algorithms are continually improved to increase the accuracy of clash detection. Yet, as pointed out by scholars, the refined clash detection algorithms still cannot effectively reduce irrelevant clashes [10], especially those caused by human errors [7].

The third method is to directly identify and filter out irrelevant clashes in the clash detection report of BIM software. One of the popular methods of identification or diagnosis on a certain domain is rule-based systems [25]. Rule-based systems, also known as rule-based expert systems, have been commonly used in many fields such as medical, engineering, manufacturing, education, etc. since the 1980s and have been proved to be effective in pattern recognition, diagnosis, decision-making, control, planning, and so on due to the transparency of knowledge reasoning and consistency of reasoning results [12]. However, despite their advantages, rule-based systems require a considerable amount of time to acquire the knowledge that is needed for reasoning. A rule-based system was proposed to automate the code checking process for green construction. Still, the researchers found that the domain knowledge is dispersed and fragmented, and rule acquisition requires human experts from many professional fields [11]. Hu et al. [7] applied the rules to identify the dependency relationship of conflicting/clashing components, thereby constructing a component-dependent network. This network can be used to identify the central components of clashes, group those repetitive clashes, and finally filter out irrelevant clashes. However, the number of irrelevant clashes being filtered out depends on the components' dependency relationships and their query algorithms, which are similar to a rule-based knowledge base, which requires a lot of effort to capture and maintain those rules. Besides, the rules developed by their study may not necessarily fit other projects.

Another method that also is popular for complex problems and does not require too many efforts on knowledge acquisition is machine learning. Machine learning algorithms use computational methods to predict results directly from historical data without relying on predetermined rules or equations on domain knowledge. Besides, the algorithms adaptively improve their performance as the number of training cases increases [13,15]. Despite its ease of identifying trends and patterns without human intervention, researchers often argued that machine learning requires a sufficiently large training dataset that allows a more complex model in order to obtain favorable results [7,15]. In the field of identifying design clashes, Hu and Castro-Lacouture [6] used a historical dataset of 204 clashes from a three-story building and implemented six different machine learning classifiers including J48-based decision trees, random forest, Jrip, binary logistic regression, naïve Bayesian, and Bayesian network to filter out irrelevant clashes. The features selected for machine learning process considered three aspects: (1) the information uncertainty level; (2) problem complexity, such as clashing objects' size, priority, materials, type, and clashing volume, and (3) contextual flexibility, such as the location, spatial relationship, and available space [6]. However, their method based on machine learning obtained an average prediction accuracy of 80%, but it required a great amount of labor to preprocess the training data. Some researchers then argued that before a sufficient number of training cases are collected, the prediction accuracy is insufficient [7].

In summary, both rule-based reasoning and machine learning have their own pros and cons to provide a solution to domain problems. The former produces a favorable result no matter how big the training data size is but often requires a great number of efforts to acquire knowledge from human experts. On the contrary, the latter does not require human efforts to prepare and formulate the domain knowledge to produce results. Still, it requires sufficient training data in order to obtain a favorable result. In the field of machine learning, many studies have proved that applying the hybrid method that combines two or more sophisticated algorithms on certain domains can obtain better results than using individual methods [14–16]. However, most studies combined two or more machine learning algorithms as their hybrid methods, but few have taken advantage of machine learning and rule-based systems from the perspectives of minimal development efforts and maximal predictive performances.

Considering the nature of clash detection, in which a large training dataset is not easy and cost-effective to collect and prepare, this study made the best use of expertise from human experts hired by the research team to both prepare the training dataset for machine learning and to be interviewed to acquire their heuristic know-how for rule-based reasoning. Based on the perspective of clash filtering, this study first used rule-based reasoning to preliminarily determine the type of clashes; subsequently, the results of the preliminary classification are added into the dataset of machine learning for training and the testing of classifiers in order to improve the prediction accuracy under a small training dataset.

3. Methodology

The knowledge acquisition of a rule-based system that takes into consideration spatial relations is not easy, and machine learning requires numerous training cases in order to obtain a reasonable accuracy. Therefore, this study proposes a hybrid method by first developing a simple rule-based system that merely takes into consideration clash attributes; then, the results of the rule-based reasoning is merged to the dataset of machine learning. This study develops the research methodology shown in Figure 1 to validate the effectiveness of this method. First, a real architectural project is selected, BIM software is used for clash detection, and then the (clash) detection report is submitted to two experts for labeling clash types. The labeling results with identical labels from the experts are selected and further adjusted to form training dataset #1; this dataset is subjected to a supervised machine learning process to obtain the "pure machine learning results". At the same time, the same experts are interviewed to acquire their knowledge of determining the types of clashes and incorporate this knowledge into rules. These rules only consider attributes of clashing components and do not delve into other deeper spatial relations with others. After implementing the rules, training dataset #1 is used in the same manner for rule-based reasoning to obtain the "pure rule-based results". Next, these results are regarded as a field of training data inserted to training dataset #1 and form training dataset #2; this dataset is again processed by the same supervised machine learning to obtain the "hybrid results". Finally, the accuracy of the three prediction results is evaluated and compared. In the following paragraphs, Section 3.1 first describes the data collection process; Section 3.2 describes the process of hiring experts to label the type of case clashes for the clash detection report; and Section 3.3 describes the selection and adjustment of the labeling results. The development of the rule-based system and its prediction results are introduced in Section 4, followed by Section 5 describing the prediction results of two similar machine learning processes.

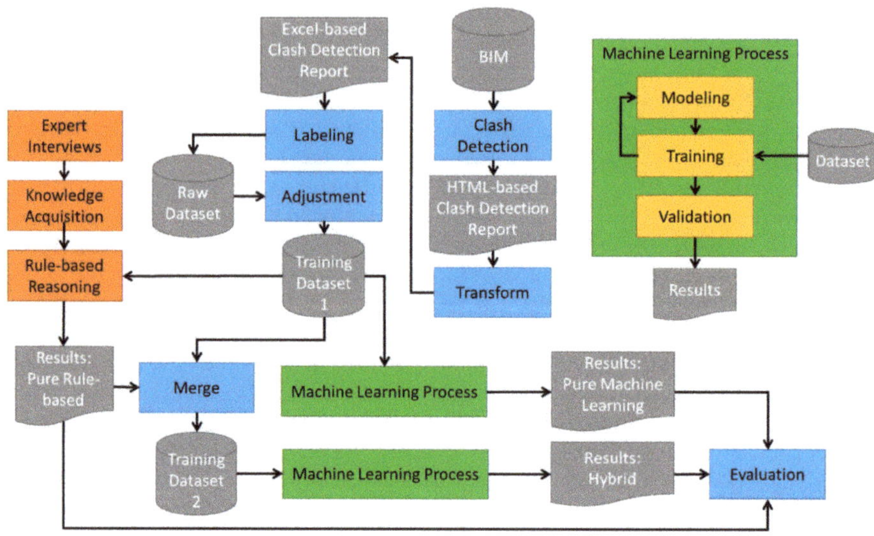

Figure 1. Process of research methodology.

3.1. Data Collection

This study used a large shopping mall with nine floors above the ground and four floors underground as the testing case. This building was considered mainly because it consisted of a large number of complicated pipes and conduits, as shown in Figure 2. The structural and MEP model of the building were extracted for clash detection through Autodesk Navisworks Manage 2017. Clashes detected by most BIM software can be hard clashes or soft clashes. Hard clashes exist when building elements have physical overlaps, whereas soft clashes occur when an element is not given the spatial tolerances it requires. Similar to most of the studies discussing clash detection mentioned in Section 2, this study only considers "hard clashes" because they have a universal definition, and therefore research results can be compared. In order to control the number of total clashes within an acceptable range for labeling work by the hired experts, this study merely selects the water supply pipes and fittings from the MEP model for clash detection against the structural model. Figure 3 shows a part of the HTML-based clash detection report. The summary table on the top records the total number of clashes in the report; the table below lists the detailed information of each conflict/clash. The detailed information includes the grid location of the conflict/clash, clash point, and distance, as well as the information of two clashing objects, including their IDs, floors, names, and types. Moreover, snapshots of two clashing objects are attached. Clicking on the thumbnails in the first column of the table allows viewing the enlarged snapshots, as shown in Figure 4. Table 1 presents the complete statistics of the clash detection report; there are a total of 415 clashes between four structural components (beams, columns, slabs, walls) and two MEP components (pipes and fittings).

Figure 2. Case study building: exterior and internal pipeline configuration of the building object.

Clash Report

Test 1	Tolerance	Clashes	New	Active	Reviewed	Approved	Resolved	Type	Status
	0.001m	107	107	0	0	0	0	Hard	OK

Image	Clash Name	Distance	Grid Location	Clash Point	Item 1				Item 2			
					Item ID	Layer	Item Name	Item Type	Item ID	Layer	Item Name	Item Type
	Clash1	-0.217	D-3 : 6F.L	x:-7373.366, y:-1108.751, z:45.685	Element ID: 1394803	6F.L	Pipe Types	Pipes: Pipe Types: S8-VP-PVC-CNS1298-B	Element ID: 1653984	6F.L	Basic Wall	Walls: Basic Wall: 15cm
	Clash2	-0.189	D-1 : 6F.L	x:-7376.020, y:-1089.845, z:45.737	Element ID: 1394872	6F.L	Pipe Types	Pipes: Pipe Types: S8-VP-PVC-CNS1298-B	Element ID: 1654008	6F.L	Basic Wall	Walls: Basic Wall: 15cm
	Clash3	-0.182	D-2 : 6F.L	x:-7373.585, y:-1097.651, z:45.657	Element ID: 1394346	6F.L	Pipe Types	Pipes: Pipe Types: S8-VP-PVC-CNS1298-B	Element ID: 1653996	6F.L	Basic Wall	Walls: Basic Wall: 15cm

Figure 3. Illustration of the clash detection report produced by Autodesk Navisworks Manage 2017.

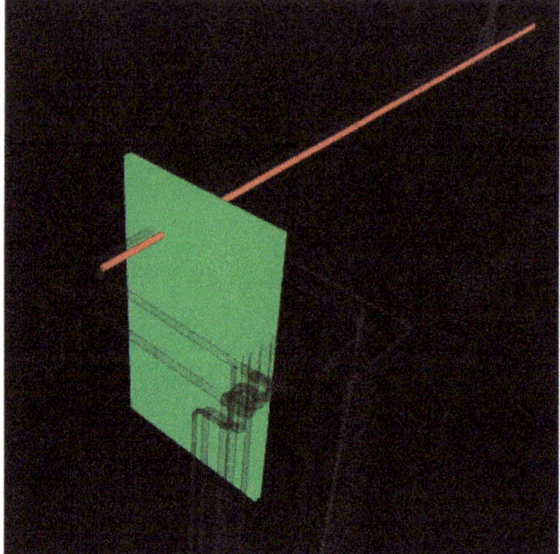

Figure 4. Snapshot of a clash produced by Autodesk Navisworks Manage 2017.

Table 1. Statistics of the clash detection report.

Clashing Item Types	Pipes	Fittings	Total
Framings	153	25	178
Walls	155	68	223
Columns	1	1	2
Slabs	12	0	12
Total	321	94	415

3.2. Labeling the Clash Types

Human experts still inevitably have different subjective judgments on the same cases, and therefore the research team hired two experts who both have more than five years of experience on clash coordination and resolution to label the clash types from the same clash detection report. Those clashes with different labels by the two experts will be excluded from the training process during the machine learning phase. Once obtaining the HTML-based clash detection report, we transform it into a spreadsheet, as shown in Figure 5, for human experts to label the clash types. Besides all the information on the clash detection report shown in Figure 3, the spreadsheet also contains a column "Clash Type" with a drop-down list to facilitate labeling clash types by the experts. The drop-down list consists of four options: serious clashes, negligible clashes, legal interventions, and unknown. These four options use a more intuitive vocabulary; at the time of subsequent analysis, these options will correspond to the four categories suggested by the literature as errors, pseudo clashes, deliberate clashes, and unknown, respectively. Serious clashes or errors are those relevant and crucial clashes that need to be carefully examined and resolved if necessary. Except for the spreadsheet containing information derived from the clash detection report, the researcher did not provide the labeling experts with other information about the building, such as CAD drawings or 3D building models. The only information for them to determine the clash types is the clash detection report mentioned in Section 3.1.

B6		▾ : ╳ ✓ f_x		
◢	**A**	**B**	**D**	**E**
1				
2	**Clash No.**	**Clash Type**	**Distance**	**Grid Location**
3	Clash1	severe clashes	-0.234	D-10 : 2F.L
4	Clash2	negligible clashes	-0.204	A-7 : 2F.L
5	Clash3	legal interventions	-0.204	E-11 : 2F.L
6	Clash4		-0.192	A-3 : 2F.L
7	Clash5	severe clashes	-0.187	E-11 : 2F.L
8	Clash6	negligible clashes / legal interventions	-0.176	E-12 : 2F.L
9	Clash7	unknown	-0.176	E-11 : 2F.L
10	Clash8		-0.174	E-11 : 2F.L
11	Clash9		-0.172	D-10 : 2F.L

Figure 5. Spreadsheet-based clash detection report for labeling of clash type by experts.

3.3. Label Adjustment

Table 2 shows a summary of the labeling results by two experts. After comparing the details, 89 clashes were found with different labels by the two experts and excluded, which means only the remaining 326 cases were used for the processing. Moreover, there was no negligible clash among the clash types labeled by both experts. An investigation reveals that most negligible clashes may occur when pipes penetrate a beam. According to the specification of reinforced concrete structures [26] published by the Chinese Society of Structural Engineers in Taiwan for the positions of legal pipes penetrating through beams, if the position of clash falls on the grid area in Figure 6, the structural behavior of the beam will not be affected. In other words, this clash can be classified as a negligible clash, or a pseudo clash. Applying this specification requires the precise dimensions of the clashing objects and the clashing position. The experts were not able to determine whether those clashes were negligible or not by merely viewing the snapshots with their naked eye. Therefore, to be on the conservative side, the experts mostly labeled those clashes with pipes penetrating through beams as "serious clashes"; a few cases were labeled as unknown.

In order to increase the granularity of the training data and improve the filtering rate of irrelevant clashes later on, researchers decided to further apply the specification as mentioned above toward the original labeling results. The research team used the Model Builder embedded by Environmental Systems Research Institute (ESRI) ArcScene to implement the specification, as shown in Figure 7, and obtain the adjusted labeling result, as shown in the rightmost column of Table 2. The original 58 serious clashes and five unknowns were adjusted as negligible clashes. After adjustment, the numbers of labeling for four clash types turn to be 100 errors, 63 pseudo clashes, 127 deliberate clashes, and 36 unknowns.

Table 2. Statistics of training and testing data.

Clash Type		Expert #1	Expert #2	Common	Adjusted
Serious clashes	Errors	159	208	158	100
Negligible clashes	Pseudo clashes	0	0	0	63
Legal intervenes	Deliberate clashes	174	148	127	127
Unknown	Unknown	82	59	41	36
Total		415	415	326	326

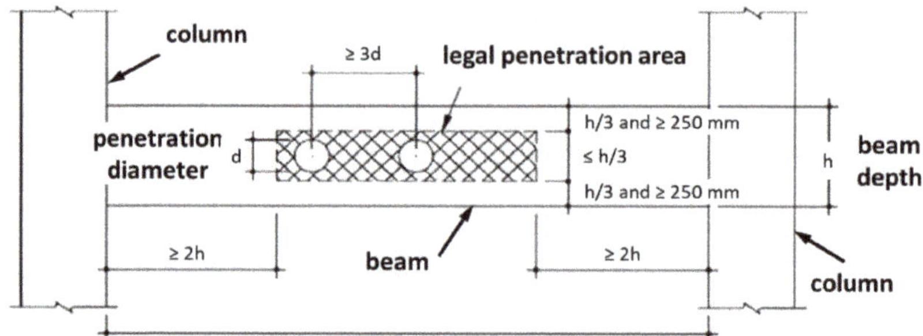

Figure 6. Legal penetrating area for structural beams [26].

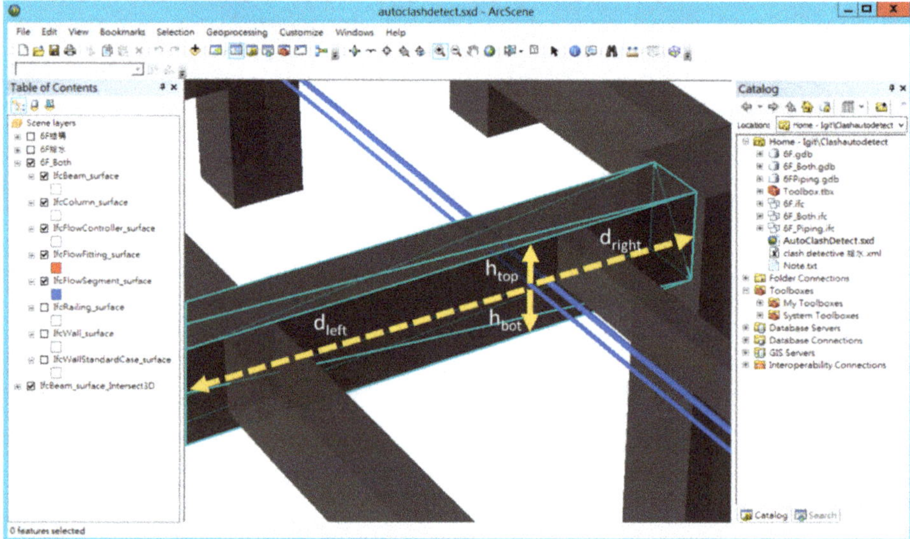

Figure 7. Automated adjustment on labeling of clash types by using ESRI (Environmental Systems Research Institute) ArcScene.

4. Rule-Based Reasoning

While directly obtaining the labeling results from the experts, we also interviewed the two hired experts to acquire their knowledge used to classify the clash types. Different from most rule-based reasoning systems [17,27] acquiring as many rules as possible, we only focus on those rules of thumb requiring facts that can be found in the clash detection report. The reason for this is that the rule-based reasoning in this study is not meant to serve as a robust method for classifying clash types; instead, it is used to serve as the catalyst to improve the prediction performance of the supervised machine learning. In addition, the clash detection report is the only reference for the experts to do their jobs. The following six rules are directly acquired after interviewing the experts:

1. Beam Rule: If the type of clashing object from the structural model is a beam, the clash type will be an "error".
2. Extended Beam Rule: If the type of clashing object from the structural model is a beam and the clash position falls within the legal area defined by the specification, the clash type will be a "pseudo clash". This rule is based on the specification we used to adjust the original labeling result mentioned in Section 3.3.

3. Column Rule: If the type of clashing object from the structural model is a column, the clash type will be an "error".
4. Slab Rule: If the type of clashing object from the structural model is a slab, the clash type will be a "deliberate clash".
5. Wall Rule: If the type of clashing object from the structural model is a wall, the clash type can be a "deliberate clash" or an "error".
6. Bearing Wall Rule: If the type of clashing object from the structural model is a wall and is not a bearing wall, the clash type will be a "deliberate clash"; otherwise, it is an "error".

Among these rules, Rule #6 requires other supporting information that the clash detection report does not provide to ensure whether the clashing wall is a bearing wall or not. Therefore, it is excluded from our final rule repository. Instead of classifying those clashing walls as "errors", the researchers simply revised Rule #5 as follows:

7. Simplified Wall Rule: If the type of clashing structural component is a wall, the clash type will be unknown.

Then, the research team applied the Rules #1–4 and #7 stated above to perform the rule-based reasoning and obtained the clash classification result, as presented in Table 3. The average accuracy rate is approximately 60% (194/326). The columns in Table 3 represent the numbers of clash types predicted using the rules, whereas the rows represent the actual clash types specified by the two experts. For example, among the 100 true errors, the rule-based reasoning correctly predicts 98 of them, and the remaining two errors are determined as unknown.

Table 3. Results of rule-based reasoning using simplified rules.

Clash Type		Predicted Labels				Total
		Errors	Deliberate	Pseudo	Unknown	
	Errors	98	-	-	2	100
True labels	Deliberate	-	12	-	115	127
	Pseudo	-	-	63	-	63
	Unknown	15	-	-	21	36
Total		113	12	63	138	326
Accuracy		98/113 (0.88)	12/12 (1.00)	63/63 (1.00)	21/138 (0.15)	194/326 (0.60)

As mentioned earlier, the aim of rule-based reasoning in this study is only to improve the outcomes of machine learning under a small training dataset. Therefore, the rules we applied did not consider deep and complex relationships among those clashing objects. As a result, the prediction accuracy (60%) tended to be low. The prediction results by rule-based reasoning here will be further included as a feature for the machine learning process that is introduced in Section 5.5.

5. Machine Learning Process

5.1. Feature Selection and Manipulation

The first task of the supervised machine learning process is to select those features of the dataset (i.e., attributes) that may contribute to problem-solving. Table 4 presents the results of the feature selection in this study. In the table, the features from the clash detection report include numeric ones, such as "Distance", "Floor-1", and "Floor-2". The values of these features are left unchanged without manipulation. The "clash point" is the coordinates with a mix of text and numbers, so the values are extracted separately and form three independent numeric features, namely, "Clash Point_x", "Clash Point_y", and "Clash Point_z". Furthermore, the features whose data types are nominal/text, such as "ItemType-1" and "ItemType-2", additionally require "one-hot encoding" to be transformed into numeric features. Therefore, the original six features found in the clash detection report finally result in the first 12 deriving features, which are shown in the last column of Table 4. These features will be used for the machine learning process to be introduced in Section 5.3 and 5.4.

Table 4. Summary of feature selection and preprocessing.

Original Feature	Description	Data Type	Example Value	Revised Feature
Distance	Length of clash point away from the component edge	numeric	−0.234	unchanged
Floor-1	The floor where the clashing component 1 is located	numeric	2	unchanged
Floor-2	The floor where the clashing component 2 is located	numeric	2	unchanged
Clash Point	Clash point coordinates	text	x: −7370, y: −1171, z: 24	Clash Point_x, Clash Point_y, Clash Point_z
ItemType-1	Component type of clashing component 1	nominal	Pipes | Fittings	Pipes; Fittings
ItemType-2	Component type of clashing component 2	nominal	Framings | Walls | Slabs | Columns	Framings, Walls, Slabs, Columns
Rule-Tag	Clash type predicted by rule-based reasoning	nominal	Errors | Pseudo clashes | Deliberate clashes | Unknown	Rule_errors, Rule_pseudo, Rule_deliberate, Rule unknown

In order to demonstrate whether rule-based reasoning can improve the predicting accuracy of machine learning, the results of rule-based reasoning mentioned in Section 4 are also added as a feature "Rule-Tag" for the machine learning process, which is addressed in Section 5.5. Since the data type of this feature is also nominal/text, "one-hot encoding" is also applied in the same manner. That makes a total of 16 features that are present in the training dataset for the machine learning process of Section 5.5.

The subsequent machine learning process is divided into two experiments. The first experiment uses the first 12 features shown in the last column of Table 4 for the training and testing of classifiers, i.e., the dataset does not contain the results of rule-based reasoning; this dataset is denoted as training dataset #1, as shown in Figure 1. Section 5.3 and 5.4 will explain this process and the results in detail. The second experiment uses all the features in the last column of Table 4; the dataset is denoted as training dataset #2. These two experiments are then compared to evaluate the impacts of rule-based reasoning on the prediction accuracy of machine learning. Section 5.5 will detail the results of this experiment.

5.2. Classification Algorithms and Parameters

This study adopted the open-source machine learning library, Scikit-Learn, as a development tool, which provides a rich set of tools and various algorithms required for classification and regression problems. As suggested by many machine learning studies [28,29], no single classifier can work best across all scenarios. The best practice of choosing a classification algorithm is to compare the performance of several learning algorithms and select the best model for the problem to be solved. Even so, the best model under this consideration may still vary regarding the nature of the training dataset we collect, the number of training cases, and the features we select for the learning process. Besides, instead of determining the best model for a particular problem, the main purpose of this study is to demonstrate the effect of the hybrid method we propose on the predictive performance. Therefore, this study implemented several classifiers using common classification algorithms including decision tree (DT) [30,31], support vector machine (SVM) [32–34], and k-nearest neighbor (k-NN) [28,35], as well as three classifiers applying ensemble learning strategies [28], and then conducted the training process using these classifiers upon two training datasets, one with the feedback from rule-based reasoning and the other without it, and compared both predictive performances. The decision tree is chosen because it works well with both numerical and categorical features and provides interpretability for decision-making [29], while SVM can generate robust results for complex classification problems [15]. k-NN, an easy-to-implement classifier that works well with a small dataset but suffers from low efficiency when the dataset grows, is used as a benchmark for DT and SVM classifiers [29]. The reason why those ensemble learning classifiers are also included in this study is that they combine multiple classifiers to have a better performance than individual classifiers alone. However, as mentioned above, we are not intending to decide which classifier best fits the domain problem, so we did not dive deep into tuning those hyper-parameters of each classifier. The results of the comparison among them will be addressed later in Section 6.

Table 5 summarizes the manipulation of training classifiers implemented by this study. To test and evaluate the prediction accuracy of different classifiers, this study randomly selected 30% of 326 cases, i.e., 98 sets, as the testing dataset; the remaining 70% was the training dataset, i.e., 228 cases. For the training of all types of classifiers, the training dataset was subjected to k-fold cross-validation (k = 5) with a 4:1 split ratio. When implementing classifiers such as kNN, SVM, and Voting, the dataset was subjected to standardization before proceeding to the learning process. Tables 6 and 7 list the modeling parameters of individual classifiers and ensemble learning classifiers, respectively.

Table 5. Summary of machine learning manipulation. K-NN: k-nearest neighbor, SVM: support vector machine.

Measures	Description
Data splitting	7:3 (228 cases for training; 98 cases for testing)
Performance metric	Error matrix (also known as Confusion matrix)
Classifiers	Decision trees, SVM, k-NN, Voting, Bagging, Random forest
Evaluation	k-fold cross-validation (k = 5)

Table 6. Modeling parameters of individual classifiers.

Classifiers	Parameters
DecisionTree	criterion = "gini", max_depth = 6
KNeighborsClassifier	n_neighbors = 3
SVMClassifier	c = 1.0, kernel = "linear" & "rbf"

Table 7. Modeling parameters of ensemble learning classifiers.

Ensemble Learning Strategy	Estimators	Parameters
VotingClassifier	DecisionTree (max_depth = 6) KNeighbors (n_neighbors = 3) SVM (kernel = "linear")	voting = "soft", weights = [5, 1, 1]
BaggingClassifier	DecisionTree (max_depth = 6)	n_estimators = 100
RandomForestClassifier	DecisionTree (max_depth = 6)	n_estimators = 100, criterion = "gini"

Furthermore, the error matrix, also known as the confusion matrix, was adopted for the evaluation, which reports the counts of the true positive, true negative, false positive, and false negative predictions of a classifier, as shown in Figure 8 [28]. Three indicators derived from the error matrix were recorded to evaluate the performance metrics against all the training classifiers, including precision, recall, and f1-score, which are defined according to Equations (1)–(3), respectively. Precision is the number of true positive results divided by the number of all the positive results returned by the classifier, and recall is the number of true positive results divided by the number of all the samples that should have been identified as positive. The f1-score is the harmonic mean of the precision and recall, with its best value at 1 and worst value at 0.

$$Precision = \frac{TP}{TP + FP} \tag{1}$$

$$Recall = \frac{TP}{TP + FN} \tag{2}$$

$$f1 - score = 2 * \frac{Precision * Recall}{Precision + Recall} \tag{3}$$

Sections 5.3 and 5.4 explain the machine learning processes using training dataset #1 for individual classifiers and ensemble learning classifiers, respectively; Section 5.5 explains the results obtained using training dataset #2 along with rule-based reasoning results for the same machine learning process.

Predicted labels

	True Positives (TP)	False Negatives (FN)
Actual labels	False Positives (FP)	True Negatives (TN)

Figure 8. Error matrix used to evaluate the performance of the trained classifiers.

5.3. Individual Classifiers of the Linear Model

1. Decision Trees (DTs) Classifier

First, the research team implemented a classifier using the DT, which can work with both numerical and categorical features and also provides interpretability for decision-making [29]. The algorithm starts at the root of the feature tree and splits the dataset on the tree node, which results in the largest information gain, IG, as shown in Equation (4) [28]. The objective function of a decision tree to optimize the training is to maximize the information gain at each node [29].

$$IG(D_p, f) = I(D_p) - \sum_{j=1}^{m} \frac{N_j}{N_p} I(D_j) \tag{4}$$

In Equation (4), f is the feature to perform the split, D_p and D_j are the datasets of the parent and jth child node, I is the measure of data applying the splitting criteria, N_p is the total number of samples at the parent node, and N_j is the number of samples in the jth child node. As stated by this equation, the information gain is the difference between the measures of splitting data of the parent node and the sum of the child node impurities. The splitting criteria we used in this study is Gini impurity (I_G), which is defined in Equation (5) [28]:

$$I_G(t) = \sum_{i=1}^{c} p(i|t)(1 - p(i|t)) = 1 - \sum_{i=1}^{c} p(i|t)^2 \tag{5}$$

Here, $p(i|t)$ is the proportion of the cases that belong to class i for a particular node t. Table 8 and Figure 9 shows one of the prediction results of the DT classifier that we implemented and trained using training dataset #1 depicted in Figure 1, or the dataset without the feedback from rule-based reasoning. Among 98 testing cases, 81 cases were correctly classified, which made an overall precision of 0.85, recall of 0.83, and f1-score of 0.83. In order to avoid bias from one test, we repeated the above training and testing process 30 times, where different training and testing cases were randomly selected for each test. The average precision, recall, and f1-score of 30 tests are 0.88, 0.87, and 0.87, respectively.

Table 8. One of the predictive results of the DT classifier using training dataset #1.

Clash Types	Precision	Recall	F1-Score	Cases
Errors	0.57	0.81	0.67	21
Pseudo clashes	0.89	0.70	0.78	23
Deliberate clashes	1.00	1.00	1.00	42
Unknown	0.75	0.50	0.60	12
Average/Total	0.85	0.83	0.83	98

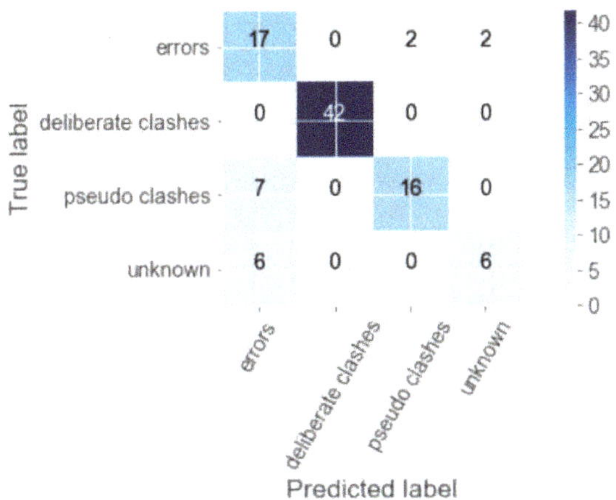

Figure 9. The error matrix of the test as shown in Table 8.

2. Support Vector Machine (SVM) Classifier

SVM is also a common machine learning classifier that is widely used in practical and theoretically domains [29]. This study implemented a classifier using the SVM algorithm because they can generate accurate and robust results for classification problems, even when training data are nonlinearly separable [29,33]. They can also be easily extended to solve nonlinear classification problems using a nonlinear "kernel" [32,34]. The optimization objective of SVM classifiers is to maximize the distance between the separating decision boundary and the training samples that are closest to this boundary [29]. A penalty can also be applied for misclassification [28]. Using training dataset #1 and repeating 30 train-then-test cycles the same as stated previously, the SVM classifier implemented by this study used a linear kernel and obtained an average f1-score of 0.79 among 30 tests and an average f1-score of 0.74 when using a radius basis function kernel.

3. K-Nearest Neighbors (k-NN) Classifier

Next, a classifier using the k-NN algorithm was implemented. k-NN was selected because it is a robust classifier that is often used as a benchmark for more complex classifiers, such as SVM and decision trees [35]. Besides, since both the number of features and size of our training dataset are not very large, this classifier does not suffer from the "curse of dimensionality" and low computation efficiency [28]. The average precision, recall, and f1-score of 30 tests for the k-NN classifier with three neighbors and a uniform weight are 0.79, 0.78, and 0.78, respectively, which is close to the results of the SVM classifier and lower than the DT classifier we implemented. This implies that the performances of both SVM and DT classifiers implemented by this study are considerably robust and trustworthy.

5.4. Multiple Classifiers by Ensemble Learning

The three individual classifiers mentioned previously could obtain a predictive precision ranging from 0.79 to 0.88. The research team further considered constructing a group of individual classifiers that can often receive a better predictive precision than any of its members, as suggested by the so-called ensemble learning [28]. The benefit of the ensemble method is that it combines different classifiers into a multiple classifier and thus has a better prediction performance than individual classifiers alone. The three ensemble learning classifiers implemented in this study are based on different strategies, including majority voting, bagging, and random forest.

1. Majority Voting Classifier

As stated earlier, during the ensemble learning, a group of individual classifiers will be involved in the prediction process. Majority voting means that we combine different types of classifiers for prediction and select the answer that is predicted by the majority of classifiers [28]. The majority voting classifier implemented by this study comprises all three individual classifiers introduced in Section 5.3. Since the DT classifier has the best predictive performance among the three, the voting weights of the DT, SVM, and k-NN classifier is assigned as 5:1:1, respectively, indicating that the prediction made by the DT classifier has five times more weight than the predictions by other two classifiers.

Table 9 and Figure 10 present one of the predictions made by the majority voting classifier that we implemented using training dataset #1, which was the one without the feedback from rule-based reasoning. Its overall predictive precision, recall, and f1-score are 0.86, 0.84, and 0.84. Again, we repeated the train-then-test cycle 30 times and obtained an average precision of 0.89, recall of 0.88, and f1-score of 0.88. The predictive performance of the majority voting classifier is close to the individual DT classifier, indicating that the ensemble learning effect is not significant. Although this result can be further improved by choosing a new set of individual classifiers or tuning the hyper-parameters such as weights, we still stopped here and moved forward to implement other ensemble learning classifiers.

Table 9. One of the predictive results of the majority voting classifier using training dataset #1.

Clash Types	Precision	Recall	F1-Score	Cases
Errors	0.61	0.81	0.69	21
Pseudo clashes	0.90	0.78	0.84	23
Deliberate clashes	1.00	0.95	0.98	42
Unknown	0.70	0.58	0.64	12
Average/Total	0.86	0.84	0.84	98

2. Bagging

Bagging is another ensemble strategy that is similar to majority voting that includes multiple classifiers and casts votes to the final decision. Instead of using different types of individual classifiers and the same training dataset to fit the classifiers, a bagging classifier only employs the same type of individual classifier and draws random samples from the training set to train the classifiers. It is a useful technique to reduce the model variance [28]. The bagging classifier we implemented employed 100 DT classifiers. Using training dataset #1 and repeating 30 train-then-test cycles, the bagging classifier obtains an average predictive performance of 0.91. Comparing with the previous individual classifiers mentioned in Section 5.3, this result moderately demonstrates the effect of ensemble learning.

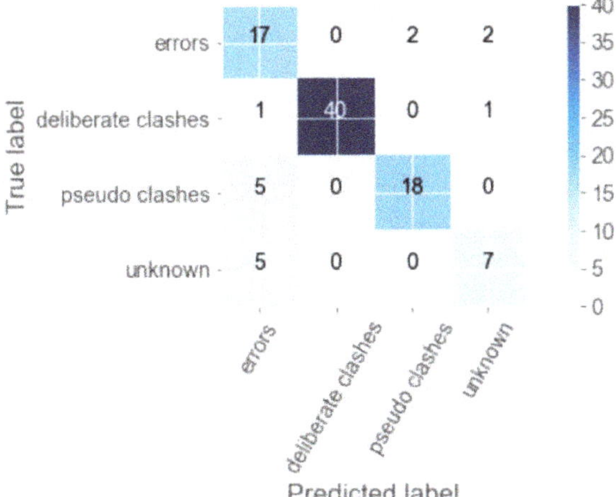

Figure 10. The error matrix of the test as shown in Table 9.

3. Random Forest

The last ensemble learning classifier we implemented is the random forest, which is a special case of bagging. Similar to bagging that only employs one type of individual classifier, which is DT, and draws random samples from the training set, the random forest also randomly selects feature subsets to fit the individual decision trees [28]. The random forest classifier we implemented also employs 100 DT classifiers and obtains a very close prediction performance of 0.89 compared with the bagging classifier, which also reflects the effect of ensemble learning.

Table 10 summarizes the average f1-scores of 30 tests using training dataset #1 for both the individual classifiers and ensemble learning classifiers implemented by this study. Table 10 also presents the average f1-scores of each clash type predicted by all classifiers.

Table 10. Summary of average f1-scores of 30 test for classifiers using training dataset #1.

Clash Types	Individual Classifiers			Ensemble Learning Classifiers		
	DT	k-NN	SVM	Voting	Bagging	Random Forest
Errors	0.804	0.698	0.713	0.829	0.871	0.838
Deliberate clashes	0.969	0.952	0.964	0.967	0.969	0.976
Pseudo clashes	0.773	0.681	0.649	0.789	0.841	0.809
Unknown	0.817	0.531	0.56	0.816	0.861	0.769
Average	0.868	0.775	0.785	0.876	0.905	0.881

5.5. Machine Learning with Feedback from Rule-Based ReasoningFormatting of Mathematical Components

According to the research process of Figure 1, we used training dataset #2 to perform the same machine learning process as described in Section 5.3 and 5.4 once we obtained the predictive results from rule-based reasoning introduced in Section 4. A feature called "Rule-tag" whose values are the corresponding clash types predicted by rule-based reasoning is inserted into training dataset #1, forming training dataset #2. Table 11 summarizes the average f1-scores of 30 tests using training dataset #2 for both the individual classifiers and ensemble learning classifiers implemented by this study. As shown in Table 11, the average f1-scores of individual classifiers using the training dataset with the feedback from rule-based reasoning range from 0.91 (both k-NN and SVM classifier) to 0.94 (DT classifier), whereas the average f1-score of ensemble learning ranges from 0.94 (random forest classifier) to 0.96 (bagging classifier). More evaluation and discussion between the results of machine learning using training dataset #1 and #2 will be detailed in Section 6.

Table 11. Summary of average f1-scores of 30 tests for classifiers using training dataset #2.

Clash Types	Individual Classifiers			Ensemble Learning Classifiers		
	DT	k-NN	SVM	Voting	Bagging	Random Forest
Errors	0.932	0.889	0.914	0.94	0.955	0.958
Deliberate clashes	0.970	0.959	0.958	0.97	0.972	0.966
Pseudo clashes	1.000	1.000	1.000	1.000	1.000	1.000
Unknown	0.766	0.577	0.515	0.823	0.844	0.746
Average	0.942	0.906	0.905	0.95	0.957	0.944

6. Results and Evaluation

6.1. Results

In summary, we first implemented a basic rule-based reasoning system, which was introduced in Section 4, for predicting the clash types according to the clash detection report generated by BIM software, and obtained the preliminary predictive accuracy rate of 60% (194/326). Next, we implemented six common classifiers (three individual classifiers and three ensemble learning classifiers) and conducted the same machine learning process twice using two training datasets: training dataset #1, derived from the clash detection report with clash types labeled by two human experts, and training dataset #2, derived from merging training dataset #1 and a feature of clash types predicted by the rule-based reasoning.

For the experiment with training dataset #1, the best predictive performance from 30 tests among three individual classifiers is obtained by the DT classifier (f1-score is 0.87), whereas the best ensemble learning classifier is the one using the strategy of bagging (f1-score is 0.91). Among the three individual classifiers, the k-NN classifier obtained the lowest f1-score, indicating that the results of both SVM and DT classifiers are considerably robust and trustworthy. The DT classifier outperforms the SVM classifiers by 10% probably because the dataset is relatively small, and the number of labels in this study is four rather than two or three, where SVM could function better [28]. The information gain of DT classifiers usually tends to be larger for the label with more cases in training data, so it may benefit from the nature of the training data we selected.

Among the three ensemble learning classifiers, there is no significant difference in their performance. Still, all of them outperform the three individual classifiers, proving the findings suggested by previous studies of machine learning [28], especially compared with the SVM and k-NN classifiers.

The experiment results with the training dataset #1 are also compared with the previous work by Hu and Castro-Lacouture [6], where six machine learning classifiers were implemented to filter out irrelevant clashes. Their best predictive performance in terms of f1-score, obtained by both random forest classifier and a rule-based classifier, Jrip, was 0.74, lower than the average f1-score of the random forest classifier in our study (0.881). This outperformance does not necessarily indicate that our classifiers are better because their classifiers may suffer from underfitting due to the smaller training dataset (204 versus 326 cases) and the larger number of features (10 versus 6 features). The other reason could also be the difference in the nature of training data, as we mentioned in the comparison between DT and SVM classifiers.

For the experiments with training dataset #2, the predictive performances of all three individual classifiers reach a favorable level of 0.91, while the level is 0.94 or higher for ensemble learning classifiers. The best performances for both individual and ensemble learning classifiers remain to be DT and bagging, respectively. The predictive performances of all the classifiers achieved improvements in terms of average f1-score, as shown in Figure 11. Taking a closer look at the comparisons, the average f1-scores of the SVM and k-NN classifiers significantly increased by approximately 15%–17% while the prediction improvement by the DT and three ensemble learning classifiers also increased by 6%–8.5%. Among those clash types predicted by machine learning classifiers, errors and pseudo clashes gained the greatest increase in two experiments, as shown in Figure 12.

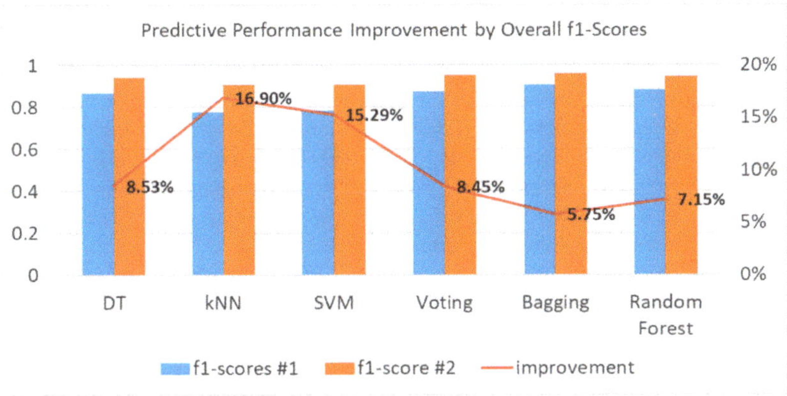

Figure 11. Predictive performance improvement by overall f1-scores.

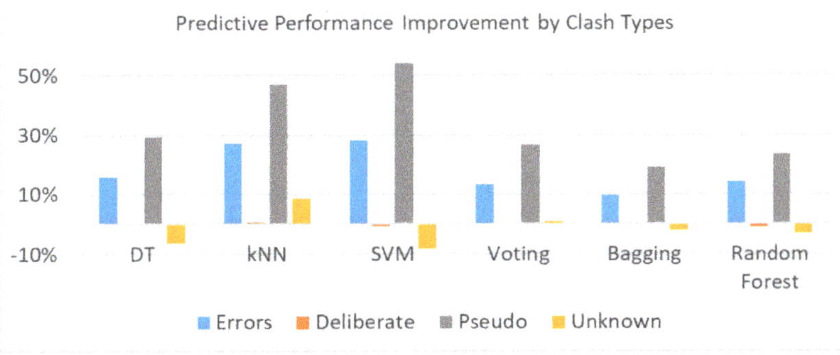

Figure 12. Predictive performance improvement by clash types.

From the results of these two experiments, we could preliminarily prove that under the condition of a small training dataset, the hybrid method proposed by this study combining rule-based reasoning and supervised machine learning can improve the predictive performance compared with using machine learning approach alone. A more in-depth evaluation will be discussed next.

6.2. Evaluation

The effect of the hybrid method on predictive performances obviously comes from the contribution of preliminary results by rule-based reasoning. Without this feedback, though it is not very accurate, the machine learning algorithms require more cases to reach a better performance. The improvement itself is not beyond our expectation, but what amazed us was that such a tiny rule base with five simple rules, possessing a relatively low predictive accuracy of 0.6, can still make contributions to those classifiers with a much higher predictive accuracy of 0.8.

Figures 13–15 illustrate the feature importance of the decision tree, bagging, and random forest classifiers. The feature importance reveals the contribution of each feature in deriving the prediction results. The left histogram of these figures shows the results using training dataset #1, whereas the right histogram uses training dataset #2. As shown in the left histograms of Figures 13–15, the most discriminative feature processed by training dataset #1 is "Framing", which implies that if the type of clashing item is a beam, it can contribute nearly 40% of correct prediction. However, among those features processed by training data with the feedback of rule-based reasoning, or training dataset #2, the rule-based tag with "errors" becomes the most contributing feature to the prediction results. The shift of the most discriminative feature between the two experiments responds and explains the above-mentioned prediction improvement. According to Table 2, the proportion of actual cases that are "errors" is around 30%, representing the second-largest label cluster in the training data and reflecting the character of DT classifiers that tend to make a prediction to those labels with more cases. The feature importance of both "framing" and "rule_errors" in our experiments may reflect the effect of the beam rule, which was introduced in Section 4. Besides, the feature "rule_pseudo" also outperforms other features. These two features could be the reason why adding the feedback of rule-based reasoning can improve the prediction performance of machine learning. On the contrary, the label with the largest cluster, "deliberate clashes", does not appear in feature important ranking list, which needs more investigation in the future.

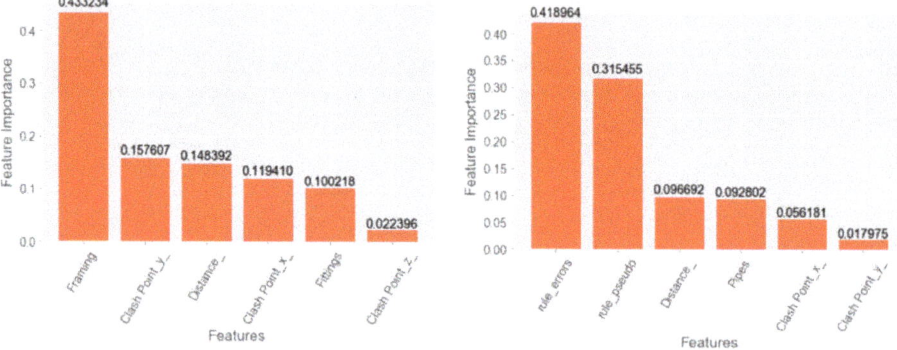

Figure 13. Feature importance suggested by the DT classifier.

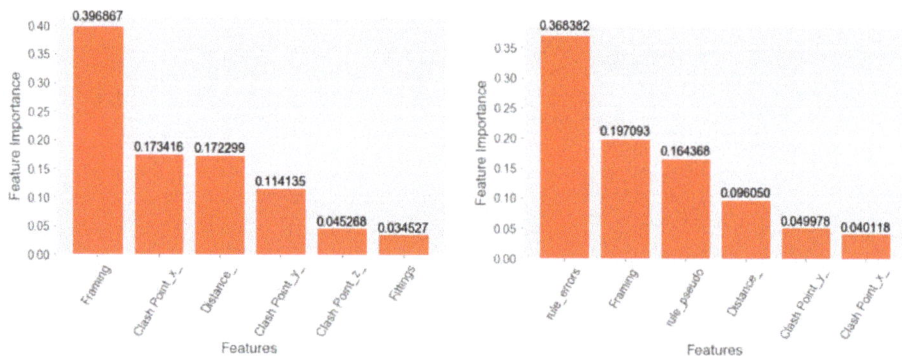

Figure 14. Feature importance suggested by the bagging classifier.

Figures 16 and 17 show the learning curves of six classifiers in the two experiments. As shown in Figure 16, the training curves and validation curves converge with a large gap, indicating that the models processed by training dataset #1 may suffer from a small degree of overfitting. This problem is significantly improved in the second experiment when the feedback of rule-based reasoning is added in the training data. According to Figure 17, all the curves converge within a small gap with an f1-score higher than 0.90.

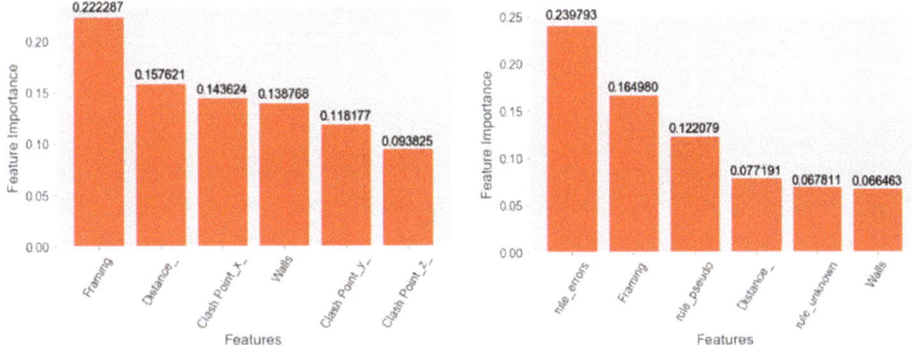

Figure 15. Feature importance suggested by the random forest classifier.

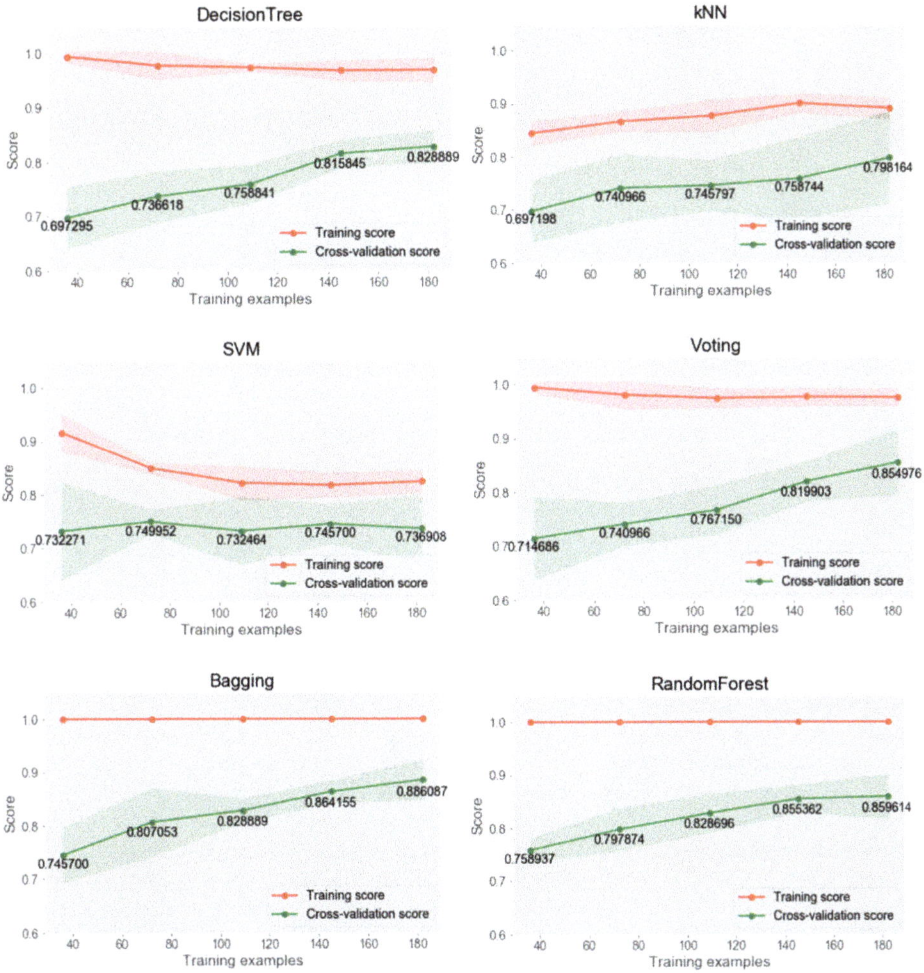

Figure 16. Learning curves of classifiers using training dataset #1.

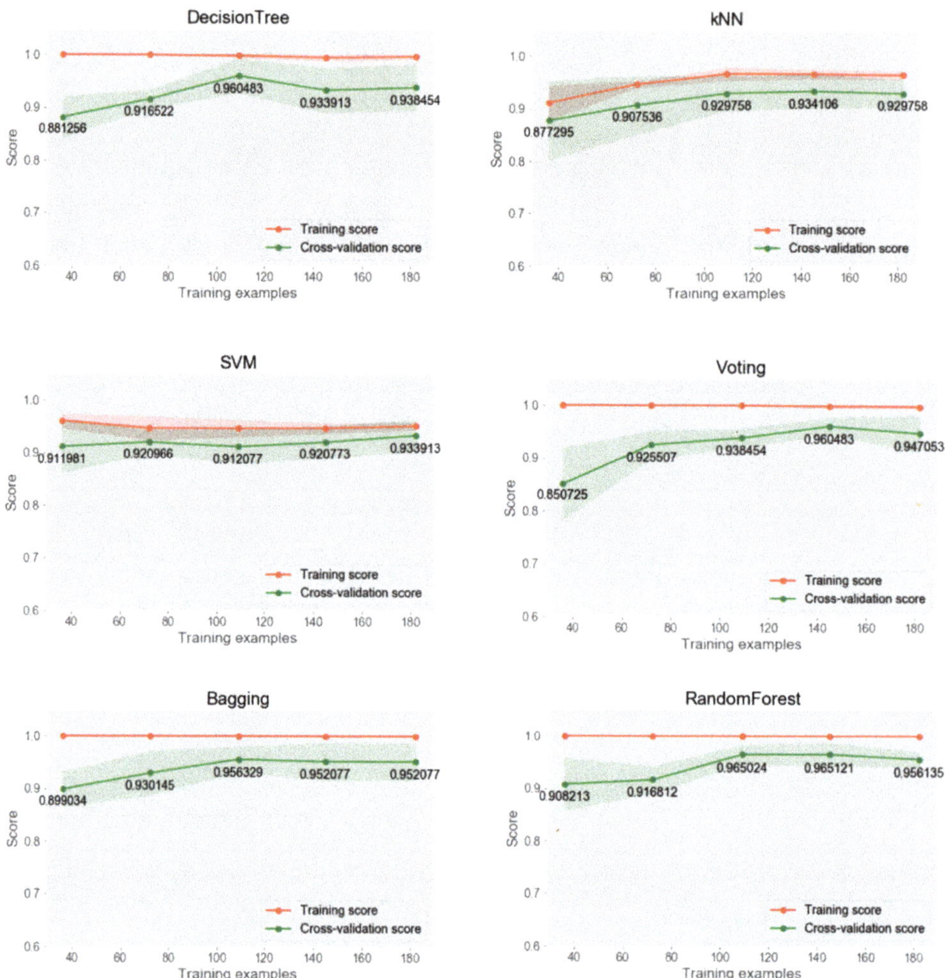

Figure 17. Learning curves of classifiers using training dataset #2.

Simply looking at the numbers, it may indicate that the models processed by training dataset #2 can achieve a satisfactory f1-score with a low bias and variance. Nevertheless, the predictive performance may be still under the influence of the nature of the training dataset we collect. For instance, among those clashes in our training data, "errors" accounted for nearly 50% of all the 326 cases. Histograms of feature importance shown in Figures 13–15 reveal the influence of this distribution on final prediction. It can be reasonably expected that the predictive performance of the same models may vary from one training dataset to another. More experiments with different training data from similar architectural projects are still needed.

Nonetheless, in light of the average f1-score improvement shown in Figure 11, the hybrid method combining the feedback from rule-based reasoning and machine learning still can be regarded as a favorable approach to enhance machine learning under a small training dataset.

Considering the ultimate goal of this study that BIM managers can benefit from the automatic clash classification of clash detection reports, both a high filtering rate of irrelevant clashes and a low misclassification rate, especially for errors, are expected. The filtering rate of irrelevant clashes we are referring to is the proportion of pseudo and deliberate clashes correctly identified among all 98 testing cases, while the error misclassification rate is the proportion that the actual errors are wrongly classified as pseudo or deliberate clashes. From the 30 tests of 98 randomly selected testing cases, we obtained an average filtering rate of 50%–60%. For the experiment of processing training dataset #1, both individual and ensemble learning classifiers wrongly classified 9.5% of errors into pseudo clashes or deliberate clashes. This number decreased to 0%–4.8% in the second experiment of processing training dataset #2. It shows that the hybrid method also had a positive effect on reducing the misclassification rate of errors.

7. Conclusions

Previous studies stated that the clash detection reports produced by most BIM software are prone to present a huge number of clashes, many of which belong to irrelevant or ignorable clashes. Manually filtering out serious clashes from the long list in the clash detection report is both time and cost consuming; thus, automatic filtering out those irrelevant clashes by algorithms is a crucial need for the current industry.

This study proposed a hybrid method that combines simple rule-based reasoning and supervised machine learning to automatically filter out irrelevant clashes from those conflicts detected by BIM software. The experiment results showed that the hybrid method can obtain a rise in the prediction accuracy of machine learning by 15%–17% for individual classifiers and 6%–8.5% for both DT and ensemble learning classifiers. The proposed method conquered the difficulties of purely developing a rule-based system considering complex relationships or merely implementing a machine learning to filter out irrelevant clashes. The former requires lots of effort for knowledge acquisition, while the number of cases collected often limits the latter's performance. It indicated that when the predictive accuracy of the conventional supervised machine learning for design clash classification is unfavorable and more training cases cannot be collected shortly, adding a feature of prediction results obtained by rule-based reasoning to the original training dataset provides an alternative to improve the prediction performance. However, the extent of improvement may depend on how well the rule-based reasoning performs. In other words, there exists a trade-off between the accuracy improvement and efforts to acquire domain knowledge when implementing the rule-based reasoning system.

The ultimate goal of identifying clash types is to resolve those errors or serious clashes before the construction phase to avoid delays and costs incurred. The resolution of serious clashes is the most important task worthy of time and effort. Even though the average predictive accuracy we obtained by the hybrid method is as high as 95%, we still conducted an analysis of the misclassification of serious clashes by our method, where actual serious clashes are wrongly classified as pseudo or deliberate clashes. One of 30 tests with 98 cases from testing dataset #2, the serious clashes misclassification rate by the bagging classifier is up to 11% (out of actual serious clashes in testing cases). How to reduce and avoid the misclassification of serious clashes remains one of the important issues in future work.

Since the training data used to conduct the machine learning process contains only clashes between structural and piping components, the current models and their predictive results can only be applied to those clash detection reports with a similar setting. The training data needs to include more MEP components, such as ducts, conduits, fire alarm devices, or lighting devices, to extend the practical value of this study. Several individual classifiers also have overfitting issues. More training data are required for more experiments in the future. Another issue of this study is that labeling the dataset highly relied on manual work. Future studies can consider applying unsupervised machine learning based on those labeled training data to build up larger training data.

Author Contributions: Funding acquisition (W.Y.L.); Conceptualization (W.Y.L.); Methodology (W.Y.L.); Data Collection (Y.-H.H.); Labeling (Y.-H.H.); Label Adjustment (W.Y.L.); Knowledge Acquisition (Y.-H.H.); Rule-based reasoning (W.Y.L.); Machine Learning Process (W.Y.L.); Validation (Y.-H.H.); Visualization (W.Y.L.); Writing—Original draft (W.Y.L.); Writing—Review and editing (Y.-H.H.).

Funding: This research was funded by the Ministry of Science and Technology, Taiwan, under grant MOST 107-2221-E-035-039.

Acknowledgments: The authors would like to express their sincere gratitude to the editor and the anonymous reviewers who significantly enhanced the contents of the study with their insightful comments. We also would like to thank Uni-edit (www.uni-edit.net) for editing and proofreading this manuscript.

Conflicts of Interest: The authors declare no conflict of interest. The funders had no role in the design of the study; in the collection, analyses, or interpretation of data; in the writing of the manuscript, or in the decision to publish the results.

References

1. Lopez, R.; Love, P.E.; Edwards, D.J.; Davis, P.R. Design error classification, causation, and prevention in construction engineering. *J. Perform. Constr. Facil.* **2010**, *24*, 399–408. [CrossRef]

2. Han, S.; Lee, S.; Pena-Mora, F. Identification and quantification of non-value-adding effort from errors and changes in design and construction projects. *J. Constr. Eng. Manag.* **2011**, *138*, 98–109. [CrossRef]

3. Won, J.; Lee, G. How to tell if a BIM project is successful: A goal-driven approach. *Autom. Constr.* **2016**, *69*, 34–43. [CrossRef]

4. Pärn, E.A.; Edwards, D.J.; Sing, M.C. Origins and probabilities of MEP and structural design clashes within a federated BIM model. *Autom. Constr.* **2018**, *85*, 209–219. [CrossRef]

5. Van den Helm, P.; Böhms, M.; van Berlo, L. IFC-based clash detection for the open-source BIMserver. In Proceedings of the International Conference on Computing in Civil and Building Engineering, Nottingham, UK, 30 June–2 July 2010; Nottingham University Press: Nottingham, UK, 2010; p. 30. Available online: http://www.engineering.nottingham.ac.uk/icccbe/proceedings/pdf/pf91.pdf (accessed on 1 May 2019).

6. Hu, Y.; Castro-Lacouture, D. Clash relevance prediction based on machine learning. *J. Comput. Civ. Eng.* **2018**, *33*, 04018060. [CrossRef]

7. Hu, Y.; Castro-Lacouture, D.; Eastman, C.M. Holistic clash detection improvement using a component dependent network in BIM projects. *Autom. Constr.* **2019**, *105*, 102832. [CrossRef]

8. Leite, F.; Akinci, B.; Garrett, J., Jr. Identification of data items needed for automatic clash detection in MEP design coordination. In Proceedings of the Construction Research Congress 2005: Building a Sustainable Future, Seattle, WA, USA, 5–7 April 2009; pp. 416–425.

9. Ciribini, A.L.C.; Ventura, S.M.; Paneroni, M. Automation in construction implementation of an interoperable process to optimise design and construction phases of a residential building: A BIM pilot project. *Autom. Constr.* **2016**, *71*, 62–73. [CrossRef]

10. Akponeware, A.O.; Adamu, Z.A. Clash detection or clash avoidance? An investigation into coordination problems in 3D BIM. *Buildings* **2017**, *7*, 75. [CrossRef]

11. Jiang, S.; Wu, Z.; Zhang, B.; Cha, H.S. Combined MvdXML and semantic technologies for green construction code checking. *Appl. Sci.* **2019**, *9*, 1463. [CrossRef]

12. Fernandez-Millan, R.; Medina-Merodio, J.A.; Plata, R.; Martinez-Herraiz, J.J.; Gutierrez-Martinez, J.M. A laboratory test expert system for clinical diagnosis support in primary health care. *Appl. Sci.* **2015**, *5*, 222–240. [CrossRef]

13. Ziolkowski, P.; Demczynski, S.; Niedostatkiewicz, M. Assessment of failure occurrence rate for concrete machine foundations used in gas and oil industry by machine learning. *Appl. Sci.* **2019**, *9*, 3267. [CrossRef]

14. Wong, T.T. A hybrid discretization method for naïve Bayesian classifiers. *Pattern Recognit.* **2012**, *45*, 2321–2325. [CrossRef]

15. Hoshyar, A.N.; Rashidi, M.; Liyanapathirana, R.; Samali, B. Algorithm development for the non-destructive testing of structural damage. *Appl. Sci.* **2019**, *9*, 2810. [CrossRef]

16. Chou, J.S.; Pham, A.D. Hybrid computational model for predicting bridge scour depth near piers and abutments. *Autom. Constr.* **2014**, *48*, 88–96. [CrossRef]

17. Wang, L.; Leite, F. Formalized knowledge representation for spatial conflict coordination of mechanical, electrical and plumbing (MEP) systems in new building projects. *Autom. Constr.* **2016**, *64*, 20–26. [CrossRef]

18. Mehrbod, S.; Staub-French, S.; Mahyar, N.; Tory, M. Beyond the clash: Investigating BIM-based building design coordination issue representation and resolution. *J. Inf. Technol. Constr.* **2019**, *24*, 33–57. Available online: http://www.itcon.org/2019_03-ITcon-Mehrbod.pdf (accessed on 12 June 2019).

19. Hartmann, T. Detecting design conflicts using building information models: A comparative lab experiment. In Proceedings of the CIB W78 2010: 27th International Conference, Cairo, Egypt, 16–18 November 2010; pp. 16–18. Available online: http://itc.scix.net/pdfs/w78-2010-57.pdf (accessed on 20 April 2019).

20. Gijezen, S. *Organizing 3D Building Information Models with the Help of Work Breakdown Structures to Improve the Clash Detection Process*; VISICO Center, Univ. of Twente: Enschede, The Netherlands, 2010.

21. Palmer, I.J.; Grimsdale, R.L. Collision Detection for Animation Using Sphere-trees. *Comput. Graph. Forum* **1995**, *14*, 105–116. [CrossRef]

22. Hubbard, P.M. Approximating polyhedra with spheres for time-critical collision detection. *ACM Trans. Graph.* **1996**, *15*, 179–210. [CrossRef]

23. Klosowski, J.T.; Held, M.; Mitchell, J.S.B.; Sowizral, H.; Zikan, K. Efficient collision detection using bounding volume hierarchies of k-DOPs. *IEEE Trans. Vis. Comput. Graph.* **1998**, *4*, 21–36. [CrossRef]

24. Gottschalk, S.; Lin, M.C.; Manocha, D.; Hill, C. Obbtree: A Hierarchical Structure for Rapid Interference Detection. Available online: http://gamma.cs.unc.edu/SSV/obb.pdf (accessed on 28 July 2019).

25. Leo Kumar, S.P. Knowledge-based expert system in manufacturing planning: State-of-the-art review. *Int. J. Prod. Res.* **2019**, *57*, 4766–4790. [CrossRef]

26. Chinese Society of Structural Engineers. *The Manual for Structural Reinforcement in Reinforced Concrete Buildings*; Technology Books Co., Ltd.: Taipei, Taiwan, 2011; ISBN 9789576554964.

27. Solihin, W.; Eastman, C. Classification of rules for automated BIM rule checking development. *Autom. Constr.* **2015**, *53*, 69–82. [CrossRef]

28. Raschka, S. *Python Machine Learning*; Packt Publishing: Birmingham, UK, 2015.

29. Bilal, M.; Oyedele, L.O.; Qadir, J.; Munir, K.; Ajayi, S.O.; Akinade, O.O.; Owolabi, H.A.; Alaka, H.A.; Pasha, M. Big Data in the construction industry: A review of present status, opportunities, and future trends. *Adv. Eng. Inform.* **2016**, *30*, 500–521. [CrossRef]

30. Pietrzyk, K. A systemic approach to moisture problems in buildings for mould safety modelling. *Build. Environ.* **2015**, *86*, 50–60. [CrossRef]

31. Desai, V.S.; Joshi, S. *Application of Decision Tree Technique to Analyze Construction Project Data, in: Information Systems, Technology and Management*; Springer: Berlin/Heidelberg, Germany, 2010; pp. 304–313.

32. Liu, H.B.; Jiao, Y.B. Application of genetic algorithm-support vector machine (ga-svm) for damage identification of bridge. *Int. J. Comput. Intell. Appl.* **2011**, *10*, 383–397. [CrossRef]

33. Mahfouz, T.; Jones, J.; Kandil, A. A machine learning approach for automated document classification: A comparison between SVM and LSA performances. *Int. J. Eng. Res. Innov.* **2010**, *2*, 53–62.

34. Dehestani, D.; Eftekhari, F.; Guo, Y.; Ling, S.; Su, S.; Nguyen, H. Online support vector machine application for model based fault detection and isolation of HVAC system. *Int. J. Mach. Learn. Comput.* **2011**, *1*, 66. [CrossRef]

35. Ur-Rahman, N.; Harding, J.A. Textual data mining for industrial knowledge management and text classification: A business oriented approach. *Expert Syst. Appl.* **2012**, *39*, 4729–4739. [CrossRef]

Article

A Two-Stage Building Information Modeling Based Building Design Method to Improve Lighting Environment and Increase Energy Efficiency

Sha Liu [1] and Xin Ning [2,*]

1   Faculty of Infrastructure Engineering, Dalian University of Technology, Dalian 116024, Liaoning, China
2   School of Investment and Construction Management, Dongbei University of Finance and Economics, Dalian 116025, Liaoning, China
*   Correspondence: ningxin@dufe.edu.cn

Received: 23 July 2019; Accepted: 25 September 2019; Published: 29 September 2019

Abstract: Buildings are one of the largest energy consumers in the world, and have great energy saving potential. Thermal systems and lighting systems take most of the energy in a building. Comparing with the optimization solutions developed for a thermal system, the research of improving the lighting system is insufficient. This study aims to improve the lighting environment and reduce the energy by optimizing the building design, which has the largest potential for cutting energy economically compared with the other stages in the life cycle of a building. Although many approaches have been developed for building design optimization, there is still one big problem obstructing their successful practices, in that the designers who take the responsibility of making building designs are not experts in building physics, thus they are not capable of calculating the most appropriate parameters and operating the professional software to optimize their designs. Therefore, this study proposes a user-friendly method for designers to improve building designs. Firstly, Building Information Modeling (BIM) and particle swarm optimization algorithm are applied to build an intelligent optimal design search system. The optimized design from this system can largely use daylighting for internal illumination and save energy. Secondly, different types of lighting control systems are compared and the one which can save maximal energy is added to the selected optimal design. A case study demonstrates that optimized designs generated by the proposed design method can save large amounts of life cycle energy and costs, and is effective and efficient.

Keywords: building design; building performance simulation; energy conservation; building information modeling

---

1. Introduction

Buildings ultimately account for 40% of the world's energy [1]. To reduce the energy consumed by buildings has great impact on mitigating the greenhouse effect. Comparing with the other stages in the life cycle of a building, the design stage has the largest potential for energy conservation with the least extra expenses. Factors which have significant influences on building energy consumption, e.g., types and dimensions of building materials, building envelope, and structure, can all be carefully decided during building design. It has been estimated that the design stage determines about 70% of the environmental impacts of the entire life cycle [2]. Therefore, making a reasonable and sustainable building design is beneficial for both the project owners and the environment.

1.1. Daylight for Indoor Illumination

The Heating, Ventilation and Air Conditioning (HVAC) system and lighting system are the two largest energy consumers in a building [3–5]. Many efforts have been put into improving thermal performance of buildings for energy reduction. However, how to make a proper building design to optimize the indoor lighting environment and saving energy still needs to be discussed.

Daylight, which best matches the human visual response, is considered as the best substitution for artificial light in the indoor environment. It provides better color rendering and a visual environment in which occupants can properly see objects [6]. Apart from that, daylighting is a more environmentally-friendly approach compared to artificial lighting. Zain-Ahmed et al. [7] found that at least 10% of building energy can be saved using daylight alone. As such, using daylight to effectively create a comfortable indoor light environment is necessary for both occupant health [8,9] and building energy conservation [10].

A good building design, particularly the envelope design, is the basic solution for getting good use of daylight. By adjusting the parameters of design factors, sufficient daylight can be led into the room to take place for artificial light, thus that a part of the energy consumed by artificial lighting will be saved, and the indoor visual comfort will be improved as well [11]. Krarti et al. [12] developed a simplified energy estimation model to analyze the energy reduction by daylighting. The model considered four parameters: Building geometry, glazing types, window areas, and geographic locations. Results showed that glazing type and window area had more significant impacts on energy conservation. Ochoa et al. [13] investigated the thermal and visual performance of buildings with different window sizes, to search for a design that can achieve the objectives of low energy consumption and high visual comfort. Although the key design factors are confirmed, it is difficult to decide the parameter value of each factor to achieve the optimal lighting performance. The direct way is the trial-and-error method. Researchers tried every possible parameter value to search for the one which can take the best performance [14–16]. This type of methods performs well in the cases of small-scale buildings or in which only a few design factors are considered. When the building scale becomes larger, they cannot afford the large amount of computations. Then simulation tools are applied for lighting performance prediction to reduce the calculative burden [17–20]. Simulation tools can largely shorten the calculation time and improve the accuracy of the results, thus designers can quickly make a good design with an optimized lighting environment. Actually, a good design must be well performed on not only the lighting aspect, but also others such as the thermal environment and energy consumption. Besides, for the project owners, project cost is a crucial factor to be carefully controlled during the design. Therefore, some other methods are required to find a balance in the problems with multi-objectives. Mangkuto et al. [21] tried a graphic optimization method to search for the designs with the minimum annual lighting energy demand and to fulfill the indoor visual requirement. Evolutionary multi-objective optimization algorithms, such as NSGA-II and PSO-HJ, are more popularly applied to achieve the win-win goal of both the energy-efficiency and visual comfort of buildings [22–24]. These optimization methods are usually combined with thermal and lighting simulation engines, e.g., EnergyPlus, Radians, and Daysim, thus that the calculation speed and the result precision can be promoted to an all-time high.

A lighting control system is another way to improve the illumination environment. There are mainly three kinds of lighting control modes, i.e., manual light switching (MLS) (on/off), automatic light switching (ALS) (on/off), and automatic light dimming (ALD) (on/off) [25]. The lighting control systems can adjust the room illuminance as the lighting situation changes to avoid glare or light deficiency. Studies has proved that lighting control can provide a more comfortable visual environment [26] and save up to 70% of the energy of a lighting system [27–30].

1.2. Barriers of Making a Good Building Design

Making an energy-efficient design with good building performance is the most cost-effective way to keep the project sustainable over its entire life cycle. It requires the designers to put more effort into the building physics and structure during design. However, for the designers, it is difficult to achieve the objectives above. The barriers are as follows.

(1)   Most of the designers are not experts in the area of building physics, thus they are not able to calculate the quantity of energy consumption and the levels of thermal and lighting performance of their building designs [31]. To solve this problem, designers have to follow the standards of the green building handbooks, such as Leadership in Energy and Environmental Design (LEED) and Building Research Establishment Environmental Assessment Method (BREEAM). However, the standards in these handbooks are the minimum thresholds for general buildings, thus they cannot fulfill the requirements of each specific building, which leads to the incomplete optimization on building performance [32].
(2)   In order to procure reliable and accurate optimization results quickly, simulation tools for thermal performance and lighting performance are necessary. Since designers do not understand every parameters of building physics and structure, they do not know how to preset the reasonable parameter values and operate the software successfully [33,34]. In addition, it will take a long time for designers to learn how to operate the simulation software, which is not only a burden for them, but also slows down the design work.

1.3. Objectives

The aim of this study was to overcome the barriers stated above and propose a user-friendly building design method, which can take advantages of both building design optimization and lighting control system to save energy as well as improve the indoor lighting environment. The specific objectives of this study are as follows:

(1)   The first objective was to improve the lighting environment in buildings by fully using daylight instead of artificial lighting and by applying an appropriate lighting control system.
(2)   The second objective was maximizing energy savings by optimizing the initial building design.

This study can help designers make optimal building designs, which can fulfill the requirements of the environment, clients and occupants, and raise the design work efficiency, thus, to increase the success rate of sustainable building projects.

2. Two-Stage Building Information Modeling (BIM) Based Lighting Design Method

The proposed lighting design method consists of two stages: Building design optimization and lighting control system selection. To maximize the efficiency of the process, BIM was selected as a computing platform to integrate the database, simulation, and computation modules. Figure 1 illustrates the schematic structure of the proposed BIM based building design method.

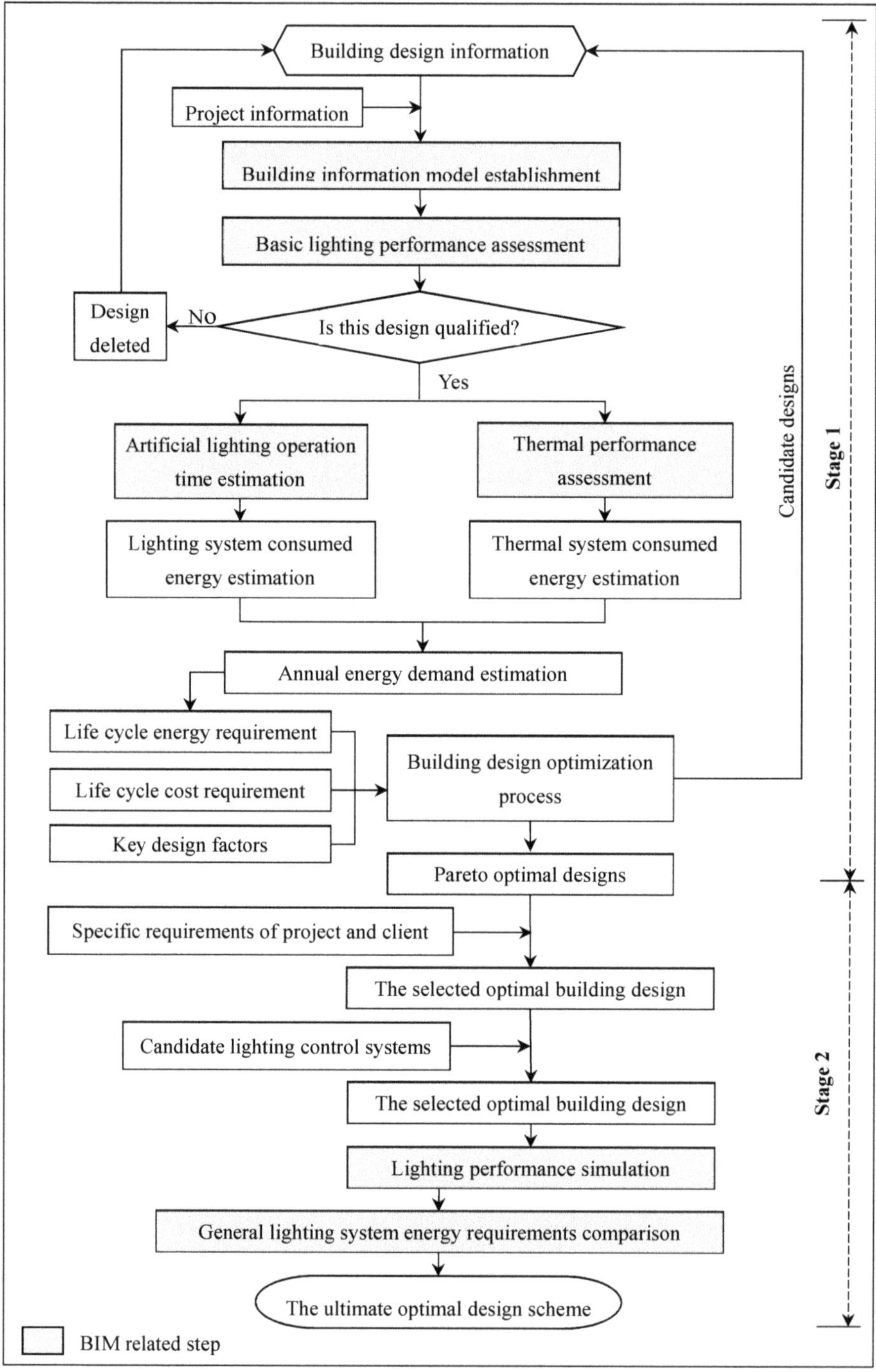

Figure 1. Schematic structure of the proposed building information modelling (BIM) based building design method.

2.1. Stage 1: Building Desing Optimziation

2.1.1. Building performance simulation

This proposed lighting design method was based on a completed initial building design scheme, which integrated specific client requirements and the project environment. Complete information about the initial design was entered into the BIM system to establish a three-dimensional (3D) building information model. Information related to the project, such as geographical coordinates and the project location's climate, must also be entered into the model.

To ensure all candidate designs could qualify the minimum lighting standard, the first step was to do a basic lighting performance assessment. Lighting performance of a building envelope can be simulated using BIM, and the ones which cannot achieve the daylight factor of 2% will be eliminated.

Several studies have found that heating, ventilation and air conditioning (HVAC), and lighting systems consume the most energy in a building [35–37]. The goal of this study was to both improve lighting performance and save energy. As such, both thermal and lighting performance must be discussed. Therefore, when receiving the qualified design, the BIM simulation module was applied to generate the estimates of artificial lighting time and the assessments of thermal performance.

For a lighting system, the operation time of the lighting facilities was estimated based on daylight autonomy (DA) at a specific illumination level. An analysis grid with a specific dimension was generated to cover the test area; test points were simulated for DA within the workday. Considering that there were plenty of test points, thus the average DA $(\overline{DA})$ of all these points was used at this step for the estimates. Consequently, the annual total operation time of lighting facilities (T_{lit}) can be calculated using Equation (1):

$$T_{lit} = (1 - \overline{DA}) \times t_{wrk} \tag{1}$$

In this expression, t_{wrk} is the annual total working time of the test area.

Based on T_{lit}, annual total energy required by the lighting system (E_{lit}) can be estimated using Equation (2):

$$E_{lit} = \sum_{n=1}^{N} (T_{lit} \times P_{lit})_n \tag{2}$$

In this expression, $n = 1, 2, 3, \ldots, N$ is the number of lighting facilities in the test area; and P_{lit} is the power of the lighting facilities.

For the thermal system, the thermal performance of the building envelope was evaluated based on the energy provided to the HVAC system to maintain thermal neutrality. Therefore, the monthly cooling load and heating load were estimated using the thermal simulation engine of the Ecotect Analysis, which was produced by Autodesk Inc. Then, the amount of energy consumed by the thermal system yearly (E_{ter}) could be computed using the coefficients of performance (COP) for the chiller and heating pump, as shown in Equation (3):

$$E_{ter} = \sum_{k=1}^{12} \left(\frac{E'_c}{e_c} + \frac{E'_h}{e_h} \right)_k \tag{3}$$

In this expression, $k = 1, 2, 3, \ldots, 12$ is the month number; E'_c and E'_h are the monthly loads for cooling and heating, respectively; and e_c and e_h are the COPs for the chiller and heat pump, respectively.

The annual total energy demand of the design (E_{sum}) is the sum of the energy consumed by the lighting system and thermal system; as such, it can be calculated using Equation (4):

$$E_{sum} = E_{lit} + E_{ter} \tag{4}$$

After the simulation process, the annual total energy demand of lighting and thermal systems are entered into the optimization module for further computation.

2.1.2. Building Design Optimization

In this study, a particle swarm optimization (PSO) algorithm was selected to generate the optimal design. This is because it has the advantages of evolutionary algorithms (i.e., inherent parallelism and the ability to exploit similarities in solutions through recombination [38]), and had the characteristics of directed mutation, population representation and operators, and others. The revised PSO algorithm used in this study ensures global convergence at a higher speed.

As mentioned above, the two objectives of the optimization module were to minimize the project's life cycle energy (LCE) consumption and life cycle cost (LCC). The effects of the design and demolishment stages on the LCE and LCC were negligible, and data on these factors were difficult to obtain. As such, the life cycle of this study only included the construction and operation stages. Therefore, the LCE can be calculated using Equation (5):

$$LCE = E_{con} + E_{opr} = \sum_{j=1}^{J} Q_j \times f_j + E_{sum} \times M \tag{5}$$

In this expression, E_{con} and E_{opr} represent the energy consumed at the construction stage and operation stage, respectively; $j = 1, 2, 3, \ldots, J$ is the number of building materials used in the design; Q_j is the quantity of the jth material; f_j is the energy intensity of the jth material; and M is the operation time for the project.

Similarly, the LCC was composed of the construction cost and the operation cost. The LCC reflects cash flows at different stages, requiring the calculation of the time value of money. Consequently, all LCC elements were converted to the present value. The actual LCC was computed using Equation (6):

$$LCC = C_{con} + C_{opr} = \sum_{j=1}^{J} (Q_j \times p_j) + \sum_{m=1}^{M} (C_{opr})_m \times (i+1)^{-m} \tag{6}$$

In this expression, C_{con} and C_{opr} are the costs of the construction stage and operation stage, respectively; p_j is the price of the jth material; and i is the real interest rate for a compounding period, which considered the effect of inflation.

Decision variables were another important aspect of optimization. In this study, design factors, which significantly influenced lighting and thermal performances, were set as decision variables. Many studies have demonstrated that wall types, glazing types, and window-to-wall ratio were the most significant design factors [39–42]. Artificial lighting types was also an important factor in illumination and energy consumption [25,43]. Therefore, these 4 design factors were selected as the decision variables in the optimization module.

For clients, project costs were a crucial element impacting decisions. As such, the constraints set in the optimization module included both the construction cost and the design's LCC, which cannot exceed their respective budgets.

The output of the proposed optimization system was a Pareto-optimal solution set, which contained several optimal design schemes. The solutions were non-dominated with each other; as such, the client must select the solution that can fulfill most of the project requirements for the following processes.

2.2. Stage 2: Lighting Control System Selection

The second stage of this study was to select an appropriate lighting control mode for the selected optimal building design. In this study, 3 popular lighting control modes were evaluated as candidates for improving lighting performance [44,45]. The first mode was a manual light switching (MLS), where people must manually turn the lights on/off, depending on indoor brightness preferences. The second mode was automatic light switching (ALS) with an occupancy sensor that automatically turns the lights on or off depending on whether someone is in the room. The third mode was an automatic light dimming with occupancy sensor (ALD). Similar was ALS, the occupancy sensor controlled the operating conditions, however, the lights dim instead of turning off when no one was in the room.

When receiving the selected optimal design scheme, the LCE and LCC of the design with 3 lighting control modes had to be estimated, respectively. The ultimate optimized design was generated by comparing the results of the 3 scenarios.

3. Case Study

The case study for this research was a lecture room in the east Teaching Building of Dalian Minzu University. The goal was to validate the effectiveness and efficiency of the proposed BIM based building design method. The room had an area of 126.36 mq and was located at the northeast of the building on the second floor. The distance from the floor to roof was 4.5 m; the ceiling height was 3.5 m. There were 8 windows in the east wall, one narrow window in the north wall, and two doors in the south wall. The narrow window in the north wall was mainly aesthetic, and as such, was not considered during the design optimization process. Figure 2 shows the room's outline. There were a total of 20 lights in the room. Table 1 shows the detailed information about the initial design. To facilitate the application of the proposed design method, it was assumed that all the building materials were purchased in Liaoning Province. The building's lifetime was calculated as 50 years.

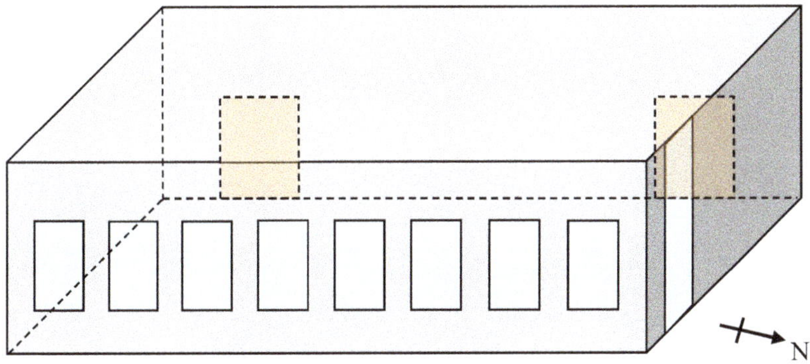

Figure 2. Outline of the lecture room.

Table 1. Initial design of the case.

Design Factor	Variable Description
Window-to-wall ratio	0.19
Glazing type	6 mm double clear float glass with 6 mm air gap
Wall type	300 mm reinforced concrete wall + 50 mm EPS board +30 mm cement and sand (1:3) screed + 10 mm plaster
Artificial light type	T8 fluorescent light, 30W

The BIM software used in this study for the simulations was Ecotect Analysis 2011 (Autodesk Inc., San Rafael, Calif., the USA, 2011). The software established a 3D building information model, with the location and climate information. The workday schedule was set as 6:00 to 18:00. When calculating the total annual energy for the room, the year was set as 365 days, with no holidays during the year.

In the lighting performance simulation module, the analysis grid was set at a height of 750 mm above the floor, with 496 uniformly distributed points. The standard illumination at the workplace was 500 lux; as such, the artificial light operation time was be calculated based on $\overline{DA500}$, and the lights were controlled manually.

The city of Dalian in Liaoning Province of China, has a warm temperate continental monsoon climate. Therefore, both heating and cooling were required. In this case, a water-cooled chiller with a COP of 4.7 was used. The indoor design temperature was set at 26 °C in the summer and 18 °C in the winter, without night setback or setup.

Additional study design details were as follows. Based on the charging standard of electricity in Dalian, the electricity tariff was set at 0.8 CNY/kWh. The real interest rate was counted to be −6.45%, based on the inflation rate. The maximum construction cost and the maximum LCC were set at CNY ¥50,000 and CNY ¥15,000,000, respectively, based on interviews with project owners and project managers.

After a series of lighting performance and thermal performance simulations, simulation outputs were used to estimate the annual energy demand. This result was entered into the building design optimization system. Key parameters were set as follows: The cognitive parameter (c1) and social parameter (c2) were both set at 1.49445; the minimum inertia weight (wi) was 0.4; the maximum inertia weight (wa) was 0.9; the population size (N) was 50; the maximum iteration number (maxit) was 100; the maximum repository size (Anum) was 50; the grid inflation parameter (α) was 0.01; the leader selection pressure parameter (β) was 10; and the grid sum per dimension (gridnum) was 50.

Table 2 presents detailed information about the decision variables; these were also important design factors. Energy intensities of the materials were collected from [46] and determined based on [47–51].

Table 2. Information of decision variables.

Decision Variable		Content	Unit Cost (CNY/mq)	Energy Intensity (MJ/mq)
Window-to-wall ratio	γ	(0, 0.9)		
Glazing type	G1	6 mm clear glass +12 mm argon gap +6 mm clear glass	400	1345.84
	G2	6 mm low-e glass +12 mm argon gap +6 mm clear glass	700	1937.60
	G3	6 mm tint glass +6 mm argon gap +6 mm clear glass	500	1549.60
Wall type	W1	300 mm reinforced concrete wall +50 mm Polystyrene (EPS) board +10 mm cement and sand (1:3) screed +10 mm plaster	143.67	564.50
	W2	300 mm aerated concrete block wall +50 mm EPS board +10 mm cement and sand (1:3) screed +10 mm plaster	125	370.40
	W3	300 mm lightweight concrete block wall +50 mm EPS board +10 mm cement and sand (1:3) screed +10 mm plaster	114.68	370.70
Artificial light type	L1	T8 fluorescent light, 30W	50	
	L2	T5 fluorescent light, 14W	38	
	L3	T8 LED light, 18W	65	

After the iterations were completed, a Pareto frontier was generated (see Figure 3). Table 3 provides detailed information about the optimal designs. These design solutions are non-dominated with each other, making it difficult to distinguish the best one. The designer selects the optimum design, based on the client's preferences and specific requirements. Of the solutions, although the first one was non-dominated with the initial design, it was not further considered because of its larger LCE. The fifth design in Table 3 was selected as the optimum solution for the process that follows.

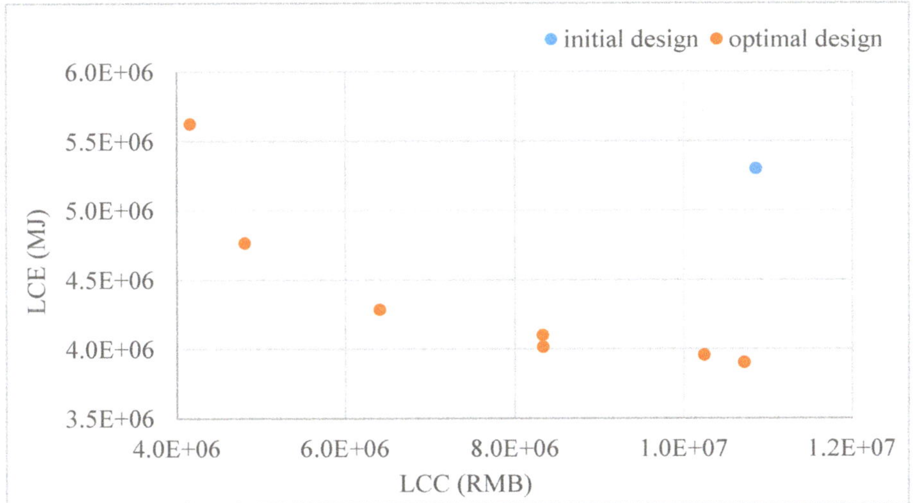

Figure 3. Pareto frontier of case study.

Table 3. Information about optimal designs.

No.	Window-to-Wall Ratio	Glazing Type	Wall Type	Artificial Light Type	LCC (CNY ¥)	LCE (MJ)
1	0.900	1	1	2	4,158,961.07	5,624,928.382
2	0.748	2	2	1	4,811,119.35	4,765,746.660
3	0.201	3	3	2	6,410,224.77	4,285,789.883
4	0.139	2	3	2	8,332,691.86	4,098,387.582
5	0.137	1	3	2	8,336,950.97	4,014,879.593
6	0.100	2	2	2	10,245,469.86	3,956,413.052
7	0.085	1	2	2	10,716,024.16	3,902,559.339

As introduced in Section 2.2, the LCEs and LCCs of the fifth optimal design were estimated, using the 3 lighting control modes. The annual energy consumption of the fifth design with ALS can be calculated using Ecotect Analysis. Considering that the MLS had high randomness, and Ecotect cannot provide precise values based on the ALD, the values for these 2 modes were determined based off statistical results from experiments and related studies [44,45]. Table 4 provides the performances of the 3 modes.

Table 4. Information about the fifth design with three lighting control modes.

Lighting Control Mode	LCC (CNY ¥)	LCE (MJ)
MLS	8,429,845.70	4,068,169.843
ALS	8,364,098.08	4,015,142.393
ALD	8,384,345.69	4,028,560.143

When comparing the performances of the 3 lighting control modes, ALS had the lowest LCC and LCE. Consequently, the ultimate optimal design scheme was the fifth design with the automatic light switching with occupancy sensor.

4. Discussion

This study developed a method that can support significant savings in both energy and cost. The LCC and the LCE of the Pareto optimal designs (except the first design) generated at Stage 1 were compared with the LCC and LCE of the initial design; Figure 4 shows the results. This outcome illustrates that most designs performed better than the initial design: The LCC savings ranged from 1.25% to 55.66% and the LCE savings ranged from 10.10% to 26.39%. As a result, this proposed method can help achieve the goal of developing high-quality designs.

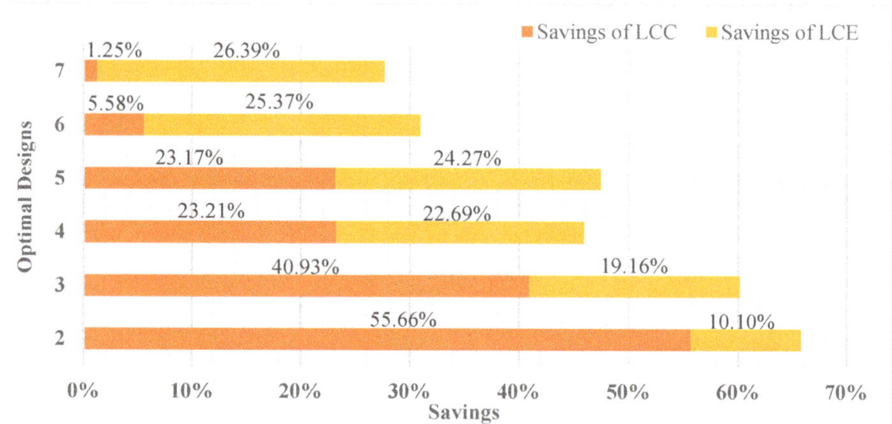

Figure 4. Life cycle cost (LCC) and life cycle energy (LCE) savings from the optimal designs compared with the initial design.

At Stage 2, the best fit lighting control mode was added to the selected optimal design. Table 5 shows that compared with the initial design, the ultimate optimal design can save 22.92% of the LCC and 24.26% of the LCE. The savings of the ultimate optimal design required a little more LCC and LCE than the fifth design. This is because in fifth design assumed that the occupancy can turn the lights on/off in time when the indoor illumination cannot/can be up to the standard of 500 lux. In contrast, for the ultimate optimal design, the lights were controlled using occupancy sensor switches, which have extra energy consumption and costs. The lighting control mode in the fifth design cannot be realistically achieved. As such, a lighting control mode is necessary in the proposed lighting design method.

Table 5. Performances of the initial design and the ultimate optimal design.

Design Scheme	LCC (CNY ¥)	Saving of LCC	LCE (MJ)	Saving of LCE
Initial	10,851,157.42	-	5,301,347.349	-
Ultimate optimal	8,364,098.08	22.92%	4,015,142.393	24.26%

In the proposed lighting design method, the computer generated most of the procedures. The computing time required 82 h, faster than with traditional design methods. Further, during the process, designers only needed to build the model into the simulation software, select the design factors, and enter the pre-set values into the program. This method required less professional knowledge of architectural physics, reducing the difficulty of selecting an excellent sustainable building design scheme. These factors demonstrated the efficient, precise, and user-friendly characteristics of this method.

5. Conclusion

This study provides a holistic building design method for improving the indoor lighting environment and saving energy. The approach has a goal of making full use of daylight to substitute for artificial light and reduce energy. This reduces energy and creates a comfortable indoor lighting environment for occupants. The proposed method can achieve these objectives by changing the key design factors of the initial design scheme to improve building performance. With the support of BIM and PSO, the optimal design procurement can be processed using the computer. This avoids the problems of heavy workloads, a narrow scope of optimal solutions, low accuracy, and other problems. When daylight cannot support enough brightness, an appropriate lighting control mode is selected to monitor the artificial lights working time and adjust the brightness. This improves energy efficiency and extends the lifetime of the lights.

This case study demonstrated the modeling approach that is easy for designers. Further, the results demonstrate that the proposed design method can largely save energy and costs within a building's life cycle. The entire operation process is easy to conduct. Specific requirements of designers and clients can be set by changing the value range of corresponding parameters before running the program. Therefore, the two stage BIM-based building design method is user-friendly, effective, and efficient tool for both designers and clients to procure optimal designs. This yields both environmental and economic benefits.

Author Contributions: Conceptualization, S.L.; methodology, S.L.; software, X.N.; validation, S.L. and X.N.; formal analysis, S.L.; investigation, X.N.; resources, S.L.; data curation, S.L.; writing—original draft preparation, S.L.; writing—review and editing, X.N.; visualization, X.N.; supervision, X.N.; project administration, S.L. and X.N.; funding acquisition, S.L. and X.N.

Funding: This research was funded by National Natural Science Foundation of China (Grant Nos. 71801029, 71501029), Program for Innovative Talents in Liaoning Province of China 2018, and the Program for Excellent Talents in Dongbei University of Finance and Economics (Grant No. DUFE2017R11).

Conflicts of Interest: The authors declare no conflict of interest.

References

1. Shaikh, P.H.; Nor, N.B.M.; Nallagownden, P.; Elamvazuthi, I.; Ibrahim, T. A review on optimized control systems for building energy and comfort management of smart sustainable buildings. *Renew. Sustain. Energy Rev.* **2014**, *34*, 409–429. [CrossRef]

2. Rebitzer, G. Integrating life cycle costing and life cycle assessment for managing costs and environmental impacts in supply chains. In *Cost Management in Supply Chains*; Physica: Heidelberg, Germany, 2002; pp. 128–146.

3. Wei, Y.; Zhang, X.; Shi, Y.; Xia, L.; Pan, S.; Wu, J.; Han, M. A review of data-driven approaches for prediction and classification of building energy consumption. *Renew. Sustain. Energy Rev.* **2018**, *82*, 1027–1047. [CrossRef]

4. Futrell, B.J.; Ozelkan, E.C.; Brentrup, D. Bi-objective optimization of building enclosure design for thermal and lighting performance. *Build. Environ.* **2015**, *92*, 591–602. [CrossRef]

5. Yoon, Y.R.; Moon, H.J. Energy consumption model with energy use factors of tenants in commercial buildings using Gaussian process regression. *Energy Build.* **2018**, *168*, 215–224. [CrossRef]

6. Lim, Y.W.; Kandar, M.Z.; Ahmad, M.H.; Ossen, D.R.; Abdullah, A.M. Building façade design for daylighting quality in typical government office building. *Build. Environ.* **2012**, *57*, 194–204. [CrossRef]

7. Zain-Ahmed, A.; Sopian, K.; Othman, M.Y.H.; Sayigh, A.A.M.; Surendran, P.N. Daylighting as a passive solar design strategy in tropical buildings: A case study of Malaysia. *Energy Convers. Manag.* **2002**, *43*, 1725–1736. [CrossRef]

8. Hwang, T.; Kim, J.T. Effects of indoor lighting on occupants' visual comfort and eye health in a green building. *Indoor Built Environ.* **2011**, *20*, 75–90. [CrossRef]

9. Shishegar, N.; Boubekri, M. Natural light and productivity: Analyzing the impacts of daylighting on students' and workers' health and alertness. In Proceedings of the International Conference on Health, Biological and Life Science (HBLS-16), Istanbul, Turkey, 18–19 April 2016.

10. Cammarano, S.; Pellegrino, A.; Lo Verso, V.R.M.; Aghemo, C. Daylighting design for energy saving in a building global energy simulation context. *Energy Procedia* **2015**, *78*, 364–369. [CrossRef]

11. Al-Khatatbeh, B.J.; Ma'bdeh, S.N. Improving visual comfort and energy efficiency in existing classrooms using passive daylighting techniques. *Energy Procedia* **2017**, *136*, 102–108. [CrossRef]

12. Krarti, M.; Erickson, P.M.; Hillman, T.C. A simplified method to estimate energy savings of artificial lighting use for daylighting. *Build. Environ.* **2005**, *40*, 747–754. [CrossRef]

13. Ochoa, C.E.; Aries, M.B.C.; van Loenen, E.J.; Hensen, J.L.M. Considerations on design optimization criteria for windows providing low energy consumption and high visual comfort. *Appl. Energy* **2012**, *95*, 238–245. [CrossRef]

14. Ashrafian, T.; Moazzen, N. The impact of glazing ratio and window configuration on occupants' comfort and energy demand: The case study of a school building in Eskisehir, Turkey. *Sustain. Cities Soc.* **2019**, *47*, 1–14. [CrossRef]

15. Reinhart, C.F.; Mardaljevic, J.; Rogers, Z. Dynamic daylight performance metrics for sustainable building design. *Leukos* **2006**, *3*, 7–31. [CrossRef]

16. Hee, W.J.; Alghoul, M.A.; Bakhtyar, B.; Elayeb, O.K.; Shameri, M.A.; Alrubaih, M.S.; Sopian, K. The role of window glazing on daylighting and energy saving in buildings. *Renew. Sustain. Energy Rev.* **2015**, *42*, 323–343. [CrossRef]

17. Huang, Y.C.; Lam, K.P.; Dobbs, G. A scalable lighting simulation tool for integrated building design. In Proceedings of the Third National Conference of IBPSA-USA, Berkeley, CA, USA, 30 July–1 August 2008.

18. Lagios, K.; Niemasz, J.; Reinhart, G. Animated building performance simulation (ABPS)—Linking rhinoceros/Grasshopper with Radiance/Daysim. In Proceedings of the Forth National Conference of IBPSA-USA, New York, NY, USA, 11–13 August 2010.

19. Kota, S.; Haberl, J.S.; Clayton, M.J.; Yan, W. Building information modeling (BIM)- based daylighting simulation and analysis. *Energy Build.* **2014**, *81*, 391–403. [CrossRef]

20. Lee, J.W.; Jung, H.J.; Park, J.Y.; Lee, J.B.; Yoon, Y. Optimization of building window system in Asian regions by analyzing solar heat gain and daylighting elements. *Renew. Energy* **2013**, *50*, 522–531. [CrossRef]

21. Mangkuto, R.A.; Rohmah, M.; Asri, A.D. Design optimization for window size, orientation, and wall reflectance with regard to various daylight metrics and lighting energy demand: A case study of buildings in the tropics. *Appl. Energy* **2016**, *164*, 211–219. [CrossRef]

22. Zhai, Y.; Wang, Y.; Huang, Y.; Meng, X. A multi-objective optimization methodology for window design considering energy consumption, thermal environment and visual performance. *Renew. Energy* **2019**, *134*, 1190–1199. [CrossRef]

23. Vera, S.; Uribe, D.; Bustamante, W.; Molina, G. Optimization of a fixed exterior complex fenestration system considering visual comfort and energy performance criteria. *Build. Environ.* **2017**, *113*, 163–174. [CrossRef]

24. Zhang, A.; Bokel, R.; van den Dobbelsteen, A.; Sun, Y. Optimization of thermal and daylight performance of school buildings based on a multi-objective genetic algorithm in the cold climate of China. *Energy Build.* **2017**, *139*, 371–384. [CrossRef]

25. Linhart, F.; Scartezzini, J. Minimizing lighting power density in office rooms equipped with Anidolic Daylighting Systems. *Sol. Energy* **2010**, *84*, 587–595. [CrossRef]

26. Hua, Y.; Oswald, A.; Yang, X. Effectiveness of daylighting design and occupant visual satisfaction in a LEED Gold laboratory building. *Build. Environ.* **2011**, *46*, 54–64. [CrossRef]

27. Li, D.H.W.; Lam, J.C. An investigation of daylighting performance and energy saving in a daylit corridor. *Energy Build.* **2003**, *35*, 365–373. [CrossRef]

28. Yun, G.Y.; Kim, H.; Kim, J.T. Effects of occupancy and lighting use patterns on lighting energy consumption. *Energy Build.* **2012**, *46*, 152–158. [CrossRef]

29. Li, D.H.W.; Lam, T.N.T.; Wong, S.L. Lighting and energy performance for an office using high frequency dimming controls. *Energy Convers. Manag.* **2006**, *47*, 1133–1145. [CrossRef]

30. Doulos, L.; Tsangrassoulis, A.; Topalis, F. Quantifying energy savings in daylight responsive systems: The role of dimming electronic ballasts. *Energy Build.* **2008**, *40*, 36–50. [CrossRef]

31. Geyer, P. Systems modelling for sustainable building design. *Adv. Eng. Inform.* **2012**, *26*, 656–668. [CrossRef]

32. Chen, X.; Yang, H.; Lu, L. A comprehensive review on passive design approaches in green building rating tools. *Renew. Sustain. Energy Rev.* **2015**, *50*, 1425–1436. [CrossRef]

33. Negendahl, K. Building performance simulation in the early design stage: An introduction to integrated dynamic models. *Autom. Constr.* **2015**, *54*, 39–53. [CrossRef]

34. Loonen, R.C.G.M.; Favoino, F.; Hensen, J.L.M.; Overend, M. Review of current status, requirements and opportunities for building performance simulation of adaptive facades. *J. Build. Perform. Simul.* **2017**, *10*, 205–223. [CrossRef]

35. Sun, B.; Luh, P.B.; Jia, Q.; Jiang, Z.; Wang, F.; Song, C. Building energy management: Integrated control of active and passive heating, cooling, lighting, shading, and ventilation systems. *IEEE Trans. Autom. Sci. Eng.* **2013**, *10*, 588–602. [CrossRef]

36. Jeon, J.; Lee, J.; Seo, J. Application of PCM thermal energy storage system to reduce building energy consumption. *J. Therm. Anal. Calorim.* **2013**, *111*, 279–288. [CrossRef]

37. Wong, S.L.; Wan, K.K.W.; Lam, T.N.T. Artificial neural networks for energy analysis of office buildings with daylighting. *Appl. Energy* **2010**, *87*, 551–557. [CrossRef]

38. Zitzler, K.; Deb, K.; Thiele, L. Comparison of multiobjective evolutionary algorithms: Empirical results. *Evol. Comput.* **2000**, *8*, 173–195. [CrossRef]

39. Yu, W.; Li, B.; Jia, H.; Zhang, M.; Wang, D. Application of multi-objective genetic algorithm to optimize energy efficiency and thermal comfort in building design. *Energy Build.* **2015**, *88*, 135–143. [CrossRef]

40. Evins, R. Multi-level optimization of building design, energy system sizing and operation. *Energy* **2015**, *90*, 1775–1789. [CrossRef]

41. Kim, H.; Stumpf, A.; Kim, W. Analysis of an energy efficient building design through data mining approach. *Automat. Constr.* **2011**, *20*, 37–43. [CrossRef]

42. Attia, S.; Gratia, E.; Herde, A.D.; Hensen Jan, L.M. Simulation-based decision support tool for early stages of zero-energy building design. *Energy Build.* **2012**, *49*, 2–15. [CrossRef]

43. Pan, Y.; Yin, R.; Huang, Z. Energy modeling of two office buildings with data center for green building design. *Energy Build.* **2008**, *40*, 1145–1152. [CrossRef]

44. Roisin, B.; Bodart, M.; Deneyer, A.; D'Herdt, P. Lighting energy savings in office using different control systems and their real consumption. *Energy Build.* **2008**, *40*, 514–523. [CrossRef]

45. Galasiu, A.D.; Newsham, G.R.; Suvagau, C.; Sander, D.M. Energy saving lighting control systems for open-plan offices: A field study. *Leukos* **2007**, *4*, 7–29. [CrossRef]

46. *Inventory of Carbon and Energy*, 2nd ed.; University of Bath: Bath, UK, 2011.

47. Asif, M.; Davidson, A.; Muneer, T. Embodied energy analysis of aluminium-clad windows. *Build. Serv. Eng. Res. Technol.* **2001**, *22*, 195–199. [CrossRef]

48. Syrrakou, E.; Papaefthimiou, S.; Yianoulis, P. Eco-efficiency evaluation of a smart window prototype. *Sci. Total Environ.* **2006**, *359*, 267–282. [CrossRef] [PubMed]

49. Prakash, O.; Kumar, A. Annual Performance of a modified greenhouse dryer under passive mode in no-load conditions. *Int. J. Green Energy* **2015**, *12*, 1091–1099. [CrossRef]

50. Chen, T.Y.; Burnett, J.; Chau, C.K. Analysis of embodied energy use in the residential building of Hong Kong. *Energy* **2001**, *26*, 323–340. [CrossRef]

51. Hammond, G.P.; Jones, C.I. Embodied energy and carbon in construction materials. *Proc. Inst. Civ. Eng.-Energy* **2008**, *161*, 87–98. [CrossRef]

Article

Automatically Generating a MEP Logic Chain from Building Information Models with Identification Rules

Ya-Qi Xiao [1], Sun-Wei Li [2] and Zhen-Zhong Hu [1,2,*]

[1] Department of Civil Engineering, Tsinghua University, Beijing 100084, China; gowtxyq@163.com
[2] Graduate School at Shenzhen, Tsinghua University, Shenzhen 518055, China; li.sunwei@sz.tsinghua.edu.cn
* Correspondence: huzhenzhong@tsinghua.edu.cn

Received: 29 March 2019; Accepted: 27 May 2019; Published: 29 May 2019

Abstract: In mechanical, electrical, and plumbing (MEP) systems, logic chains refer to the upstream and downstream connections between MEP components. Generating the logic chains of MEP systems can improve the efficiency of facility management (FM) activities, such as locating components and retrieving relevant maintenance information for prompt failure detection or for emergency responses. However, due to the amount of equipment and components in commercial MEP systems, manually creating such logic chains is tedious and fallible work. This paper proposes an approach to generate the logic chains of MEP systems using building information models (BIMs) semi-automatically. The approach consists of three steps: (1) the parametric and nonparametric spatial topological analysis within MEP models to generate a connection table, (2) the transformation of MEP systems and custom information requirements to generate the pre-defined and user-defined identification rules, and (3) the logic chain completion of MEP model based on the graph data structure. The approach was applied to a real-world project, which substantiated that the approach was able to generate logic chains of 15 MEP systems with an average accuracy of over 80%.

Keywords: BIM; MEP; logic chain

---

1. Introduction

Mechanical, electrical, and plumbing (MEP) systems mainly include the heating, ventilation and air conditioning (HVAC), electric, electronic, plumbing, and anti-power subsystems. These systems are groups of components that, when connected, provide specific building services. MEP relationships are necessary to decrease the difficulty and time to locate components during operation and maintenance. In locating the component, the task can be decomposed into two pieces, i.e., topological analyses and logical inferences. The former focuses on the topological relationships of the MEP components, and the latter on their logic chain.

If equipment or terminals are topologically connected, they are always spatially linked to each other through the elements, such as pipe fittings or segments. However, the functional relationships among the topologically-connected components are not apparent. On the other hand, the logic chain shows the upstream and downstream connections for each MEP component, which builds up the functional relationships of the MEP system. The functional relationship is useful as the upstream components usually control the downstream ones. For example, the HVAC system consists of several equipment, and the fan certainly controls its downstream counterparts, such as ducts, elbows, etc. The typical logic relationship is shown in Figure 1.

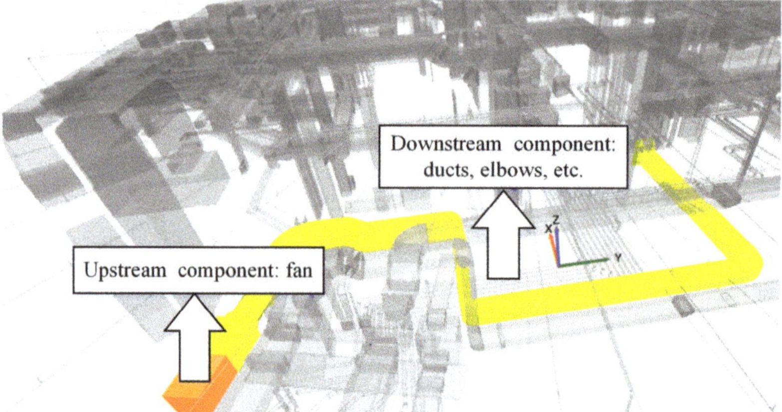

Figure 1. Typical logic chain in the subsystem of HVAC.

Setting-up the logic chain to show the upstream and downstream connections among MEP components is important for improving the efficiency of facility management (FM). For example, the logic chain can help locate the upstream valve of a leaking pipe in an emergency. However, on-site FM personnel currently rely on document-based blueprints or their experiences to locate such equipment. Consequently, some of the equipment cannot be located in time because of the complexity in the MEP network and the manpower limitation. Locating the equipment and retrieving its FM-related information are repetitive and time-consuming tasks for both technicians and facility managers.

Building information modeling/model (BIM) technology has been heralded as a facilitator to improve the FM efficiency by integrating the FM related information [1]. The application of BIM to FM is beneficial because it provides FM personnel with the opportunity to manipulate and utilize the information associated with the 3D objects and, thus, FM tasks can be conducted in a more scientific, reasonable, and orderly way [2]. These improvements are accrued during the generation and management of facilities' digital specifications and characteristic data [3]. Nevertheless, although the BIM technology enables data integration within the building lifecycle, incongruence between the supply of and demand for has also proved that insufficient valuable information is considered one of the key obstacles of BIM-based FM [4]. A survey demonstrates that more than 80% of the FM time is consumed in finding relevant information that is often disregarded by designers [5]. Enriched as-built BIM delivered to facility managers can save time for information acquisition, thus, it is suggested that the logic tree of the objects should be implemented to facilitate the management of various components within the MEP model [6]. However, manually setting up the logic chain of the MEP components requires tremendous workloads [7]. Hence, a unified logical information rule is necessary to build an accurate MEP logic tree, which benefits not only the operation and maintenance (O and M) management but also the emergency responses associated with the specific MEP component. Substantiated by the facility managers, making use of the MEP model containing the appropriate logic information could save up to 30% work force in locating the equipment and emergency decision-making compared with the traditional MEP database [8].

Existing modeling standards, however, do not include clear definition of the logic chain and, hence, existing modeling tools do not support quick generation of relevant logic information. Survey results also illustrate that, because of lacking relevant information, manual input can take up to two years to put MEP-related data into computerized maintenance management systems after the handover stage [4]. Worse still, associating components with correct data in various databases manually is an error-prone process. In addition, inadequate data integration is a frequently observed issue amongst building information modelers due to the differences in syntax, schema, or semantics [4]. The Industry

Foundation Classes (IFC) [9] standard, an open and object-oriented 3D vendor-neutral data file format (Volker 2011), supports data sharing and exchange between the construction and O and M phases of the building lifecycle [10]. However, the logic information within MEP systems is not explicitly considered in the widely used IFC standard even though spatial topology between components can be presented in property sets according to the IFC schema.

To solve these problems, this study proposes an automatic approach to generate the logic chain of the MEP system from BIMs with the help of pre-defined or user-defined identification rules. The rest of the paper is organized as follows. A literature review concerning both the spatial topological analysis and logical inference through domain query language is presented in Section 2. Afterwards, three key processes to automatically generate the MEP logic chain are discussed thoroughly in Section 3, including the parametric and nonparametric spatial topological analysis within MEP models, the pre-defined and user-defined identification rules of logic chain in MEP systems and the logic chain completion of MEP models. Section 4 contains a description of the application of the proposed approach to a real project. Finally, conclusion remarks are presented in Section 5.

2. Existing Studies

The application of BIM technology provides not only the convenience for MEP engineers to browse the 3D model of the specific MEP component, but also an integrated information platform to facilitate the management tasks. However, the generation of the MEP logic chain, which forms the essential foundation of MEP management during operation and maintenance period, is still a tedious and error-prone task even though BIM technology is widely adopted nowadays. This paper proposes an approach to automatically generate the MEP logic chains. The approach mainly relies on geometric information and query language. The relationship between the completion of logic chain and these two foundations is shown in Figure 2. Topological analysis through geometric information provides a path for the generation of the logic chain, because the upstream and downstream components are usually located in the same system and spatially connected to each other. A query language attached rule selector and rule engine can transform the domain knowledge into computer readable language for the completion of the MEP logic chain. Therefore, the literature review consists of two parts: spatial topological analysis and logical inference through a domain query language.

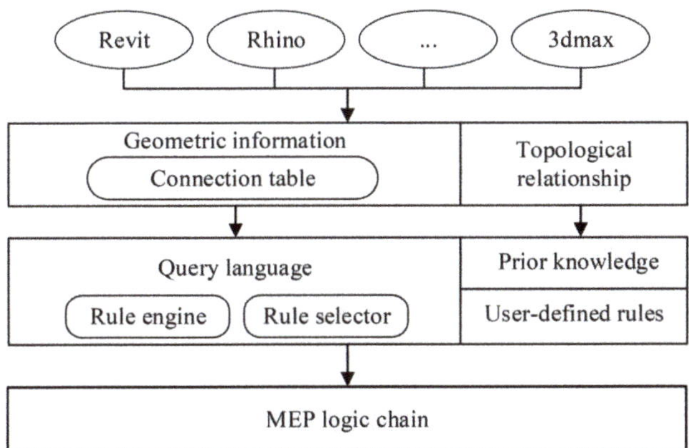

Figure 2. The foundations of automatically generating MEP logical relationships. MEP: Mechanical, Electrical, and Plumbing.

2.1. Spatial Topological Analysis

The spatial topology analysis relies on the geometric information of the MEP model. IFC can describe spatial topology objects, such as ducts and pipelines, as well as abstract concepts, such as spaces, organizations, relations, and processes [11]. For BIM applications in the O and M phase an extension of IFC that includes FM related information and repair information of MEP assets [12–14]. has been established. IFC-based BIMs can store various types of geometric data. Conventionally, the 3D object is identified as either a surface model or a solid model in the BIM. The solid model can further be divided into swept solid and CSG (constructive solid geometry) representations. Apart from geometric property of objects, 3D models need spatial relations (or topological relations) of components to facilitate complicated analysis and decision-making, e.g., building object classification [15]. Existing BIMs can be directly used to estimate spatial relations between 3D objects for spatial queries or further analysis. Some algorithms [16] have been proposed to automatically suggest topological relationships of building components from 3D CAD (Computer Aided Design) models. Generally, existing analytical BIM studies focus on information extracted from completed 3D models [17]. Particularly, based on the representation of 3D objects, the following topological relations were computed: adjacency, separation, containment, intersection, and connectivity. However, they are still limited in the cases of simple building configurations. Topological [18] and directional [19] relations can also be extracted from an IFC-represented BIM. Other proposed method can estimate topological relations for spatial queries based on boundary representations (B-rep) of BIMs [20]. B-rep is the most common surface-based representation, in which a shape is described by a set of surface elements (its surface boundary) along with connectivity information to describe the topological relationships between the elements. Previous works mainly focus on the architecture elements, such as the space, wall, and so on. However, most MEP components are represented in line structures such as air ducts and pipelines, which means topological analysis can be optimized with its specific swept solid representation. Different from the aforementioned work, this paper explores methods to complete the spatial relations (especially connections in MEP system) relying on both B-rep and swept solid represented models to improve the efficiency and accuracy of topological information generation.

2.2. Logical Inference through a Domain Query Language

On the other hand, domain query languages that specify the lexicon and syntax to generate correct topological information statements [21] have been used in BIM for the integrated object query based on attribute value conditions, including shape [22]. A domain-specific language for BIM can provide a more direct access to the required data, as it is tailored to the domain model and it hides the details of the internal data [23]. MEP knowledge is always presented in nature language and, thus, it must be transformed into a query language to become potential for automatically generating the logic chain. Sensors, a kind of MEP equipment, can optimize retrieval and analysis methods according to the characteristics of monitoring information collected by them, such as the types of monitoring information (humidity, temperature, etc.), monitoring time, monitoring range, etc. Some scholars [24] utilized a logic-based modeling language TCOZ (Timed Communicating Object Z) to describe constraints of the sensor imposed by its environmental conditions or relations with other sensors. Integration of spatial operations into standardized structured query language (SQL) queries makes the BIM data much more visible for wide ranges of query capabilities and decision-making processes [25]. Apart from the query language, some visual programming languages, including QL4BIM and VCCL, represent a new, more intuitive mechanism for data retrieval [26]. During the logic chain generation process, the prior knowledge provides relevant information that helps to guide and accelerate the processes. Specifically, the pieces of equipment and valves in MEP system have a relationship with neighboring components. In the cases of industrial plants, such information can be derived from the piping and instrumentation diagrams (P and IDs). These diagrams, which are the overall documents used to define the industrial process, contain logic information about their functional interrelationships [27]. There has been a growing consensus as to how prior knowledge can guide and facilitate the process of reconstruction of an existing large-scale facility. Prior knowledge is also used for 3D reconstruction of as-built industrial instrumentation models from laser-scan data [28]. Incorporation of prior knowledge into the reconstruction process provides supports for the automated procedures instead of the manual input [29]. However, the above-mentioned studies related to prior knowledge are not considered in the design and facility management of public buildings, therefore there are still some challenges for generate MEP logic tree during O and M phases. To automatically generate the MEP logic chain, this paper defines some descriptions to specify the prior knowledge of MEP systems in public buildings.

Current BIM adoptions in MEP engineering have been concentrated on modeling [30,31], measurement [32], collision detection [33] or collaborative design [34]. Several studies focused on MEP logic chains and the implementation of BIMs for facilitating FM tasks. Some prototype systems [35] have been developed to improve FM based on BIM and prior knowledge, including the reconstruction of 3D models from laser-scan data and a 3D CAD database in the O and M phase [24]. A multi-scale BIM management system was also developed for FM of large public buildings, containing the upstream and downstream relationship management of MEP components [31] and a method of generating schematic diagrams from 3D models of MEP systems was proposed. A cross-platform O and M management system was developed using the MEP-related information in the as-built model to run routine O and M tasks and to effectively response to MEP-related emergencies. Another research [8] studied logic relation retrieval for MEP systems based on the Revit database. With regard to the data mining applications of MEP systems, an object-oriented data capture approach was also developed for MEP information models [36].

2.3. Discussion

Figure 2 summarizes and compares some closely related studies on relationship information establishment with the proposed approach. According to Figure 2, some methods were proposed for MEP logical relationships generation by 3D point clouds, but not IFC files or 3D CAD models. Some other studies focused on typological analysis with BIM or CAD model of the AEC industry, but they did not consider the specific characteristics of MEP components and MEP knowledge, which contributed to the logical relationships of MEP model. These approaches attempted to explore the intrinsic value of MEP components and to represent this information in a more rational and user-friendly format, which could be beneficial to a variety of FM practices. Although studies shown in Table 1 introduce the attempts of generating relationship information, the analysis of spatial topology and use of prior knowledge still requires a more intelligent and automatic approach. Otherwise, the generation of the logic chain for MEP systems is a repetitive, time-consuming and error-prone task for either repair technicians or equipment managers. In this study, a BIM-based topological analysis method, a query language identification and a MEP logical query engine within MEP systems are proposed to generate the logic chain automatically to improve the efficiency of FM. In the proposed approach, the connection tables are the results from spatial topology analysis, which provide the foundation for the generation of the logic chain; and identification rules are extracted from prior practical knowledge or user-defined rules, providing the foundation for logical inferences. The connection tables and MEP knowledge are two foundations of the MEP logical query engine for generating the MEP logic chains automatically.

Table 1. Comparison of methods for generating MEP relationships.

Class	Relationship Modeling		Relationship Completion Mode		Platform		Geometric Data Source		
Sub-Class	Spatial	Logic	Manual	Automatic	Business Software	Self-Developed Engine	IFC Files	3D CAD Model	3D Point Clouds from Laser-Scan Data
Son et al. [28]	•	•	•	•	•	•			•
Hu et al. [7,35]		•	•			•		•	
Nguyena et al. [16]	•			•	•			•	
Borrmann et al. [19]	•			•		•	•	•	
Chen et al. [8]		•	•		•			•	
Proposed approach	•	•		•		•	•	•	

3. An Approach to Generate the MEP Logic Chain from BIMs

This section proposes a three-step approach to generate the MEP logic chain as shown in Figure 3.

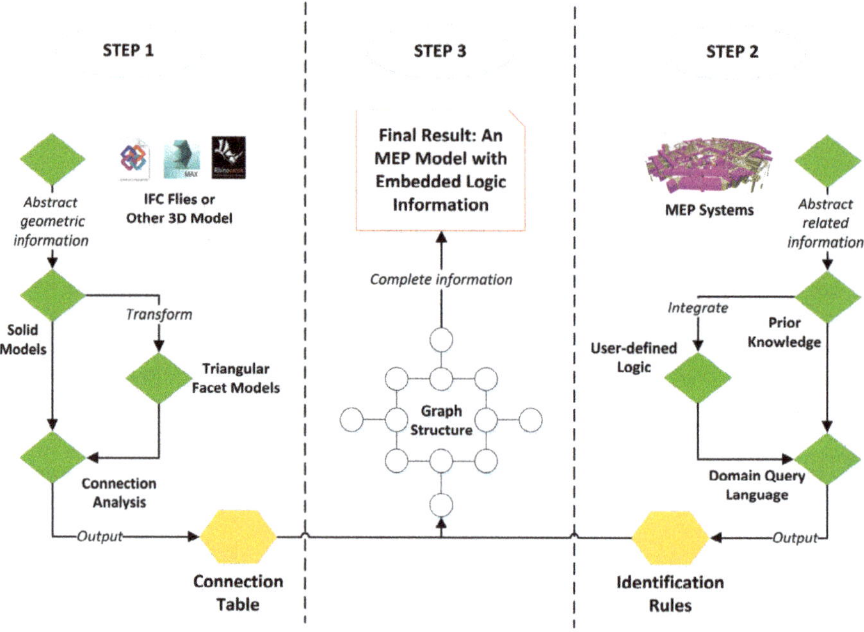

Figure 3. Flow diagram of the automatic completion mechanism.

The basis of the approach is that the components spatially connected to each other are usually located in the same system. Within this system, it can be further determined that whether they are at the same logic level or closely connected as upstream or downstream components, through prior knowledge or user-defined identification rules. The following three processes utilize the MEP information and some pre-defined knowledge to identify the logic relations among various MEP components.

The first step discussed in Section 3.1 analyses the geometric information of MEP components stored in either IFC or other 3D formats. If the information contained in the existing model is sufficient to assess spatial relations, the relevant parameters in the IFC file are extracted for analysis. If the solid model does not contain the required spatial information, B-rep representations could also be used for topological analysis. B-rep representations will be transformed into triangular facet models for spatial triangle intersection judgments. These two methods will both output a connection table to be used in the following steps. A connection table here refers to a table structure that stores the spatial relation in forms of MEP component pair records indicating the spatial connections in corresponding subsystems.

The second step discussed in Section 3.2 involves the transformation of typical MEP systems and custom information requirements to generate the identification rules. To generate the MEP logic chain, engineers must figure out some logic identification rules to specify which kind of components is upstream or downstream of a specific component. These rules are further divided into two parts. One part is based on the prior experience knowledge, and the other is based on the user-defined logic. The former extracts the upstream and downstream relationships from typical compositions of various MEP systems as a template, while the latter integrates user-defined rules of logic relations according to the characteristics of MEP components as the extended rules for automatic generation. The above two rules utilize a query language to represent their information requirements in formal statements, which are then parsed in the third step.

The third step discussed in Section 3.3 is the retrieval of MEP logics, and completion of the logic chains. The upstream component of a given MEP subsystem will first be determined by identification rules. After creating a graphic structure based on the upstream terminal component and connection table, automatic completion algorithms will perform a traversal searching for a component that meets the requirements of inspections and add upstream or downstream components as extended attributes to represent their logics, an algorithm for selective traversal which uses a tag for each node in the MEP logic tree. [37]

3.1. Spatial Topological Analysis within MEP Models

Different design tasks require different types of topological information. For example, architects pay more attention to the space of the building, whereas MEP engineers focus on the connections between pipelines and air ducts. Due to the lacking logic chain among components in most models delivered by contractors, the upstream or downstream connections of a given component within the MEP system cannot be extracted directly for FM tasks. Therefore, the geometric information of MEP components in BIMs is extracted for analysis to obtain the connectivity between MEP components as the basis for subsequent analyses. There are two kinds of geometric representations, i.e., parametric and nonparametric. Parametric representations describe a shape using a model with pre-defined parameters. For example, a cylinder as a swept solid may be represented by its radius, axis, and the start and end points. In addition, a cylinder can also be represented in nonparametric form, such as in triangles-based B-rep format. Accordingly, spatial topological analysis would be divided into two parts as show in Figure 4, i.e., the IFC-based parametric topological analysis and nonparametric topological analysis.

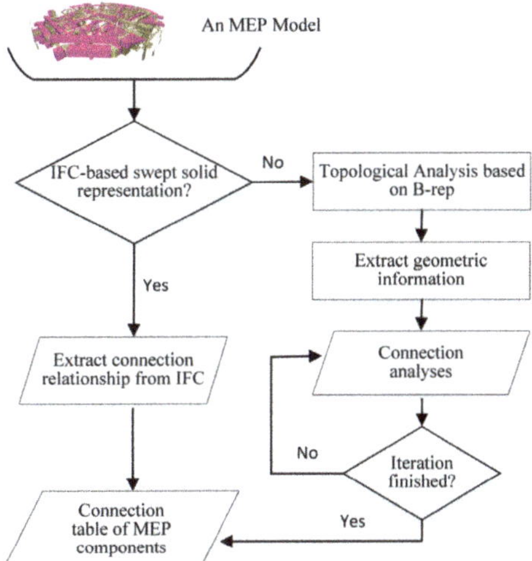

Figure 4. BIM-based topological analysis within MEP systems.

3.1.1. The IFC-Based Parametric Topological Analysis Algorithm

In the IFC schema, object entities are subtypes of the abstract entity IfcObject. Objects that appear in a spatial or geometric context are described by abstract entities of type IfcProduct. As a subtype of IfcProduct, MEP products are described by the abstract entity IfcElement, which can be connected to other IFC elements. Then the objectified relationship entity IfcRelConnectsElements is used to describe connections between elements. Specific subtypes of IfcElement entities are used to further distinguish the semantic meaning of a MEP product, including specific object entities to describe HVAC elements, plumbing elements, etc. In some well-developed BIMs, the connection information can be extracted directly by IfcDistributionPort, IfcDistributionElement and other tools. As shown in Figure 5, according to IFC schema, MEP components are represented by IfcDistributionElement and its subclasses such as IFCFan and IFCPipeFitting, etc. The connection relationship between these components is represented through IfcDistributionPort. According to the port between the component and connection form, IfcDistributionPort connects to the equipment by IfcRelConnectsPortToElement. The connection relationship between IfcDistributionPort is defined by IfcRelConnectsPorts, and the direction of the upstream and downstream relationship is defined through the two reverse attributes (ConnectedFrom and ConnectedTo) of IfcDistributionPort.

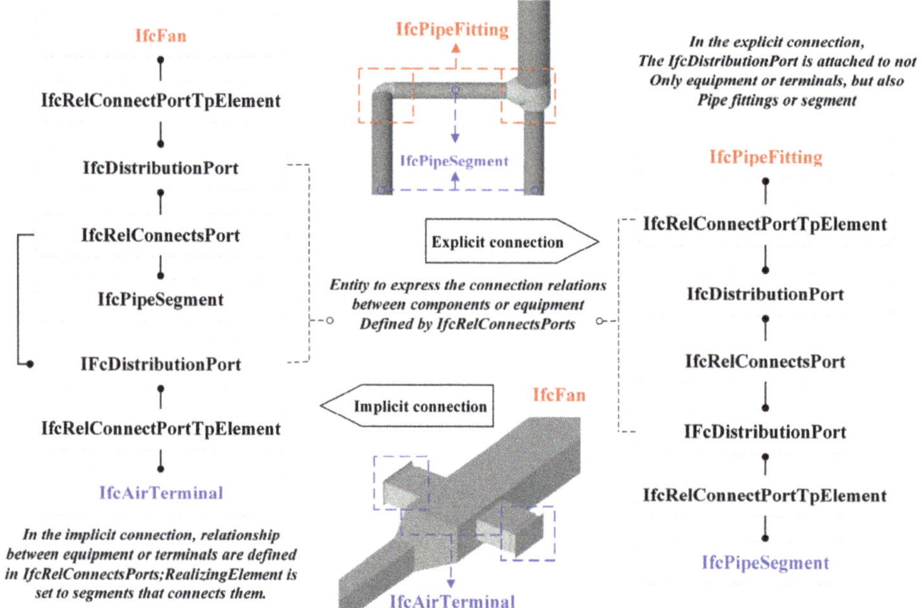

Figure 5. Expression of MEP connection information in IFC.

There are two descriptions of connective information in MEP subsystems, implicit connection and explicit connection. For implicit connection, IfcDistributionPort only relies on the MEP equipment or terminal component. If these two equipment or terminals are connected, they are defined in IfcRelConnectsPorts and then the RealizingElement attribute value of the relationship is set to pipe fittings or segments that connect them while, in the explicit connection, the IfcDistributionPort is attached to not only equipment or terminals, but also pipe fittings or segments. The equipment and terminals must be connected via the IfcRelConnectsPorts with pipe fittings or segments that connect them, ignoring the attribute value of the RealizingElement.

Furthermore, geometric information can also be used for topological analysis even if connection information is ignored in BIM model. For example, a pipe is parametrically represented in diverse BIM tools, as a "SweptSolid", IFC model defines the cross section, the starting position of the extrusion, the direction of the extrusion and its length. The pipe's Cartesian coordinate value will also be illustrated in the local coordinate system. Using this information, two pipes can be assessed as connected if the cross-section parameters of the starting point of one component is approximately equal to the ones at the end point of another component.

3.1.2. The Nonparametric Topological Analysis Algorithm

There are many other nonparametric data formats available for MEP modeling, such as triangle-based B-rep representations. For example, if these objects are imported through rendering-oriented software such as 3ds Max, or shape-oriented design tools, such as CATIA or Sketch Up. Moreover, some BIM platforms focus on efficient 3D display to help real-time management, and they will transform the parametric models to triangles to enhance the OpenGL or DirectX render efficiencies. Therefore, it is necessary to propose the nonparametric topological analysis algorithm.

Generally, pipe fittings, equipment, and short pipes have different kinds of connections between each other. The connections between segments are usually in one-to-one form, while for fittings and segments are usually in one-to-many form. For example, a reducing tee connects three pipes. Considering the magnitude of errors buried in MEP modeling, a deviation threshold, i.e., 10 mm between outlets of MEP segments, fittings or equipment can be determined as touching. The algorithm to determine the relation between components is depicted as follows.

The test can return the result of spatial analysis of triangles, which distinguish between touching and intersecting triangles. If a pair of triangles from A and B intersect, this also indicates an intersection of the two interiors of the objects under examination. It is possible that A covers B, or vice versa. A and B can, nevertheless, still touch each other. To disprove the inside relationships, the algorithm is realized by means of ray tests. Each vertex of a triangle is used as a starting-point for rays. The rays are emitted from the selected vertex and the number of intersections with the B is counted. An even intersection number indicates that the triangle is located outside, whereas an odd number indicates that the triangle is inside another.

After iterating all components and spaces, the connection tables of the MEP components and the space are ready for the next step. The connection tables can also be used to record the connection components or equipment. Each connection pair indicates that the two components are connected in space, the connection type and the name of components. For example, DS-SS- <E1, E2> indicates that entity E1 is connected to entity E2, both are segments in the duct system. Other typical connection types include FS, ES, which stand for fittings or equipment connected to segment, respectively. Moreover, different types can then be used for analysis of upstream or downstream relations.

3.2. Identification Rules of Logic Chain in MEP Systems

The identification rules refer to domain query rules for identifying the logic relations of MEP components with attribute conditions. As mentioned above, except for the geometric relations among MEP components, specific rules are also needed for identifying which component is the upstream or downstream of the specific component to generate the logic chain. Generally speaking, the methods to establish such rules are extracted from MEP knowledge. This section discusses two kinds of such rules, i.e., the typical prior knowledge rule and the user-defined logic rule. Moreover, a domain-specific query language with pre-defined lexicon and syntax is represented in a formal way to figure out the relationship between upstream and downstream components.

3.2.1. The Typical Prior Knowledge Rule

In each MEP system, countless components formulate a logic web with intertwined relationships. Taking HVAC system as an example, it consists of several circles with different functions. As shown in Figure 6, in most cases, an HVAC system with a chilled water air-handling unit is a common design for general commercial buildings or office buildings. In such a design, the supply air of each floor comes from its engine room, where the air-handling unit consists of cooling coils, filters, and supply fans. The chilled water in the frozen coil is provided by the central chilled water plant in the building and releases heat by cooling tower heat rejection. The cold air produced during refrigeration is circulated to each room by cooling water. The entire system consists of relevant equipment and various types of pipelines and valves. Chillers and air-handling units can be viewed as a whole ignoring their internal logic. Considering the practical requirements of FM and the characteristics of existing HVAC control systems, the upstream and downstream logics between ducts or pipes play important roles in emergency management. For air circulation, the fan can be taken as the ultimate upstream terminal and its downstream components involve the damper, duct and air terminal box. Similarly, an evaporator cooler can be regarded as the upstream terminal component of the water circulation, and its downstream components involve the water pump, valve, and pipe.

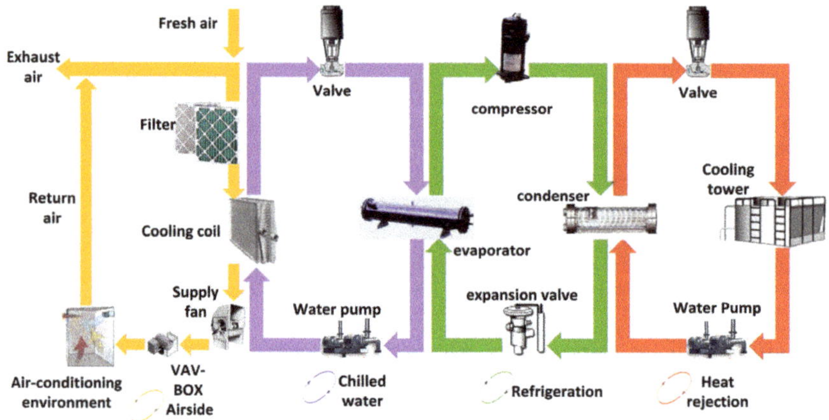

Figure 6. Four circles of HVAC systems.

Domain schemas of IFC files provide entity classification in different MEP subsystems. This research firstly extracted the logic chain of IfcElement types corresponding to several common MEP subsystems from the knowledge accumulated along with field engineering practices, literatures and manuals of each MEP subsystem. These typical logic chains are summarized in Figure 7.

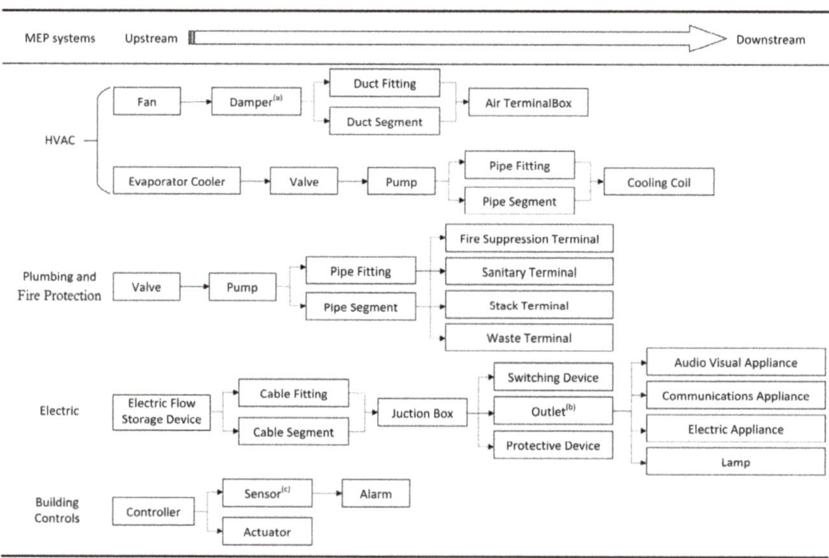

Note: (a) The type of Damper includes balancing damper, control damper, fire damper, smoke damper and gravity damper; (b) The type of Outlet includes audio visual outlet, communications outlet, power outlet, data outlet and telephone outlet; (c) The type of Sensor includes fire sensor, flow sensor, heat sensor, moisture sensor, etc.

Figure 7. Typical logic relationships of Element types within several MEP subsystems.

Take HVAC system as an example, proportional control is one of the control forms in HVAC systems. The valve or damper is positioned in intermediate positions in proportion to the response to slight changes in the controlled condition. A modulating valve controls the amount of chilled water entering a coil so that cool supply air is just sufficient to match the load at a desired setpoint. If the temperature goes beyond the setpoint, the on- and off-times change according to the temperature differences. Thus, according to the control knowledge, if a valve is spatially connected to a cooling coil in HVAC systems, the former one can be determined as the upstream component of latter components. Moreover, a valve entity can be the upstream component of ether pipe fitting or pipe segment which connects them. Particularly, pipe segments connected to each other can be regarded in the same logic level, which means they have the same upstream and downstream components. According to the tabulated rules, the MEP models with correct entity classifications are easy to be placed in the logic chain in most systems.

3.2.2. The User-Defined Logic Rule

According to different MEP design or FM requirements, pre-defined rules stipulated according to prior experience do not always work. Therefore, this approach proposes a user-defined logic rule as a supplementary to pre-defined rules. To define those rules, a domain query language with lexicon and grammar of their statements should be introduced first. The query, for example, may ask for the position of the valve on the supply water pipe connected to the reheat coil in the terminal box [38,39]. The contextual information items used in the query language for HVAC systems can be categorized into three general groups: entities, properties, and relationships. The relationship between entities is simplified as follows: A is the upstream component of B, A is the downstream component of B, or NULL (means without any relationship).

Each entity specifies a type of object that has a set of instances in MEP systems. Hence, in a rule statement, the name of an entity represents all instances with the same name. Each entity contains several property sets, each of which contains a number of properties, such as the component name, ID, or spatial geometry information. This information provides lexicon of upstream and downstream logic rules. Numbers or strings can typically represent the values of these properties. The predicates of these values shown in Table 2 were identified according to SPARQL.

Table 2. Predicates reused from SPARQL for the inspection rules.

Predicate	Syntax	Description
=	v1 = v2	True if value v1 equals to v2
Contains	v1 contains v2	true if value v2 is a substring of value v1
Connects	v1 connects v2	true if value v2 is topologically connected to value v1

Each property of an entity instance has a set of values that describe the instance. The values can be individual or an array of either a primitive, such as number, character and Boolean, or instances of other entities. Since the values of properties describe the characteristics of the instances, they can be used to specify the target instances. For example, to retrieve the instances of all temperature sensors in the data model, the users can specify the query statements to identify the instances of sensors that have their property "MeasurementType" with value equaling to "Temperature", such as (Sensor. MeasurementType, =, "Temperature").

Although the actual relationships between components can be highly complex, the relationship in the user-defined identification rules considered by the algorithm is either "UpstreamOf" or "DownstreamOf". The upstream and downstream relationships are not interconnected in space, unlike the functional relationships. For example, the valve as a logic upstream component controls its downstream components, such as pipes, and the axial fan as an upstream component will affect the working state of its downstream components, including outlets.

The syntax of a query language specifies the grammar for the logic chain statements. It provides guidelines about how the statements should be parsed and interpreted. After defining the information items in a specific logic chain, the syntax of object-oriented query language rules is extended to express the entity, property and upstream and downstream logic relations. Table 3 provides examples for each definition.

Table 3. Description and examples of the syntax.

Definition	Syntax	Example	Explanations
1	E	Valve	E represents an entity type that defines the scope of an entity to which an inspection rule is applied.
2	$E_U.P$	Valve.Name	E_U represents an inspection rule, and P is an attribute of the relevant component; then, $R_{UE}.P$ is the specified attribute value of the upstream component examined by the inspection rule.
3	$R_{UE} = (E_U.P, \theta, v)$	(Valve.Name, contains, "XX")	E_U represents an inspection rule, P is an attribute of the relevant component, θ is a predicate, and v is the value; then, $R_{UE} = (R_{UE}.P, \theta, v)$ is an inspection rule for all instances in UE that have values for property P that satisfy the predication.
4	(R_{UE}, R, R_{DE})	((Valve.Name, contains, "XX"), UpstreamOf, (Pipe.Name, contains, "XX"))	If R_{UE} and R_{DE} are two queries for instances and R provides a relationship (UpstreamOf or DownstreamOf) of these two groups of instances, then (R_{UE}, R, R_{DE}) is a complete inspection rule for the generated part of the MEP logic chain.

3.3. Logic Chain Completion of MEP Models

Following the steps articulated above, the connection tables and identification rules can be obtained. The connection tables extracted from geometric data provide topological relationships between components. Each component is considered as a node and an edge is then developed between two nodes if their represented components are spatially connected. In addition to the topological relationship, attributes of specific node in different types of systems contribute to generate the logic chain. Identification rules extracted from prior knowledge or user-defined rules are then transformed into computer readable language. The logic query engine retrieves attributes' information according to the identification rules for establishing the logic chain. An example of an HVAC subsystem is discussed below reveals the mechanism of the MEP logical query engine, as shown in Figure 8.

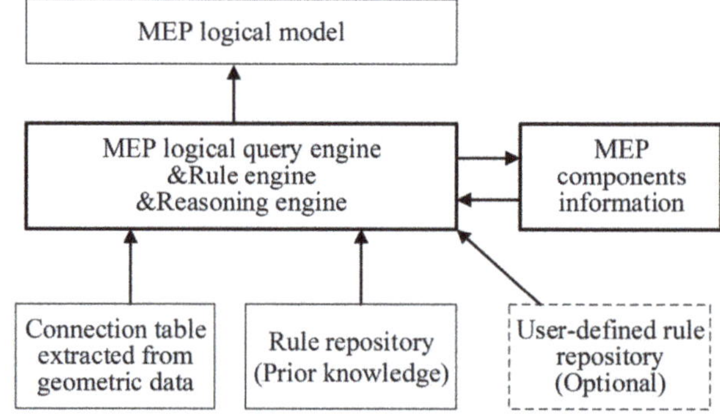

Figure 8. The workflow of logical relationships completion of MEP model.

3.3.1. Graph Structure of the MEP Topology

When modeling MEP systems by Revit or other BIM software, the component usually contains extended properties including system type, system classification, and system name. Taking an ordinary duct as an example, these properties can be specified as "air duct system", "supply air", and "supply air 8". Accordingly, this approach firstly parses the model (which can be exported in IFC format) to obtain the hierarchy relationships between MEP systems. The first-level node refers to the system type such as duct or pipe systems pre-defined in Revit. The second-level node refers to the system classification, such as supply air, return air and exhaust air in the duct system. The third-level node refers to the system name that can be defined by users. Some other attributes extracted from MEP model are shown in Table 4.

The list of attributes is assigned to each node. This list acts as the semantic feature of the system topology. It makes the element do have a "meaning" inside the system.

In a subsystem, the proposed approach determines the upstream terminal components by analyzing the identification rules. Considering that there may be some circles within MEP systems, this study selects the graph structure and uses the adjacency list as the data structure. Comparing with the search in the tree structure, graph search has an advantage that whenever the algorithm explores a new node, it marks it as visited. In this study, the node can be marked as an upstream component or downstream component when being visiting.

Taking an HVAC loop subsystem as an example, the graph in Figure 9b shows a simplified supply air system with 13 components, where E1 is the air handing unit, E8, E10, and E13 are VAVs, and the remainders are ducts. After performing the spatial topology analysis discussed in step 1, the connection table is shown in Figure 9a. Each component pair indicates that the two components are spatially connected, where <E1, E2> indicates that air handing units E1 and duct E2 are connected. If entity E1 is selected manually or determined automatically to be the upstream terminal component by identification rules, the algorithm can generate a directed graph as shown in Figure 9b, and E1 will be selected as the start node for the generation of the logical relationships. The completed inverse adjacent list is shown in Figure 9c. Each list describes the set of input nodes of a vertex in the graph.

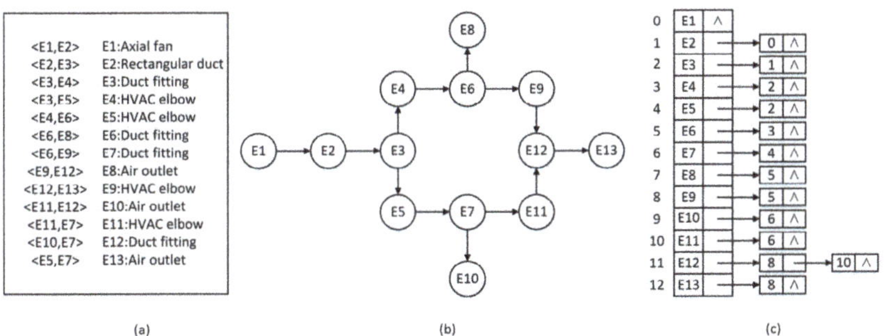

Figure 9. Data structure of MEP subsystem's components. (**a**) Connection table for illustrating topological relationships. (**b**) Graph structure of the MEP subsystem. (**c**) Inverse adjacent list of MEP components.

Table 4. Example of the attributes list.

Attributes	Types	Brief Description
Element Id	Number	Unique Id in BIM model for every element.
Element Location	Coordinate Point	The element location point.
Space Located	String	The spaces of this element located.
System classification	String	The system classification includes supply air, return air, etc.
System name	String	The name of system which elements belong to, such as supply air, return air, etc.
Flow	Number	Flow in and out of a diffuser or duct.
Start Point	Coordinate Point	Start point of a duct. For duct only.
End Point	Coordinate Point	End point of a duct. For duct only.

3.3.2. MEP Logical Query Engine

After generating a directed graph of the MEP subsystem, the breadth search for the directed graph will be performed with each identification rule to append the MEP logic information. The semantic analysis of identification rules will output the requirements of the upstream and downstream components, including the entities, properties, relationships and operations. The search sequence of Figure 9 Note (b) is E1, E2, E3, E4, E5, E6, E7, E8, E9, E10, E11, E12, and E13. For each component meeting the downstream component requirements, the first satisfying upstream component of the same rules searched in the reverse direction of the graph will be defined as the upstream component, and the logic information will be added to the model. An enriched MEP information model with a logic chain can then be generated when the entire subsystem completes the traversal of all identification rules. In this example, the fan is the upstream component of the duct and the duct is the upstream component of the air terminal box. An inspection of the components according to the first rule indicates that E2, E3, E4, E5, E6, E1, E7, E9, E11, and E12 are the downstream components of E1. When considering the second rule, E8, E10, and E13 should be the downstream components of E6, E7, and E12, respectively.

3.3.3. The Enriched MEP Logic Chain Representation

The above-generated MEP logic information will finally be added in the BIM. After generating the attributes list, another issue concerned is how to show the logic information of duct and elements in HVAC systems for checking and tracing purpose. System topology with logic information is a way to present the MEP model in a simple manner such that the system architecture can be easily understood. A typical MEP services layout consists of air ducts, water pipes, equipment, cable containment, etc. Taking a typical layout plan for HVAC air-side services (see above half of Figure 10) as an example, it is too complicated for engineers to know how the system is designed and how it works.

Obtaining the upstream and downstream elements is inconvenient with complex HVAC lines in 2D drawings and requires professional skills. The large size and dual-line representation of the air duct and equipment result in such congested drawings. The bottom half of Figure 11 shows a system logic representation of an HVAC subsystem. The duct fittings and equipment are denoted by 3D models to identify their location and connectivity. When considering BIM integrated with logic information, the MEP models provide the upstream and downstream component list of selected nodes. In user interface, the logic management panel provides the upstream and downstream component list of selected nodes. The downstream list of axial fans includes several ducts, elbows, fire dampers and transitions; meanwhile the upstream component of the duct is the axial fan. There is only one upstream component for each MEP component except for the upstream terminal one. Thus, the axial fan has many downstream components but no upstream components, and the duct has an upstream component but no downstream component.

The connection relationship and MEP logic information will finally be added in the FM model which can be described by IFC. The IfcRelConnectsElements as an objectified one-to-one relationship provides the generalization of the connectivity between elements. The concept of two elements being spatially or logically connected could be described independently from the connecting elements. The spatial connectivity may contain related geometric representation of the connected entities. As shown in Figure 11, In this case the geometrical constraints of the connection are provided by the optional relationship to the IfcConnectionGeometry. The connection geometry is provided as a point, curve, or surface within the local placement coordinate systems of the connecting elements. If the logic connection is developed by the above algorithm based on connection relationship, the connection geometry can be replaced by logic information within the extended property set which conclude the extended properties of RelUpstreamElement and RelDownstreamElement.

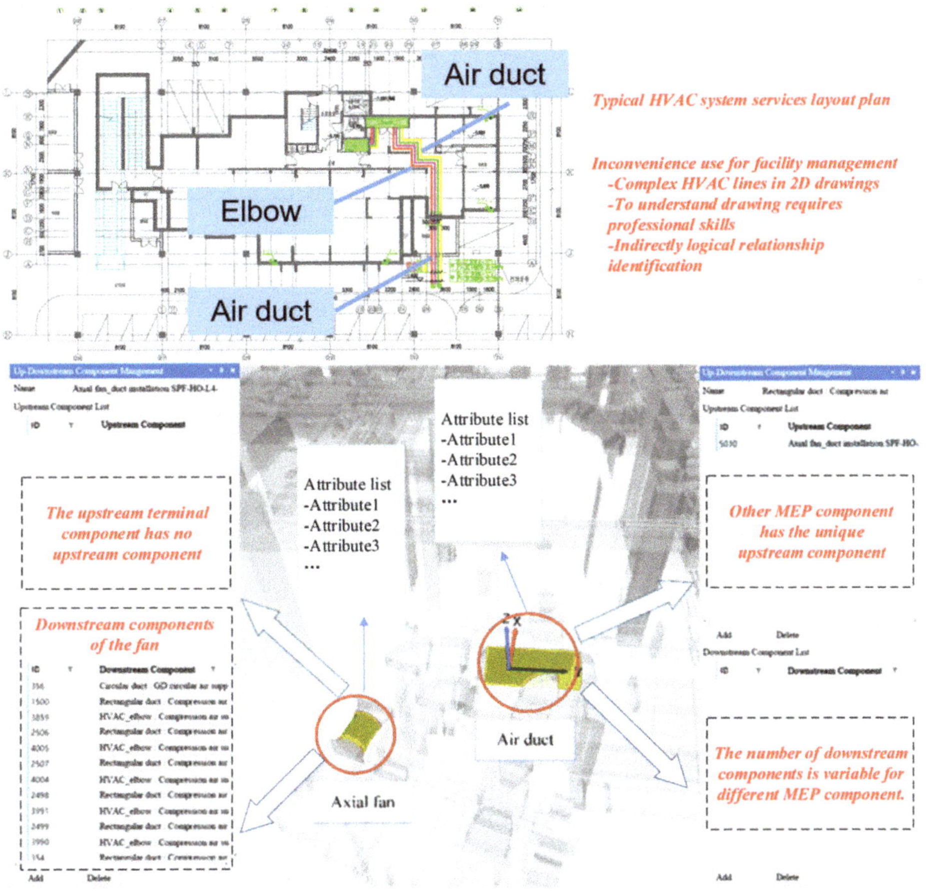

Figure 10. The HVAC system layouts.

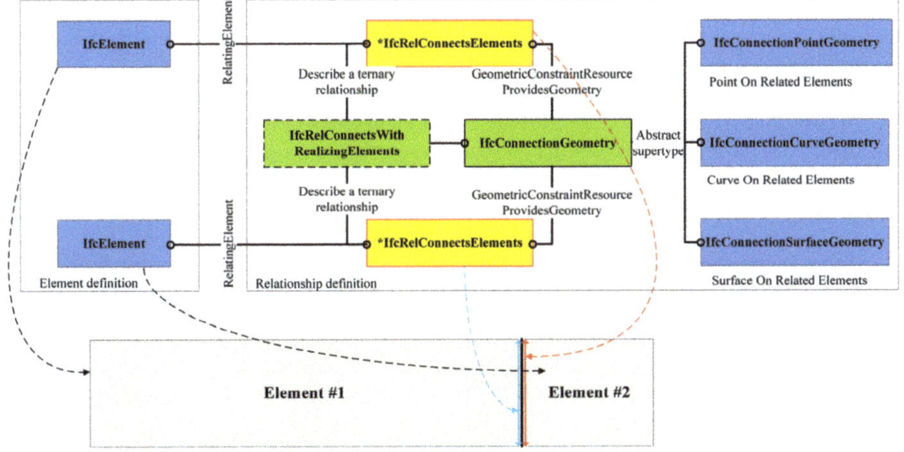

Figure 11. The EXPRESS-G representation of element connection in BIM.

4. A Case Study

4.1. Introduction to the Project

The China Resources headquarters building is around 400 m high with a construction area of 267,137 m^2. It has four floors below and 66 floors above ground. The MEP systems contain HVAC, electric, plumbing, anti-power, and anti-water. The structural and MEP models were created in Revit, and the MEP model was used as a case study (Figure 12). Related information of MEP components (e.g., suppliers, manufacturing date, code) were continuously input during the modeling and construction stage for O and M staff to complete the MEP logic chain among components based on the automatic completion approach proposed in this study. The principles of the upstream and downstream relations were as follows. In the following part, the duct system was illustrated in detail.

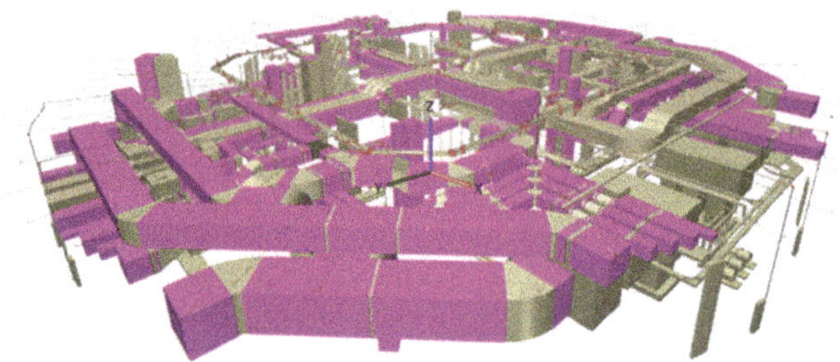

Figure 12. The MEP model for the 5th floor of the Chinese resource headquarters building.

4.2. Application Details of the Approach

Modeling of the MEP systems was performed mainly on Revit. In addition, other software such as 3dx Max was used for design and evaluation of the accuracy of the MEP model. One of the differences between them was that the geometric data of IFC files exported by Revit included axes information for spatial relationship analysis. After extracting the system type and system classification of MEP systems from IFC files, the supply air subsystem was selected to complete its logic chain. First, the approach determined the upstream terminal component in the subsystem by identification rules and users, as shown in Figure 13. The connection table was then automatically generated, and a set of components represented a spatially connected path by the same color. As shown in Figure 13, the fan is the upstream component of fire dampers, ducts, elbows, and transitions. Outlets are the downstream components of transitions. The axial fan is the upstream terminal component, as well as the start node of the directed graph traversal.

The method was also verified in the fire sprinkler system, where the riser is the upstream component of valves, pipes, transitions, and elbows, while sprinklers are the downstream components of transitions. The pipes include feed mains, cross mains, and branch lines. Most of the logic chain were generated based on prior knowledge while several user-defined rules, as shown in Table 5, were added as a supplement before retrieving the fire sprinkler system.

Table 5. Part of inspection rules of MEP subsystem.

No.	Inspection Rules of A Supply Air Subsystems
1	(Component.Name, contains, "Fan"), UpstreamOf, (Valve.Name, contains, "FAD")
2	(Component.Name, contains, "Fan"), UpstreamOf, (Duct.Name, contains, "duct")
3	(Component.Name, contains, "Fan"), UpstreamOf, (Duct.Name, contains, "elbow")
4	(Component.Name, contains, "Fan"), UpstreamOf, (Duct.Name, contains, "Transition")
5	(Duct.Name, contains, "Transition"), UpstreamOf, (Duct.Name, contains, "outlet")

Table 5. *Cont.*

No.	Inspection Rules of A Fire Sprinkler Subsystems
1	(Component.Name, contains, "Riser"), UpstreamOf, (Pipe.Name, contains, "Valve")
2	(Component.Name, contains, "Riser"), UpstreamOf, (Pipe.Name, contains, "pipe")
3	(Component.Name, contains, "Riser"), UpstreamOf, (Pipe.Name, contains, "elbow")
4	(Component.Name, contains, "Riser"), UpstreamOf, (Pipe.Name, contains, "Transition")
5	(Duct.Name, contains, "Transition"), UpstreamOf, (Pipe.Name, contains, "sprinkler")

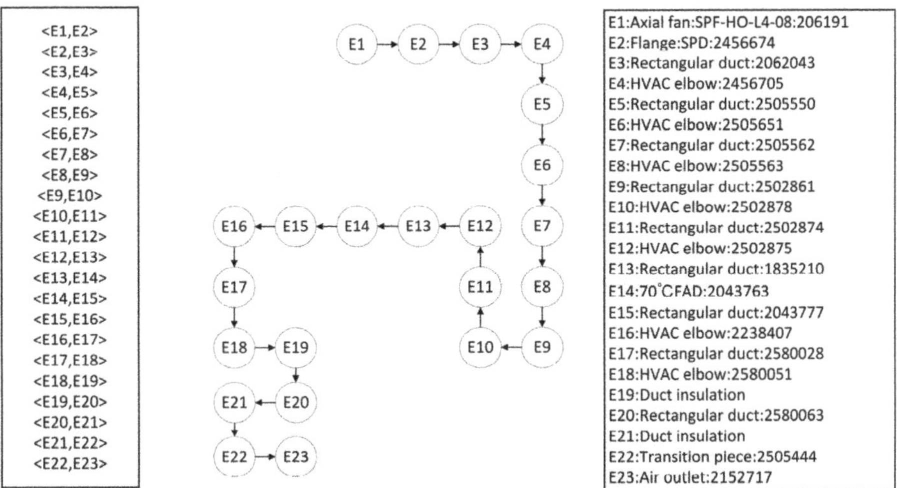

Figure 13. Data structure of a supply air system.

4.3. Evaluation and Discussion

A total of 15 subsystems in HVAC, including fresh air, air supply and exhaust air systems, and 10 subsystems in fire sprinkler systems, including wet pipe sprinkler systems and dry pipe sprinkler systems were described and discussed. Each subsystem contains 20–40 components, with a total number of 763 components in the HVAC and fire sprinkler system. Table 6 summarizes the number of omissions and errors that occurred. The statistical results demonstrate that there are one or more omissions in the process for generating the connection table in six subsystems and one error in the process for generating the logic model in four subsystems.

Table 6. Performance of the algorithms developed for generating MEP logic models.

Subsystem Name	Number of Components	Generate Connection Table Process		Generate Logic Model Process	
		Omissions (Pairs)	Accuracy (%)	Errors (Pairs)	Accuracy (%)
Return air 1	23	0	100	0	100
Return air 2	19	0	100	1	96.8
Return air 3	28	3	88.9	2	92.6
Return air	70	3	95.7	3	95.7
Supply air 1	32	1	96.8	0	100
Supply air 2	34	0	100	0	100
Supply air 3	41	0	100	0	100
Supply air 4	32	2	96.7	0	100
Supply air 5	23	0	100	0	100

Table 6. *Cont.*

Subsystem Name	Number of Components	Generate Connection Table Process		Generate Logic Model Process	
		Omissions (Pairs)	Accuracy (%)	Errors (Pairs)	Accuracy (%)
Supply air 6	26	0	100	0	100
Supply air	188	3	98.4	0	100
Intake air 1	21	1	95	0	100
Intake air 2	19	0	100	0	100
Intake air 3	29	5	82.1	2	92.9
Intake air 4	24	2	91.3	1	95.7
Intake air 5	32	0	100	0	100
Intake air 6	35	0	100	0	100
Intake air	160	8	95	3	98.1
Wet pipe 1	23	0	100	0	100
Wet pipe 2	28	0	100	0	100
Wet pipe 3	40	1	97.5	2	100
Wet pipe 4	34	1	97.1	1	100
Wet pipe 5	23	0	100	0	100
Wet pipe 6	30	0	100	0	100
Wet pipe	178	2	98.9	3	98.3
Dry pipe 1	20	0	100	0	100
Dry pipe 2	28	0	100	0	100
Dry pipe 3	14	0	100	2	85.7
Dry pipe 4	33	1	97.0	0	100
Dry pipe 5	28	0	100	0	100
Dry pipe 6	44	2	95.5	2	95.5
Dry pipe	167	3	98.2	0	100

From this case study, the following conclusions can be drawn.

1. In 15 subsystems, both the connection table and logic model were generated with 100% accuracy, demonstrating that the approach could generate the right logic chain.

2. There were ten subsystems with 1–5 omissions when generating the connection tables, and the accuracy rates remained higher than 80%. The spatial connection was misjudged for the following reasons in these cases:

    i. There were deviations in the position of the component during modeling. When the model was observed under magnification, there was no connection between components that were, in fact, connected to each other in practical engineering.

    ii. A portion of the information was missing or incorrect in the modeling process. System classification and system name attributes were missing or missing in certain subsystem components' properties, which influenced the inspection of components.

    iii. The intersection judgment of space triangles was not sufficiently accurate. Because the triangular size was small, improper handling of truncation errors for intermediate calculations considerably affected on the results. Therefore, the algorithm should be improved.

Among the three types of problems, missing or incorrect information during modeling occurred most frequently, followed by misjudgments of triangle intersections and then deviations in the component positions during modeling. During the modeling process, information about the system classification and system name were generally inputted in commercial software. Designers or modelers were likely to ignore the accuracy of relevant property information, which, when combined with the lack of relevant modeling standards, led to the problem of missing or incorrect information. If two components of a subsystem were spatially connected to each other but the model assessed the connection incorrectly, the two components were distinguished as shown as Figure 14. In addition, the connection table could be manually added with two adjacent components to correct such omissions so that a complete connected path could be obtained.

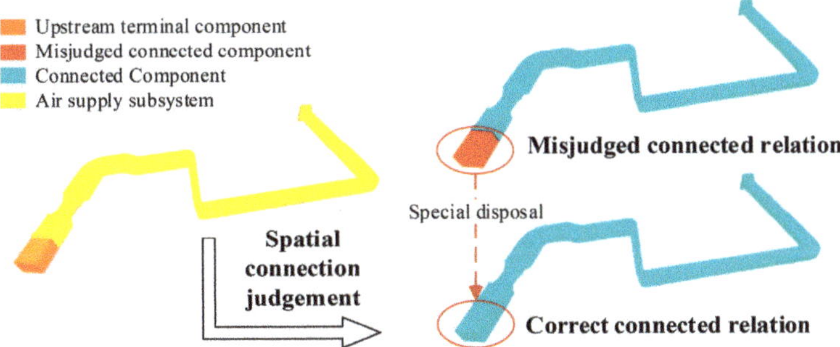

Figure 14. Results of spatial connection judgement within an MEP subsystem.

3.   There were four subsystems with less than two errors when generating logic models and the accuracy rates remained higher than 90%. An analysis of the results demonstrated that the values of properties were ambiguous in terms of their semantic meanings. For example, the fire damper was one of the valves in the HVAC system. However, if users used "valve" to represent the damper instead of using "damper", the algorithm was unable to match the valve description with the fire damper component. Such problems were corrected by adding and deleting upstream and downstream components manually.

The results also indicated that this approach could achieve the automatic completion of logic chain. As discussed above, one limitation of this approach is because there was no unified modeling standard for MEP systems, the omission of certain necessary data in the model would deduce the accuracy.

5. Conclusions and Future Work

Implementations of BIM technology can provide visualization and integrated information models for FM. However, the logic relationships between MEP components are often ignored not only during the modeling process but also in the widely adopted IFC standard. Considering that manual generation of the logic chain is time and labor consuming, this study proposes an approach to generate the logic chain for MEP systems from BIM with identification rules.

The approach consists of three steps which focus on the geometric information of MEP models, identification rules for logic inferences and the data structure of MEP components, respectively. The assessment of IFC-based spatial relationships is mainly based on 3D solid models or triangular facet models. The identification rules of MEP systems contain typical prior knowledge and user-defined logic rules. After determining the connection table, classification rules, and the terminal component, the graph data structure can be generated to represent the relationships between MEP components. Finally, the logic is generated by performing traversal searching for each connected pair of components with an appropriate identification rule.

The feasibility and accuracy of the approach were tested in an MEP model of an actual project. The results indicated that the approach was capable of generating logic chains for different subsystems and the accuracy of generating both connection tables and logic model exceeded 80% in 15 subsystems. Some problems, such as mismatching, occurred during the tests while they were discussed to be solved by adding connections or logic information manually.

Future works will focus on improving the accuracy of the approach and exploring efficient control methods of energy conservation based on the MEP logic chain and related control theories of MEP systems. Based on the experiences gained from this application, more service-oriented functions can be developed for FM depending on a thorough exploration and a comprehensive application of BIM and related information technologies.

Author Contributions: Conceptualization: S.-W.L. and Z.-Z.H.; data processing: Y.-Q.X.; funding acquisition: Z.-Z.H.; methodology: Y.-Q.X.; software: Y.-Q.X.; supervision: S.-W.L. and Z.-Z.H.; validation: Y.-Q.X. and Z.-Z.H.; visualization: Y.-Q.X.; writing—original draft: Y.-Q.X.; writing—review and editing: Z.-Z.H. and S.-W.L.

Funding: This research is supported by the National Key R&D Program of China (no. 2017YFC0704200), the National Natural Science Foundation of China (no. 51778336 and no. 51478249), and the Tsinghua University-Glodon Joint Research Centre for Building Information Model (RCBIM).

Acknowledgments: The authors also thank Pine Liu (Carnegie Mellon University) and Kincho H. Law (Stanford University) for their valuable comments and contributions.

Conflicts of Interest: The authors declare no conflict of interest.

References

1. Akbarnezhad, A.; Ong, K.C.G.; Chandra, L.R. Economic and environmental assessment of deconstruction strategies using building information modeling. *Autom. Constr.* **2014**, *37*, 131–144. [CrossRef]
2. Akcamete, A.; Akinci, B.; Garrett, J.H. Potential utilization of building information models for planning maintenance activities. In Proceedings of the 13th International Conference on Computing in Civil and Building Engineering, Nottinham, UK, 30 June–2 July 2010; pp. 151–158.
3. Chotipanich, S. Positioning facility management. *Facilities* **2004**, *22*, 364–372. [CrossRef]
4. Shen, W.; Hao, Q.; Mak, H.; Neelamkavil, J.; Xie, H.; Dickinson, J.; Thomas, R.; Pardasani, A.; Xue, H. Systems integration and collaboration in architecture, engineering, construction, and facilities management: A review. *Adv. Eng. Inform.* **2010**, *24*, 196–207. [CrossRef]
5. Parn, E.A.; Edwards, D.J.; Sing, M.C.P. The building information modelling trajectory in facilities management: A review. *Autom. Constr.* **2017**, *75*, 45–55. [CrossRef]
6. Becerik-Gerber, B.; Jazizadeh, F.; Li, N.; Calis, G. Application Areas and Data Requirements for BIM-Enabled Facilities Management. *J. Constr. Eng. Manag.* **2012**, *138*, 431–442. [CrossRef]
7. Hu, Z.-Z.; Tian, P.-L.; Li, S.-W.; Zhang, J.-P. Bim-Based Integrated Delivery Technologies for Intelligent Mep Management in the Operation and Maintenance Phase. *Adv. Eng. Softw.* **2018**, *115*, 1–16. [CrossRef]
8. Chen, S.; Zheng, J.; Bi, K.; Zhang, P. BIM-Based Logic Retrieval of MEP Systems. *J. Eng. Manag.* **2016**, *30*, 132–137.
9. BuildingSMART. IFC4 Release Candidate 4. Retrieved from BuildingSMART. 2013. Available online: http://www.buildingsmart-tech.org/ifc/IFC2x4/rc4/html/index.html (accessed on 24 September 2015).
10. Kassem, M.; Kelly, G.; Dawood, N.; Serginson, M.; Lockley, S. BIM in facilities management applications: A case study of a large university complex. *Built Environ. Proj. Asset Manag.* **2015**, *5*, 261–277. [CrossRef]
11. Lee, Y.C.; Eastman, C.M. Logic for ensuring the data exchange integrity of building information models. *Autom. Constr.* **2018**, *93*, 388–401. [CrossRef]
12. Hassanain, M.A.; Froese, T.M.; Vanier, D.J. Development of a maintenance management model based on IAI standards. *Artif. Intell. Eng.* **2001**, *15*, 177–193. [CrossRef]
13. Krijnen, T.; Beetz, J. An IFC schema extension and binary serialization format to efficiently integrate point cloud data into building models. *Adv. Eng. Inform.* **2017**, *33*, 473–490. [CrossRef]
14. Cho, G.H.; Won, J.S.; Kim, J.-U. The Extension of IFC Model Schema for Geometry Part of Road Drainage Facility. *J. Korea Acad. Ind. Cooper. Soc.* **2013**, *14*, 5987–5992. (In Korea) [CrossRef]
15. Brilakis, I.; Lourakis, M. Toward automated generation of parametric BIMs based on hybrid video and laser scanning data. *Adv. Eng. Inform.* **2010**, *24*, 456–465. [CrossRef]
16. Nguyena, T.H.; Oloufa, A.A.; Nassar, K. Algorithms for automated deduction of topological information. *Autom. Constr.* **2005**, *14*, 59–70. [CrossRef]
17. Bruno, S.; De Fino, M.; Fatiguso, F. Historic building information modeling: Performance assessment for diagnosis-aided information modeling and management. *Autom. Constr.* **2018**, *86*, 256–276. [CrossRef]

18. Nepal, M.P.; Staub-French, S.; Zhang, J. Deriving construction features from an IFC model. In Proceedings of the Annual Conference–Canadian Society for Civil Engineering, Québec, QC, USA, 10–13 June 2008.

19. Borrmann, A.; Rank, E. Specification and implementation of directional operators in a 3D spatial query language for building information models. *Adv. Eng. Inform.* **2009**, *23*, 32–44. [CrossRef]

20. Daum, S.; Borrmann, A. Processing of Topological BIM Queries using Boundary Representation Based Methods. *Adv. Eng. Inform.* **2014**, *28*, 272–286. [CrossRef]

21. Bertino, E.; Negri, M.; Pelagatti, G.; Sbattella, L. Objectoriented query languages: The notion and the issues. *IEEE Trans. Knowl. Data Eng.* **1992**, *4*, 223–237. [CrossRef]

22. Kang, T.W. Object composite query method using IFC and LandXet ML based on BIM linkage model. *Autom. Constr.* **2017**, *76*, 14–23. [CrossRef]

23. Mazairac, W.; Beetz, J. BIMQL—An open query language for building information models. *Adv. Eng. Inform.* **2013**, *27*, 444–456. [CrossRef]

24. Dong, J.S.; Feng, Y.; Sun, J. *Sensor Based Designs for Smart Space*; Technical Report; Computer Science Department, National University of Singapore: Singapore, 2004.

25. Solihin, W.; Eastman, C.; Lee, Y.C. A simplified relational database schema for transformation of BIM data into a query-efficient and spatially enabled database. *Autom. Constr.* **2017**, *82*, 367–383. [CrossRef]

26. Preidel, C.; Daum, S.; Borrmann, A. Data retrieval from building information models based on visual programming. *Vis. Eng.* **2017**, *5*, 18–32. [CrossRef]

27. Thambirajah, J.; Benabbas, L.; Bauer, M.; Thornhill, N.F. Cause-and-effect analysis in chemical processes utilizing XML, plant connectivity and quantitative process history. *Comput. Chem. Eng.* **2009**, *33*, 503–512. [CrossRef]

28. Son, H.; Kim, C.; Kim, C. 3D reconstruction of as-built industrial instrumentation models from laser-scan data and a 3D CAD database based on prior knowledge. *Autom. Constr.* **2015**, *49*, 193–200. [CrossRef]

29. Tang, P.B.; Huber, D.; Akinci, B.; Lipman, R.; Lytle, A. Automatic reconstruction of as-built building information models from laser-scanned point clouds: A review of related techniques. *Autom. Constr.* **2010**, *19*, 829–843. [CrossRef]

30. Liu, H.; Singh, G.; Lu, M.; Bouferguene, A.; Alhussein, M. BIM-based automated design and planning for boarding of light-frame residential buildings. *Autom. Constr.* **2018**, *88*, 235–249. [CrossRef]

31. Yarmohammadi, S.; Castro-Lacouture, D. Automated performance measurement for 3D building modeling decisions. *Autom. Constr.* **2018**, *93*, 91–111. [CrossRef]

32. Leite, F.; Akcamete, A.; Akinci, B. Analysis of modeling effort and impact of different levels of detail in building information models. *Autom. Constr.* **2011**, *20*, 601–609. [CrossRef]

33. Ciribini, A.L.C.; Mastrolembo Ventura, S.; Paneroni, M. Implementation of an interoperable process to optimise design and construction phases of a residential building: A BIM Pilot Project. *Autom. Constr.* **2016**, *271*, 62–73. [CrossRef]

34. Wang, L.; Leite, F. Process Knowledge Capture in BIM-Based Mechanical, Electrical, and Plumbing Design Coordination Meetings. *J. Comput. Civ. Eng.* **2016**, *30*, 04015017. [CrossRef]

35. Hu, Z.Z.; Zhang, J.P.; Yu, F.Q. Construction and facility management of large MEP projects using a multi-Scale building information model. *Adv. Eng. Softw.* **2016**, *100*, 215–230. [CrossRef]

36. Li, W.; Fernanda, L. Knowledge Discovery of Spatial Conflict Resolution Philosophies in BIM-Enabled MEP Design Coordination Using Data Mining Techniques: A Proof-of-Concept. In Proceedings of the ASCE International Workshop on Computing in Civil Engineering, Los Angeles, CA, USA, 23–25 June 2013; pp. 419–426. [CrossRef]

37. Driscoll, J.R.; Lien, Y.E. A selective traversal algorithm for binary search trees. *Commun. ACM* **1978**, *21*, 445–447. [CrossRef]

38. Djuric, N.; Novakovic, V.; Frydenlund, F. Heating system performance estimation using optimization tool and BEMS data. *Energy Build.* **2008**, *40*, 1367–1376. [CrossRef]

39. Liu, X.S.; Akinci, B.; Bergés, M.; Garrett, J.H., Jr. Domain-Specific Querying Formalisms for Re-trieving Information about HVAC Systems. *J. Comput. Civ. Eng.* **2014**, *28*, 40–49. [CrossRef]

Article

Combined MvdXML and Semantic Technologies for Green Construction Code Checking

Shaohua Jiang [1,*], Zheng Wu [1], Bo Zhang [1] and Hee Sung Cha [2]

[1] Department of Construction Management, Faculty of Infrastructure Engineering, Dalian University of Technology, Dalian 116024, China; wuzheng0709@gmail.com (Z.W.); zhangbo5951@163.com (B.Z.)

[2] Department of Architectural Engineering, Ajou University, Suwon 16499, Korea; hscha@ajou.ac.kr

* Correspondence: shjiang@dlut.edu.cn; Tel.: +86-411-8470-7482

Received: 31 January 2019; Accepted: 27 March 2019; Published: 8 April 2019

Abstract: The construction process plays a key role in sustainable development of the environment. With the concept of sustainable construction being put forward in the world, some countries released green construction standards to strengthen the requirements in the construction phase. Green construction code checking needs to integrate semantic information embedded in green construction standards and model information involved in Industry Foundation Classes (IFC) and/or Model View Definition (MVD), which are generated separately and lead to difficulty in information integration for green construction code checking. At present, the existing code-checking methods cannot be directly used for green construction. Related practitioners need an efficient and convenient method for green construction code checking urgently. To ameliorate this situation, this research proposes an innovative approach to organize, store, and re-use green construction knowledge by combining mvdXML and semantic technologies. The code checking of green construction is classified into four types based on the difficulty level to meet the requirements of the clauses in green construction standard. Depending on the characteristics of each inspection type, mvdXML or semantic technology is adopted for the appropriate inspection type. This paper demonstrates the deployment and validation of such automated checking procedures in a case study. Based on these experiences, a detailed discussion about the identified issues is provided as the starting point for future research.

Keywords: green construction; code checking; mvdXML; semantic technology

1. Introduction

As people pay more and more attention to the living environment, the concept of green construction is becoming widespread [1]. According to statistics, a great level of pollution and waste occur in the different phases of a building's life cycle [2]. Most of these activities take place in the construction phase; thus, green construction is a very important strategy of sustainable development [3]. For example, waste generated in construction and demolition processes comprised around 50% of the solid waste in South Korea in 2013 [4]. Many cases show that the code checking based on building information modeling (BIM) is an effective means to reduce the amount of construction waste, since it is mainly generated due to improper design and unexpected changes in the design and construction phases. In response to this reality, some countries such as America and China released green construction-related standards and codes. The definitions of green construction in different countries are similar. In China, green construction is defined as "under the premise of guaranteeing the basic requirements of quality and safety, the construction activities which should be adopted to maximize resource conservation and minimize the negative impact on the environment through scientific management and technological progress, and to achieve the goal of four savings (energy, land, water, and materials) and environmental protection", which is specified in the Code for Green Construction of Building (GB/T50905-2014) issued by the Ministry of Housing and Urban-Rural Development of the People's Republic of China (MOHURD) in 2014.

The evaluation of green construction is becoming more and more complex; however, traditional evaluation processes need a lot of manpower, materials, and time, which will undoubtedly increase the unreliability and bring uncertainty to the quality of construction project. Therefore, construction industry practitioners put forward urgent requirements for a more efficient approach of code checking of green construction.

Some code-checking cases are shown in other researchers' papers. For example, Won et al. showed that BIM-based validation could prevent 4.3–15.2% of construction waste that might have been generated without using BIM [4]. However, the requirements of green construction were not fully considered. Therefore, existing approached cannot be directly used for green construction code checking, which results in an imperative need for innovative technology to improve the current condition.

Green construction code checking requires distinct domain knowledge and building model information. Model view definition (MVD), proposed by buildingSMART, defines a subset of the IFC schema needed to satisfy data exchange requirements, and it helps software vendors develop IFC import and export features that allow project participants to share and exchange BIM information. It is a kind of handshake protocol that allows expected meaningful requirements, implementation, and results. The method to publish the concepts and associated rules is called mvdXML [5]. Ontology and semantic web technology offer an opportunity to enable domain knowledge to be represented semantically. Semantic Query Web Rule Language (SQWRL) is a Semantic Web Rule Language (SWRL)-based query language that provides SQL(Structured Query Language)-like operators for extracting information from ontologies in Web Ontology Language (OWL), which is also widely used for code checking [6,7]. The goal of the mvdXML and semantic technologies is not only to put forward an automated knowledge framework, but also to offer the impetus to gear up the Architecture, Engineering and Construction (AEC) industry toward greater interoperability through the deployment of the BIM-based code checking.

However, both mvdXML and semantic technologies have limitations. This paper combines mvdXML and semantic technologies to transfer green construction standards to computer-interpretable code. This approach allows researchers to target critical needs across application areas and to solve the problems of code checking.

2. Research Background

Three levels of BIM implementation are proposed in the Bew–Richards maturity model, in which a code-checking process is implemented in Level 2 and can be continually developed with more functions in Level 3 [8]. Many studies were conducted on code checking, which can be categorized into four classes based on the adopted approaches. In this section, the importance of green construction, particularly green construction code checking, is summarized, and previous code-checking approaches are reviewed [9–11].

2.1. Importance of Green Construction

With the acceleration of the urbanization process and the rapid development of the social economy, the business scope of the construction industry is constantly expanding. At the same time, the adverse impact of waste and pollution caused by construction activities on the environment cannot be ignored, especially in the construction stage. In recent years, people's awareness of environmental protection and resource conservation increased; thus, the concept of sustainable development was widely acknowledged. Green construction is the embodiment of the concept of sustainable development in the construction industry. Only by paying attention to the dissemination of the green construction concept and the application of green construction technology can the sustainable development of the construction industry be realized.

2.2. Green Construction Code Checking

Knowledge and rules are two critical parts in the green construction code-checking process. The knowledge related to the green construction code and the project-specific information required by green construction code checking are scattered and fragmented. The rules in the green construction code-checking process should be interpreted systematically and made tractable, whereby complete green construction rule interpretation involves knowledge from human experts in many professional fields. Therefore, the exact requirements for green construction code checking are often much more complex than other standards. Because much information relevant to green construction standard is not available in the BIM model, the IFC standard needs to be physically extended to express the missing information. As not everything in the BIM can be defined at the initial stage, some missing, evolving, or additional information can be inserted into the mvdXML, allowing the mvdXML to include the original IFC information and the extended information.

Some scholars systematically analyzed code-checking-related systems, concept libraries, query languages, reasoners, and model view definitions [12,13]. From their analysis, we can see that there are many kinds of rules and concepts supported by the information included in different levels of detail (LOD). In traditional rule checking or code checking, LOD 300 is assumed in general, but green construction code checking needs higher LOD. Before green construction code checking, the BIM model should be modeled as specific assemblies, with complete fabrication, assembly, and detailing information, in addition to precise quantity, size, shape, location, and orientation. Non-geometric information of the model elements can also be attached, which includes model details and elements that represent how building elements interface with various systems and other building elements with model view. Requirements for code checking should match with all kinds of green construction-related information in BIM. In this case, LOD 400 is expected for green construction code checking.

2.3. Code-Checking Approaches in Construction Industry

There are four types of code-checking approaches, i.e., code checking based on existing BIM software, code checking based on object-oriented approach, code checking based on logical rules, and code checking approach based on semantic web. Their characteristics and limitations are summarized in this section.

The code-checking approach based on existing BIM software and code checking based on an object-oriented approach are usually used for simple quantitative model checking, and they both have less scalability. The code-checking approach based on logical rules and the code-checking approach based on semantic web are both widely used. MvdXML technology is a code-checking approach based on logical rules. Fahad et al. compared mvdXML technology and semantic web technology to deal with abstract concepts or physical objects, and pointed out that each method has its pros and cons and can be used according to the requirements of the research project [14]. The mvdXML method is an appropriate choice for simple logical rules. In the implementation of mvdXML method, the BIM Collaboration Format (BCF) is used to report identified issues. BCF is an open standard originally proposed by buildingSMART to enable workflow communication between different applications. Unlike traditional text-based reports, mvdXML method links communication to model view, and it can also be easily implemented by visualization applications. The code-checking approach based on semantic web can perform complicated semantic inferences which cannot be conducted in the mvdXML approach; therefore, it is also widely used.

Based on the particularity and requirements of green construction code checking, we combined mvdXML technology and semantic web technology to conduct green construction code checking. Therefore, we mainly introduce the latter two approaches here.

2.3.1. Code Checking Based on Logical Rules

The most common language in interpreting the natural language is first-order predicate logic. Choi et al. created links between the Standard for the Exchange of Product Model Data (STEP) and Extensible Markup Language (XML) formats to develop a standardized automated inspection system to share architectural drawings and document information [15]. MvdXML is an XML format file including MVD information. BIM Rule Language (BIMRL) was developed by SQL to support code checking, which is a query and manipulation language used for managing the rule data [16]. Nawari et al. demonstrated the construction plan with the BIM model and used the model-checking software to achieve automatic inspection of the normative terms, while the specification and standards were in monotonous and rigid formats [17]. Nawari et al. pointed out that the implementation of the automatic specification check is based on two main parts; one is to convert building code to computable rules, i.e., Smart Codes, the other is to use the BIM model to ensure the level of detail [18]. Ilhan et al. proposed a framework for providing an integrated platform to facilitate the green code generation. The proposed model helps design team-assigned sustainable properties and helps assess sustainability performance of a project during the design stage so that decisions can be made on time for Building Research Establishment Environmental Assessment Methodology (BREEAM) certification [19]. Ciribini et al. implemented an interoperable IFC-based process to support code checking through four-dimensional (4D) BIM; an auto-matching between BIM objects and construction activities was achieved [20]. Zhang et al. developed algorithms that automatically analyze a building model to detect safety hazards and suggest preventive measures to users for different cases involving fall-related hazards. A rule-based engine was implemented on the top of a commercially available BIM platform to show the feasibility [21]. Kasim et al. presented a generic approach for automating BIM compliance, checking with standards and best practice with a focus on sustainability-related standards. A methodology was presented to provide a reusable solution for compliance checking within a similar context. The proposed approach was generic in nature, thus allowing implementation in different fields of engineering including electrical engineering, energy, water, and electronic engineering. In this approach, the RASE (Requirement, Applicability, Selection, Exception) methodology was utilized to extract requirements from sustainability-based regulations, with the goal of converting them into coded rules for execution by a rule engine [22]. Lee et al. suggested a robust MVD validation process using a modularized validation platform that evaluates an IFC instance file according to diverse types of rule sets of MVDs. This research employed the MVD of the precast concrete domain using the identified rule logic and checking structures to extract rule templates that consisted of 10 types of rule logic [23]. Dimyadi et al. proposed a formalized way to capture these requirements using a graphical notation to represent code rules in a machine- and human-readable language, which allowed unambiguous description of the requirements. Such unambiguous description could be understood by all the participants in the implementation efforts [24,25]. Zhang et al. proposed an automated compliance checking system for automatically extracting information from construction regulatory textual documents and transforming them into logic clauses using semantic Natural Language Processing (NLP) and logic inference, which focused on presenting quantitative requirements and the corresponding algorithms. The system could be extended to support the checking of different types of existential requirements containing certain building elements in rules [26–28].

2.3.2. Code Checking Based on Semantic Web

There are some building codes computerizing systems based on ontology-based approaches and semantic web information. Ontology-based approaches focus on formalizing conformance requirements, which are conducted under knowledge extraction and semantic mapping [29]. Semantic web information is about how to enhance the IFC model using description language, which is based on a logic theory. Pauwels et al. took advantage of the logical basis of Resource Description Framework (RDF), using semantic web technologies to support data interoperability and integrate relationships between semantic information and model information [30,31]. The semantic web was built in a semantic network on the World Wide Web (WWW). In semantic web, objects and logical relationships between objects are displayed by graph. The Resource Description Framework (RDF) is a standard model for data interchange on the web [32,33]. RDF is used for representing the graph structure, which provides expressions and statements to describe relationships between resources. RDF extends the linking structure of the web to use Unique Resource Identifiers (URIs) to name the relationship between things, as well as the two ends of the link. Using this simple model, it allows structured and semi-structured data to be mixed, exposed, and shared across different applications. RDF is used in Web Ontology Language (OWL) to describe more complex ontologies [34,35].

The ontological approach was also adopted by International Code Council (ICC) through the development of SMARTcodes in 2006. ICC establishes an International Energy Conservation Code (IECC) dictionary, which is a method of communication between regulations and building models [36]. In addition to IECC, Bakillah et al. proposed a new system that supports geospatial data retrieval from multiple and complementary sources [37]. The system uses the SQWRL to support inference with complex queries and to enable combination of complementary and heterogeneous data. Zhong et al. explored an ontology-based semantic modeling approach of regulation constraints and proposed a meta model for construction quality inspection and evaluation, i.e., CQIEOntology. Based on CQIEontology, the regulation constraints were modeled into OWL axioms and SWRL rules. Using these OWL axioms and SWRL rules, the regulation provisions imposed on construction quality inspection can be translated into a set of inspection tasks, before being associated with the specific construction tasks [38]. Lu et al. extracted information from construction safety regulations and represented constrain rules with SWRL. The construction safety-checking processes were implemented in the JESS (Java Expert Shell System) engine [39]. Ding et al. took advantage of the strength of BIM, ontology, and semantic web technology to establish an ontology-based framework for construction risk knowledge management in a BIM environment [40]. Lee et al. proposed an ontology-based framework which would help accurately recognize domain knowledge and appropriate requirements for developing reusable concept modules [41]. Zhang et al. proposed a pragmatic method to check real-world-scale BIM models, which transforms BIM models into a well-defined OWL model. Rules were formalized by a structured natural language (SNL) designed intentionally to describe building regulations. The checking engine was based on Simple Protocol and RDF Query Language (SPARQL) queries on OWL models, which can effectively improve the time efficiency and deal with large-scale applications [42]. De Farias et al. combined the ifcOWL ontology with SWRL rules and performed a more intuitive and flexible extraction of building views than the MVD approach [43].

The objective of this study was to propose an approach to organize, store, and re-use green construction knowledge. Therefore, considering the particularity of green construction standards and the limitation in current code-checking approaches, in this paper, BIM, logical rule, ontology, and semantic web query language were integrated to conduct green construction code checking. The conceptual model ontology, work item ontology, and construction condition ontology for green construction were built, while the missing information for green construction code checking was added using IfcDoc, and various kinds of green construction rules were checked using an approach combining mvdXML and semantic technologies. Users can modify the parameters of each rule to match their local regulations using the approach proposed in this paper. Once the rule structure is defined, it is available for multiple code-checking projects.

3. Classification of Code Checking for Green Construction

With the development of green building certification, some national organizations and institutions further infiltrated the issues related to the construction phase to make up for the insufficiency of the requirements regarding green construction in green building certification. To reflect the environmental problems and guide the environmental management on the construction site, at present, some countries such as America and China released green construction-related standards and codes.

The code checking of green construction standards can be categorized into the following four types based on the difficulty level to meet the requirements of the clauses.

Class 1: Code checking based on simple defined attribute values in IFC schema;
Class 2: Code checking based on the extension of IFC entities;
Class 3: Code checking based on simple conceptual reasoning;
Class 4: Code checking based on construction condition reasoning.

This paper takes green construction standards and codes in China as an example.

In China, sustainability was widely debated in the construction industry in recent years. In order to strengthen the "green" development during the construction phase, MOHURD organized research efforts to conduct green construction research, issuing the "Evaluation Standard for Green Construction of Buildings" in 2010 (GB/T506040), which was promulgated to promote green construction, and releasing the "Code for green construction of building" in 2014 (GB/T50905-2014), which provides directional guidance for green construction.

According to the above classification of code checking, some example clauses in GB/T50905-2014 are shown in Table 1.

Table 1. Classification of various kinds of clauses in GB/T50905-2014.

Classification of Clauses	Types of Inspection	Example Clauses in GB/T50905-2014
Class 1	Code checking based on simple defined attribute value in IFC schema.	&3.2.1 Building material storage site should be no more than 500 km from construction site.
Class 2	Code checking based on the extension of IFC entities.	&4.0.5 A database of building materials should be established and a building material with good green performance should be adopted.
Class 3	Code checking based on simple conceptual reasoning	&9.2.4 Cast-in-place concrete raw materials should be mixed in construction site.
Class 4	Code checking based on construction condition reasoning	&9.2.7 When spraying materials on site, the ambient temperature should be 10–40 °C

In the case study, for each type of rule, an example from these standards with logic formulas was given using mvdXML and semantic technologies to specify its semantics.

4. Methodology of Green Construction Code Checking

Each method has its own characteristics and should be used according to requirements of the green construction specification. XML and mvdXML data can be transformed into OWL/RDF description language as the rule base of ontology (see Figure 1), which allows mvdXML and semantic technologies to be integrated for code checking. Based on this concept, an approach combining mvdXML and semantic technologies for code checking of green construction is introduced below (see Figure 2).

It is necessary to follow some specific steps for high efficiency and performance of code checking.

(1) Data source. The green construction standard is selected as one data source, which can be converted into XSLT (Extensible Stylesheet Language Transformations)/XML based on rule expression and classification of clauses. Model information in IFC and/or MVD is selected as another data source, which can be expressed in Excel and transformed to mvdXML form using the mvdXML rule transformer by data preprocessing [44]. The missing information for green construction code checking can be added in mvdXML using IfcDoc.

(2) Knowledge base. The conceptual model ontology, work item ontology, and construction condition ontology for green construction are built in this part. The ontologies in the knowledge base consist of facts and axioms.

(3) Rule base. The rule base includes customized rules that are used in conjunction with the content in the knowledge base for the JESS inference. The customized rules are based on the model information in IFC and/or MVD. The rules in this study were established in Protégé 3.4, and the grammar format was based on RDF [45].

(4) JESS inference. The facts and the rule base are matched to obtain inference results through JESS inference engine. After the transformation from the JESS rule language to RDF/OWL, the inferred facts can be used to update the knowledge base.

(5) Query. To meet the need of query function, the result needs to be transformed into a computer-readable format. The converted data representation is in the form of a triplet, which consists of resources, attributes, and values that can be represented by the concept and relationship of ontology and can be queried by SQWRL or SWRL. Finally, the report of code checking for green construction can be produced using Hypertext Markup Language (HTML).

In addition, the IFC model can be checked by the mvdXML ruleset and a BCF report is generated using a BCF generator; then, the issues can be shown in Revit using BCFier with a BIM view [46].

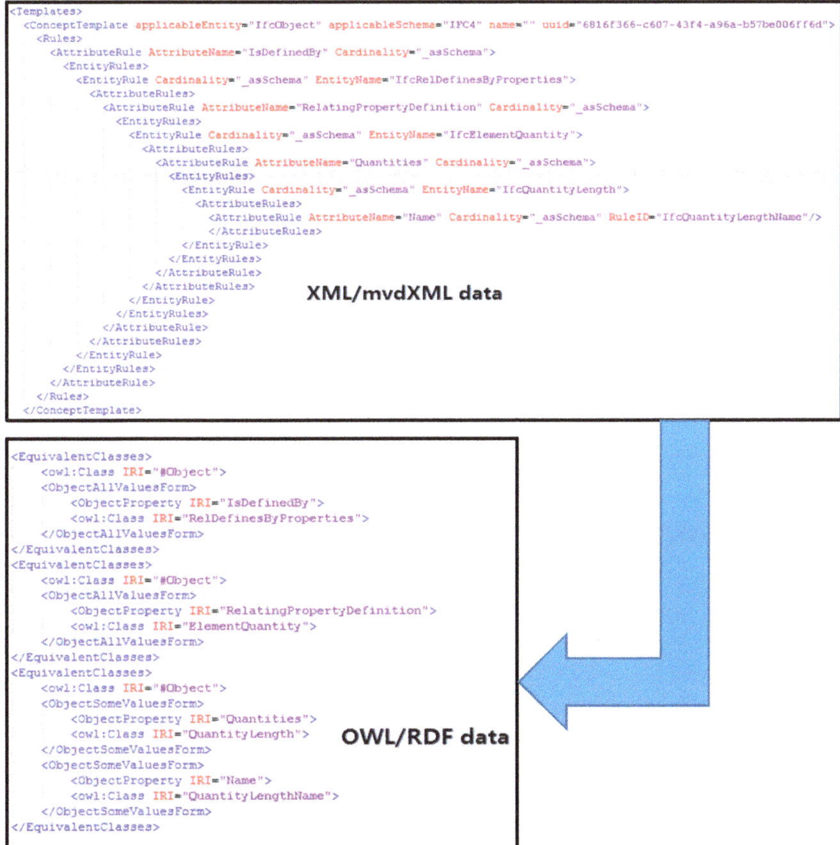

Figure 1. Transform between XML/mvdXML and OWL/RDF data.

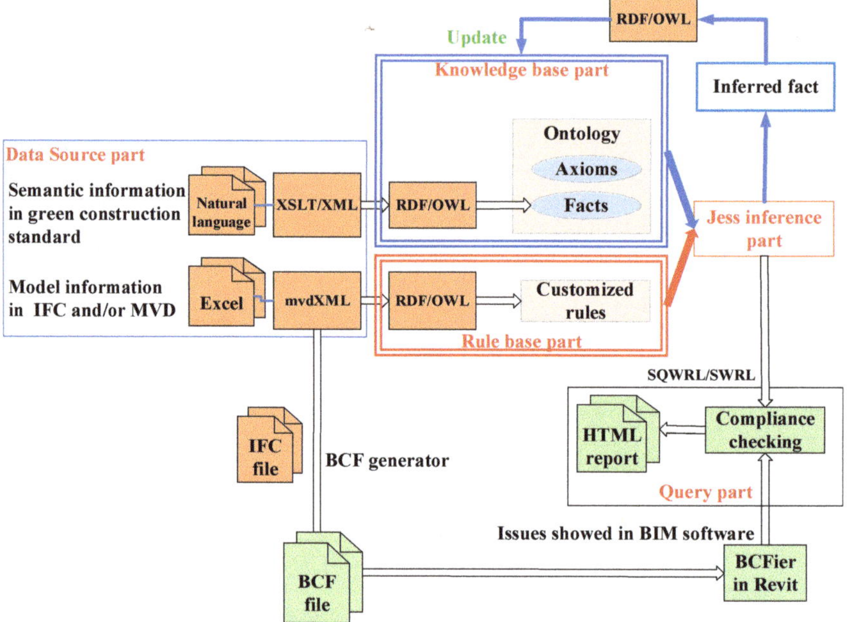

Figure 2. Workflow of code checking.

4.1. MvdXML Technology for Code Checking

Model view definitions (MVDs) are encoded in a format called mvdXML, which defines allowable values at particular attributes of particular data types [47]. For example, an MVD may require a wall providing a fire rating, a classification according to Table 22 of OmniClass, and information required for structural analysis, such as the elastic modulus of materials. In simple cases, mvdXML may define a single attribute on a single data type, while more complex cases may consist of graphs of objects and collections which can be defined by semantic technology. To check the rules in mvdXML, in general, four steps are needed: (1) generating IFC models using BIM-related software and model information in IFC and/or MVD expressed in Excel; (2) transforming Excel to mvdXML form using an mvdXML rule transformer by data preprocessing and inserting the information which is missing, evolving, or new in the mvdXML using IfcDoc. The IfcDoc tool is used to generate documentation of the IFC specification itself, the baseline MVD ConceptTemplates, and concepts that are part of the IFC specification. It, therefore, allows quickly expanding the generic MVD concepts to cover the specific requirements and business rules that are introduced by a set of exchange requirements; (3) evaluating and comparing IFC and mvdXML files [45]; (4) reading the output of the BCF generator using BIM software.

A checker was developed based on the open standard mvdXML as the format for structuring checking rules and the BIM Collaboration Format (BCF) to issue reports of the checking process. Two BIM operational standards, i.e., mvdXML and BCF were used for green construction code checking [46].

4.1.1. Data Source

The data source was an Excel spreadsheet. Figure 3 gives an overview of the data source. The data source contains rules that can validate if an object (e.g., window) contains certain properties and quantity parameters (e.g., self-closing). The property and quantity rules are often used in model checking.

Information Item	Required	IFC Support
Object:Window		
Subtitle: Existence object parameters		
Rule type:SCHEMADEFINITION(Properties)		
SelfClosing	Yes	IfcWindow->IfcPropertySingleValue.Name=SelfClosing
FireRating	Yes	IfcWindow->IfcPropertySingleValue.Name=FireRating
IsExternal	Yes	IfcWindow->IfcObject.IsDefinedBy.IfcRelDefinesByProperties.RelatingPropertyDefinition.IfcPropertySet.HasProperties.IfcPropertySingleValue.Name=IsExternal
Rule type:SCHEMADEFINITION(Quantities)		
Height	No	IfcWindow->IfcObject.IsDefinedBy.IfcRelDefinesByProperties.RelatingPropertyDefinition.IfcElementQuantity.Quanties.IfcQuantityLength.Name=Height
Width	No	IfcWindow->IfcQuantityLength.Name=Width
Thickness	No	IfcWindow->IfcObject.IsDefinedBy.IfcRelDefinesByProperties.RelatingPropertyDefinition.IfcElementQuantity.Quanties.IfcQuantityLength.Name=Thickness
Volume	No	IfcWindow->IfcObject.IsDefinedBy.IfcRelDefinesByProperties.RelatingPropertyDefinition.IfcElementQuantity.Quanties.IfcQuantityVolume.Name=Volume
Area	No	IfcWindow->IfcQuantityArea.Name=Area
Object:Door		
Subtitle: Existence object parameters		
Rule type:SCHEMADEFINITION(Properties)		
SelfClosing	No	IfcDoor->IfcPropertySingleValue.Name=SelfClosing
FireRating	No	IfcDoor->IfcObject.IsDefinedBy.IfcRelDefinesByProperties.RelatingPropertyDefinition.IfcPropertySet.HasProperties.IfcPropertySingleValue.Name=FireRating
IsExternal	No	IfcDoor->IfcObject.IsDefinedBy.IfcRelDefinesByProperties.RelatingPropertyDefinition.IfcPropertySet.HasProperties.IfcPropertySingleValue.Name=IsExternal
Rule type:SCHEMADEFINITION(Quantities)		
Height	No	IfcDoor->IfcObject.IsDefinedBy.IfcRelDefinesByProperties.RelatingPropertyDefinition.IfcElementQuantity.Quanties.IfcQuantityLength.Name=Height
Width	No	IfcDoor->IfcQuantityLength.Name=Width
Thickness	No	IfcDoor->IfcObject.IsDefinedBy.IfcRelDefinesByProperties.RelatingPropertyDefinition.IfcElementQuantity.Quanties.IfcQuantityLength.Name=Thickness
Volume	No	IfcDoor->IfcObject.IsDefinedBy.IfcRelDefinesByProperties.RelatingPropertyDefinition.IfcElementQuantity.Quanties.IfcQuantityVolume.Name=Volume
Area	No	IfcDoor->IfcQuantityArea.Name=Area
LoD 200 – Approximate Geometry		
Rule type:SCHEMADEFINITION(Properties)		
SelfClosing	No	IfcDoor->IfcPropertySingleValue.Name=SelfClosing
FireRating	No	IfcDoor->IfcObject.IsDefinedBy.IfcRelDefinesByProperties.RelatingPropertyDefinition.IfcPropertySet.HasProperties.IfcPropertySingleValue.Name=FireRating
IsExternal	No	IfcDoor->IfcObject.IsDefinedBy.IfcRelDefinesByProperties.RelatingPropertyDefinition.IfcPropertySet.HasProperties.IfcPropertySingleValue.Name=IsExternal
Rule type:SCHEMADEFINITION(Quantities)		
Height	No	IfcDoor->IfcObject.IsDefinedBy.IfcRelDefinesByProperties.RelatingPropertyDefinition.IfcElementQuantity.Quanties.IfcQuantityLength.Name=Height
Width	No	IfcDoor->IfcQuantityLength.Name=Width
Area	No	IfcWindow->IfcQuantityArea.Name=Area
Volume	No	IfcWindow->IfcQuantityVolume.Name=Volume
Thickness	No	IfcWindow->IfcObject.IsDefinedBy.IfcRelDefinesByProperties.RelatingPropertyDefinition.IfcElementQuantity.Quanties.IfcQuantityLength.Name=Thickness
Object:Wall		
Subtitle: Existence object parameters		
Rule type:SCHEMADEFINITION(Properties)		
FireRating	No	IfcWall->IfcPropertySingleValue.Name=FireRating
LoadBearing	No	IfcWall->IfcPropertySingleValue.Name=LoadBearing
IsExternal	No	IfcWall->IfcObject.IsDefinedBy.IfcRelDefinesByProperties.RelatingPropertyDefinition.IfcPropertySet.HasProperties.IfcPropertySingleValue.Name=IsExternal
Rule type:SCHEMADEFINITION(Quantities)		
Thickness	No	IfcWall->IfcQuantityLength.Name=Thickness
Area	No	IfcWall->IfcQuantityArea.Name=Area
Volume	No	IfcWall->IfcObject.IsDefinedBy.IfcRelDefinesByProperties.RelatingPropertyDefinition.IfcElementQuantity.Quanties.IfcQuantityVolume.Name=Volume

Figure 3. Excel data source.

The data source described in Figure 3 consists of three columns. The first column "Information Item" classifies the rule type and can be used to append a name to each rule. The second column "Required" specifies whether the rule should be converted by the mvdXML rule transformer. Each row should specify if the rule should be transformed into mvdXML format. Thus, "Yes" means that the mvdXML rule transformer should convert the rule to mvdXML, and "No" means the opposite. The third column "IFC Support" is a path that is interpreted by the mvdXML rule transformer. The section "IFC Support" elaborates how to create and adjust IFC support path.

4.1.2. MvdXML Rule Transformer

The mvdXML rule transformer is a tool for the generation of the mvdXML ruleset (see Figure 4). MvdXML rules are based on the open mvdXML standard. The open mvdXML standard ensures easy access and extensions of the ruleset by users. The mvdXML rule transformer can generate rules for all rule types defined in mvdXML schema. If there is some information that is missing, evolving, or new in the mvdXML, IfcDoc allows additional content to be created in IFC baseline, which can be transformed to mvdXML files. IfcDoc is divided into four submenus as follows:

(1) Documentation: content only affecting documentation output such as terms, references, and examples.

(2) Schema: content describing underlying data structures such as entities, enumerations, selections, and defined types added within a schema (data schemas, computer interpretable listings, alphabetical listings, and inheritance listings in IfcDoc).

(3) User data: content describing dynamic data structures such as property sets and quantity sets, added within a schema (data schemas, computer interpretable listings, alphabetical listings, and inheritance listings in IfcDoc).

(4) Model views: content describing model view definitions such as templates (concepts in IfcDoc), model views and exchanges (scope in IfcDoc), and concepts (underneath entities within data schemas, computer interpretable listings, alphabetical listings, and inheritance listings in IfcDoc).

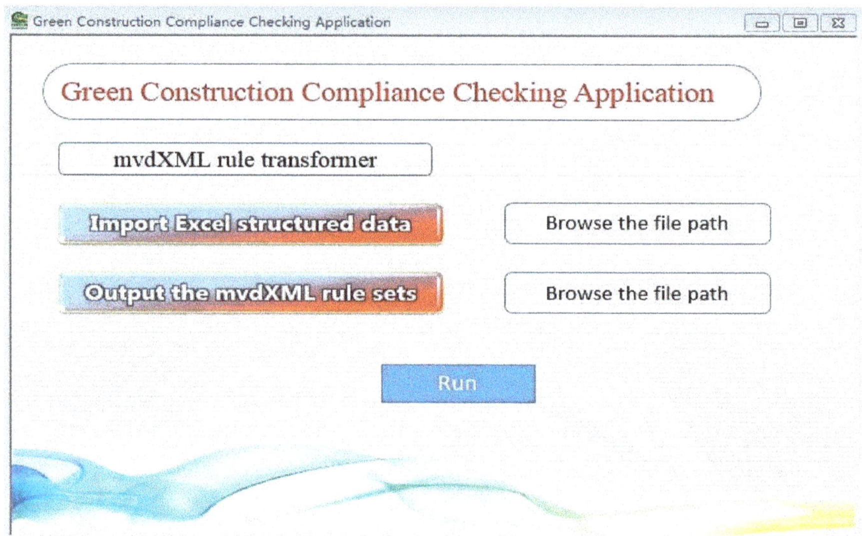

Figure 4. MvdXML rule transformer.

In this paper, the information related to green construction standards not defined in the existing IFC was extended in the case study.

4.1.3. MvdXML Rules

To use the mvdXML rule transformer, basic knowledge about the structure of mvdXML files is required. The <ConceptTemplate> element defines the structure that should be complied with by related concepts (see Figure 5). For instance, Concept Template is able to assign IfcPropertySingleValues to all subtypes of an IfcObject. Concept Template defines rules for the entire IfcObject based on IFC2X4. The elements between <Rules> define the path to the ApplicableEntity (i.e., IfcPropertySingleValueName). The universally unique identifier (UUID) of the Concept Template is an essential element which is generated to relate concepts to the Concept Template. For instance, Figure 6 is the mvdXML snippet that shows the definition of the "external flag" concept of "external loadbearing wall", and Figure 7 is the mvdXML snippet that shows the definition of "applicability". The name of ConceptRoot element can be used to search for the name of a concept [47,48]. Three parts are relevant in the ConceptRoot definition.

(1)   ApplicableRootEntity and the applicability element (see Figure 6) define all IfcWall instances that fulfill the applicability constraints.
(2)   The <TemplateRules> element defines the availability of the property Pset_WallCommon.IsExternal (see Figure 7).
(3)   The <Requirements> element (see Figure 6) defines the usage for a set of exchangeRequirements (e.g., the exchangeRequirement with the ID 00000023-0000-0000-0000-000000000352).

This part of ConceptRoot defines the condition under which the requirement test has to be applied to an IFC instance. Some requirements are shown in Table 2. The example shown in Figure 6 defines the applicableRootEntity of IfcWall, which includes all subtypes of IfcWall. Additionally, the <Applicability> element specifies constraints of Pset_WallCommon.IsExternal and Pset_WallCommon.LoadBearing.

```xml
</ConceptTemplate>
<ConceptTemplate applicableEntity="IfcObject" applicableSchema="IFC4" name="" uuid="0f7f7621-dc0a-4ab8-8118-64ead2ee3cca">
  <Rules>
    <AttributeRule AttributeName="IsDefinedBy" Cardinality="_asSchema">
      <EntityRules>
        <EntityRule Cardinality="_asSchema" EntityName="IfcRelDefinesByProperties">
          <AttributeRules>
            <AttributeRule AttributeName="RelatingPropertyDefinition" Cardinality="_asSchema">
              <EntityRules>
                <EntityRule Cardinality="_asSchema" EntityName="IfcElementQuantity">
                  <AttributeRules>
                    <AttributeRule AttributeName="Quantities" Cardinality="_asSchema">
                      <EntityRules>
                        <EntityRule Cardinality="_asSchema" EntityName="IfcQuantityVolume">
                          <AttributeRules>
                            <AttributeRule AttributeName="Name" Cardinality="_asSchema" RuleID="IfcQuantityVolumeName"/>
                          </AttributeRules>
                        </EntityRule>
                      </EntityRules>
                    </AttributeRule>
                  </AttributeRules>
                </EntityRule>
              </EntityRules>
            </AttributeRule>
          </AttributeRules>
        </EntityRule>
      </EntityRules>
    </AttributeRule>
  </Rules>
</ConceptTemplate>
```

Figure 5. Example of ConceptTemplate.

```xml
<ConceptRoot uuid="00000023-0000-0000-2000-000000029085" name="External flag : Predefined Properties for Walls : External loadbearing wall" applicableRootEntity="IfcWall">
  <Definitions>
    <Definition>
      <Body><![CDATA[External flag : Example for checking properties defined by IFC (prefix Pset) : External loadbearing wall]]></Body>
    </Definition>
  </Definitions>
  <Applicability>
  <Concepts>
    <Concept uuid="00000023-0000-0000-0000-000000029085" name="External flag">
      <Definitions>
        <Definition>
          <Body lang="en"><![CDATA[External flag : Example for checking properties defined by IFC (prefix Pset) : External loadbearing wall]]></Body>
        </Definition>
      </Definitions>
      <Template ref="00000000-0000-0000-0001-000000000001"/>
      <Requirements>
        <Requirement applicability="import" exchangeRequirement="00000023-0000-0000-0000-000000000352" requirement="mandatory"/>
      </Requirements>
      <TemplateRules operator="and">
    </Concept>
  </Concepts>
</ConceptRoot>
```

Figure 6. MvdXML snippet.

```xml
<ConceptRoot uuid="00000023-0000-0000-2000-000000029085" name="External flag : Predefined Properties for Walls : External loadbearing wall" applicableRootEntity="IfcWall">
  <Definitions>
    <Definition>
      <Body><![CDATA[External flag : Example for checking properties defined by IFC (prefix Pset) : External loadbearing wall]]></Body>
    </Definition>
  </Definitions>
  <Applicability>
    <Template ref="00000000-0000-0000-0001-000000000001"/>
    <TemplateRules operator="and">
      <TemplateRules operator="or">
        <TemplateRule Parameters="Set[Value]='Pset_WallCommon' AND Property[Value]='IsExternal' AND Value[Value]=TRUE"/>
        <TemplateRules operator="and">
          <TemplateRules operator="nor">
            <TemplateRule Parameters="Set[Value]='Pset_WallCommon' AND Property[Value]='IsExternal'"/>
          </TemplateRules>
          <TemplateRules operator="or">
            <TemplateRule Parameters="T_Set[Value]='Pset_WallCommon' AND T_Property[Value]='IsExternal' AND T_Value[Value]=TRUE"/>
          </TemplateRules>
        </TemplateRules>
      </TemplateRules>
      <TemplateRules operator="or">
        <TemplateRule Parameters="Set[Value]='Pset_WallCommon' AND Property[Value]='LoadBearing' AND Value[Value]=TRUE"/>
        <TemplateRules operator="and">
          <TemplateRules operator="nor">
            <TemplateRule Parameters="Set[Value]='Pset_WallCommon' AND Property[Value]='LoadBearing'"/>
          </TemplateRules>
          <TemplateRules operator="or">
            <TemplateRule Parameters="T_Set[Value]='Pset_WallCommon' AND T_Property[Value]='LoadBearing' AND T_Value[Value]=TRUE"/>
          </TemplateRules>
        </TemplateRules>
      </TemplateRules>
    </TemplateRules>
  </Applicability>
```

Figure 7. <Applicability> definition of a requirement.

Table 2. Examples of requirements.

Related Concept	Requirements
External loadbearing wall	Show checking results for all ConceptRoots that fulfill applicability test of "external loadbearing wall"
External flag	Test all child node requirements of external flag.
Standard walls	Test for all "standard walls" including "acoustic rating", "external flag", and "load bearing flag"

In Figure 8, the existence of the property Pset_WallCommon.IsExternal is checked. This example is very specific as it tests a property that is already part of the applicability test. If the exchangeRequirement is mandatory, then this test should always pass as it is already required in the applicability test (external walls are identified by this property). If the exchangeRequirement is "excluded", then this test should always fail.

```
<ConceptRoot uuid="00000023-0000-0000-2000-000000029085" name="External flag : Predefined Properties for Walls : External loadbearing wall" applicableRootEntity="IfcWall">
  <Definitions>
  <Applicability>
  <Concepts>
    <Concept uuid="00000023-0000-0000-0000-000000029085" name="External flag">
      <Definitions>
        <Template ref="00000000-0000-0000-0001-000000000001"/>
        <Requirements>
        <TemplateRules operator="and">
          <TemplateRules operator="or">
            <TemplateRule Parameters="Set[Value]='Pset_WallCommon' AND Property[Value]='IsExternal'"/>
            <TemplateRules operator="and">
             <TemplateRules operator="nor">
               <TemplateRule Parameters="Set[Value]='Pset_WallCommon' AND Property[Value]='IsExternal'"/>
             </TemplateRules>
             <TemplateRules operator="or">
               <TemplateRule Parameters="T_Set[Value]='Pset_WallCommon' AND T_Property[Value]='IsExternal'"/>
             </TemplateRules>
            </TemplateRules>
          </TemplateRules>
        </TemplateRules>
      </Concept>
  </Concepts>
</ConceptRoot>
```

Figure 8. <Requirements> definition example.

An mvdXML file contains a broad set of entities and types of ruleset because an mvdXML file contains multiple <ConceptTemplates>. Moreover, each <ConceptTemplate> has multiple referring concepts, which allows integrating a variety of rules and entities in an mvdXML file. After the mvdXML ruleset is completed, the mvdXML checker can check IFC model by rules.

4.1.4. IFC Support Path

The path for IFC support is essential for the creation of a ruleset. The mvdXML rule transformer processes a path like "IfcWindow->IfcObject.IsDefinedBy. IfcRelDefinesByProperties. RelatingPropertyDefinition.IfcPropertySet.HasProperties.IfcPropertySingleValue.Name=IsExternal". The path consists of an applicable object. IFC4 specification requires parameter value. The applicable IfcObject can be an object described in IFC4 or previous schema, for instance, IfcDoor, IfcWindow, IfcWall, or IfcColumn. The rule can easily be applied to a different object by changing the applicable IfcObject, for instance, replacing IfcWindow with IfcDoor.

The elements are separated with operators. The applicable IfcObject is in front of the "->" operator. The part between the "->" operator and "=" operator is specified according to IFC4. Each instance within the IFC4 specification path is separated using a "."; after the "=" operator, the required parameter value is specified.

The second element of the IFC support path is the specification path according to IFC 4. The mvdXML rule transformer is based on IFC4, an example of IFC documentation can be found in Figure 9. The example specifies property sets for IfcObject [49].

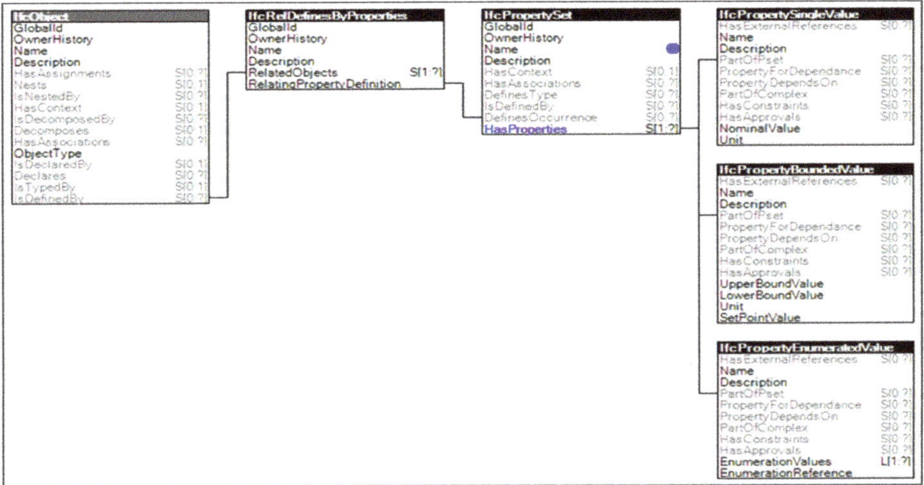

Figure 9. IFC support path.

4.1.5. BCF Report Generator

The IFC model can be checked by the mvdXML ruleset. The IFC objects and attributes from the instance file can be extracted by the developed mvdXML file. Depending on rule types in mvdXML, these values are checked to evaluate their existence, quantity, content, uniqueness, and conditional dependency. The BCF report generator executes the mvdXML checking on the IFC building model (see Figure 10).

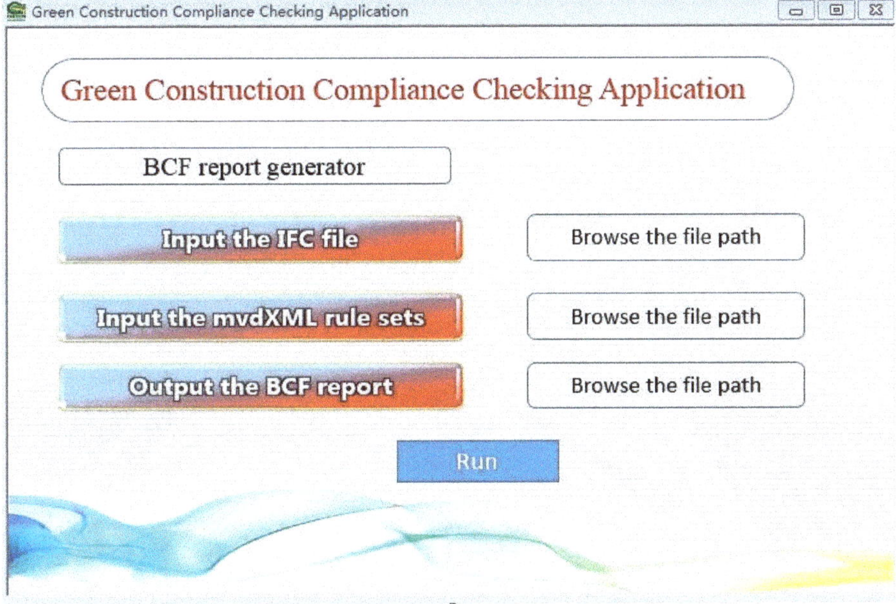

Figure 10. BCF report generator.

4.2. Semantic Technology for Code Checking

4.2.1. Knowledge Base

The knowledge base consists of three types of ontologies, i.e., conceptual model ontology, work item ontology, and construction condition ontology. The purpose of this section is to construct a systematic and computer-readable green construction knowledge base for green construction code checking. A project is composed of a series of work items, and conceptual models can be further classified by work items, which are defined on a project-by-project basis. The conceptual model of green construction is divided into three levels, i.e., basic concept, core concept, and practicability concept. In order to build a better green construction knowledge base, it is necessary to implement inference mapping among conceptual model ontology, work item ontology, and construction condition ontology.

Conceptual Model Ontology

Conceptual model ontology is to extract related concepts from building standards, construction documents, and BIM models, before using the ontology modeling method to build the relationship model of green construction concepts. Through analyzing the relationship between the green construction concepts, the relationship can be described by rules. During green construction code checking, when the information is input to the inference engine, the chosen concepts represent a series of potential requirements. These concepts are regarded as a series of hierarchical concepts, and some semantic relationships exist between concepts. The implementation of inference rules can infer and identify the necessary concepts and their relationships.

Therefore, through the analysis of the characteristics of green construction, the conceptual model ontology includes two parts: (1) structure of concepts; (2) definition of semantic relationships.

(1)   Structure of concepts

The first task of conceptual model ontology construction is to define the classification of concept structure and hierarchy structure related to green construction. Based on the previous studies and the characteristics of green construction, this paper defines the relevant attributes of the main classes and their sub-classes including project classification, building product, building feature, etc.

(2)   Definition of semantic relationships

When the relationship between concepts is defined, the relationship should be associated with richer semantics. The semantic relationship describes the relationships between concepts in ontology, which can be divided into hierarchical and non-hierarchical [50,51]. The relationship between classes can be divided into internal and external. Through the analysis of the definition of semantic relations, the semantic relations in green construction field are mainly divided into two kinds: hyponymy (also expressed as superclass–subclass) and association. Association can be mainly divided into synonymy relationships (also expressed as Equivalent (x, y)), antonymy relationships (also denoted as Disjoint (x, y)), and meronymy relationships (also expressed as Whole-Part (x, y)). The following is a brief introduction to these relationships:

(1)   Equivalent (x, y) is used to describe the similarity between two concepts.
(2)   Disjoint (x, y) is used to describe the relationship of independence or antinomy between two concepts.
(3)   Whole-Part (x, y) is used to describe the meronymy relationship between two concepts.

Work Item Ontology

A work item is the smallest unit of work defined for building elements in a construction project. The work item can be applied to separate phases from design to operation. The classification of work items is according to national standards for project coding and project naming. A construction process may consist of one or more sub-processes, and a sub-process may consist of a collection of work items for multiple building elements. To establish the ontology of work items, the proposed ontology needs to know the corresponding work items and the corresponding amount of resources necessary. Therefore, the construction of work item ontology can facilitate green construction inspection.

Existing work item ontologies, such as FreeClassOWL ontology, are built based on datasets from the European building and building materials market [52]. By describing the building products and services, they can satisfy a wide range of products, suppliers, warehouses, and other related construction product search needs. More than 88 million triplet business data, 81 product brands, 19 distributors, 56,360 product types, and 1,783,798 products are described in the FreeClassOWL Ontology. Therefore, the work item ontology is huge in terms of scale. During the construction process, participants who have rich knowledge and experience need to be gathered to define the ontology model. The work item ontology proposed in this section is composed of four basic entities: actor, knowledge, resource, and schedule. The basic structure of work item ontology is shown in Figure 11. The actor consists of organization and personnel, the organization is composed of government and company, and personnel is composed of the design, construction, operation, maintenance, and other personnel who assume a variety of roles. Knowledge is generated or updated from each construction process or event. Resource consists of workforce, equipment, material, specification, etc. Construction material is classified according to classification Table 41 of OmniClass. Schedule is the construction schedule, including the actual schedule and planning schedule.

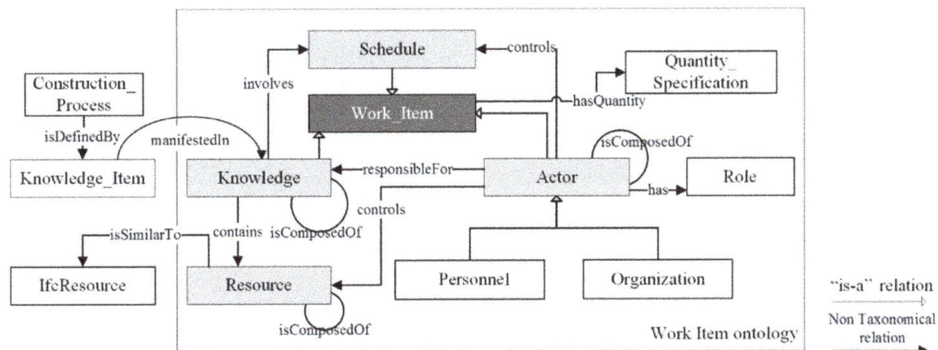

Figure 11. Work item ontology.

Construction Condition Ontology

After building the work item ontology, the product, process, and personnel involved in the ontology can be classified in detail. To realize the information description of the work items about the specific construction conditions, it is very necessary to build mapping relationships between the work item ontology, conceptual model ontology, and construction condition ontology.

Based on the work item ontology and concept model ontology, this paper uses logical inference technology to deduce corresponding concepts or work items through the proposed construction condition ontology. The information in the ontology of construction condition comes from the BIM model (see Figure 12), evaluation criteria, historical data, etc., which include the construction environment, construction machinery, construction sites, construction workers, building models under construction, etc. (see Figure 13). In construction condition ontology, the corresponding work items can be found according to the defined inference rules. For example, specific material information extracted from IFC or a construction document is converted into OWL/RDF format, which can be found in the resource of the construction condition ontology.

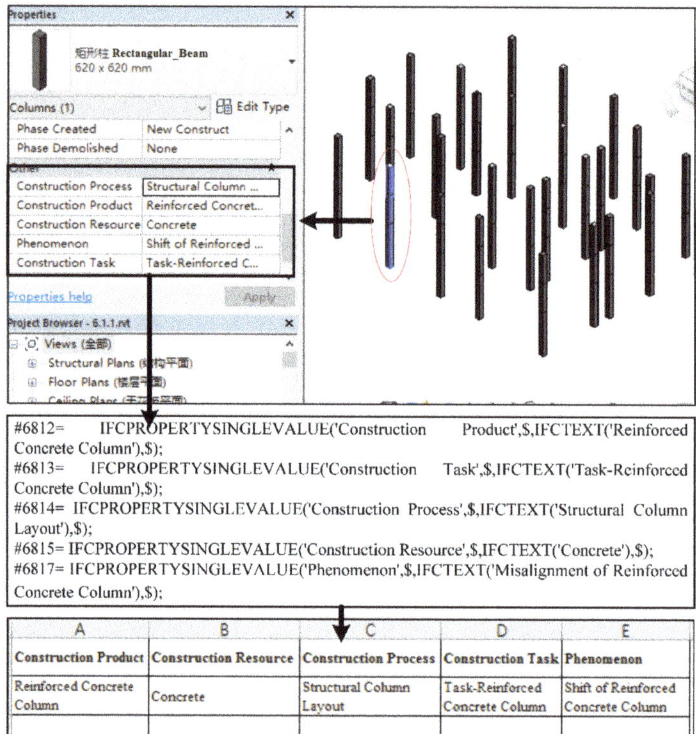

Figure 12. Extraction of construction information from BIM model.

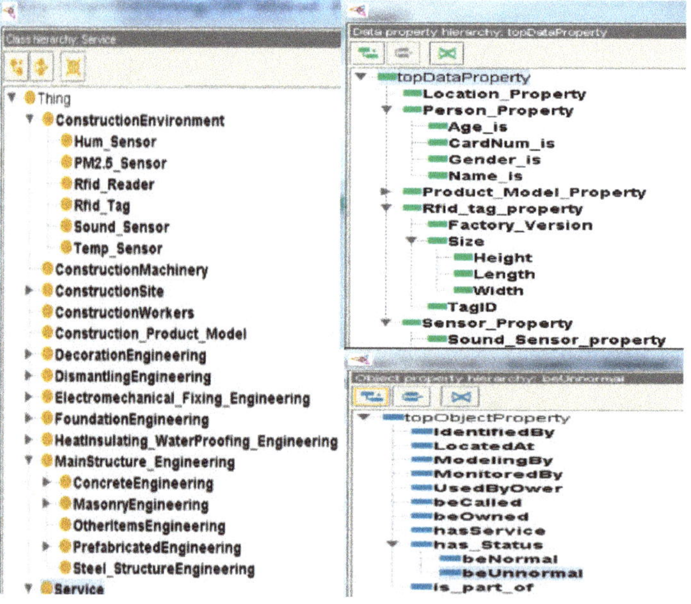

Figure 13. Construction condition ontology.

The needed information to build a green construction knowledge base mainly comes from documents, developers, construction enterprises, governmental agencies, and operation enterprises. Many persons are required to conduct data analysis, extract required information, and build ontology, which takes a lot of manpower and time. Therefore, due to the particularity of green construction code checking, the work of ontology building will be a long-term process.

4.2.2. Semantic Inference

SWRL is used to define related semantic rules during the inference phase; however, SWRL is a language that does not rely on any inference engine and cannot be used without inference tools. Therefore, OWL and SWRL need to be converted into available rules for inference. In the process of ontology inference, JESS is used as an inference engine and consists of a fact base and a rule base. As the most mature inference engine for SWRL, JESS is used in many fact-based systems and is capable of handling ontology-based SWRL rules. In this paper, the JESS rule engine is used to transform OWL and SWRL rules into JESS facts, and it is used to match the rule base and fact base to implement inference. Some specific steps for ontology inference are as follows:

Step 1: In the inference layer, XSLT/XML is used to convert conceptual model ontology, work item ontology, construction condition ontology, and their corresponding inference rules into JESS knowledge for query.

Step 2: The JESS inference engine is used to match the rule base and fact base during the inference process. The inference result is converted to RDF for output using XSLT/XML.

Step 3: Fact and rule bases are stored in the JESS repository.

Some classes and their attribute relationships are defined in Protégé (see Figure 14).

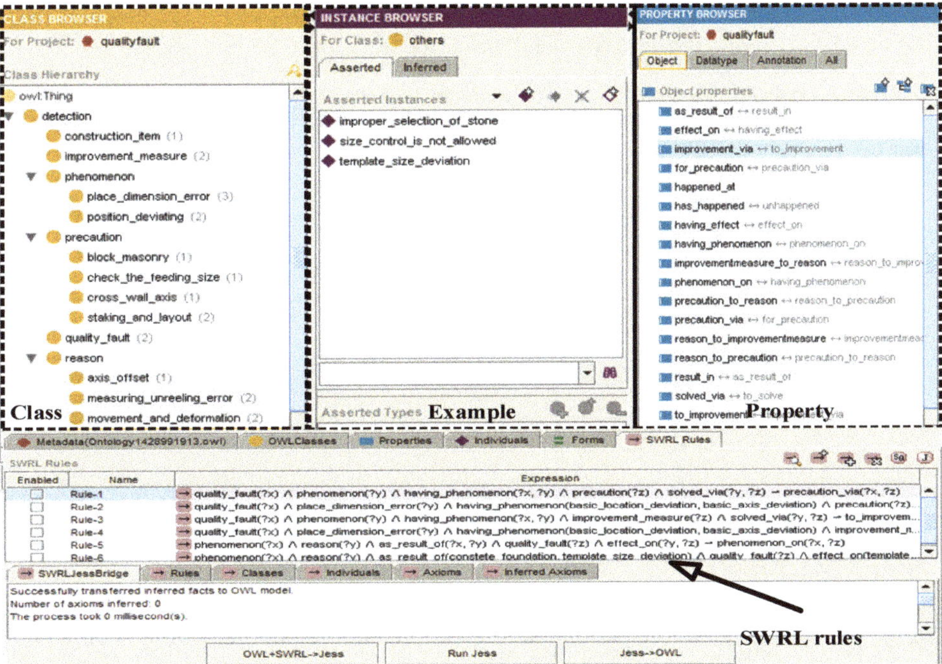

Figure 14. Some classes and instances built by Protégé.

The JESS inference engine can only interpret the JESS code; thus, the structured presentation languages OWL and SWRL need to be transformed into the JESS-encoded form using the OWL2Jess and SWRL2Jess translators. Both translators are built-in plug-ins of Protégé to integrate heterogeneous information between OWL, SWRL, and JESS. IFC files can be converted from models saved in ISO STEP (Standard for the Exchange of Product model data). The models are specified in EXPRESS and represented by OWL. EXPRESS is a standard data modeling language for product data. Table 3 shows the data type mapping relationships between EXPRESS, OWL, and JESS.

Table 3. The data type mapping relationships between EXPRESS, OWL, and JESS.

EXPRESS	OWL	JESS
real, integer, number, etc.	owl: real, xsd: integer, xsd: decimal, owl: rational, etc	jess.ru.integer, jess.ru.float, jess.ru.long
string	xsd: string	jess.ru.string
boolean	xsd: boolean	The atoms "TRUE" and "FALSE"

The constructions of conceptual model ontology, work item ontology, and construction condition ontology are beneficial for the green construction knowledge sharing and reuse. OWL can express the concept of knowledge of green construction. SWRL can describe inference rules in ontologies. Through the combination of SWRL and the JESS inference engine, inference results of green construction code checking can be easily obtained.

5. Case Study of Green Construction Code Checking

In this paper, some types of rules and the corresponding clauses in GB/T50905-2014 are shown as examples.

Class 1: Code checking based on simple defined attribute value in IFC schema.

9.3.5 Extra item: staff dormitory must meet the requirements of at least two square meters per person.

mvdXML rule:

IfcPhysicalQuantity.IfcPhysicalSimpleQuantity.IfcQuantityArea.AreaValue>2*4 (there are four people in every room in this case).

SQWRL rule:

```
Room(?r) ˆ hasWidth(?r, ?width) ˆ hasHeight(?r, ?height) ˆ
    swrlm:eval(?area, "width * height", ?width, ?height)
    -> sqwrl:select(?area).
```

Class 2: Code checking based on the extension of IFC entities.

6.2.1 General item: fly ash, slag admixture, and other new materials should be used during the construction period.

The fly ash and slag admixture must be inserted into IFC baseline, because they are not defined in existing IFC (see Figure 15).

mvdXML rule:

IfcObject.IsDefinedBy.IfcRelDefinesByProperties.RelatingPropertyDefinition.IfcMaterialDefinition.HasProperties.IfcMaterialProperties.Name=Flyash

IfcObject.HasAssocialtions.IfcRelAssociatesMaterial.RelatingMaterial.IfcMaterial.Material.Name= slag.

After the checking is executed, the mvdXML checker captures each generated issue in a BCF report which mainly includes a markup file and a viewpoint file. The markup file is where most of the information about an issue resides. In a markup file, information about the model, about the issue, and about each comment on the issue can be obtained. The generated issue comments in the markup file contain the description of the "concept" defined in the mvdXML file to make domain users understand the requirement that is violated. The viewpoint file contains all the information about the camera, the elements in the view, and the software used. BIM software can be used to find and analyze the generated issues from the mvdXML checker. The BCF report can be opened with BCFier in Revit (see Figure 16).

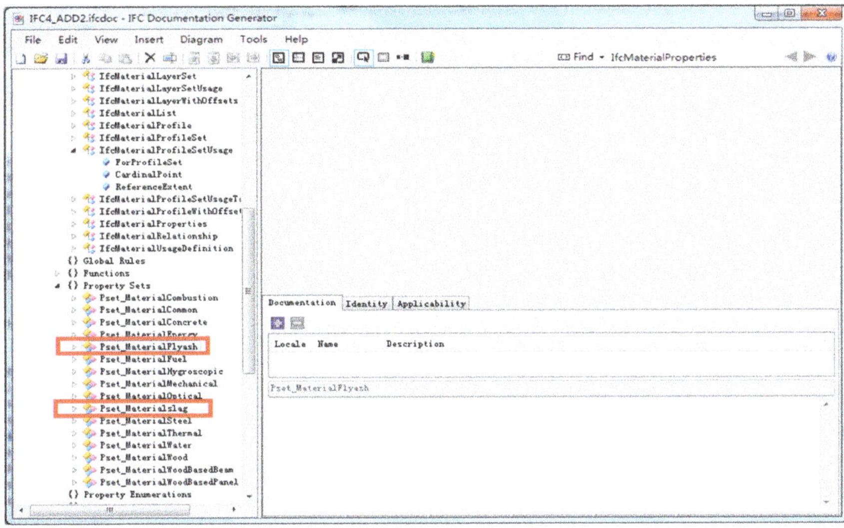

Figure 15. Insert the new properties in IFC.

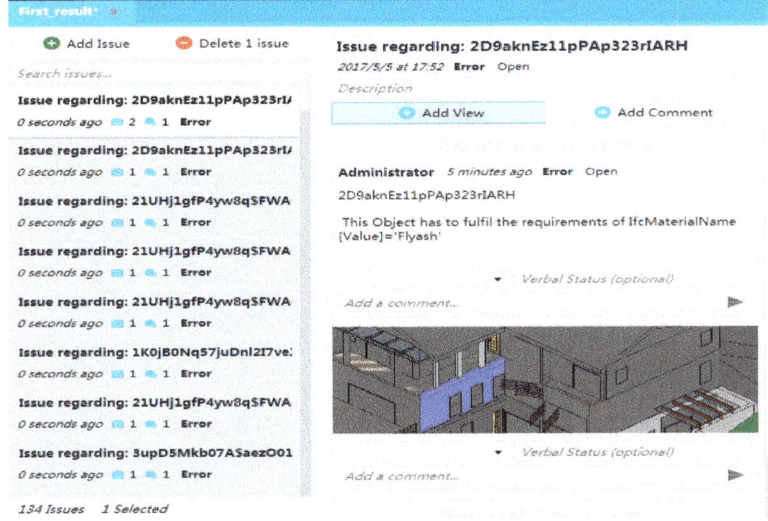

Figure 16. The BCF report of the code checking.

Class 3: Code checking based on simple conceptual reasoning.

9.2.4 Extra item: Cast-in-place concrete raw materials should be mixed in construction site.

The constructed ontology model and rules are stored in the knowledge base and rule base, respectively. They are integrated to conduct inference in the JESS inference part. The inference result from the JESS inference engine is shown in Figure 17. The result shows that "cast-in-place concrete" is composed of materials "cement", "sand", "stone", and "water". The "Rectangular_Beam" and its attributes are defined in the conceptual model ontology, while "cement", "sand", "stone", and "water" are defined in the work item ontology. The inference process is related to construction condition knowledge. Therefore, the realization of this inference process is based on the concept model ontology, work item ontology, and construction condition ontology.

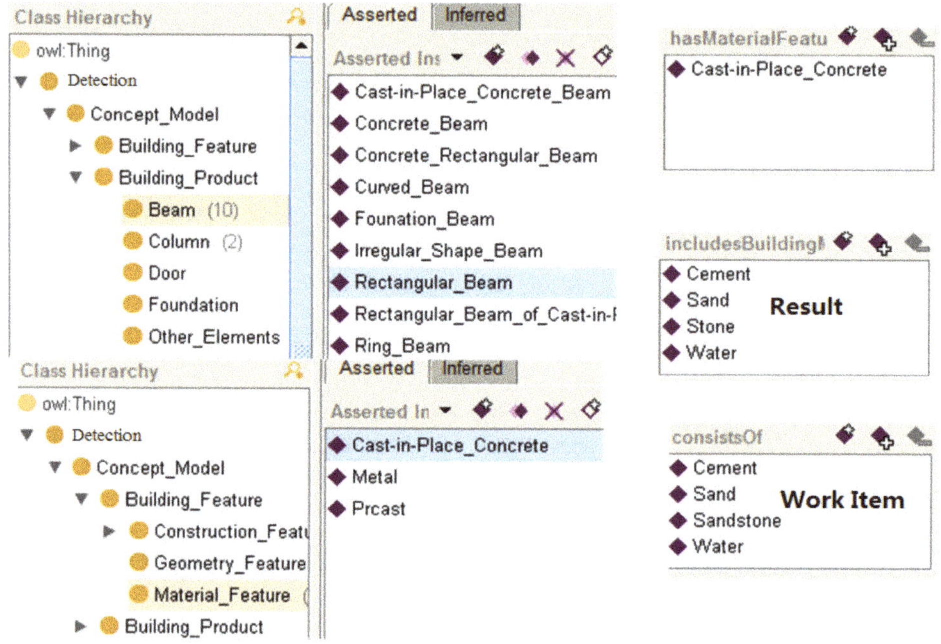

Figure 17. Inference result.

SQWRL rule:

Cast_In_Place_Concrete(?c) ˆ hasComponents(?c, ?z) ˆ hasMaterials(?z, ?d) -> sqwrl:select(?d) ˆ sqwrl:orderBy(?d).

Class 4: Code checking based on construction condition reasoning.

3.3.2 Extra item: Construction field noise emission at daylight should not exceed 70 dB (A) and should not exceed 55 dB at night (A).

This kind of clause is closely related to the construction site environment. The noise monitoring in the construction process is embodied in the construction condition ontology. The PM2.5_Sensor and Sound_Sensor are set in ontology (see Figure 18). The diagram of classes and properties in construction condition ontology is shown in Figure 19.

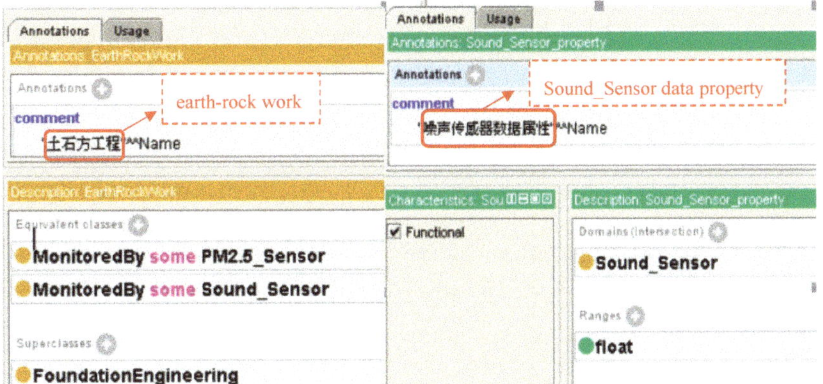

Figure 18. The sensor set in construction condition ontology.

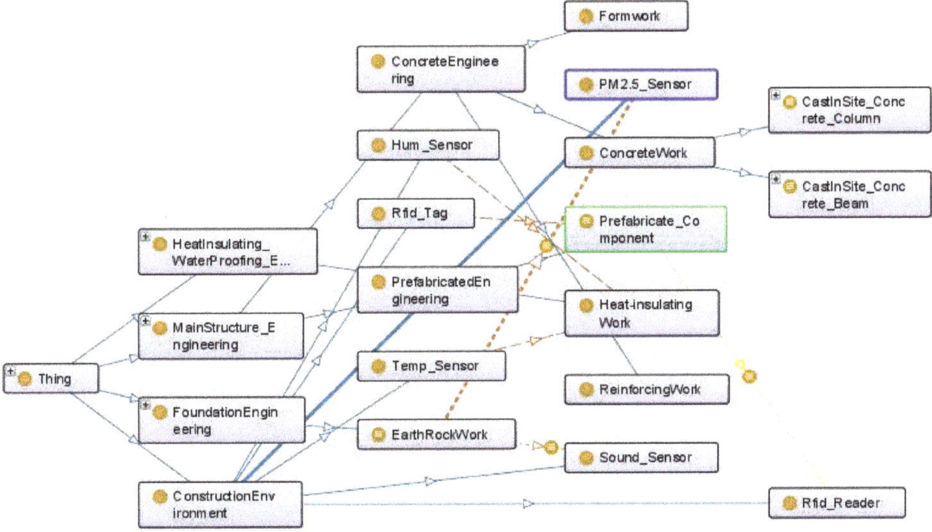

Figure 19. The diagram of classes and properties in construction condition ontology.

For example, if the entry place of the construction site is the fixed position of the Radio-frequency identification (RFID) reader, the component, material, and the person after passing the RFID tag will have the position information, which can be used for ontology inference, along with other attributes, to judge the execution of the service. The construction site will be equipped with noise sensors to collect noise data; when the noise data exceeds the threshold specified in green construction code, an early warning will be triggered automatically. The overrun alarm for noise values is shown in Figure 20. The rule expressions are as follows:

SWRL rule1

> (?Rfid_Tag t1:beOwneredBy ?Prefabricate_Component)
> (?Rfid_Tag t1:IdentifiedBy ?entrance_Rfid_Reader)
> -> (?Prefabricate_Component t1:locatedAt ?Entrance).

SWRL rule2

        (?Prefabricate_Component t1:locatedAt ?Entrance)
->   (?Service t1:beCalled t1:Register).

SWRL rule3

        (EarthRockWork t1:MoniterdBy ?Sound_Sensor)
        (?Sound_Sensor t1:hasStatus t1:is_unnormal)
->   (?Service t1:beCalled t1:Alarm).

SQWRL rule

Sound_Sensor (?p) ˆ hasValue(?p, ?s) ˆ swrlb:greaterThan(?s, 70) -> unnormal(?p).

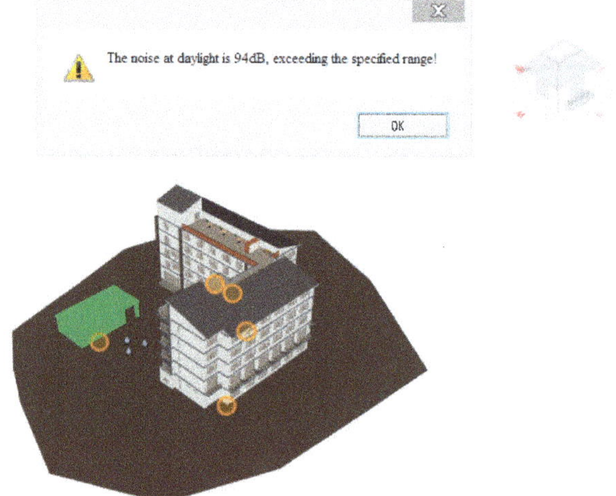

Figure 20. The overrun alarm for noise values.

Finally, all the code checking results can be summarized to a HTML report (see Figure 21).

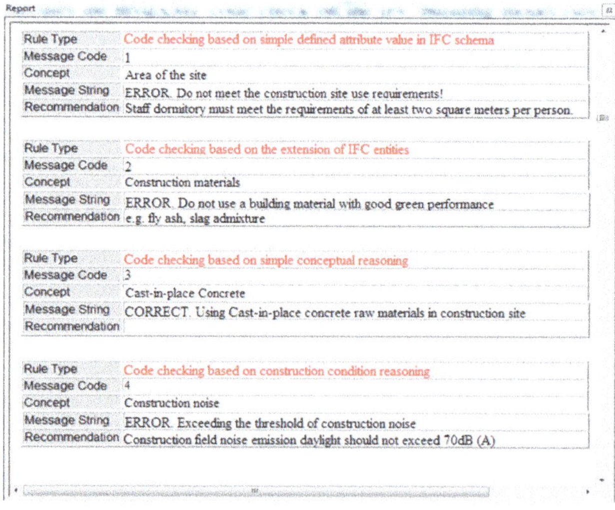

Rule Type	Code checking based on simple defined attribute value in IFC schema
Message Code	1
Concept	Area of the site
Message String	ERROR. Do not meet the construction site use requirements!
Recommendation	Staff dormitory must meet the requirements of at least two square meters per person.

Rule Type	Code checking based on the extension of IFC entities
Message Code	2
Concept	Construction materials
Message String	ERROR. Do not use a building material with good green performance
Recommendation	e.g. fly ash, slag admixture

Rule Type	Code checking based on simple conceptual reasoning
Message Code	3
Concept	Cast-in-place Concrete
Message String	CORRECT. Using Cast-in-place concrete raw materials in construction site
Recommendation	

Rule Type	Code checking based on construction condition reasoning
Message Code	4
Concept	Construction noise
Message String	ERROR. Exceeding the threshold of construction noise
Recommendation	Construction field noise emission daylight should not exceed 70dB (A)

Figure 21. The HTML report of code checking.

6. Conclusions and Perspectives

Although professionals in the AEC domain realize the importance of green construction, there is no efficient and accurate evaluation method for green construction. Traditional green construction evaluation is usually completed by manual inspection, with low efficiency and high error rate. Therefore, construction industry practitioners put forward urgent requirements for a more efficient and convenient approach to conduct green construction code checking.

Based on the particularity of green construction standards and the limitation of existing code-checking approaches, this paper combines mvdXML technology and semantic web technology to organize, store, and re-use green construction knowledge. In our approach, BIM, logical rule, ontology, and semantic web query language are integrated to conduct green construction code checking. These techniques are used for automated information transformation and extension, knowledge base construction, logical inference, and information query. The automation is facilitated by semantic-based and logic-based representations, which is general and flexible. The approach is composed of five main parts: data source, knowledge base, rule base, JESS inference, and query.

In this paper, code checking of green construction standards was classified into four classes based on the difficulty level to meet the requirements of the clauses, and the particularity of green construction code checking was analyzed. Based on the above classification and analysis, several types of green construction rules were checked. The proposed BIM-based code-checking approach for green construction was validated by a case study.

The proposed approach not only focuses on quantitative requirements, but also supports the checking of other types of requirements, such as checking based on semantic information of BIM and checking based on project criteria, owner requirement, etc.

7. Limitations and Future Work

The proposed approach offers a solution to code checking for green construction. However, there are some limitations in this approach, and there is still some work to be done in future research. Firstly, the scoring rules of existing green construction standards are relatively rough, which results in weak operability; thus, a more detailed standard is expected to be issued to improve green construction inspection. Secondly, although the approach proposed in this paper enables making various rules, parameters that are not predefined cannot be checked without being encoded into programming language. Thirdly, automated regulatory code checking requires automated extraction of requirements from regulatory textual documents, and then the extracted requirements need to be transformed into a formalized format that enables JESS inference, which still requires human judgement and cannot reach the level of artificial intelligence. Information extraction (IE) in code checking is a challenging task that requires complex analysis and processing of text. Some researchers started using deep learning algorithms like NLP (Natural Language Processing) to solve this kind of problem, but it is still imperfect. The field of automatic code checking is emerging, and it will continue to offer more help in the future.

Author Contributions: S.J. conceived the idea for the paper, oversaw its development and supervised all aspects of this research. Z.W. designed the methodology of code checking. B.Z. analyzed and classified code checking of green construction. H.S.C. conducted the case study and analyzed the results. All authors have reviewed and approved the final manuscript.

Funding: This research was funded by the National Key R&D Program of China grant number 2016YFC0702107.

Conflicts of Interest: The authors declare no conflict of interest.

References

1. Cole, R.J. Energy and greenhouse gas emissions associated with the construction of alternative structural systems. *Build. Environ.* **1998**, *34*, 335–348. [CrossRef]
2. Wong, J.K.W.; Zhou, J. Enhancing environmental sustainability over building life cycles through green BIM: A review. *Autom. Constr.* **2015**, *57*, 156–165. [CrossRef]

3. González, M.J.; García Navarro, J. Assessment of the decrease of CO2 emissions in the construction field through the selection of materials: Practical case study of three houses of low environmental impact. *Build. Environ.* **2006**, *41*, 902–909. [CrossRef]

4. Won, J.; Cheng, J.C.P.; Lee, G. Quantification of construction waste prevented by BIM-based design validation: Case studies in South Korea. *Waste Manag.* **2016**, *49*, 170–180. [CrossRef] [PubMed]

5. IFC Model View Definition Format. Available online: http://www.buildingsmart-tech.org/downloads/accompanying-documents/formats/mvdxml-documentation/MVD_Format_V2_Proposal_080128.pdf (accessed on 17 December 2018).

6. Connor, M.O.; Das, A. SQWRL: A query language for OWL. In Proceedings of the International Conference on Owl: Experiences and Directions, Chantilly, VA, USA, 23–24 October 2009; pp. 208–215.

7. Fudholi, D.H.; Maneerat, N.; Varakulsiripunth, R.; Kato, Y. Application of protégé, SWRL and SQWRL in fuzzy ontology-based menu recommendation. In Proceedings of the ISPACS 2009—2009 International Symposium on Intelligent Signal Processing and Communication Systems, Kanazawa, Japan, 7–9 December 2009; pp. 631–634.

8. BIM Levels Explained. Available online: https://www.thenbs.com/knowledge/bim-levels-explained (accessed on 19 November 2018).

9. Ismail, A.S.; Ali, K.N.; Iahad, N.A. A Review on BIM-based automated code compliance checking system. In Proceedings of the IEEE International Conference on Research and Innovation in Information Systems, ICRIIS, Langkawi, Malaysia, 16–17 July 2017.

10. Solihin, W.; Eastman, C. Classification of rules for automated BIM rule checking development. *Autom. Constr.* **2015**, *53*, 69–82. [CrossRef]

11. Eastman, C.; Lee, J.; Jeong, Y.; Lee, J. Automatic rule-based checking of building designs. *Autom. Constr.* **2009**, *18*, 1011–1033. [CrossRef]

12. Dimyadi, J.; Solihin, W.; Hjelseth, E. Classification of BIM-based Model checking concepts. *J. Inf. Technol. Constr.* **2016**, *21*, 354–370.

13. Krijnen, T.; Berlo, L.A.H.M. Van Methodologies for requirement checking on building models: A technology overview. In Proceedings of the the 13th International Conference on Design and Decision Support Systems in Architecture and Urban Planning, Eindhoven, The Netherlands, 27–28 June 2016; pp. 1–11.

14. Fahad, M.; Bus, N.; Andrieux, F. Towards Mapping Certification Rules over BIM Towards Mapping Certification Rules over BIM. In Proceedings of the the 33rd CIB W78 Conference, Brisbane, Australia, 31 October–2 November 2016.

15. Choi, J.; Kim, I. An Approach to Share Architectural Drawing Information and Document Information for Automated Code Checking System. *Tsinghua Sci. Technol.* **2008**, *13*, 171–178. [CrossRef]

16. Lee, Y.C.; Eastman, C.M.; Lee, J.K. Validations for ensuring the interoperability of data exchange of a building information model. *Autom. Constr.* **2015**, *58*, 176–195. [CrossRef]

17. Nawari, N. The Challenge of Computerizing Building Codes in a BIM Environment. *Comput. Civ. Eng.* **2012**, *1*, 285–292.

18. Nawari, N. SmartCodes and BIM. In Proceedings of the Structures Congress 2013, Pittsburgh, PA, USA, 2–4 May 2013; pp. 928–937.

19. Ilhan, B.; Yaman, H. Green building assessment tool (GBAT) for integrated BIM-based design decisions. *Autom. Constr.* **2016**, *70*, 26–37. [CrossRef]

20. Ciribini, A.L.C.; Mastrolembo Ventura, S.; Paneroni, M. Implementation of an interoperable process to optimise design and construction phases of a residential building: A BIM Pilot Project. *Autom. Constr.* **2016**, *71*, 62–73. [CrossRef]

21. Zhang, S.; Teizer, J.; Lee, J.K.; Eastman, C.M.; Venugopal, M. Building Information Modeling (BIM) and Safety: Automatic Safety Checking of Construction Models and Schedules. *Autom. Constr.* **2013**, *29*, 183–195. [CrossRef]

22. Kasim, T.; Li, H.; Rezgui, Y.; Beach, T. Automated Sustainability Compliance Checking Process: Proof of Concept. In Proceedings of the 13th International Conference on Construction Applications of Virtual Reality (ConVR), London, UK, 30–31 October 2013; pp. 11–21.

23. Lee, Y.C.; Eastman, C.M.; Solihin, W.; See, R. Modularized rule-based validation of a BIM model pertaining to model views. *Autom. Constr.* **2016**, *63*, 1–11. [CrossRef]

24. Dimyadi, J.; Solihin, W.; Preidel, C.; Borrmann, A. Towards code compliance checking on the basis of a visual programming language. *J. Inf. Technol. Constr.* **2016**, *21*, 402–421.

25. Dimyadi, J.; Solihin, W.; Solihin, W.; Eastman, C. A knowledge representation approach in BIM rule requirement analysis using the conceptual graph. *J. Inf. Technol. Constr.* **2016**, *21*, 370–402.

26. Zhang, J.; El-Gohary, N.M. Semantic NLP-Based Information Extraction from Construction Regulatory Documents for Automated Compliance Checking. *J. Comput. Civ. Eng.* **2016**, *30*, 04015014. [CrossRef]

27. Zhang, J.; El-Gohary, N.M. Integrating semantic NLP and logic reasoning into a unified system for fully-automated code checking. *Autom. Constr.* **2017**, *73*, 45–57. [CrossRef]

28. Zhang, J.; El-Gohary, N.M. Automated Information Transformation for Automated Regulatory Compliance Checking in Construction. *J. Comput. Civ. Eng.* **2015**, *29*, B4015001. [CrossRef]

29. Jiang, L.; Leicht, R.M. Supporting automated Constructability checking for formwork construction: An ontology. *J. Inf. Technol. Constr.* **2016**, *21*, 456–478.

30. Pauwels, P.; Terkaj, W. EXPRESS to OWL for construction industry: Towards a recommendable and usable ifcOWL ontology. *Autom. Constr.* **2016**, *63*, 100–133. [CrossRef]

31. Pauwels, P.; de Farias, T.M.; Zhang, C.; Roxin, A.; Beetz, J.; De Roo, J.; Nicolle, C. A performance benchmark over semantic rule checking approaches in construction industry. *Adv. Eng. Inform.* **2017**, *33*, 68–88. [CrossRef]

32. RDF 1.1 Semantics. Available online: https://www.w3.org/TR/rdf11-mt/ (accessed on 17 December 2018).

33. RDF 1.1 XML Syntax. Available online: https://www.w3.org/TR/2014/REC-rdf-syntax-grammar-20140225/ (accessed on 17 December 2018).

34. OWL 2 Web Ontology Language Document Overview (Second Edition). Available online: https://www.w3.org/TR/owl2-overview/ (accessed on 18 December 2018).

35. Bouzidi, K.R.; Fies, B.; Faron-Zucker, C.; Zarli, A.; Thanh, N. Le Semantic Web Approach to Ease Regulation Compliance Checking in Construction Industry. *Future Internet* **2012**, *4*, 830–851. [CrossRef]

36. Do, S.L.; Shin, M.; Baltazar, J.C.; Kim, J. Energy benefits from semi-transparent BIPV window and daylight-dimming systems for IECC code-compliance residential buildings in hot and humid climates. *Solar Energy* **2017**, *155*, 291–303. [CrossRef]

37. Bakillah, M.; Mostafavi, M.A.; Liang, S.H.L. Enriching SQWRL Queries in Support of Geospatial Data Retrieval from Multiple and Complementary Sources. In *Proceedings of the Sixth International Workshop on Semantic and Conceptual Issues in GIS*; Castano, S., Vassiliadis, P., Lakshmanan, L.V.S., Lee, M.L., Eds.; Springer: Florence, Italy, 2012; pp. 241–250.

38. Zhong, B.T.; Ding, L.Y.; Luo, H.B.; Zhou, Y.; Hu, Y.Z.; Hu, H.M. Ontology-based semantic modeling of regulation constraint for automated construction quality compliance checking. *Autom. Constr.* **2012**, *28*, 58–70. [CrossRef]

39. Lu, Y.; Li, Q.; Zhou, Z.; Deng, Y. Ontology-based knowledge modeling for automated construction safety checking. *Saf. Sci.* **2015**, *79*, 11–18. [CrossRef]

40. Ding, L.Y.; Zhong, B.T.; Wu, S.; Luo, H.B. Construction risk knowledge management in BIM using ontology and semantic web technology. *Saf. Sci.* **2016**, *87*, 202–213. [CrossRef]

41. Lee, Y.C.; Eastman, C.M.; Solihin, W. An ontology-based approach for developing data exchange requirements and model views of building information modeling. *Adv. Eng. Inform.* **2016**, *30*, 354–367. [CrossRef]

42. Zhang, H.; Zhao, W.; Gu, J.; Liu, H.; Gu, M. Semantic Web Based Rule Checking of Real-World Scale BIM Models: A Pragmatic Method. *Semant. Web*. 2017. Available online: http://www.semantic-web-journal.net/system/files/swj1621.pdf (accessed on 29 March 2019).

43. de Farias, T.M.; Roxin, A.; Nicolle, C. A rule-based methodology to extract building model views. *Autom. Constr.* **2018**, *92*, 214–229. [CrossRef]

44. Schwabe, K.; König, M.; Teizer, J. BIM Applications of Rule-based Checking in Construction Site Layout Planning Tasks. In Proceedings of the 33rd International Symposium on Automation and Robotics in Construction (ISARC), Auburn, AL, USA, 18–21 July 2016; pp. 209–217.

45. mvdXML: Specification of a Standardized Format to Define and Exchange Model View Definitions with Exchange Requirements and Validation Rules. Available online: http://www.buildingsmart-tech.org/downloads/accompanying-documents/formats/mvdxml-documentation/mvdXML_V1-0.pdf (accessed on 18 December 2018).

46. Van Berlo, L.; Krijnen, T. Using the BIM Collaboration Format in a Server Based Workflow. *Procedia Environ. Sci.* **2014**, *22*, 325–332. [CrossRef]
47. Zhang, C.; Beetz, J.; Weise, M. Interoperable validation for IFC building models using open standards. *J. Inf. Technol. Constr.* **2015**, *20*, 24–39. [CrossRef]
48. IFC Model Checking Based on mvdXML 1.1. Available online: https://core.ac.uk/display/80693771 (accessed on 18 December 2018).
49. Industry Foundation Classes—Version 4—Addendum 2. Available online: http://www.buildingsmart-tech.org/ifc/IFC4/Add2/html/ (accessed on 18 December 2018).
50. Khan, S.; Safyan, M. Semantic matching in hierarchical ontologies. *J. King Saud Univ. Comput. Inf. Sci.* **2014**, *26*, 247–257. [CrossRef]
51. Punuru, J.; Chen, J. Extraction of Non-hierarchical Relations from Domain Texts. In Proceedings of the 2007 IEEE Symposium on Computational Intelligence and Data Mining, Honolulu, HI, USA, 1–5 April 2007; pp. 444–449.
52. Radinger, A.; Rodriguez-Castro, B.; Stolz, A.; Hepp, M. BauDataWeb: The Austrian building and construction materials market as linked data. In *Proceedings of the 9th International Conference on Semantic Systems—I-SEMANTICS '13*; ACM Press: New York, NY, USA, 2013; pp. 25–32.

Article

Automated Generation of Daily Evacuation Paths in 4D BIM

Kyungki Kim [1] and Yong-Cheol Lee [2,*]

[1] Department of Construction Management, University of Houston, 4734 Calhoun Road #111, Houston, TX 77204-4020, USA; kkim38@central.uh.edu

[2] Bert S. Turner Department of Construction Management, Louisiana State University, 3315E Patrick F. Taylor Hall, Baton Rouge, LA 70803, USA

* Correspondence: yclee@lsu.edu; Tel.: +1-225-578-5483

Received: 22 February 2019; Accepted: 18 April 2019; Published: 29 April 2019

Featured Application: The proposed approach can potentially become one of path finding applications or add-ins of Building Information Modeling (BIM) tools. This application utilizes detailed daily or weekly construction plans to generate all possible evacuation paths of crews.

Abstract: Spatial movements of workers and equipment should be carefully planned according to project plans. In particular, it is crucial for workers' safety to prepare emergency evacuation paths according to changing construction site configurations and construction progress. However, creating evacuation paths for all crews for each day can be an extremely labor-intensive task if it is done manually. Consequently, in most construction projects, evacuation plans are not provided to managers and crews throughout the entire construction. Even state-of-the-art technologies do not suggest ways to generate evacuation paths according to changing progresses presented in 4-Dimensional Building Information Model (4D BIM). This research proposes a framework to automatically analyze, generate, and visualize the evacuation paths of multiple crews in 4D BIM, considering construction activities and site conditions at the specific project schedule. This research develops a prototype that enables users to define parameters for pathfinding, such as workspaces, material storage areas, and temporary structures to automatically identify the accessible evacuation paths. This prototype shows the secured evacuation paths in the 4D BIM environment and allows the users to organize the automatically generated evacuation paths. A case study using the BIM model of a real construction project involved in this paper demonstrates the potential of the proposed method.

Keywords: BIM; safety; path planning; A-Star Searching; evacuation

1. Introduction

Throughout a dynamically changing construction project, managing limited spaces of a construction site is imperative to seamlessly facilitate linearly and cross-linking planned construction activities and operations. A project manager or an associated industry professional must carefully create a site logistic and safety plan that explicitly illustrate secured work spaces and accessible pathways according to construction phases and various moving paths for workers, equipment, and material delivery. In addition, before initiating daily work on a construction site, work crews in diverse domains must be trained and educated regarding the evacuation paths, emergency exits, and secured spaces so that they can rapidly escape in time when emergency situations occur. In a tool-box meeting where each work team shares daily work agendas, work crews are supposed to be educated about daily task details, designated work locations, relevant safety requirements, emergency manuals, and evacuation path information. However, such construction space management and path planning require labor-intensive and time-consuming tasks because construction site conditions are generally varied by numerous factors, such as construction schedules, associated work locations, major equipment, temporary structures, and others. Especially, interactively creating and updating emergency evacuation plans according to the changing circumstances may require significant time and effort. Usable evacuation paths should be identified, secured, and shared with related work crews and construction managers through construction management tools or Building Information Modeling (BIM) software.

Despite this importance, there are several challenges in interactive evacuation path planning because of the complexity of modern construction projects and manual planning practices. A construction environment is generally formed by building objects (e.g., walls, columns, floors) and various non-building objects (e.g., workers, equipment, and temporary structures). Since the status and locations of these objects dynamically change even more often than on a daily basis, it is challenging for individual work crews or construction managers to perceive all the building and non-building changes in order to identify optimal evacuation paths. The lack of identification of secured pathways on site frequently results in additional cost growth and schedule delays during the construction phase. In particular, this unforeseeable condition and working environment of the construction site is one of the primary factors affecting workers' safety. When it comes to safety, it is challenging to predict construction accidents and emergency situations during construction. To provide against such possibilities, a well-developed emergency and evacuation plan representing the existing construction conditions is mandatory. Even though OSHA standards [29 CFR 1910.38(a)] entail the emergency action plan (EAP), the interactive evacuation plans during the construction phase still require further study. In current practices, construction site managers and domain professionals use their past experience and intuitive understanding about changing workplace conditions when planning moving paths of workers, logistics, and vehicles [1]. This relies on subjective judgment that can produce suboptimal or unsafe paths.

With the advancement of data representation and information technology, BIM has increasingly developed to fulfill the evolving demands of architecture, engineering, construction, and facility management (AEC-FM) industries. The remarkable gist of this BIM technology is the direct integration of a geometry and its corresponding information. This integration enables domain professionals to intuitively manage design and construction resources on the BIM platform and to virtually analyze the construction processes and operations in accordance with the given 3D model. One of the promising BIM applications is 4-Dimensional Building Information Model (4D BIM) that merges construction schedule information into a 3D BIM environment. 4D BIM is considered one of the most advanced ways to digitally represent, not only dynamically changing construction sites, but also constantly shifting construction working conditions through the construction timeline. In addition, since BIM generally entails resources and information of relevant objects and relationships, professionals in the construction industry are able to integrate indispensable construction data into one consolidated platform and manipulate them for distinct purposes during the construction phase. With its time-dependent and realistic visualization, various moving paths (e.g., evacuation, site entrance, material delivery) can be planned on top of the 4D BIM platform.

However, the benefits of current 4D BIM applications and analysis are still limited to the visualization of expected construction site conditions. Since 4D BIM can explicitly represent construction processes with associated resources based on a planned schedule, it is a desirable approach to plan work procedures, work crews' paths, and evacuation scenarios according to a planned schedule. However, it can be extremely labor-intensive for project managers to manually analyze complex 4D BIM to establish path plans for multiple work crews and different purposes (entrance, material delivery, evacuation) on a daily basis. Unfortunately, even though several 4D BIM applications have been proposed for daily safety hazard identification [2,3], path planning within 4D BIM has been overlooked.

From this perspective, there is a need for BIM-based automation that generates evacuation paths throughout the construction processes. This research presents a method to automatically generate rational evacuation paths of multiple crews using information in 4D BIM. The analyzed paths according to dynamic site conditions and changing construction progress will be valuable assets for field managers and the workforce to secure guaranteed pathways and accessible spaces.

This paper is organized as follows. The literature review section presents a review of state-of-the-art approaches in the area of 4D BIM-based automated analysis and BIM-based construction path generation. A point of departure, a research objective, and a study scope are presented. The next section presents the framework and algorithms of the proposed evacuation path generation in 4D BIM. The case study section presents a validation in BIM of a real-world construction project and the benefit section contains the expected applications and advantages. The conclusions and discussion section presents the conclusion, limitations, and suggestion for future studies.

2. Literature Review

2.1. Automated Construction Analysis Using 4D BIM

In the past, construction scheduling and planning was conducted using the Critical Path Method (CPM) and the Computer-Aided Drawing (CAD). Despite wide uses of the CPM and CPM-based scheduling techniques (Kim et al. 2015; Senouci and Hassan 2008), these approaches have a critical limitation that is the absence of visualization of construction processes. BIM incorporated CPM activities into 3D building objects so that expected construction progresses can be visualized in a timeline of a construction schedule. In many studies, 4D BIM demonstrated its usefulness in construction planning [4], jobsite safety analysis [5], and constructability checking [6].

Taking advantage of rich digital information in BIM, diverse research studies presented methods to automatically conduct several types of construction analyses. Akinci et al. [7] created various types of spaces in 4D BIM for automated workspace conflict identification. Jongeling et al. [8] integrated spatial workflows of multiple crews in a construction project to quantitatively analyze potential productivity losses due to proximity among them. Zhang et al. [3] presented a rule-based safety hazard identification system focusing on falling hazards. Kim and Teizer [9] automatically generated scaffolding objects in 4D BIM, Kim et al. [2] identified potential safety hazards related to the scaffolding in 4D BIM, and the automated safety hazard checking tool was used to assist in decision making [10,11].

2.2. Automated Generation of Moving Paths in BIM

In terms of algorithms for path planning and egress path finding, diverse approaches to predicting and analyzing the conditions of available paths and spaces on a construction site have been studied and developed. Soltani et al. [1] presented the applications of path planning algorithms (Dijkstra, A*, and Genetic Algorithms) to generate optimized paths in terms of travel distance, safety risks, and visibility. On a grid-based site representation, three additional layers of visibility, hazard, and distance were incorporated and the weighted-sum of multi-criteria scores was used for optimization. While this research successfully incorporated critical issues (safety and productivity) into path planning, it has limitations: (1) Dynamically changing spatial-temporal conditions cannot be analyzed. (2) Unsafe locations need to be specified manually. Unsafe locations appear and disappear during construction. Considering the two limitations above, the implementation of the proposed approach may require excessive user-defined inputs. One research used room boundaries to define spaces and implement the analysis relying on architectural information [12]. However, it may lead to wrong results because site configuration dynamically changes upon installation and the dismantlement of components and rooms may not exist until a certain time. In terms of safe egress time research, one paper contains a Medial Axis Transform (MAT)-based granular evacuation modeling framework [13] that describes a safe egress time and the density-based evacuation model. Using Voronoi tessellation of a set of points, this study represents paths and its attributes such as widths and areas. The implementation utilizes the geometry and topology of Industry Foundation Classes (IFC) BIM models. Instead of using MAT or Straight Medial Axis Transform (S-MAT) that define a geometric network, this paper using Voronoi tessellation of a set of points could implement density-based analysis such as flow analysis and path widths and areas. When it comes to the geometry checking applications of a BIM model, various commercial and open source applications have been developed to support the automated validation. The Solibri Model Checker® (SMC), which is a java-based BIM application, verifies an IFC instance file regarding rule sets defined by a user [14]. This commercial application supports diverse types of geometry checking such as object existence, space relations, circulation, fire code exits, path distance checking, and space program checking [15]. Since users can manipulate rule templates embedded in SMC, the IFC instance model can be flexibly validated according to user-defined rules [16]. One research paper also applied the spatial queries of a relational database for evaluating BIM models [15]. In addition, diverse types of BIM-based validation and their challenges regarding the accuracy and interoperability of BIM data have been studied using semantic modular and automated rule-based checkin [17–20]. One paper also demonstrates rule logic according to diverse BIM data validation scenarios according to BIM data exchange standards [21]. In the area of evacuation analyses, Wang et al. [22] and Rüppel et al. [23] illustrate the evacuation path finding using BIM models and virtual reality systems. These papers also describe A* and Dijkstra algorithms for analyzing an egress path, but do not address the interactive evacuation analysis using a 4D BIM model that represents different site conditions according to construction phases.

The investigated research and papers indicate that even though 4D BIM is broadly used, evacuation path planning has not been properly incorporated into 4D BIM. In addition, diverse pathfinding algorithms that have been developed give an insight for developing an interactive evacuation path planning using 4D BIM as a platform.

3. Objective and Scope

The primary objective of this paper is to investigate the frameworks for generating and visualizing evacuation plans in 4D BIM and to evaluate its feasibility and practicability throughout a case study. Evacuation path planning needs to take into account various factors associated to available exits, construction activities, equipment, material stacks, and others. Since these objects cannot be easily defined in the early planning stages, these factors should be generated as user-defined parameters according to the particular project requirements and site conditions. With the assumption that these user-input factors are correctly defined, this paper presents the framework and development of the evacuation planning using 4D BIM. For evaluating the accessible and secured evacuation path, this framework adopts the A* algorithm, which is one of path finding and graph traversal methods calculating the shortest and directed path among multiple nodes. Even though there are three path planning algorithms including A*, Dijkstra, and Genetic Algorithm, the A* algorithm can generate more logical and optimal paths than the other two algorithms [1]. Therefore, this paper implements the A* algorithm on top of the 4D BIM platform to identify an evacuation path using project information and user inputs. To assess its accuracy and feasibility, the case study of the real construction project using its BIM model for evacuation path planning is included in this paper. Out of the entire construction period, four critical construction phases were selected to provide the associated project's 4D BIM and user-defined parameters for the pathfinding implementation.

The application of the accurate pathfinding algorithm using user-defined parameters to 4D BIM is critical because daily-changing construction work and site conditions should be evaluated by a robust pathfinding method to generate the daily egress paths. The analyzed daily evacuation paths and secured space information are expected to be shared in daily pre-task tool-box meetings with work crews so they can recognize the accessible and secured pathways to quickly get out of the emergent or dangerous areas. In addition, the analyzed results can be significantly useful to dispatch a rescue team to the right place throughout a secured pathway in the case of construction accidents or emergency situations.

4. Framework and Algorithms for Automated Generation of Evacuation Paths in 4D BIM

This section presents a framework and implementation of the proposed evacuation path planning in 4D BIM.

4.1. Proposed Framework

Figure 1 illustrates the proposed framework comprising five steps that are "updating 4D BIM", "preparing user input", "generating evacuation paths", "reviewing and selecting paths", and "distributing the information in selected paths to related crews".

Step 1: Update 4D BIM: An updated 4D BIM is an essential system input that presents current progresses and configuration of the construction site. There have been research studies (Mani et al. 2009) to track construction progresses automatically and reflect the results in 4D BIM. However, automated progress tracking is out of the scope of this research. This research assumes that the input 4D BIM properly reflects construction progresses, configuration, and shape of a current construction site. Therefore, users of this system should manually track current construction progresses and reflect them in 4D BIM.

Step 2: Prepare user input: Using the case study of a real construction project, the authors identified key parameters that can impact the calculation of possible pathways within the interactive BIM-based path planning system. The parameters are used to develop rule-checking features for finding out the shortest path and a shelter within a building or a construction site. Even though the listed variables are not the formalized criteria or standardized requirements for calculating evacuation paths, the variables identified from a real construction project are essential for implementing the proposed 4D BIM-based evacuation planning. The list of parameters will be a great initiator that can establish a generalized list of variables in future research that must be considered to assess evacuation paths using 4D BIM. However, because of varied processes and requirements of construction projects, the formalization of such variables and key factors impacting pathfinding were not conducted in this research. The considerable parameters of evacuation planning that the authors identified from the case study involve the following factors:

- Exits that allow work crews to escape from a construction structure and a jobsite
- Secured spaces that can prevent structure collapse, fire, or flammable gas
- Conditions of paths such as accessible or movable paths (e.g., A rebar work location is not an ideal pathway)
- Workspaces of onsite work crews that can impact site congestion and the flows
- Existing structures and objects that can block the pathway of workers
- Interference of workspaces that are crosslinking planned
- Stacked materials that are supposed to be instantly used for construction works
- Various types of equipment being used near construction tasks
- Temporary structures such as scaffolding

The information about construction activities, equipment, materials, and site conditions are generally documented by a field manager or a superintendent on a daily basis to monitor work progresses. This research was conducted with the assumption that such information can be added as user-defined parameters on 4D BIM models to represent the most recent construction site conditions.

Step 3: Generate all available evacuation paths: Based on the two system inputs prepared in step 1 and step 2, the BIM-based path planning system automatically identifies and quantitatively evaluates all available evacuation paths between work locations and exists. In this research, to identify the shortest evacuation path, a distance was considered as a primary criterion.

Step 4: Review and select evacuation paths: In this step, superintendents and crews' qualitative reviews automatically generated evacuation paths and selected the paths to be used during construction.

Both the path visualization in 4D BIM and the quantitative evaluation of all available evacuation paths assist in the decision making process for the most accessible and secure path. To account for possible spatial conflicts and situations not reflected in 4D BIM, it is desired for multiple crews and managers to review and select evacuation paths together as part of their pre-task planning.

Step 5: Distribute selected paths to crews: After the evacuation paths are selected, the information can be communicated with related work crews in various channels, including printed papers or hand-held computers. Using wireless communication, any changes regarding evaluation paths or other safety relevant issues can be immediately distributed to all the work crews and field managers.

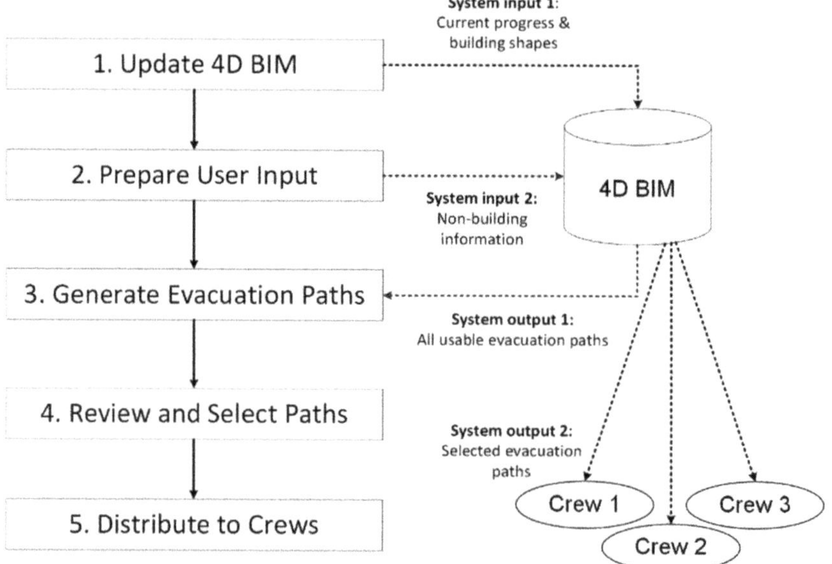

Figure 1. Framework for evacuation path planning in 4D BIM.

4.2. System and Algorithm Development

4.2.1. Custom 4D BIM Platform for Path Planning

To implement the path finding algorithm, this research employed the 4D BIM platform that the authors customized to import information extracted from Autodesk Revit and manipulate it according to a pre-designated schedule and construction working conditions (Figure 2). A plug-in was developed using Revit Application Programming Interface (API) to automatically extract necessary geometric and non-geometric information from a building model created in Revit. The developed custom 4D BIM platform reads the output XML files for path planning. The custom platform has a user-interface for schedule integration and site component creation.

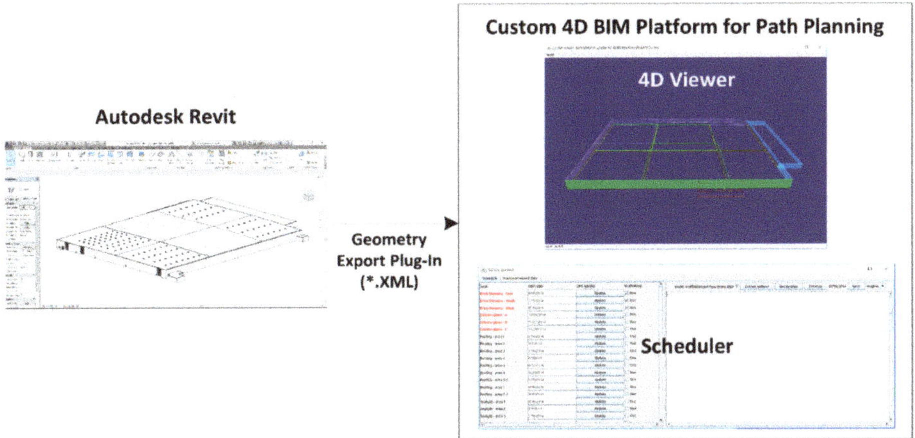

Figure 2. Custom 4D BIM platform for proposed path planning integration.

4.2.2. A* Searching Application for Path Finding

This study utilizes the A* search algorithm to calculate the shortest path from workers' locations to secured exits in a construction site. Associated job spaces were defined and searched for by drawing a rectangle grid mapped on the entire area of the job site, with 10 feet of node spacing. For each BIM model of four scenarios, the grid nodes were configured according to the distinct site conditions of the particular work day. The nodes in the grid consist of four types: A start node, an exit node, a blocked node (non-navigable), and a regular (non-blocked and navigable) node. Start nodes and exit nodes denote the workers' and exit locations, respectively. Blocked nodes correspond to inaccessible regions such as within 10 ft of installed columns, the inside areas of the steel stack and rebar stack, or areas directly underneath ongoing roofing activities including a scaffolding structure. Non-blocked nodes correspond to all other accessible regions.

Based on the navigation grid, A* path planning finds an optimal path between a start node and an exit node that passes through non-blocked nodes and avoids blocked nodes. It starts with a single node (a start node) and keeps expanding the path by adding nodes until it generates a complete path to the exit node. As shown in Figure 3, 16 immediate neighbors of the current node (as shown in Figure 3) were considered to expand the path.

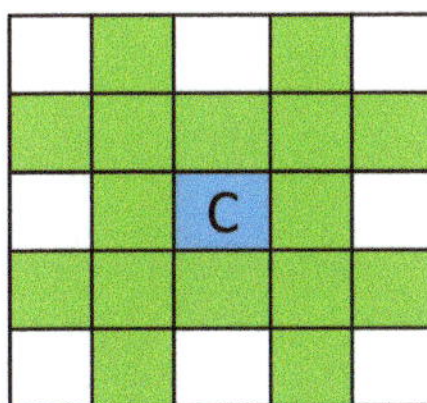

Figure 3. A grid node with neighbors.

From the 16 neighbors, blocked nodes and previously explored nodes are removed and the rest of the nodes are evaluated based on the heuristic function F(n).

$$F(n) = G(n) + H(n)$$

where G(n) = distance from start node to node n, H(n) = straight distance from node n to exit node.

5. A Case Study and Implementation

To investigate the feasibility and applicability of evacuation path planning, this authors utilized the BIM model of the real construction project (Figure 4a) and implemented the proposed process shown in Figure 1.

5.1. Step 1: Update 4D BIM

The first step updated the 4D BIM according to the current status of the project. Due to the large amount of missing information in BIM prepared by the general contractor, four sub-BIM models were extracted and work conditions according to a project schedule were created in coordination with the general contractor. Figure 5 shows the entire schedule of the facility construction involving the work processes of a foundation, a structure, a skin, Mechanical Electrical and Plumbing (MEP), and a roof.

5.2. Step 2: Prepare User Input

The second step is to generate non-building components, such as workspaces, temporary structures, and exits. Four milestones that represent the status and progress of each construction activity in the associated domains were defined and used to generate the four sub-BIM models. For the easy organization of work processes and orders, the general contractor of this construction project split the work zone into seven areas (Figure 4b). For example, Scenario Model 1 involves building and site conditions of construction on May 20th 2014, which illustrate foundation work in area 3 and steel structure work in area 2 and 3. The construction work activities and site conditions on May 20th are diagrammed in Figure 6. Each sub-BIM model contains information and its properties pertaining to a work crew location, a work zone, a material stack, a completed structure, an exit, and other factors that can impact generation of evacuation paths.

(a) 3D View   **(b) Work zones**

Figure 4. (a) BIM model and (b) work zones of case study project.

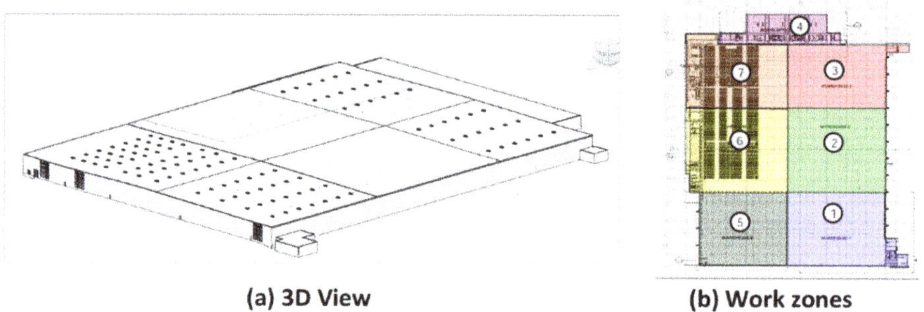

			4	5	6	7	8	9	10	11	12
Foundations - Distribution Center	4/7/2014	7/23/2014									
Structure - Distribution Center	5/5/2014	8/22/2014									
Skin	6/16/2014	12/2/2014									
MEP	4/7/2014	1/1/2015									
Roofing	6/16/2014	10/2/2014									

Model 1 (05/20/14)
- Foundation area 3
- Structure area 2, 3

Model 2 (07/2/14)
- Foundation area 6
- Structure area 5, 6
- Masony east side
- Roofing area 2

Model 3 (09/25/14)
- Masony west side
- Roofing area 7, 7-2

Model 4 (11/25/14)
- Glass area A, B, C
- MEP area 1

Figure 5. Construction schedule and four milestones.

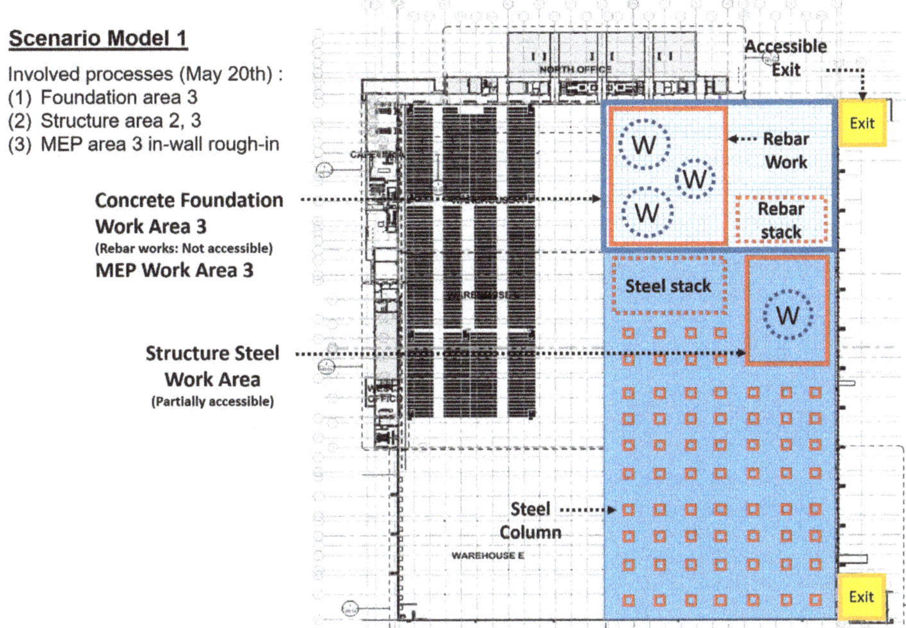

Figure 6. Scenario Model 1: Work status and site conditions.

Scenario Model 1 shows the following three working processes: (1) Foundation work in the area 3, (2) structure work in area 2 and 3, and (3) MEP work in area 3. Each domain activity is color-coded on the floor plan to efficiently represent the types of construction work. On May 20th, area 3, having a concrete foundation task with a reinforcing bar (rebar) installation, is only accessible from the nearest exit in yellow. In other words, rebar workers can use the exit in the upper side of the floor plan, but structure steel workers cannot use it for the evacuation path. Thus, the accessible exit for steel workers is the one located in the bottom of the plan. In addition, each work activity entails associated materials. The location and the area of materials such as a rebar stack is an imperative component for evaluating real-time evaluation planning. However, this manuscript does not involve any formularized and generalized information about work types, areas, and considerable parameters because construction projects generally involve significantly distinct and unique project types, program requirements, tasks, and a schedule that result in different site conditions. For the real-time evacuation analysis, this information can be manually tailored and updated by a BIM modeler according to a project and site information.

Scenario Model 2 shows the following five working processes (Figure 7): (1) Foundation work in area 6, (2) structure work in area 5 and 6, and (3) masonry work in the east side, (4) MEP work in area 1 and 6, and (5) roofing work in area 2. On July 2nd, the roofing work area containing scaffolding cannot be used as an accessible evacuation pathway because of the shallow path spaces, the complex scaffolding structure, and the low illumination level of the area.

Scenario Model 3 in Figure 8 shows the following three working processes: (1) Masonry works in the west side, (2) MEP work in area 3 and 5, and (3) roofing work in area 7 and 7-2.

Scenario Model 4 in Figure 9 shows the following two working processes: (1) Masonry work in the east and west sides and (2) MEP work in area 1.

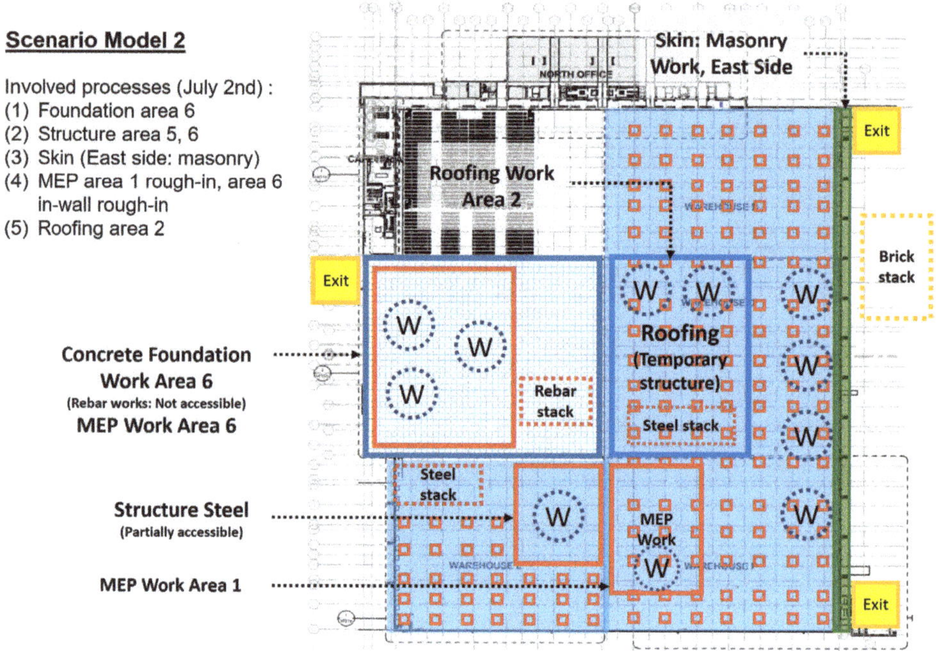

Figure 7. Scenario Model 2: Work status and site conditions.

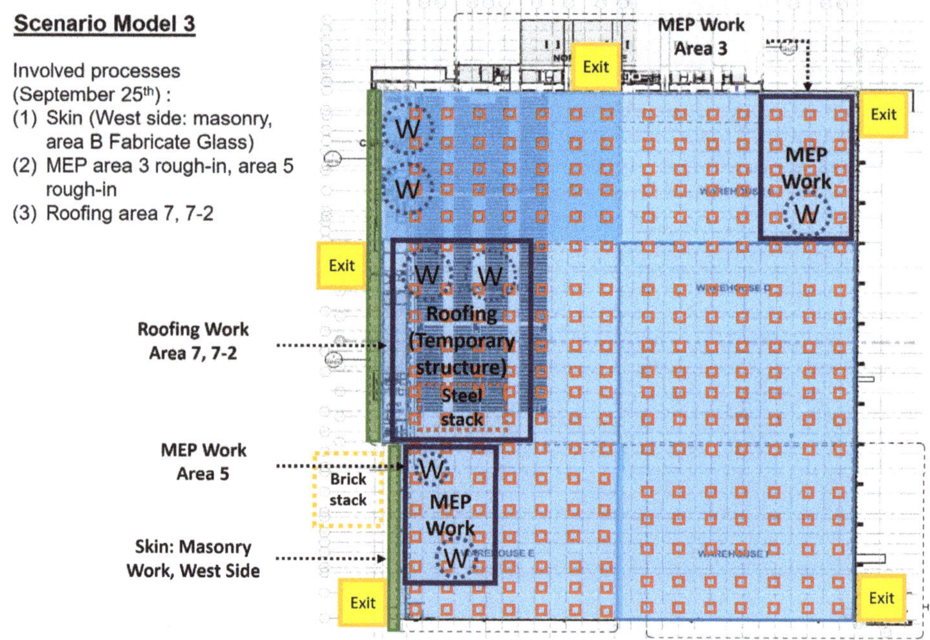

Figure 8. Scenario Model 3: Work status and site conditions.

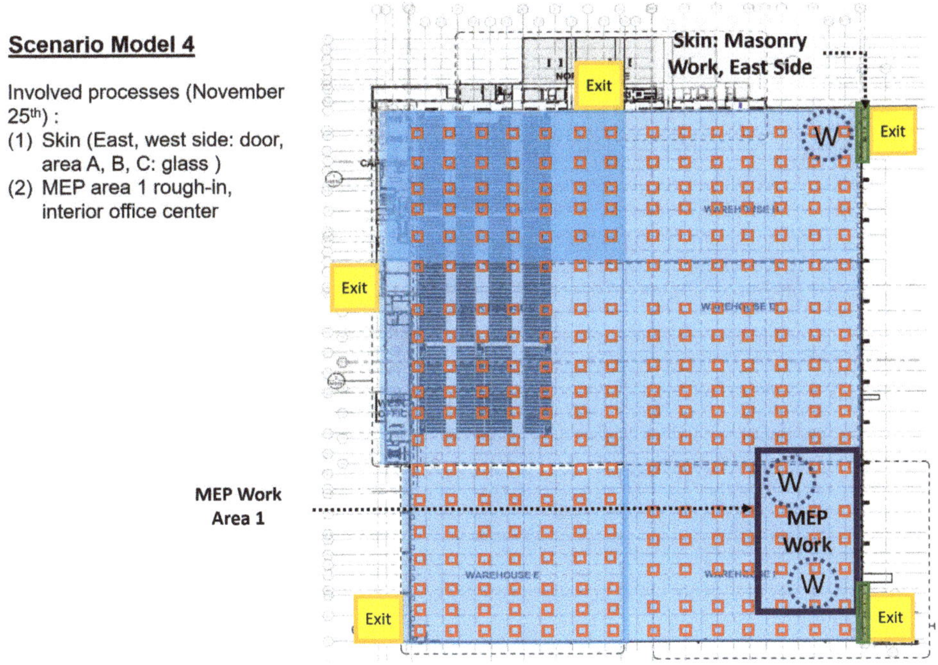
Scenario Model 4
Involved processes (November 25th) :
(1) Skin (East, west side: door, area A, B, C: glass)
(2) MEP area 1 rough-in, interior office center
Skin: Masonry Work, East Side
Exit
Exit
Exit
Exit
Exit
W
W
W
MEP Work
MEP Work Area 1

Figure 9. Scenario Model 4: Work status and site conditions.

5.3. Step 3 and 4: Generate All Available Evacuation Paths and Select Paths for Distribution

As shown in the four scenarios, work areas, task types, and accessible exits are varied according to construction progress, a schedule, a weather, a logistics, site conditions, or diverse variables. To secure evacuation paths of a changing construction site, automated and dynamic path planning and identification are required for generating a daily evacuation plan. This paragraph illustrates the detailed implementation and its processes. The construction site environment and work activities related to the four key scenarios in Section 5.1 were reflected in the developed path planning platform. Then, this system is designed to automatically identify and evaluate accessible evacuation paths. Figure 10, Figure 15, Figure 16, and Figure 17 present the results of the four scenarios in the four different views (site conditions presented in 4D BIM, a navigation grid, all available paths, and selected paths).

Figure 11 illustrates the site condition of scenario 1 including work locations, structural columns, exits, etc. As shown in Figure 12, a navigation grid of 10 feet granularity was created. The navigation grid presents accessible and inaccessible places within the construction site. Based on the site conditions (Figure 11) and a grid (Figure 12), all available evacuation paths were identified and evaluated in terms of a total distance (Figure 13). Finally, one or two evacuation paths were selected by users as shown in Figure 14.

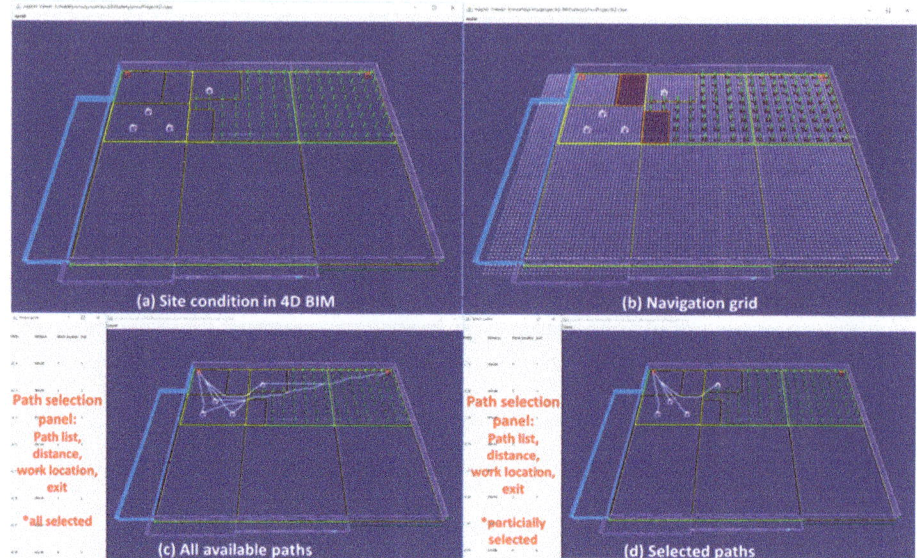

Figure 10. Scenario 1 (**a**) site condition, (**b**) navigation grid, (**c**) all available paths, and (**d**) selected paths.

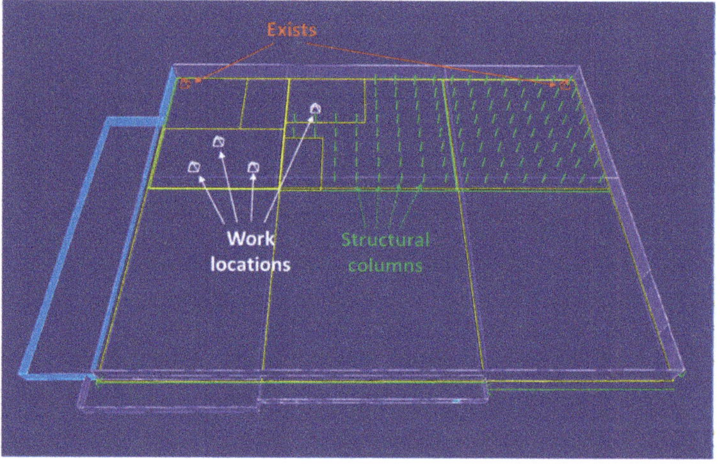

Figure 11. Site condition of scenario 1 reflected in the prototype system.

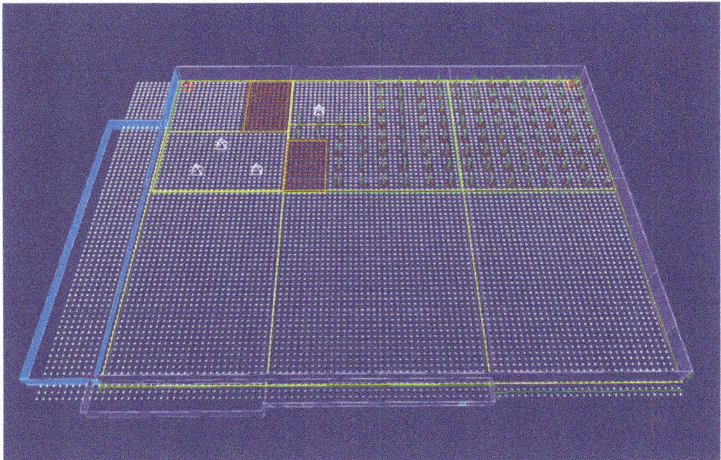

Figure 12. Scenario 1 navigation grid with accessible and inaccessible nodes.

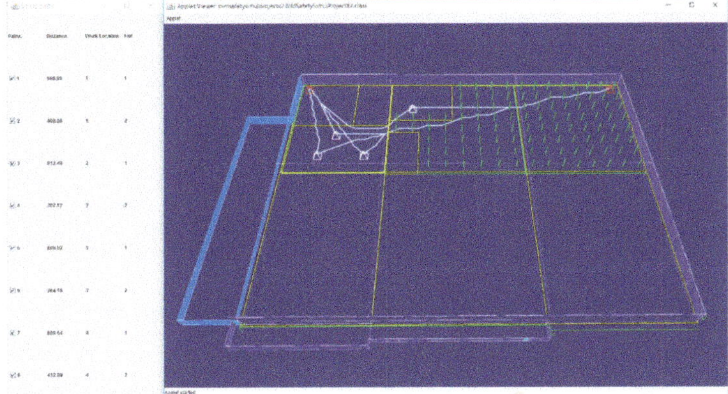

Figure 13. All evacuation paths found from scenario 1.

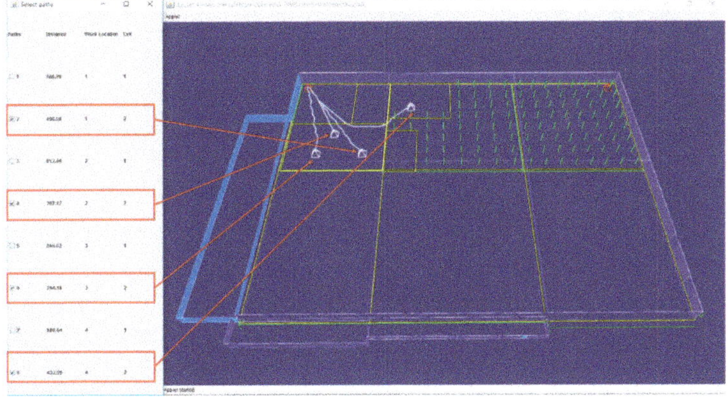

Figure 14. Selected evacuation paths found from scenario 1.

Figure 15, Figure 16, and Figure 17 show the results from scenario 2, 3, and 4, respectively. Figure 18 shows paths from roofing work locations. This case study assumed that crews working for roofing generally go down to the first floor level through the opening in the middle of the roofing area. For the four scenarios, the developed system successfully generated rational evacuation paths for each work location by properly considering the construction site conditions and explicitly represent secured circulation paths connecting to the exits. Even though final decisions on determining the shortest path can also be made automatically, based on calculated distances, the proposed approach puts its objective to assist the decision making process of domain professionals with the automatically generated pathways.

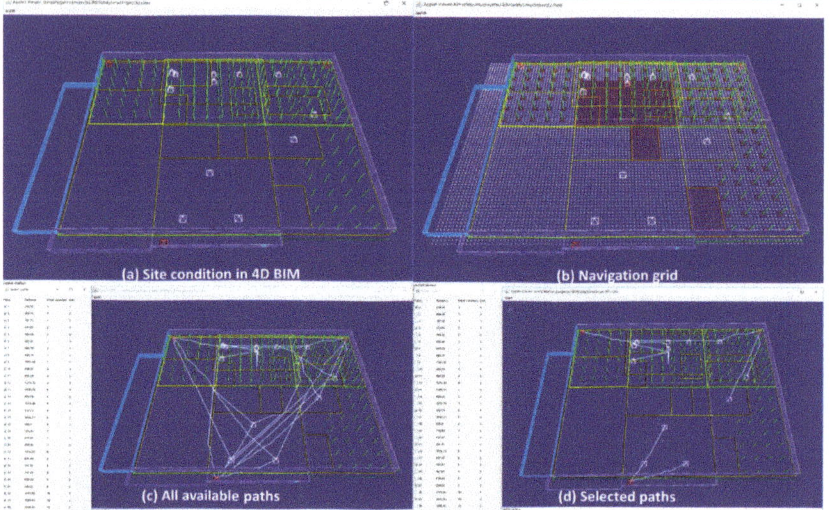

Figure 15. Scenario 2 (**a**) site condition, (**b**) navigation grid, (**c**) all available paths, and (**d**) selected paths.

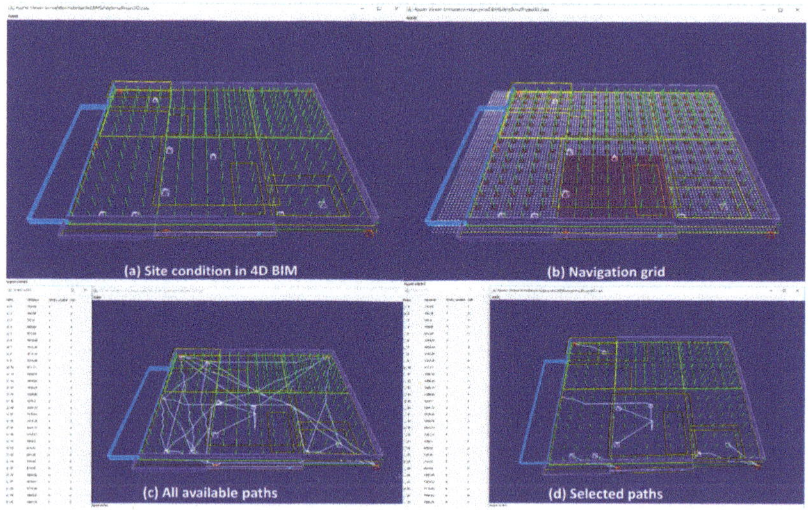

Figure 16. Scenario 3 (**a**) site condition, (**b**) navigation grid, (**c**) all available paths, and (**d**) selected paths.

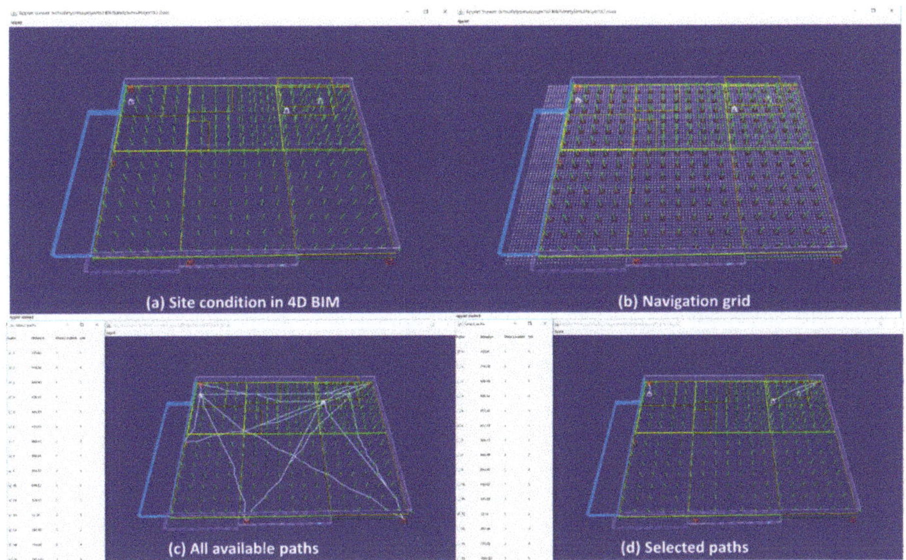

Figure 17. Scenario 4 (**a**) site condition, (**b**) navigation grid, (**c**) all available paths, and (**d**) selected paths.

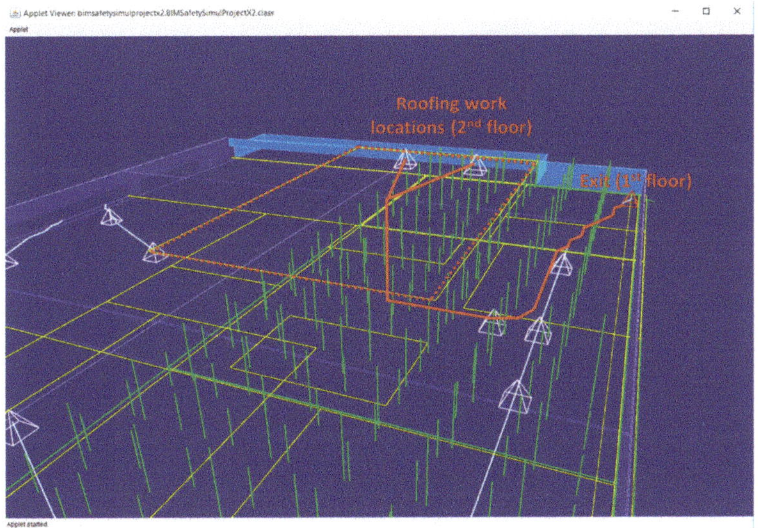

Figure 18. Evaluation paths from roofing work locations in scenario 2.

5.4. Step 5: Distribute Selected Paths to Crews

After generating all the evacuation paths of the four project milestones, certain evaluation paths were selected based on the project-specific review as discussed in Section 5.3. The research team provided information on the selected evacuation paths to the general contractor for distribution to the work crews of the project. Although there was no incident of safety hazards that required evacuation from the construction site, the general contractor was able to provide evacuation paths that were generated based on BIM containing up-to-date project status as well as construction site conditions that are impacted by non-building components, such as workspaces, temporary structures, etc.

5.5. Benefits and Potentials

The proposed approach, which uses the customized 4D BIM platform, has potential for any commercial 4D BIM tool such as Synchro or Vico. The pathfinding algorithm and process can be integrated into any 4D BIM application to generate egress paths and circulation paths. The key benefit of 4D BIM-based path planning is the automated generation of daily, weekly, or monthly updatable and changeable evacuation plans that reflect project progress, planned work schedules, and site conditions. Users can define work zones, schedules, work orders, and other required parameters on the 4D BIM platform to flexibly manage BIM models and analyze possible pathways. The accessible pathways, secured spaces, and installed exits shown in 4D project models that can be shared and trained in every day's tool-box meetings, will enhance emergency preparedness planning and help a rapid reaction in any emergency situations or construction accidents. In particular, in emergency situations, such as a collapse or fire on site, a dispatched rescue team can quickly recognize the site conditions using the mapped 4D models and determine the shortest and secured pathway to save labors from the site. From a project management perspective, the daily updated geometric representation of work zones and a construction site will improve logistics planning, task sequence management, and material stack/equipment movement planning by incorporating path planning and visualization capabilities into 4D BIM and daily pre-task planning. Since a construction project has to organize a series of tasks, numerous labors, diverse materials within a limited jobsite space, a site logistic, and work planning, using 4D BIM and its automated circulation path finding features will play a pivotal role in the project management of diverse construction projects. Furthermore, an explicit illustration of the existing conditions of project tasks and the available movement paths of equipment or material, will facilitate more effective construction project management and planning.

6. Discussion and Conclusions

Establishing evacuation paths for multiple work crews based on changing construction site conditions requires a significant amount of manual and labor-intensive tasks. Most of construction projects do not provide situation-specific evacuation plans for workers through the construction. To resolve this challenge, this research proposed a framework to automatically generate available evacuation paths for multiple crews using project information mapped in 4D BIM. To account for construction site components, such as workspaces, exists, and temporary structures, that are difficult to model in 4D BIM in advance, the developed prototype system requires users to manually create these components as part of a daily pre-task planning. A case study on four real-world construction scenarios successfully generated rational evacuation paths for multiple work crews.

Even though this research shows valuable results pertaining to the dynamic and automated evacuation path planning, there are multiple limitations that should be addressed. (1) The first limitation is a manual 4D BIM update. Since the proposed approach uses current geometric conditions as an essential system input, 4D BIM should reflect the construction site's conditions accurately. However, identifying construction progresses and reflecting the current status is a tedious process. Due to this importance, many research studies attempted to automate the process of updating BIM and its embedded information. However, this is out of the scope of the research presented in this paper. (2) The second limitation is a manual user input for non-building components in construction sites. Other than building geometric information, non-building elements, such as workspaces, temporary structures, major tools, storage areas, need to be identified and modeled in 4D BIM environment by the users. To achieve this, construction, VDC, or BIM managers of a construction project should coordinate with work crews and specify their work locations accurately in the system. (3) The third limitation is the reliance for situation-specific heuristic functions. Depending on the project sizes, spatiotemporal conditions, and preferences of users, different search objectives and heuristic functions should be designed. In the case study of this research, we used the total moving distance as the path finding objective. However, for construction projects with different conditions and requirements, objectives and heuristic functions should be designed according to project-specific conditions. In order to address these challenges, future research should involve the investigation of the key factors that should be taken into account by the implementation framework of evacuation path finding. In addition, a well-structured user interface for developing a 4D BIM and manipulating site conditions is required for accurately identifying accessible paths and practically adopting research findings to construction projects. The mobile-based approach to share and disseminate real-time evacuation paths to labors and rescue teams will be further studied to bolster this egress path planning research and construction safety environment.

Author Contributions: Conceptualization, K.K. and Y.-C.L.; software development, K.K.; validation, Y.-C.L.; original draft preparation, K.K. and Y.-C.L.

Funding: This research received no external funding.

Acknowledgments: The authors appreciate Sajiva Pradhan a research assistant at University of Houston for her assistance in implementing A* searching algorithm during the case study.

Conflicts of Interest: The authors declare no conflict of interest.

References

1. Soltani, A.R.; Tawfik, H.; Goulermas, J.Y.; Fernando, T. Path planning in construction sites: Performance evaluation of the dijkstra, a*, and GA search algorithms. *Adv. Eng. Inform.* **2002**, *16*, 291–303. [CrossRef]
2. Kim, K.; Cho, Y.K.; Zhang, S. Integrating work sequences and temporary structures into safety planning: Automated scaffolding-related safety hazard identification and prevention in BIM. *Autom. Constr.* **2016**, *70*, 128–142. [CrossRef]
3. Zhang, S.; Teizer, J.; Lee, J.-K.; Eastman, C.M.; Venugopal, M. Building Information Modeling (BIM) and Safety: Automatic Safety Checking of Construction Models and Schedules. *Autom. Constr.* **2013**, *29*, 183–195. [CrossRef]
4. Kang, J.H.; Anderson, S.D.; Clayton, M.J. Empirical Study on the Merit of Web-Based 4D Visualization in Collaborative Construction Planning and Scheduling. *J. Constr. Eng. Manag.* **2007**, *133*, 447–461. [CrossRef]
5. Sulankivi, K.; Kähkönen, K.; Mäkelä, T.; Kiviniemi, M. *4D-BIM for Construction Safety Planning*; CIB: Manchester, UK, 2010.
6. Koo, B.; Fischer, M. Feasibility Study of 4D CAD in Commercial Construction. *J. Constr. Eng. Manag.* **2000**, *126*, 251–260. [CrossRef]
7. Akinci, B.; Fischen, M.; Levitt, R.; Carlson, R. Formalization and Automation of Time-Space Conflict Analysis. *J. Comput. Civ. Eng.* **2002**, *16*, 124–134. [CrossRef]
8. Jongeling, R.; Kim, J.; Fischer, M.; Mourgues, C.; Olofsson, T. Quantitative analysis of workflow, temporary structure usage, and productivity using 4D models. *Autom. Constr.* **2008**, *17*, 780–791. [CrossRef]
9. Kim, K.; Teizer, J. Automatic design and planning of scaffolding systems using building information modeling. *Adv. Eng. Inform.* **2014**, *28*, 66–80. [CrossRef]
10. Kim, K.; Cho, Y.K.; Kim, K. BIM-Based Decision-Making Framework for Scaffolding Planning. *J. Manag. Eng.* **2018**, *34*, 4018046. [CrossRef]
11. Kim, K.; Cho, Y.; Kim, K. BIM-Driven Automated Decision Support System for Safety Planning of Temporary Structures. *J. Constr. Eng. Manag.* **2018**, *144*, 4018072. [CrossRef]
12. Lee, J.K.; Eastman, C.M.; Lee, J.; Kannala, M.; Jeong, Y.S. Computing walking distances within buildings using the universal circulation network. *Environ. Plan. B Plan. Des.* **2010**, *37*, 628–645. [CrossRef]
13. Swieboda, W.; Krauze, A.; Nguyen, H.S. A MAT-based granular evacuation modeling framework. In Proceedings of the 2014 Federated Conference on Computer Science and Information Systems, Warsaw, Poland, 7–10 September 2014; pp. 337–342. [CrossRef]
14. Ghannad, P.; Lee, Y.C.; Dimyadi, J.; Solihin, W. Automated BIM data validation integrating open-standard schema with visual programming language. *Adv. Eng. Inform.* **2019**, *40*, 14–28. [CrossRef]
15. Lee, Y.-C.; Eastman, C.M.; Lee, J.-K. Validations for ensuring the interoperability of data exchange of a building information model. *Autom. Constr.* **2015**, *58*, 176–195. [CrossRef]
16. Solihin, W.; Eastman, C.; Lee, Y.-C. Multiple representation approach to achieve high-performance spatial queries of 3D BIM data using a relational database. *Autom. Constr.* **2017**, *81*, 369–388. [CrossRef]

17. Lee, J.K.; Eastman, C.M.; Lee, Y.C. Implementation of a BIM domain-specific language for the building environment rule and analysis. *J. Intell. Rob. Syst.* **2015**, *79*, 507–522. [CrossRef]

18. Lee, Y.-C.; Eastman, C.M.; Solihin, W.; See, R. Modularized rule-based validation of a BIM model pertaining to model views. *Autom. Constr.* **2016**, *63*, 1–11. [CrossRef]

19. Solihin, W.; Eastman, C.; Lee, Y.-C. Toward robust and quantifiable automated IFC quality validation. *Adv. Eng. Inform.* **2015**, *29*, 739–756. [CrossRef]

20. Lee, Y.C.; Solihin, W.; Eastman, C.M. The Mechanism and Challenges of Validating a Building Information Model regarding data exchange standards. *Autom. Constr.* **2019**, *100*, 118–128. [CrossRef]

21. Lee, Y.C.; Eastman, C.M.; Solihin, W. Logic for ensuring the data exchange integrity of building information models. *Autom. Constr.* **2018**, *85*, 249–262. [CrossRef]

22. Wang, B.; Li, H.; Rezgui, Y.; Bradley, A.; Ong, H.N. BIM based virtual environment for fire emergency evacuation. *Sci. World J.* **2014**, *2014*, 589016. [CrossRef] [PubMed]

23. Rüppel, U.; Abolghasemzadeh, P.; Stübbe, K. BIM-based immersive indoor graph networks for emergency situations in buildings. In Proceedings of the 2010 International Conference on Computing in Civil and Building Engineering, Nottingham, UK, 30 June–2 July 2010.

Article

Taking Advantage of Collective Intelligence and BIM-Based Virtual Reality in Fire Safety Inspection for Commercial and Public Buildings

Daxin Zhang [1,†], **Jinyue Zhang** [2,*,†], **Haiming Xiong** [3], **Zhiming Cui** [4] **and Dan Lu** [4]

1   Research Institute of Architectural Design and Urban Planning, Tianjin University, Tianjin 300073, China; dzhang@ttjl.org
2   Trimble Joint Laboratory for BIM, Department of Construction Management, Tianjin University, Tianjin 300072, China
3   Guangxi Hualan Engineering Management Co., Ltd., Nanning 530011, China; haiming@ttjl.org
4   Tianjin Sanpintiangong Construction Technology Inc., Tianjin 3000101, China; zhiming.cui@sureknow.com (Z.C.); dan.lu@sureknow.com (D.L.)
*   Correspondence: jinyuezhang@tju.edu.cn; Tel.: +86-22-2740-2272
†   Daxin Zhang and Jinyue Zhang have equal contribution to this paper.

Received: 25 October 2019; Accepted: 21 November 2019; Published: 24 November 2019

Abstract: Commercial and public buildings are more vulnerable to fires because of their complex use functions, large number of centralized occupants, and the dynamic nature of the use of space. Due to the large number of these types of buildings and the limited availability of manpower, annual fire inspections cannot ensure the continuous compliance of fire codes. A crowdsourcing application, iInspect, is proposed in this paper to harvest collective intelligence in order to conduct mass inspection tasks. This approach is supported by building information modeling (BIM) based virtual reality (VR) and an indoor real-time localization system. Based on the International Fire Code and 27 fire inspection checklists compiled by various local authorities, a generic list of inspection items suitable for iInspect is proposed, along with a reputation-based monetary incentive model. A prototype of iInspect was created for Android mobile phones, and a case study was performed in an office building in Tianjin, China, for verification of this crowdsourcing inspection approach.

Keywords: fire safety inspection; building information modeling; real-time location system; smartphone; crowdsourcing

---

1. Introduction

As a major cause of injury and death across the world, fire is an important hazard that deserves special attention in risk management for buildings. A total of 237,000 fires were reported in 2018 in China, resulting in 1407 deaths, 798 injuries, and direct property losses of about 3.67 billion Chinese Yuan [1]. Some severe fires have devastating consequences. In Table 1, which lists the major fires in China over the last two decades [2], one can find that most of the severe fires that resulted in a large number of casualties occurred in commercial or public buildings.

The rapid and continuous economic development in China over the last four decades has led to large-scale urbanization. The intensive use of land in urban areas has greatly increased the prevalence of large-scale buildings, mainly complex commercial buildings such as shopping malls as well as public buildings such as hospitals. These types of buildings are deemed to have a higher risk of fire. First, the use of the interior spaces in these buildings is more complex. For example, a large commercial complex has multiple business types such as shops, restaurants, and cinemas. Second, due to the large scale and capacity of these buildings, the large number of occupants is concentrated in specific areas and is not easy to evacuate effectively in the event of a fire. Third, the highly dynamic nature of these buildings makes fire prevention management more difficult. For example, it is not easy to track and manage the storage of flammable goods in shops.

Table 1. Major structural fires in China over the past two decades.

Structure Type (Location)	Date of Accident	Deaths/Injuries	Probable Cause(s)
Shopping mall (Tianjin)	30 June 2012	10/16	Short-circuit in the storage room located on the first floor.
Apartment building (Shanghai)	15 November 2010	58/71	Welding in the maintenance of building's external façade in violation of safety regulations.
Commercial building (Jilin)	5 November 2010	19/24	Short-circuit in the storage room located on the first floor; some fire extinguishing equipment not available or out of operation.
Dance club (Shenzhen)	21 September 2008	44/58	Setting off fireworks on an indoor stage; fire escape route did not follow fire safety regulations and was too narrow.
Hospital (Liaoyuan)	15 December 2005	40/95	Short-circuit in the electrical room; some fire extinguishing equipment out of operation.
Hotel (Shantou)	10 June 2005	31/21	Short-circuit in the kitchen; some fire exits were locked; the water pressure of the fire hydrant was insufficient.
Iron mine (Heibei)	20 November 2004	68/0	Welding slag burned wood supports in the mine tunnel.
Shopping mall (Jilin)	15 February 2004	54/70	A cigarette end in the temporary storage room was not extinguished.
Shopping mall (Luoyang)	25 December 2000	309/7	Welding operation conducted in violation of safety regulations; a number of exits and escape routes were blocked or locked.

The design, construction, and operation of numerous building service systems are increasingly sophisticated, especially for large-scale commercial and public buildings. In order to reduce the number of fires, fire safety has been at the heart of legislation on building regulations [3]. Fire safety codes are often incorporated into local building codes and have been formulated by legislators following the advice of politicians and industry experts. There are also many safety-related requirements in current occupancy standards. Even if the building's systems strictly follow all regulations, however, one cannot assume the systems will run forever without creating any safety hazards. Dieken [4] states that in fire situations, one third of safety barriers are not in proper working order because of improper inspection or inadequate testing or maintenance. As such, building inspection is an important process for ensuring compliance with all safety regulations in everyday operation.

A building fire inspection is usually performed by a fire inspector, who makes a professional judgment about whether a building meets the requirements in a building fire code. Best inspection practices aim at identifying and correcting problems before they can have a negative impact on the building or the general public. A fire inspection program has significant consequences for building owners and property managers as well as the insurance companies that assume some risk or liability for the building, even if the inspection process is not be appreciated or even noticed by the people who use the building.

Numerous research studies acknowledge the significance of fire safety inspection, and Table 2 lists some examples of studies in the literature. A popular topic of research related to fire safety inspection is the functionality of fire safety facilities. Sobral and Ferreira [5] proposed a qualitative approach of a fault tree analysis to assess the availability of a fire pump system, taking into account the specificity of dormant systems and the importance of the frequency of inspections or tests of these systems. Sobral and Ferreira [6] also stressed the importance of tests, inspections, and maintenance operations in the context of a fire sprinkler system. They proposed a methodology based on international standards and supported by test/inspection reports to adjust the frequency of these actions according to the level of degradation of the components and with regard to safety. In 2004, Sierra et al. [7] conducted a field study in Spain, visiting 164 hotels and analyzing the status of their fire protection systems.

Along with advances in information and communication technologies, better approaches are adopted for fire safety inspection simulation and education. Han et al. [8] proposed a simulation training system for fire safety inspection to be used in colleges and fire brigades, based on three key technologies: three-dimensional (3D) geographic information systems, 3D modeling technology, and visualizations of human-computer interaction. Lee et al. [9] investigated the effectiveness of using an on-line training program for improving awareness of fire prevention in hospitals and found that visual-based on-line training can greatly improve healthcare workers' knowledge about fire prevention and evacuation. Zhang et al. [10] proposed an architecture for a smart system that provides customized instructions for an occupant to evacuate a building in the event of a fire; the system incorporates 3D building models with the support of 3D building models and a real-time location system (RTLS). Some researchers have focused on inspection criteria. Lo and a co-worker discussed how a building safety inspection system should accommodate buildings constructed in accordance with previous prescriptive standards by a fire risk assessment to determine the safety index [11] and from the index to rank the priority of improvement actions [12]. Wen and Fan [13] analyzed the catastrophic behavior of a flashover and developed a more appropriate way to determine the inspection criteria for flashover in compartment fires.

Associated with the criteria of fire safety inspection is a fire safety assessment. For large-scale commercial buildings, Liu et al. [14] proposed a fire risk assessment system based on the fire characteristics and the state of maintenance of fire equipment in the buildings. The proposed assessment system focused on evaluating the safety performance of the fire protection systems in the buildings. The Central Business District of Dares Salaam City carried out a study to assess urban fire risk with respect to public awareness on the use of firefighting facilities and preparedness in the event of the outbreak of fire [15]. The policy and making of laws and regulations with regard to fire safety inspection is also an active topic of research. In collaboration with the Atlanta Fire Rescue Department, Madaio et al. [16] developed the Firebird framework to help municipal fire departments identify and prioritize fire inspections for commercial properties by using machine learning, geocoding, and information visualization. Shao et al. [17] conducted inductive analysis and a survey questionnaire to discuss the integration of inspection and reporting systems for fire safety equipment and public building security. This study attempted to clarify the ambiguity between firefighting and building safety and to provide initial results to support amendments to related laws and regulations.

Table 2. Examples of research studies related to fire safety inspection.

Category	Studies in the Literature
Fire safety facilities	Sobral and Ferreira [5], Sobral and Ferreira [6], Sierra et al. [7]
Simulation and education	Han et al. [8], Lee et al. [9], Zhang et al. [10]
Inspection criterion	Lo [11], Lo and Cheng [12], Wen and Fan [13]
Safety assessment	Liu et al. [14], Kachenje et al. [15]
Policy and law making	Madaio et al. [16], Shao et al. [17]

Existing research related to fire inspection does not consider continuous compliance to fire codes. Fire inspections are usually scheduled periodically (for example, on an annual basis). On one hand, good inspection programs require a significant amount of investment in human resources to conduct the inspection, and the inspectors are presumed to know the myriad code requirements, procedures, and processes. In reality, fire inspection programs may be inadequately staffed, may not be supported with a strong enough level of commitment by the local authority, or simply may not exist. On the other hand, some items require inspection at more frequent intervals because of the dynamic features of their status. For example, exit doors could be improperly locked during the daily operation for security purposes and are only unlocked temporarily during inspections. Another example is the storage of highly hazard contents that may be present during the normal daily operation of a facility.

To focus on the inspection of items in commercial and public buildings that have dynamic features, this study proposes a social approach that combines collective knowledge from the general public with building information modeling (BIM)-based virtual reality (VR) technology to identify possible issues related to fire safety. Based on this approach, a prototype mobile app called iInspect was developed. With the help of an indoor real-time location system (RTLS), iInspect notifies occupants when they are near an item in the building that is fire safety—related and could be inspected (for example, a fire door). If a person has some spare time and would like to perform an inspection, the app will display the VR scene on the user's mobile phone to help him/her locate the items to be inspected and will provide a checklist and some additional information such as images of the correct status of the item and common problems or issues with the item. App users can perform an inspection and upload a report along with the necessary photos for verification purposes. The office of the appropriate authority will be notified and can take the necessary action(s) according to the severity of the issue.

In the following section of the paper, we will review related concepts including crowdsourcing and collective intelligence, BIM and VR, and RTLS. The system design for iInspect will be presented in terms of framework design, inspection content, and incentive mechanisms for social participation. A case study is conducted to examine and verify the usability of this approach, and the limitations of the system and possible uses of this approach in other inspection-related areas will be discussed.

2. Related Work

2.1. Crowdsourcing and Collective Intelligence

Crowdsourcing is a portmanteau to describe how businesses use the Internet to outsource work to the crowd. After studying more than 40 definitions of crowdsourcing in the scientific and popular literature, Estellés-Arolas and González-Ladrón-de-Guevara [18] summarized crowdsourcing as a type of participative online activity in which a party proposes to the public via a flexible open call the voluntary undertaking of a task. Crowdsourcing in the modern conception is an information technology (IT)–mediated phenomenon, meaning that a form of IT is always used to create and access crowds of people [19]. Members of the public submit their work, money, knowledge, and/or experience to the initiating party and are compensated with monetary prizes or with recognition of their contribution. Usually the crowdsourced result is given back to the public as a kind of response, especially when there no monetary compensation is involved [20].

By properly using a crowdsourcing method, one can manage large jobs with thousands of workers or do small jobs that require just a single person. According to the problem being solved, crowdsourcing can be classified into following four types [21]:

- Crowdcontests: A party tasks a crowd with proposing ideas about a given topic, for example, logo design, software testing, or other creative projects. This type of crowdsourcing can identify the best worker for a given job, as many people are proposing or creating items. The initiating party rewards only a single contributor.
- Macrotasks: A party tasks a crowd with a specific job, for example, software application development or thesis editing. The initiating party selects only one person/team to carry out the entire job, and the hired party is compensated for the task.
- Microtasks: A party tasks a crowd with many small units of a job that are a part of a larger and more complicated project that typically requires a large workforce: for example, tagging a million photos or transcribing many medical records. These small jobs may or may not require human intelligence. All contributors are given recognition as a kind of reward.
- Crowdfunding: In contrast to the three types mentioned above, crowdfunding requires a monetary contribution from the crowd. This has been a very popular approach to raise money through

social networks, especially for startup companies or innovative ideas that need money to realize them. Contributors can receive goods or services or they can give money as a donation.

The crowdsourcing work involved in this research is a type of microtask. In order to compensate for the lack of an adequate fire inspection workforce, the key point in this study is to hire the crowd to inspect many commercial and public buildings against fire safety-related requirements. In this case, the huge job of inspecting many buildings at more frequent intervals is divided into many small tasks of inspecting only one or two items in a given building. This makes a seemingly impossible goal more achievable. To accomplish the crowdsourcing, there is a need to create a platform that can engage interested people to participate in the fire inspection process. Contributors can be compensated by cash (if the initiating party has a budget) or by acknowledgement (or any other in-kind offering).

Collective intelligence (CI), known as the wisdom of the crowd, refers to the collective opinion of a group of individuals rather than that of a single expert [22]. Crowdsourcing usually generates CI, which provides better results than can be achieved by hiring one or a few experts. A typical example is Wikipedia, an online, collaboratively created, and maintained encyclopedia—for this type of project, it is impossible to achieve the desired quality in a traditional editing paradigm [23]. CI is not new to the Information Age, but the rapid expansion of mobile Internet has largely facilitated the development of social information sites that rely on collective human knowledge, such as Yahoo! Answers, Quora, Stack Exchange, and similar sites. A crowdsourcing mobile inspection app, such as the iInspect app proposed in this study, can better take advantage of collective knowledge. For example, a mistakenly reported violation could be flagged as incorrect by other users in order to make the inspection results more accurate.

2.2. BIM Modeling and Visualization Using VR

BIM is a "modeling technology and associated set of processes to produce, communicate, and analyze building models" used for the design, construction, and routine operation of buildings [24]. BIM has been used to improve product delivery, and systems based on BIM have been adopted to perform tasks such as clash detection in the design and construction industry [25], visualize and optimize schedules in four dimensions (4-D) [26], and to prepare quantity take-offs during the pre-construction phase [27]. As a BIM model can encapsulate large amounts of data and information accumulated in the design and construction phases, it is a good starting point for facility management. Many concurrent facility management systems such as ARCHIBUS use BIM to support the visualization of data and information in 3D spaces, for example, allowing users to click on a pressure gage in the model to access information such as the brand/model, installation date, purchase price, warrantee information, maintenance schedule, and other information. In this research, any information that is relevant to fire safety inspection is incorporated into the BIM model of the building and is associated with each piece of fire safety equipment.

For crowdsourced inspections, it is necessary to take BIM models of commercial and public buildings to a cloud-based service that can enable visualization on mobile phones. The online visualization of BIM models is different from that employed in models running on local workstations used in the design and construction phases. An online BIM model must be significantly more lightweight (in terms of memory footprint and central processing unit usage) than a locally stored BIM model that may be more than 300 Mb in size. Until Internet service on 5G mobile phones is widely available, lightweight BIM data will be required to the model for mobile phones in order to improve the user experience. Moreover, all mobile phones do not use the same operating system (OS) and an online BIM model must be compatible with various OSs. For example, smartphones used by people in a building may use Apple iOS or Google Android, while the workstations in the fire safety management office might be running on a Microsoft Windows or Apple macOS operating system. It is imperative for the transfer of information on these platforms to be seamless.

The system developed in this study uses the Forge Viewer platform to enable BIM visualization on the cloud. This platform was formerly known as Large Model View, since it allowed the handling of

3D models that were too large to run on Autodesk Design Review. Forge Viewer converts information from a simple vector format (SVF) to Web Graphic Library (WebGL) format for display on an Internet browser; this allows people with mobile devices that use any OS to view information without having to install a plug-in. WebGL is an application programming interface (API) that allows users to create 3D graphics in their device's web browser, regardless of the software package used and without having to pay royalty fees [28], and some researchers have used this API to visualize BIM models online. For example, Xu et al. [29] developed a methodology for creating BIM model visualizations in three dimensions that combines the Industry Foundation Classes (IFC) data model using WebGL technology. In addition, Chen et al. [30] developed a cloud-based system architecture that enables very large BIM models to be viewed, stored, and analyzed for purposes such as facility management.

In order to help the building occupants to easily spot fire safety-related facilities, iInspect incorporates VR scenes along with the BIM model. Typically, VR is a simulated experience that allows users to become fully immersed in a virtual environment that could be similar to or completely different from the real world. Traditionally, a specially designed room with multiple large screens was required to create a VR environment. Currently, standard VR systems use head-mounted devices such as HTC Corporation's VIVE system to simulate a user's presence in a virtual environment [31]. In contrast to immersive VR, which is not practical to use for an app, the VR used in the proposed iInspect system is VR photography, which is the interactive viewing of a 360 °C circle or a photograph. Instead of using image stitching technology, hardware improvement has made the capture of VR photos much easier by using 360 °C cameras having two or more lenses. An Insta360 two-lens camera was used in this research to generate VR photos at places where some fire safety–related facilities are located that could potentially be inspected by iInspect users. Figure 1 shows the BIM model view and VR view of the perspective of a corridor in a commercial building.

(a)   (b)

Figure 1. (**a**) Building information modeling (BIM) model view of a corridor. (**b**) Virtual reality (VR) photograph of the same corridor from the same perspective.

2.3. Indoor Real-Time Location System

To locate iInspect users in real time, a real-time location system (RTLS) was used. iInspect will notify the user if any fire safety-related facilities are in nearby locations, and the user can view the information on his/her mobile phone and decide whether or not to perform an inspection. Accurate indoor RTLS is needed to inspire the crowd to effectively inspect fire safety-related facilities, as people will be more apt to take on an inspection job if they are notified by the system when some facilities are nearby, instead of having to search for nearby facilities on their own. There are several technologies currently used in indoor RTLSs, and these technologies are discussed in greater detail in the following paragraphs.

Radio frequency identification (RFID) has been adopted in many RTLS systems because it uses an inexpensive and flexible approach to identify objects and people. RFID is commonly used in applications for indoor settings, as its range of accuracy is 1–3 m; the global positioning system (GPS) has a lower range of accuracy (5–10 m), and it is more often applied in outdoor settings. For fire evacuation management, RFID would require all occupants of a building to wear RFID tags. It might be possible for an office building in which most building occupants are regular occupants; however, the majority of occupants of a shopping mall or a hospital visit only on an occasional basis.

Ultra-wide band (UWB) technology has an accuracy that ranges from 0.1 to 0.3 m, giving it greater accuracy than RFID. In addition, its short response time makes it applicable to settings that are indoor or out-of-doors. The main drawback is the high cost for implementing a UWB solution, which can make its use unfeasible for fire evacuation management [32].

While vision analysis has advantages over RFID and UWB because no tracking devices need to be affixed to the people or objects being tracked, it does have some drawbacks. Vision analysis needs access to a voluminous amount of labeled data sets to use for training prior to implementation. In addition, efficient visual analysis is difficult to achieve in an environment that is dark or dusty, which makes it problematic for evacuations during a fire, when smoke and flames may drastically reduce the accuracy of a vision-based location system.

Other systems—such as Bluetooth low energy (BLE), wireless local area networks (WLAN), Wi-Fi, ultrasound, lasers, radar, infrared, and magnetic marker fields—can be used for indoor localization, and each technology has benefits and drawbacks. In previous research on RTLS technology [10], the authors compared various systems reported in the literature (Figure 2) in terms of the range of accuracy, the affordability of the technology, and how easy the technology would be to implement. In Figure 2, the size and position of each are relative, as they are not based on precise values.

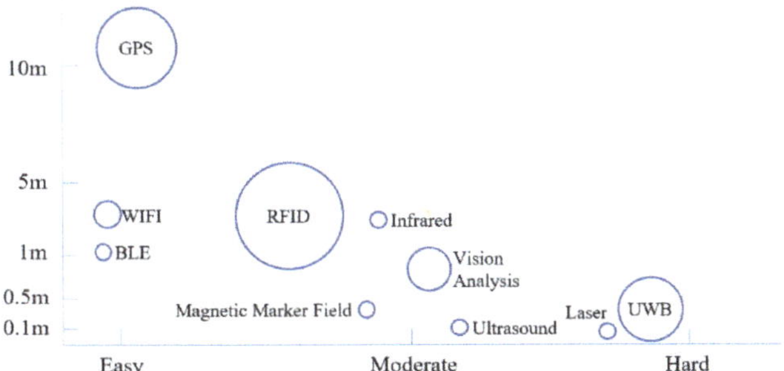

Figure 2. Comparison of major real-time location system (RTLS) technologies (where the size of each circle reflects how widely the technology has been adopted, the vertical axis indicates the range of accuracy, and the horizontal axis indicates the ease of implementation).

In this study, the proposed RTLS uses indoor localization that is based on BLE technology, which enables contextual information to be conveyed between various connected devices (such as between BLE sensors, smartphones, and computers) using a minimal amount of equipment (with regard to the size and cost of the required infrastructure) [33]. As a BLE-based system has low power requirements to operate, the tracking devices will have a long lifespan. In addition, since the BLE devices used in this system are small, they are considered to be a wearable device (it can be attached to the hardhats or safety vests of construction workers, for example). Despite its potential to be used in a wide range of areas, researchers and professionals have not considered BLE-based RTLS for very many application scenarios.

For buildings, such as the floor of the building shown in Figure 3, Bluetooth sensors can be placed in multiple locations on each floor, with a layout that depends on the room layout for each floor, as a single Bluetooth sensor will generally cover an area with a radius of 5–10 m. The Bluetooth sensors are the size of a coin and cost about US$12 each, can be used for 6–12 months before the batteries need to be replaced, and are small enough that they will not have an obvious or unappealing appearance. Batteries should be checked at six-month intervals to ensure that they are working properly. A mobile phone can be used to establish a communication link between a Bluetooth-enabled device and the surrounding sensors, and the location of the smartphone can be determined using the received signal strength (RSS) method.

Figure 3. Bluetooth low energy (BLE)-based indoor RTLS.

3. System Design for iInspect

3.1. iInspect Framework

Figure 4 illustrates the framework of iInspect, where the solid lines represent physical movement and the dashed lines show the circulation of information in cyber space. iInspect will determine which building a user is currently occupying with the assistance of the GPS module in the user's mobile phone and the geographic information system (GIS) coordinates for the building. The iInspect user is able to see the 3D spaces of the building through the cloud-based lightweight BIM model, which should be updated in a timely manner following any changes that are made to the physical building. In a sense, the BIM model is the digital twin of the physical building. While the user walks through the building, iInspect will combine the real-time location of the user and the location of fire safety-related equipment as recorded in the BIM model, and the app will notify the user if there are any facilities nearby that can be inspected.

After being notified by iInspect, a user can browse information related to the fire safety-related facilities that are associated with 3D digital components in the BIM model. There are three types of information that would be available via the app. First is basic information, which includes facility attributes (such as the type and expiration date for a fire extinguisher) and the items that can be inspected along with inspection guidelines (i.e., how to determine the operational and defective status). Second, some VR photos of the real environment will be made available via the app. Using these VR photos, iInspect users can easily find facilities to be inspected and understand the correct status and common problems. Third, if any inspection reports have been generated by other users for the facility, all iInspect users can see those reports and are able to verify the content by performing an additional inspection. All of the information would be updated periodically to reflect the latest status of each facility. Based on this information, an iInspect user could perform an inspection, upload some photographs as a kind of proof, and file an inspection report. After the inspection, a new report or an updated report will be generated and, if a defective status is noted in the report, fire safety management officers will be notified by flags of various colors (which can be used to indicate, for example, different levels of severity).

Fire safety management officers will review all reports having flags and can give feedback to the users who filed the reports regarding the validity of those flags. iInspect users will receive rewards of different kinds for valid flags they have reported. At the same time, the building manager should be notified (and issued a ticket) if there are indeed some fire safety violations. If it is deemed necessary, fire safety management officers will make a trip to the building to perform an onsite inspection.

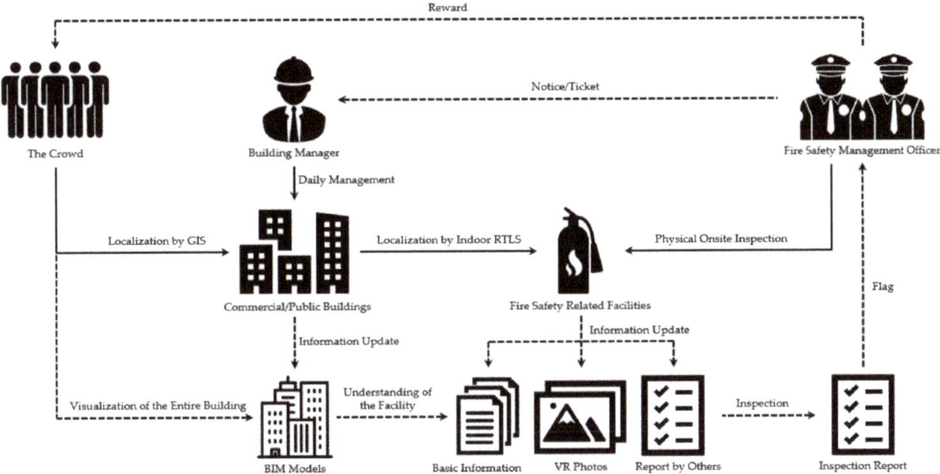

Figure 4. System illustration of iInspect.

3.2. Inspection Content

It is necessary to clearly define the inspection content for iInspect because not all fire safety related facilities in a building are accessible and able to be inspected by the general public (for example, to check if the sprinkler heads are free of grease, dust, rust, etc.). Based on the 2018 International Fire Code [34] and Fire Inspector's Guide [35], along with a systematic review of 27 fire safety inspection checklists made by different local fire departments, authors identified items suitable for iInspect; these items are summarized in Table 3.

It is noted that this list is only a subset of the fire inspection list that is used for regular annual inspection; thus, this crowdsourcing-based inspection can in no way replace a regular fire inspection. The focus in the list for iInspect are items that are publicly accessible and have a dynamic status of validation. When iInspect is applied to a specific setting, the list could be amended by local authorities to make it suitable for a particular circumstance.

Table 3. Checklist for iInspect.

Category	Items
Building access and outdoor premises	
	☐ The building address can be clearly seen from the street.
	☐ Exterior access to the building is not blocked or otherwise inaccessible.
	☐ The fire department connection is easily accessible.
	☐ The lockbox is easy to find.
	☐ Fire hydrants are easy to find and are not blocked off.
Exits and escape routes	
	☐ Exit doors are easy to identify/access and are in good working order.
	☐ Exit doors all open from the inside without the need for keys or specialized knowledge.
	☐ Fire doors are free from visible damage.
	☐ Fire doors are all self-closing and latching.
	☐ Zone maps for emergency evacuation are posted clearly.

Table 3. *Cont.*

Category	Items
Electrical systems	
	□ A minimum of 30 inches of space in front of each electrical panel is available to allow access. □ Extension cords are not in use for permanent fixtures. □ Any extension cords used for small appliances are heavy duty cords. □ Extension cords are all grounded. □ Power strips all have built-in circuit breakers.
Emergency lighting	
	□ The means of exiting the building are all appropriately lit. □ The emergency lighting units are all functional. □ Exit signs are all well illuminated and well situated.
Fire extinguishers	
	□ Fire extinguishers are within 75 feet of all areas of the building. □ Fire extinguishers are all visible and readily accessible. □ Fire extinguishers meet the necessary/specialized standards. □ Fire extinguishers are stored off the floor but at a height that is not above 5 feet. □ Fire extinguishers have inspection tags valid from the previous 12 months.
Fire alarm system	
	□ The manual pull stations for fire alarms are all visible and are readily accessible.
Fire sprinkler system	
	□ Fire sprinkler heads appear to be in good shape (not physically damaged). □ Distance of storage to sprinkler heads is no less than 18 inches.
Heat producing devices and appliances	
	□ Only UL-listed portable heaters are used—and only on a temporary basis. □ Portable electric heaters have 36 inches of space on either side. □ Gas-fired heat producing devices have 36 inches distance from combustible materials. □ Gas-fired, heat-producing devices have vents. □ Light fixtures have all combustibles cleared away from them.
Maintenance of building areas	
	□ Trash and waste are removed daily. □ Oily rags or similar materials are disposed in approved metal containers. □ Combustible materials are stored securely and orderly. □ Exits do not have combustibles stored in their pathway. □ Combustible decorations have been treated with fire retardants. □ Combustible or flammable liquids are stored in approved containers. □ Compressed gas canisters are secured standing up. □ "No Smoking" signs are installed/displayed as required.
Smoke and carbon monoxide (CO) detectors	
	□ Smoke and CO detectors are not indicating a low-battery alert (either by light or sound).

3.3. Incentive Model

People's participation and willingness to contribute are critical issues that one must take into account when developing crowdsourcing solutions. Hence, people should be given some incentives to become part of this collaborative process [36]. Incentives can be intrinsic (personal enthusiasm or altruism) or extrinsic (monetary or in kind reward) [37]. Mohammad et al. [37] argued that intrinsic incentives are more positive than extrinsic ones in terms of the quality of the crowdsourcing project. Monetary incentives can speed up the attraction of participants, however, the payment assigning method can better affect the outcome's quality than simply increasing the amount of money [38]. In all cases, insufficient incentives result in dropping out from the crowdsourcing efforts [39].

There are many cases that people on the Internet take different task for intrinsic and extrinsic incentive other than money. This is more used in knowledge sharing website (for example, Wikipedia and wikiHow) and question answering websites (for example, Quora and Stack Overflow). Different non-monetary incentive types include:

- Fun and entertainment: Since people are willing to do tasks that interest them more, making a task more entertaining will attract more people to participate.
- Personal development: Some crowdsourcing tasks can help participants develop personal skills, for example, obtaining new knowledge (Wikipedia) or learning a new language (Duolingo).
- Competition: Many crowdsourcing projects are designed to have participants compete against each other to find the best solution [40].
- Reputation and identity: As discussed by Wu et al. [41], people tend to do more tasks to earn a good reputation or gain positive attention. In this case, they will stop working if their reputation and identity are lost. Some websites (such as Stack Overflow) use points or credits to quantify reputation.
- Humanity: As a form of altruism, many people will offer help during disasters.
- Other in-kind incentives: There are many other kinds of incentives that do not fit into the categories above. For example, PhotoCity gives away free T-shirts as an incentive [42].

A carefully designed incentive system is vitally important for the success of a crowdsourcing project. A reputation-based monetary incentive model is proposed for the iInspect app developed in this research. More money could increase participation but not necessarily the quality of the results and offering less money may lead to an insufficient number of participants accepting the task. As such, participants need to be paid based on their reputation score (RS), which is earned through the quality of their work. Some researchers have designed monetary incentive models, which integrate participant history/reputation for determining task allocation and payment of rewards [43,44].

In the incentive model for iInspect, the crowdsourcing project requester, i.e., the local fire safety management office, has to evaluate all inspection reports and return a rating called a feedback rating (FR) where $FR \in \{-1, 1\}$ to participants. A rating of "-1" represents an unsatisfactory inspection report, where the inspection result contains mistakes or false fire safety violations. In contrast, a rating of "1" represents a satisfactory report, which means that violations were found to be valid. Every participant in iInspect is assigned an initial reputation score (RS) of 10; the RS is dynamically adjusted by the FR of a completed inspection task. The monetary reward (M) paid to participants is calculated as the product of a base payment (BS, which is US\$1 in this study) and RS, i.e., $M = BS * RS \mid RS > 0, FR = 1$. This means that a participant will be paid more if he/she has a higher reputation score. There are two conditions required to obtain a positive value for M. The first one is that the RS must be a positive value. If a participant has a negative value for RS, it means he/she submitted too many unsatisfactory inspection reports; in order to be paid, he/she must perform more inspections and submit satisfactory inspection reports to bring the reputation score back to a positive level. A participant will also be paid for inspection work only when a report is determined to be satisfactory (i.e., $FR = 1$). In short, after each inspection, the reputation score for a participant will be updated, and reward money will be calculated and paid based on the feedback rating.

Aside from the monetary incentives calculated using reputation scores, some non-monetary incentives are also embedded in this crowdsourcing application. First, by participating in fire safety inspection tasks, a user can develop their personal awareness and fire safety skills. This is a very important personal development and is valuable for personal safety. Second, it builds up high regard and honor in a user's social network. In this research, iInspect links a user's account with his/her account in a major social media platform in China, WeChat, which allows people to know how well their friends are performing in iInspect. A user will be very proud of completing a number of inspection tasks and having a good reputation in the system, because this means he/she contributes significantly to public safety.

4. Case Study

4.1. BIM Model and VR Scene Dvelopment

The case study conducted by the authors used an office building located in Tianjin, China, for verification of the social inspection approached proposed in this research. The floor plan of the 34th floor of the selected building is shown in Figure 5a. A BIM model was first built by importing the construction model for the building into Autodesk Revit and removing any irrelevant information on the facilities related to fire safety that were located on the 34th floor. Next, the BIM model was exported to Autodesk Forge, where a lightweight 3D model for the 34th floor of the building was rendered (as presented in Figure 5b).

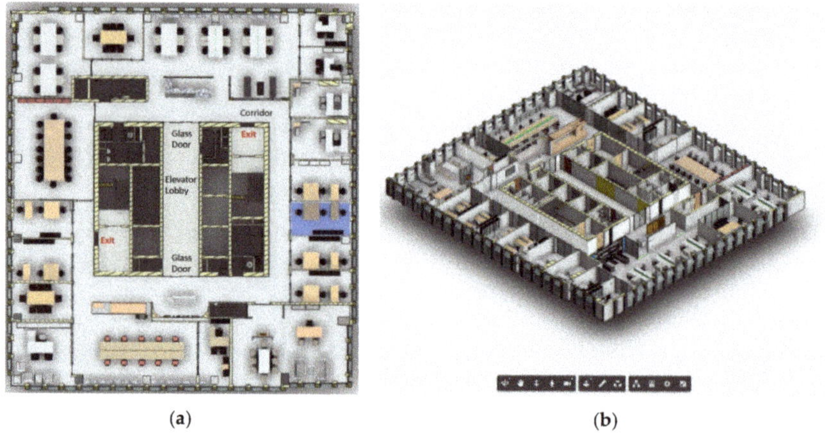

(a) (b)

Figure 5. Commercial building that was used in the case study: (**a**) Floorplan of the 34th floor of the building. (**b**) BIM model of the 34th floor as viewed in Forge (with navigation tools).

VR photos for the building, which were obtained using a dual-lens Insta360 VR camera, were collected at the following locations: two exit doors, two fire doors, floor map sign, four emergency lights, six exit signs, one fire alarm call point, two fire extinguisher boxes, sixteen sprinkler heads, four smoke detectors, and four storage locations. VR photos allow iInspect users to view the environment in a 720 °C perspective (360 °C horizontally and 360 °C vertically). The VR photos are very helpful, as they assist the participants in locating the facilities to be inspected by enabling them to compare the physical environment to the VR environment. Figure 6 shows the VR images of a fire door and a fire extinguisher box.

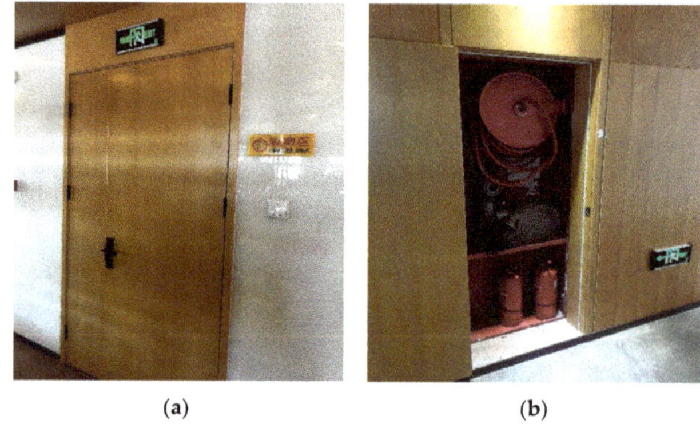

(a) (b)

Figure 6. (**a**) VR image of a fire door. (**b**) VR image of a fire extinguisher box.

4.2. Implementation of BLE-Based RTLS

Due to budget restrictions, only a single floor of the commercial building was used in the case study. Bluetooth sensors (Figure 7a) were used to implement the RTLS for the RTLS. The installation included 26 Bluetooth sensors in the 34th floor of the building (the locations of the sensors can be seen in Figure 7b), and the coverage area for each sensor had a radius of 5–10 m. A cloud server was used to determine the location of a smartphone user in the building based on the data exchanged between the smartphone and the Bluetooth sensors. The Bluetooth system considered in the case study was accurate to within a range of 1–5 m. The actual range in other settings depends on the quantity of sensors deployed and the locations of these sensors, as well as on the obstructions in the building (such as walls) and the material composition of these obstructions, the sensor power, and other factors. After this testing system was deployed, it was calibrated by comparing the coordinates for a building occupant that were calculated by the system to the actual coordinates of the occupant in the building. Since the building had a structure made of reinforced concrete, the partition walls on this floor of the building were not considered to be a serious obstruction. As tested, this system was able to achieve an accuracy of between 1.5 and 1.8 m.

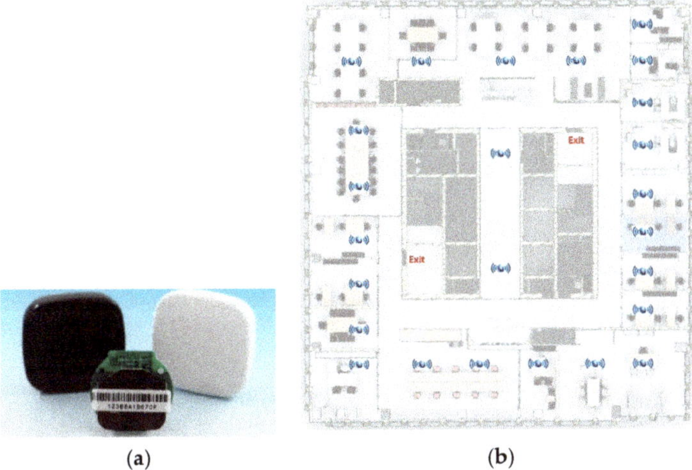

(a)   (b)

Figure 7. (a) Bluetooth sensor employed in this study and (b) layout of 26 Bluetooth sensors.

4.3. Typical Reporting Process

The prototype iInspect app (Figure 8) developed in this study for smartphones using Android did not include functions (such as account management or other functions) that were not directly related to the verification The researchers presumed that an iInspect user had previously downloaded and installed the prototype and had set up a user account. In addition, iInspect is able to use GPS information from the smartphone to determine if a user has entered a building where inspection information for the building has not previously been downloaded using the app—in such a case, the app will be able to prompt the smartphone user to download inspection information for the building from the external server to view any possible inspection tasks.

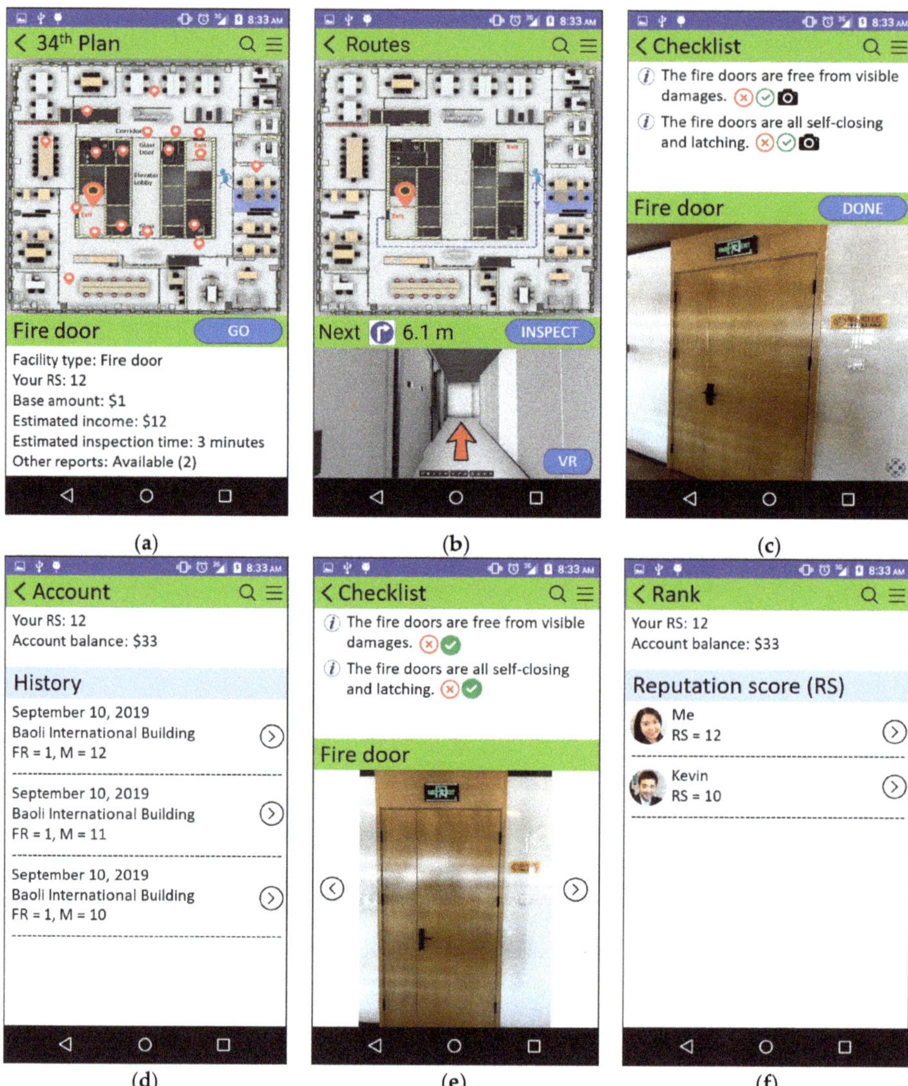

Figure 8. Prototype application of iInspect: (**a**) Possible inspection locations. (**b**) Indoor navigation to the inspection location. (**c**) Items to be inspected. (**d**) Account overview. (**e**) Inspection details. (**f**) Rank details.

When an iInspect user is near a fire safety-related facility, the system will push a notification to the user, and an overview of checkable facilities will be displayed on the phone that contains additional information about the facilities to be checked. As shown in Figure 8a, each bubble in the floor plan represents a certain type of facility available for inspection; when a bubble is selected, the basic information about the associated facility and the amount of reward money for the inspection task will displayed at the bottom of the screen.

After clicking the "Go" button, iInspect will show an interface of indoor navigation to guide the user to locate the facility, as shown in Figure 8b. The "VR" and "BIM" buttons in the navigation view allow the user to switch between the BIM view and the VR view if a VR view is available.

If the "Inspect" button is clicked, a checklist will be displayed, as shown in Figure 8c. By clicking the "i" icon in front of each item, a detailed explanation and the associated illustrations will be displayed at the bottom of the screen. If the user believes there is a violation for a given item and clicks the "cross" icon, the system will require a photograph to be uploaded as proof of the violation. If all items pass the inspection, the system will prompt the user to upload a photo of the entire facility along with a timestamp as proof of the inspection.

A participant can upload an inspection report to the server by clicking the "Done" button, and fire safety management officers will be immediately notified if any safety violation is reported. As mentioned previously, an inspection report will be rated, and the RS and monetary reward (M) will be calculated. An iInspect user can view his/her account to check the value of the RS and the balance of M, as well as the history of inspections, as shown in Figure 8d. Inspection details will be displayed (Figure 8e) after the participant clicks any inspection history record. In addition, the ranking of the user with respect to his/her friends (Figure 8f) will also be displayed, and this information is shared on social media.

5. Conclusions

Fire safety is a vitally important issue for all buildings. Annual fire safety inspections performed by the local fire safety management office is the current practice to ensure there are no violations for any fire safety-related requirements. Commercial and public buildings are deemed to be more vulnerable to fires because of their complex use functions, large number of centralized building occupants, and the dynamic nature of the use of these buildings. Ideally, there needs to be an approach to track the status of all fire safety-related facilities in a real-time manner, or at the very least to conduct safety inspections on a more frequent basis. However, the reality is that many cities lack a sufficient number of skilled fire safety inspectors to ensure continuous compliance of all fire codes for the large number of commercial and public buildings under their jurisdictions.

This paper proposed a crowdsourcing application, iInspect, which can be used to recruit members of the general public to carry out fire safety inspection tasks through the assistance of BIM + VR and indoor RTLS. iInspect assigns inspection tasks to its users when an indoor RTLS detects there are fire safety-related facilities nearby, and it employs BIM + VR technology to help the user find the facility and understand what needs to be inspected. In this way, iInspect can harvest collective intelligence, which is the key benefit of a crowdsourcing project.

Since the general public cannot access and inspect all fire safety-related facilities, this research summarized a list of items that can be easily checked by a typical building occupant. The items on the list are all accessible by the general public and have highly dynamic features, including building access and outdoor premises, exits and escape routes, electrical systems, emergency lighting, fire extinguishers, fire alarm systems, fire sprinkler systems, heat-producing devices and appliances, maintenance of building areas, and smoke and carbon monoxide (CO) detectors.

An appropriate incentive model is the key for the success of any crowdsourcing project. This research proposed a reputation-based monetary incentive model for iInspect. An iInspect user will gain a higher RS when he/she takes on more inspection tasks and produces high-quality inspection reports. The actual money an iInspect user will earn is the product of a base amount and his/her RS, which means that a user with a higher reputation score will earn more money than other users when performing the same inspection task. Some other non-monetary incentives are also considered in this research, such as the development of personal skills and altruism/honor.

A prototype application of iInspect was created for mobile devices using the Android platform, and a case study was conducted to verify the developed system by considering an office building located in Tianjin, China. A BIM model was developed along with VR photos of all major items to be inspected. An indoor RTLS was deployed using BLE technology, and an accuracy of 1.5–1.8 m was achieved. The prototype app, which runs on a mobile phone, follows the iInspect framework shown in Figure 4 and has verified major functions.

This crowdsourcing approach can be applied to many other inspection circumstances with a lack of inspectors that present a highly dynamic status of compliance. For example, the on-site safe management of a construction project is a challenge due to its dynamic nature. By using a crowdsourcing approach, every on-site worker with a smartphone has the potential to be an inspector, and any potential risk could be identified and reported in a timely manner. Similarly, for some infrastructure facilities such as highways, the spatial distribution of networks makes continuous inspection/assessment of the network condition difficult and costly. This issue will be well handled if an adequate number of infrastructure users can be recruited as defect inspectors.

There are some limitations in this research. First, indoor real-time location technology is still in its early stages and there is much room for improvement in terms of accuracy and cost. As a result, an indoor RTLS cannot be readily implemented in most buildings; thus, push notifications of nearby checkable facilities and indoor navigation could be functions that may not be available in most buildings, and users will need to find facilities on their own using the BIM model, even a 2D map, as a reference. Second, the open nature of the crowdsourcing approach presents an opportunity for individuals who exhibit antisocial behavior to participate in the inspection process. For example, people may upload incorrect inspection reports deliberately. Therefore, the quality control of such a crowdsourcing inspection system is the next logical topic of study. Third, considering that non-expert general public many have limited ability to perform inspection tasks, local authorities may use a customized version of Table 3 in the early stage of implementation. Last but not least, the effectiveness of the incentive model employed in this research needs to be validated by implementing the iInspect app on a large scale. The prototype application was tested by only a few users, and it is not possible to examine the effectiveness of the incentive model when using such a small sample size.

Author Contributions: Conceptualization, J.Z. and D.Z.; funding acquisition, J.Z.; investigation, D.Z. and J.Z.; project administration, J.Z.; software, Z.C., D.L. and H.X.; writing—original draft, D.Z. and J.Z.

Funding: This work was supported by the National Natural Science Foundation of China under Grant 71272147 and Tianjin Science and Technology Committee under Grant 15ZXHLSF00070.

Conflicts of Interest: The authors declare no conflict of interest.

References

1. Ministry of Emergency Management Fire Administration Department. Available online: http://www.119.gov.cn/xiaofang/hztj/36306.htm (accessed on 18 April 2019).
2. Wang, Z.; Zhang, X.; Xu, B. Spatio-temporal features of china's urban fires: An investigation with reference to gross domestic product and humidity. *Sustainability* **2015**, *7*, 9734–9752. [CrossRef]
3. Visscher, H.; Meijer, F.; Sheridan, L. Fire safety regulation for housing in Europe compared. *Build. Res. J.* **2008**, *56*, 215–227.
4. Dieken, D. Inspection, Testing and Maintenance of Fire Protection Systems at Electric Generating Plants. Hartford Steam Boiler Inspection and Insurance Co. Available online: www.hsb.com/thelocomotive/Story/FullStory/ST-FS-FIRESUP.html (accessed on 12 February 2019).
5. Sobral, J.; Ferreira, L.A. Availability of fire pumping systems under periodic inspection. *J. Build. Eng.* **2016**, *1*, 285–291. [CrossRef]
6. Sobral, J.; Ferreira, L. Maintenance of fire sprinkler systems based on the dynamic assessment of its condition. *Process Saf. Prog.* **2016**, *35*, 84–91. [CrossRef]
7. Sierra, F.J.M.; Rubio-Romero, J.C.; Gámez, M.C.R. Status of facilities for fire safety in hotels. *Saf. Sci.* **2012**, *50*, 1490–1494. [CrossRef]
8. Han, H.Y.; Hua, Z.G.; Ding, X.X. Design of simulation training system for fire safety inspection based on computer simulation technology. *Procedia Eng.* **2018**, *211*, 199–204. [CrossRef]
9. Lee, P.H.; Fu, B.; Cai, W.; Chen, J.; Yuan, Z.; Zhang, L.; Ying, X. The effectiveness of an on-line training program for improving knowledge of fire prevention and evacuation of healthcare workers: A randomized controlled trial. *PLoS ONE* **2018**, *13*, e0199747. [CrossRef]

10. Zhang, J.; Guo, J.; Xiong, H.; Liu, X.; Zhang, D. A framework for an intelligent and personalized fire evacuation management system. *Sensors* **2019**, *19*, 3128. [CrossRef]
11. Lo, S.M. A building safety inspection system for fire safety issues in existing buildings. *Struct. Surv.* **1998**, *16*, 209–217.
12. Lo, S.M.; Cheng, W.Y. Issues of site inspections for fire safety ranking of multi-storey buildings. *Struct. Surv.* **2003**, *21*, 79–86. [CrossRef]
13. Weng, W.G.; Fan, W.C. An inspection criterion for flashover in compartment fires based on catastrophe theory. *J. Fire Sci.* **2001**, *19*, 413–427. [CrossRef]
14. Liu, F.; Zhao, S.; Weng, M.; Liu, Y. Fire risk assessment for large-scale commercial buildings based on structure entropy weight method. *Saf. Sci.* **2017**, *94*, 26–40. [CrossRef]
15. Kachenje, Y.; Nguluma, H.; Kihila, J. Assessing urban fire risk in the central business district of Dares Salaam, Tanzania. *Jàmbá J. Disaster Risk Stud.* **2010**, *3*, 321–334. [CrossRef]
16. Madaio, M.; Chen, S.T.; Haimson, O.L.; Zhang, W.; Cheng, X.; Hinds-Aldrich, M.; Chau, D.H.; Dilkina, B. Firebird: Predicting fire risk and prioritizing fire inspections in Atlanta. In Proceedings of the 22nd ACM SIGKDD International Conference on Knowledge Discovery and Data Mining, San Francisco, CA, USA, 13–17 August 2016; pp. 185–194.
17. Shao, Y.W.; Huang, C.J.; Kao, S.F.; Wu, Y.S.; Tseng, S.C.; Chang, K.Y. Integration research on fire safety equipments and systems of inspection and reporting for the building public security. *J. Appl. Fire Sci.* **2010**, *20*, 323. [CrossRef]
18. Estelles-Arolas, E.; González-Ladrón-de-Guevara, F. Towards an integrated crowdsourcing definition. *J. Inf. Sci.* **2012**, *38*, 189–200. [CrossRef]
19. Afuah, A.; Tucci, C.L. Crowdsourcing as a solution to distant search. *Acad. Manag. Rev.* **2012**, *37*, 355–375. [CrossRef]
20. Claypole, M. Learning through Crowdsourcing is Deaf to the Language Challenge. The Guardian: London, UK. Available online: https://www.theguardian.com/education/2012/feb/14/web-translation-fails-learners (accessed on 14 February 2012).
21. Brabham, D.C. *Crowdsourcing*; The MIT Press: Cambridge, MA, USA, 2013; ISBN 9780262518475.
22. Landemore, H. Collective wisdom—Old and new. In *Collective Wisdom: Principles and Mechanisms*; Landemore, H., Elster, J., Eds.; Cambridge University Press: Cambridge, UK, 2012; ISBN 9781107010338.
23. Baase, S. *A Gift of Fire: Social, Legal, and Ethical Issues for Computing and the Internet*, 3rd ed.; Prentice Hall: Upper Saddle River, NJ, USA, 2007; pp. 351–357, ISBN 0-13-600848-8.
24. Eastman, C.; Teicholz, P.; Sacks, R.; Liston, K. *BIM Handbook: A Guide to Building Information Modeling for Owners, Managers, Designers, Engineers, and Contractors*, 2nd ed.; John Wiley & Sons, Inc.: Hoboken, NJ, USA, 2011; ISBN 9780470541371.
25. Hu, Y.; Castro-Lacouture, D.; Eastman, C.M. Holistic clash detection improvement using a component dependent network in BIM projects. *Autom. Constr.* **2019**, *105*, 102832. [CrossRef]
26. Koo, B.; Fisher, M. Feasibility study of 4D CAD in commercial construction. *J. Constr. Eng. Manag.* **2000**, *126*, 251–260. [CrossRef]
27. Monteiro, A.; Martins, J.P. Survey on modeling guidelines for quantity takeoff-oriented BIM-based design. *Autom. Constr.* **2013**, *35*, 238–253. [CrossRef]
28. Parisi, T. *WebGL: Up and Running*; O'Reilly Media, Inc.: Sebastopol, CA, USA, 2012; ISBN 144932357X.
29. Xu, Z.; Zhang, Y.; Xu, X. 3D visualization for building information models based upon IFC and WebGL integration. *Multimed. Tools Appl.* **2016**, *75*, 17421–17441. [CrossRef]
30. Chen, H.M.; Chang, K.C.; Lin, T.H. A cloud-based system framework for performing online viewing, storage, and analysis on big data of massive BIMs. *Autom. Constr.* **2016**, *71*, 34–48. [CrossRef]
31. Darbois, N.; Guillaud, A.; Pinsault, N. Does robotics and virtual reality add real progress to mirror therapy rehabilitation? A scoping review. *Rehabil. Res. Pract.* **2018**, *2018*, 6412318. [PubMed]
32. Deak, G.; Curran, K.; Condell, J. A survey of active and passive indoor localisation systems. *Comput. Commun.* **2012**, *35*, 1939–1954. [CrossRef]
33. Park, J.; Kim, K.; Cho, Y.K. Framework of automated construction-safety monitoring using cloud-enabled BIM and BLE mobile tracking sensors. *J. Constr. Eng. Manag.* **2016**, *143*, 05016019. [CrossRef]
34. International Code Council. *2017 International Fire Code*; International Code Council: Washington, DC, USA, 2017.

35. International Code Council. *Fire Inspector's Guide Based on the 2018 International Fire Code*; International Code Council: Washington, DC, USA, 2018.

36. Iyad, R.; Sohan, D.; Alex, R.; Naroditskiy, V.; McInerney, J.; Venanzi, M.; Jennings, N.R.; Cebrian, M. Global manhunt pushes the limits of social mobilization. *Computer* **2013**, *46*, 68–75.

37. Mohammad, A.; Boualem, B.; Aleksandar, I.; Motahari-Nezhad, H.R.; Bertino, E.; Dustdar, S. Quality control in crowdsourcing systems: Issues and directions. *IEEE Internet Comput.* **2013**, *17*, 76–81.

38. Goncalves, J.; Hosio, S.; Rogstadius, J.; Karapanos, E.; Kostakos, V. Motivating participation and improving quality of contribution in ubiquitous crowdsourcing. *Comput. Netw.* **2015**, *90*, 34–48. [CrossRef]

39. Manuel, C. Searching for Someone: From the "Small World Experiment" to the "Red Balloon Challenge," and Beyond. MIT Media Lab. 18 February 2016. Available online: https://medium.com/mit-media-lab/searching-for-someone-688f6c12ff42#.tlleaq622 (accessed on 25 April 2019).

40. Nam, K.K.; Ackerman, M.S.; Adamic, L.A. Questions in, knowledge in?: A study of Naver's question answering community. In Proceedings of the SIGCHI Conference Human Factors in Computing Systems, Boston, MA, USA, 4–9 April 2009; pp. 779–788.

41. Wu, F.; Wilkinson, D.M.; Huberman, B.A. Feedback loops of attention in peer production. In Proceedings of the IEEE International Conference on Computational Science and Engineering, Vancouver, BC, Canada, 29–31 August 2009; Volume 4, pp. 409–415.

42. Tuite, K.; Snavely, N.; Hsiao, D.Y.; Tabing, N.; Popovic, Z. Training experts at large-scale image acquisition through a competitive game. In Proceedings of the SIGCHI Conference Human Factors in Computing Systems, Vancouver, BC, Canada, 7–12 May 2011; pp. 1383–1392.

43. Xie, H.; Lui, J.C.; Jiang, J.W.; Chen, W. Incentive mechanism and protocol design for crowdsourcing systems. In Proceedings of the 52nd Annual Allerton Conference on Communication, Control, and Computing, Monticello, IL, USA, 30 September–3 October 2014; pp. 140–147.

44. Zhang, Y.; Schaar, M. Reputation-based incentive protocols in crowdsourcing applications. In Proceedings of the IEEE International Conference on Computer Communications, Coimbatore, India, 10–12 January 2012; pp. 2140–2148.

Article

Developing a Building Fire Information Management System Based on 3D Object Visualization

Suhyun Jung [1], Hee Sung Cha [1,*] and Shaohua Jiang [2]

[1]   Department of Architectural Engineering, Ajou University, Suwon 16499, Korea; suhyun0567@ajou.ac.kr
[2]   Department of Construction Management, Dalian University of Technology, Dalian 116024, China; shjiang@dlut.edu.cn
*   Correspondence: hscha@ajou.ac.kr; Tel.: +82-31-219-2508

Received: 18 December 2019; Accepted: 21 January 2020; Published: 22 January 2020

Abstract: In a building fire disaster, a variety of information on hazardous factors is crucial for emergency responders, facility managers, and rescue teams. Inadequate information management limits the accuracy and speed of fire rescue activities. Furthermore, a poor decision-making process, which is solely dependent on the experiences of emergency responders, negatively affects the fire response activities. Building information modeling (BIM) enables the sharing of locations of critical elements and key information necessary for effective decision-making on disaster prevention. However, it is non-trivial to integrate and link the relevant information generated during the life cycle of the building. In particular, the information requirements for building fires should be retrieved in the BIM software because most of them have spatial characteristics. This paper proposes a prototype system for a building's fire information management using three-dimensional (3D) visualization by deriving the relevant information required for mitigating building fire disasters. The proposed system (i.e., Building Fire Information Management System (BFIMS)) automatically provides reliable fire-related information through a computerized and systematic approach in conjunction with a BIM tool. It enables emergency responders to intuitively identify the location data of indoor facilities with its pertinent information based on 3D objects. Through scenario-based applications, the system has effectively demonstrated that it has contributed to an improvement of rapid access to relevant information.

Keywords: building information modeling (BIM); fire disaster; facility management

1. Introduction

More than 66% (28,012 cases) of the 42,337 fires reported in 2018 in Korea occurred in buildings and other structures. The highest proportion of fires (12,376 cases or 44.2%) was caused by "carelessness," followed by "electrical factors" (10,469 cases or 30.15%). In addition, more than 78.5% of the 12,635 "severe hazardous" cases were exacerbated by "rapid burning of combustible materials" (45%) and "delay in fire recognition and reporting" (33.5%). In particular, the most serious cases of building fires were caused by insufficient safety management facilities and inspection systems, malfunctioning of critical firefighting facilities, or insufficient fire response manuals and interventions [1].

These statistical data indicate the importance of fire disaster prevention and response management in buildings to minimize damage due to fire events because it is difficult to predict the incidence of building fires. In particular, these data highlight the critical need for management and inspection of fire safety facilities, and information about the presence of hazardous and harmful substances present in a building should be obtained in advance to enable the preparation of an adequate fire emergency response. In other words, the data point to the need for a series of integrated information management systems that can not only facilitate prevention but also advise about the steps to be taken after responding to a fire. However, the current fire disaster management system focuses on the performance inspection of newly constructed buildings and the temporary restoration of existing buildings alone [2].

To deal effectively with building fires, it is crucial to support and respond to fire sites accurately and rapidly in the early stages [3]. The first emergency responders, such as building managers, firefighters, rescuers, and other trained members, should be aware of all aspects of the situation to be able to analyze all information pertaining to the fire site, including the internal structures, facilities, and site-specific risk factors, and make the best decisions under the given circumstances [4]. However, the current fire response method is problematic in that the large amount of distributed information is not transmitted accurately and quickly in the event of a fire incident [2]. Rather, it is likely to transmit inaccurate information from a chaotic fire site. It is also important to collect information in a timely manner, which may prevent secondary damage and property loss [5]. Although the related information requests may be sent to the facility manager, information management in the operation and maintenance (O&M) stage is not carried out automatically. As a result, it is difficult to share information quickly, and the accuracy with which information is collected is limited owing to inconsistencies between the building drawings and the actual buildings, as well as the lack of location data for indoor facilities. Further, subjective judgment, which comprises the experiences of fire responders, is common, and deteriorates the safety of fire response activities. Therefore, first emergency responders should be provided with accurate location data of indoor spaces and facilities [6], and the related information needs to be integrated and maintained in the O&M stage to facilitate easy acquisition of the relevant information [7].

Recently, building information modeling (BIM) is widely adopted as a management tool that can share important information among project stakeholders by creating three-dimensional (3D) object models [8]. BIM is proposed as a powerful solution to integrate and link the information generated during the O&M stage to enable the sharing of important information at building fire sites. It is necessary to address how fire response information requirements can be managed in advance in the O&M stage. In this context, the purpose of this study was to develop a building fire information management system (i.e., Building Fire Information Management System (BFIMS) that can rapidly locate and collect critical information by visualizing these requirements for first emergency responders at building fire sites.

2. Background

Building fire safety should be managed as part of overall disaster control or mitigation measures and requires an integrated management system [9]. In particular, unlike daily-use convenience facilities, it is important to maintain fire protection facilities on a routine basis because it is difficult to detect any functional failure without the occurrence of an emergency [10]. Facility managers, who are generally responsible for supporting buildings and related services, manage information and knowledge (i.e., financial, human, and physical resources) for strategic building operations and maintenance [11]. Above all, records of O&M work history and technical documents, such as building drawings, guidelines, and manuals, are important assets for O&M work. Throughout the digitalization and computerization process, the availability and accessibility of information can be improved sharply [12]. The conventional computer-aided facility management (CAFM) involves documentation, information provision and classification, and monitoring functions related to O&M tasks [13]. The amalgamation of CAFM and BIM tools is advantageous in that it can identify the location of the building components and provide real-time information relevant to O&M via 3D object models containing large amounts of physical information. In emergencies, BIM-based facility management systems can access data in real time using BIM, enabling emergency responders to identify potential countermeasures and risk factors [14]. In spite of these advantages, in practice, two-dimensional (2D) computer-aided design (CAD) drawings, spreadsheets, or web-based information systems, all of which may store crucial information, are seldom interlinked in an effective way.

In addition, even buildings with their own respective CAFMs use them in management tasks primarily to optimize the building's operation cost. Thus, the information preparation and management for various disaster response activities are somewhat incidental. The absence of required information reduces the speed of fire responses and threatens the safety of emergency responders. Therefore, to easily obtain relevant information for effective building fire response, firefighting, and disaster prevention-related work, maintenance information should be integrated and managed in the O&M stage. In addition, in terms of information usability, an instant information output is required for each building space and internal facility. To ensure reliable information management, it is necessary to verify the quality of each data set, and data standardization should be performed to enhance the quality of information management [15]. Therefore, the facility manager can benefit from computerizing the information required in the fire response and continuously updating the changes during the building's life cycle [16].

2.1. BIM for Fire Disaster Management Systems

BIM is increasingly being employed to collect and analyze data from various research fields. The combination of computer technology, fire and safety management, facilities, and equipment has become a new trend [17]. When introducing BIM into the field of firefighting and disaster prevention, it is possible to share locations and important information pertaining to emergency decision making at fire sites in connection with the information generated in the O&M stage [18,19]. Most of the requirements for developing a fire response have spatial characteristics; therefore, they are stored in the BIM, which allows easy information access and use [14]. Wang et al. [7] proposed a BIM-based integrated system consisting of evacuation analysis and evaluation, evacuation route planning, safety education, and equipment maintenance modules. This system can be used in conjunction with a fire dynamics simulator to analyze fire safety and store information that supports safety management in web-based modules. Recognizing the limitation of CAD models in terms of urban disaster prevention, Isikdage et al. [20] proposed a model that combines the geospatial information of buildings and cities using the geographic information system and BIM. Cheng et al. [21] conducted a study on an integrated system that can be used continuously for fire prevention during the countermeasure stages using BIM and a wireless sensor network. The proposed system can display a fire situation and an occupant's location data on a screen using real-time monitoring, and it can provide bidirectional route guidance between an occupant and a rescuer using mobile interworking and a light-emitting diode guiding device.

The above studies utilized BIM as a platform for collecting real-time information, and also provided various functions, such as real-time monitoring, tracking of an occupant's position, and simulation analysis, in consideration of specific factors by linking other software or integrating sensors. However, this methodology has limitations in that it cannot actively utilize the object data of the building as required by emergency responders. In addition, they are limited with respect to system development for integrated information management of requirements, which are classified without reflecting their linkages and vary depending on the users of the information. The absence of information management with regard to the internal facilities hampers the active utilization of the object data, which is the greatest advantage of the BIM. Some studies made attempts to store and utilize object data through the BIM, but they only considered the property information of the materials affected by the fire, the physical characteristics (e.g., location and capacity of the facility), and management status (e.g., inspection history and operation status). They failed to report on the technological aspects of employing information technology integration as a method for collecting and analyzing the real-time information generated at a fire response site. Thus, in the O&M stage, management information on buildings and related fire response information, such as that pertaining to firefighting facilities, should be complemented by an information management system that can manage and preserve them in advance.

2.2. Data Management for Fire Disaster Prevention

Emergency responders often receive conflicting information at fire sites. As they should be able to accurately share and communicate reliable information in the shortest time, the quality of data collection is paramount [15]. To achieve this, a common platform should be used, and a standardized information requirement should be analyzed to establish a system database. Such a system can provide a recent status of relevant information needed for emergency responders (who are the information users). In particular, because the integrated information based on collaborative processes among emergency responders is very important [18], a common and standardized information system should be established.

Nunavath et al. [4] identified the information structure needed for firefighting teams during a search-and-rescue operation on a university campus, and using the unified modeling language (UML) model, they employed the items as useful tools for communicating with end users. In this study, they classified the general information about the university building, risk factors in the building, personal and family details, the cause and extent of the fire, and resource information about the building. Li et al. [5] classified details about the rescue team before and after arrival at the site using scenario techniques to define the required information according to the response process at the fire site. By performing one-on-one interviews with the rescue team, it was determined that they used the card game method to become a virtual player and derive necessary data according to the response process. Chen et al. [22] proposed a data model that can be used for fire response from a management viewpoint. They classified the activity into sub-activities according to the rescue response process on and off the site to establish a comprehensive classification information structure for fire response. Lee et al. [23] proposed an integrated intelligent urban facilities management approach for real-time emergency response. They attempted to integrate facility information by including a common database, facility database, and 3D geospatial database using sensor information.

Thus, there have been a number of attempts to combine facility management information in the O&M stage for effective emergency response data management. However, it is difficult to apply the information requirements derived from the previous studies as an integrated system in terms of data sharing and management because the pieces of information differ according to the methodologies or users, and they are classified without reflecting the relationships in the communication flow between emergency responders. In other words, additional complementary pieces of information that can be linked to each other in terms of information sharing are required. Therefore, it is necessary for the facility manager to integrate the pieces of information required for first emergency responders, such as the firefighting or rescue team, so that they may perform response activities directly at the fire site, and to manage them appropriately in the O&M stage.

3. Methodology

This study defines an information structure in conjunction with 3D object model for fire disaster management using BIM software, and develops a BFIMS system prototype that intuitively visualizes the information requirements at building fire response sites. To achieve the purpose of this research, the following three steps were conducted. First, the information requirements to be utilized directly or indirectly by emergency responders were derived based on previous studies related to fire disasters. Based on the derived requirements, the UML was used to establish the data structure and define the relationships among the data. Second, the Sketchup software and the programming language Ruby were used to develop a BFIMS that can intuitively identify location data and detailed information of the building space as well as internal facilities. Finally, the scenario-based case study project was applied to verify the effectiveness of the proposed system, followed by in-depth interviews with experts and practitioners.

4. Prototype Design of the BFIMS

4.1. Information Requirements

In complex and chaotic building fire situations, it is important for emergency responders to be able to collect accurate and reliable information in the shortest time, and to analyze the situation by using all available information pertaining to the site [6]. Therefore, sufficient information should be provided in the event of fire disasters involving buildings, and information reliability should be guaranteed by performing regular information management and inspection.

An information management system supports situational awareness by providing a large amount of on-site information to the site commander promptly and accurately, thus guaranteeing the reliability of the information by combining the relevant details and incorporating them into a database [24]. These information requirements were selected only for the building fire prevention and response stages for issues that were mentioned repeatedly in the classification structure proposed by previous studies, namely general information about the building, internal building information, installed facilities and equipment, and real-time fire information. The authors identified the information requirements at the building fire prevention and response stages by carrying out interviews with practitioners (e.g., building managers, fire fighters, and other rescuers.).

As depicted in Figure 1, classification criteria for information requirements related to building fire safety activities were categorized as static information. They included the building type, management history, and storage locations for dangerous property that existed before the occurrence of a fire disaster, as well as dynamic (situational) information, which is generated and collected in real time [6]. Static (non-situational) information can be computerized via collection at the O&M stage in the system in advance. In particular, most of the information requirements have spatial characteristics. While constructing an information management system that supports information transfer through continuous data management, these requirements were defined only for static information. For the purpose of this research, the static (non-situational) information breakdown structure was established, as provided in Table 1.

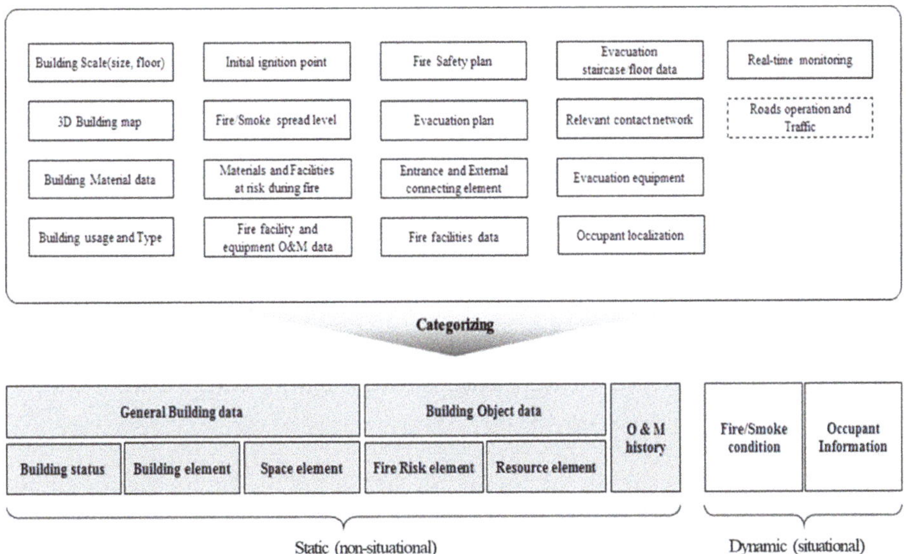

Figure 1. Categorization of the data structure based on the literature review.

Table 1. Static (non-situational) information breakdown structure for Building Fire Information Management System (BFIMS).

Level 1	Level 2	Level 3	Description	Example
Physical information	General building data	Building status	Building status: General information without forms	Building name, address, main usage, building structure and scale (number of floors and area), contact network
		Building element	Building outline composed of structural and non-structural elements that can be used in fire response	Structural elements: slabs, walls, columns, etc. Non-structural elements: doors, window, stairs, etc.
		Space element	Spatial data	Room usage, location, materials
			Vertical flow data	Location of general stairs and elevators, escalators, emergency stairs and elevator
			Externally connected elements	Main entrance, exits, rooftop, external connection zone
	Building object data	Fire risk element	Facility/area to be checked and responded to in case of fire event	Location and impact of areas vulnerable to fire; hazardous materials; facilities' locations, types, quantities
		Resource element	Fire extinguishing facilities	Location and status (type and quantity) of indoor/outdoor fire hydrants, fire extinguishers, etc.
			Firefighting activity facilities	Locations and statuses (types and quantities) of connected sprinklers, water pipe connections and facilities, etc.
			Firefighting water facilities	Location, capacity, quantity, type
			Evacuation and rescue facilities	Location and status (type and quantity) of evacuation equipment (e.g., evacuation ladders and rescue lifts) and lifesaving equipment (e.g., air respirators and heat shields)
			Fire warning facilities	Location and status of operation (fire detection, smoke detection, sensors, etc.)
Management information O&M history			Fire management personnel data	Responsible members' ID, name, role, contact data, and jurisdiction
			Maintenance history of physical information	Work details, including whether the facility is operational, updated data, and manager

UML, an object-oriented modeling language, is a useful tool and considered as a standard tool in developing a database structure in many fields [25]. A class diagram (one of the UML diagrams) illustrates an overview of the system by showing the relationships among the classes. Relationships between objects in a particular application environment and semantically responsible objects can be depicted in this manner. The object properties are specified in the class box, and the relationships between the objects are determined according to line types.

In this study, the attribute data for each object are defined as the type. For instance, the building object data defined in Table 1 are categorized as types. They include the fire risk elements and resource elements that can be used in building fire response. The resource elements are divided into fire extinguishing facilities, firefighting activity facilities, firefighting water facilities, evacuation and rescue facilities, and fire warning facilities. Each object data is a series of component data, including location data and a detailed description about the indoors of the building. The component type is used to visualize the location data more intuitively by specifying color by type. General information, which includes the component type and component data, is represented by 3D object data in association with the building shape model together with the management information.

This enables intuitive location data to be collected, and reliable information can be provided by managing O&M information according to the object. Simultaneously, it is necessary to define additional types so that the flexibility of information management can be guaranteed according to the user's convenience. For the purpose of this study, an UML diagram was constructed as a conceptual data model for BFIMS, as depicted in Figure 2.

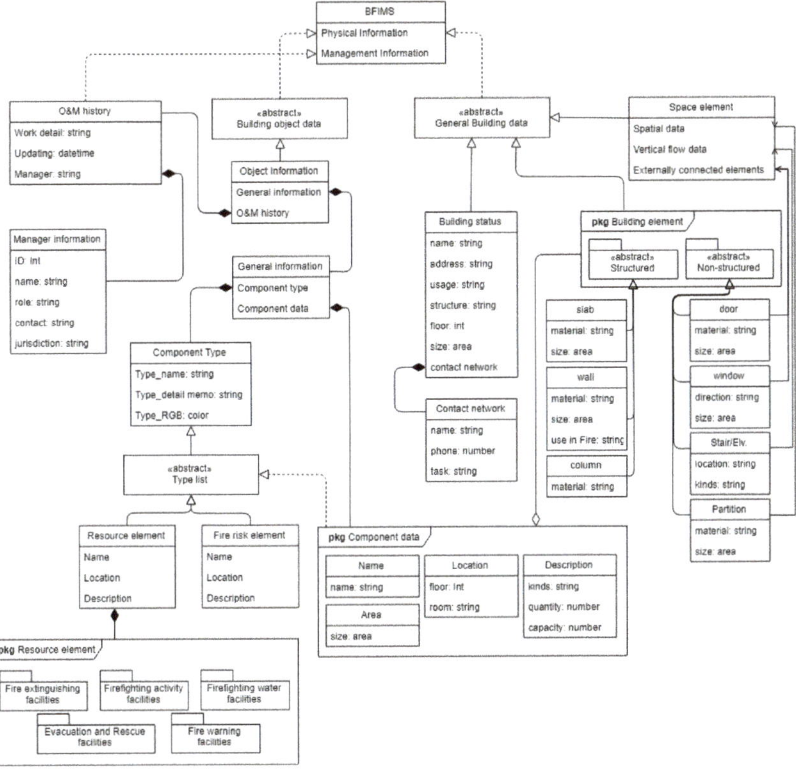

Figure 2. Unified Modeling Language (UML) diagram of conceptual data model for BFIMS.

4.2. Design of the BFIMS

The complexity of a system is determined by the plug-in, user accessibility, and functions through the platform chosen to implement and run the system. Furthermore, in terms of system accessibility, users should generally be able to easily modify and configure the database for changes according to the level of information they can process. In particular, for emergency responders, the system interface should be simple and easy to manage. It is also necessary to guarantee data processing speed by reducing large-capacity information loads [26]. Therefore, the system to be developed should be able to extract only the minimum level of information requirements for fire response activities.

Sketchup^{TM}, a conceptual 3D modeling tool, has been unsuitable for complicated building models. However, it is one of the easiest modeling programs for non-professionals to use in that it is useful for quickly constructing approximate mass models. It also requires less system storage capacity compared to other BIM software and, therefore, it can improve data access speed. Sketchup has an easy-to-follow interface and an extension (or plug-in) functionality, which can be applied to a variety of applications [27]. Further, it is easily integrated with Google Earth, making it advantageous to apply the extended concept to future urban disaster management. Sketchup is generally used as a schematic design tool in the earlier stages [28].

As a plug-in system, BFIMS was developed via Sketchup software and can further be expandable as a BIM model. The system can be installed as an add-on program to Sketchup. Ruby, which is the default language for Sketchup, was adopted as a programming language for the system.

The Sketchup plug-in program used in this study is named BFIMS. It is a system prototype for managing information that can be used in the maintenance phase for fire safety and in the fire emergency phase for fire control. As seen in Figure 3, the facility manager installs the BFIMS into the Sketchup, which contains a 3D object model with building components. Based on the information structure provided in the previous section (see Table 1), the type is determined. In this stage, any specified object data, including building components, are also stored into the system. When changes are made in the maintenance stage, the stored data are updated using the modification algorithm. In addition, visualization in color designated by the type is possible. The position of the component (object) belonging to each type in the 3D environment is obtained. The necessary information can be classified according to the corresponding procedure by outputting the details.

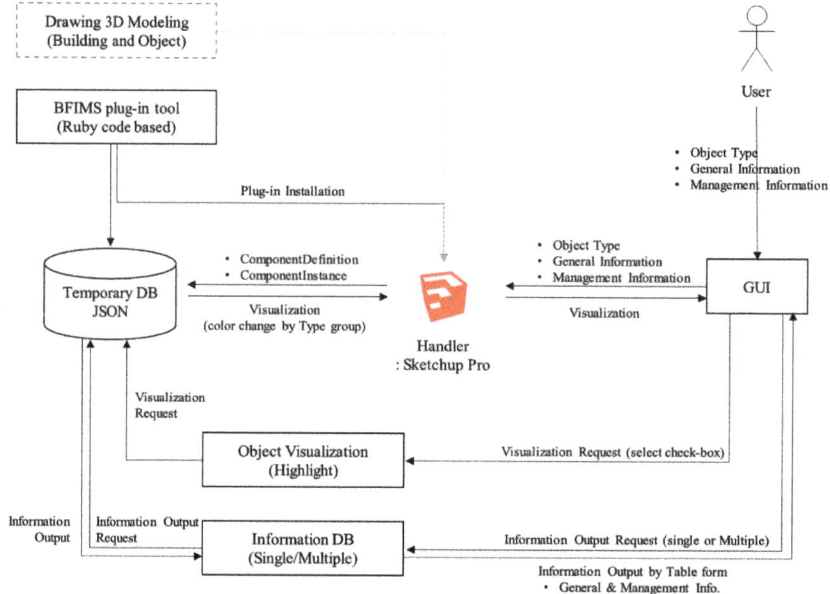

Figure 3. BFIMS architecture.

To construct a database of the 3D object, the authors developed a building model in the Sketchup environment via BFIMS. The model view is in a mass form to improve the data processing speed. The user inputs the object type as well as physical and maintenance information using the graphical user interface and the plug-in. The type information linked with each object and each relevant information are encoded and stored in Javascript Object Notation (JSON) in the attribute of the model. At this time, the type information given to the object is stored in the ComponentDefinition entity, and the information given with regard to each object is stored in the ComponentInstance entity. Sketchup visualizes the single or grouped multi-object information selected by the user. As a result of the data processing, it can also retrieve the detailed information in a real-time basis.

5. Prototype Development and Scenario Analysis

5.1. Prototype Implementation of the BFIMS

The BFIMS consists of four modules: (1) "Management" module for setting by type; (2) "object" module for object data input; (3) "information" module; and (4) "report" module for object data output (see Figure 4a). The BFIMS uses graphical elements to convey information to end users instantly (see Figure 4b). In the object module, a particular type is set as a metadata for grouping and visualizing individual object information. The type is interlined with the values used in JSON (lines 34–49 in Figure 5) by designating the name and describing the type and the colors for visual representation of the grouped objects (lines 7–23 in Figure 5). It is grouped by the type list in the checkbox format. The type consists of fire risk elements and resource elements as depicted in Figure 4.

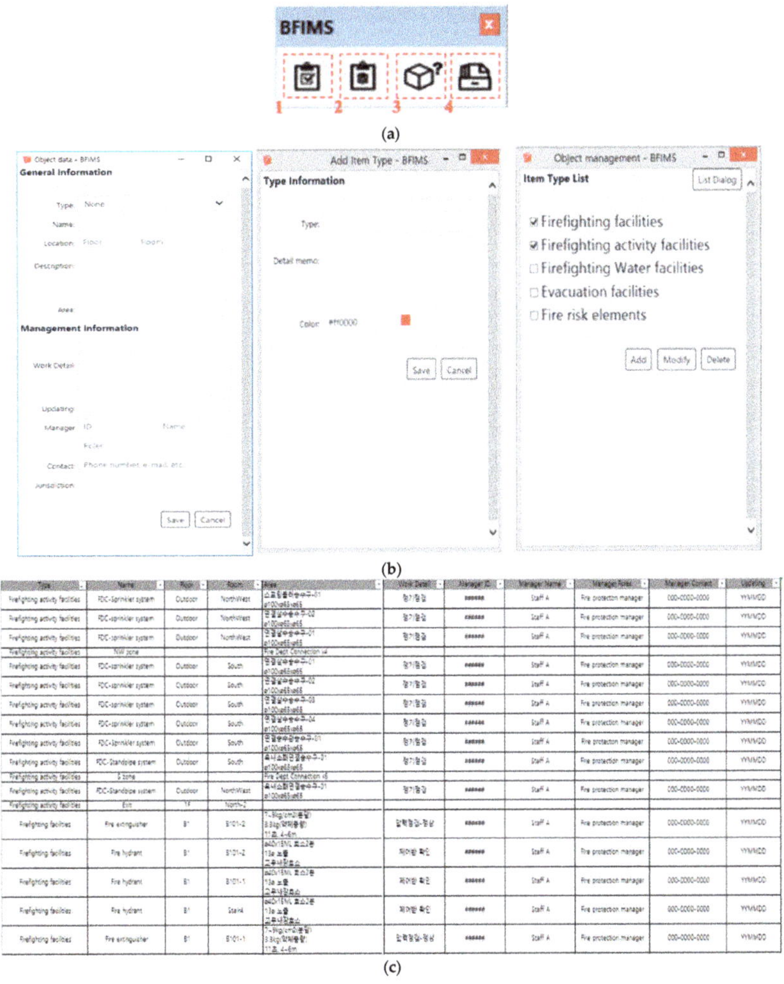

Figure 4. Examples of the BFIMS screenshot. (**a**) Graphic User Interface (GUI) module icons in BFIMS (1: Management; 2: Object; 3: Information; 4: Report); (**b**) GUI screenshots in BFIMS (left: type input; center: type selection; right: data input); (**c**) Fire facilities data derived from BFIMS

```ruby
 6      require 'bfims/constants.rb'
 6
 7  def addItemType(item)
 8      model = Sketchup.active_model
 9      itemTypes = getItemTypes()
10      m = itemTypes.max_by { |element|
11          element[:id]
12      }
13      id = (m ? m[:id] + 1 : 1)
14      itemTypes.push({
15          :typename => item[:typename],
16          :memo => item[:memo],
17          :range => item[:range],
18          :color => item[:color],
19          :id => id
20      })
21      itemTypes = JSON.generate(itemTypes)
22      model.set_attribute(DataKey, ItemTypesKey, itemTypes)
23  end
24
25  def editItemType(item)
26      model = Sketchup.active_model
27      itemTypes = getItemTypes().map { |e|
28          e[:id] == item[:id] ? item : e
29      }
30      itemTypes = JSON.generate(itemTypes)
31      model.set_attribute(DataKey, ItemTypesKey, itemTypes)
32  end
33
34  def getItemTypes()
35      model = Sketchup.active_model
36      a = model.get_attribute(DataKey, ItemTypesKey, '[]')
37      a = JSON.parse(a)
38      itemTypes = []
39      a.each { |e|
40          itemTypes.push({
41              :typename => e['typename'],
42              :memo => e['memo'],
43              :range => e['range'],
44              :color => e['color'],
45              :id => e['id']
46          })
47      }
48      return itemTypes
49  end
50
```

Figure 5. System code: data input.

Typically, the object model created in Sketchup can be stored such that each component has its own location and physical information, but there are limits to the storage of user-recognizable location data and maintenance information. In the "object" module, the user creates a particular object as a Sketchup component, and then sets the category to which each object belongs by selecting the type defined in the "management" module as well as the relevant attribute data, such as location and maintenance history data (lines 16–20 in Figure 6). Objects with the selected type can be visualized in the color designated by selecting the check box in the type list, and are classified by type (Figure 7a). In addition, objects that are distinguished by color can be visualized in a 3D environment, which can show instant spatial relationships and their locations.

```
8     $dataToolShown = false
9   def showDataTool()
10      return if $dataToolShown
11      webDialog = UI::WebDialog.new('Object data - BFIMS', false, nil,
12          400, 770, 200, 200, false)
13      html = Sketchup.find_support_file('bfims/dialogs/data-dialog.html', 'Plugins')
14      webDialog.add_action_callback('syncItemTypes') { |dlg, args|
15          model = Sketchup.active_model
16          itemTypes = model.get_attribute(DataKey, ItemTypesKey, '[]')
17          dlg.execute_script("setItemTypes(#{itemTypes})") }
18      webDialog.add_action_callback('syncObjectData') { |dlg, args|
19          selection = Sketchup.active_model.selection
20          next if selection.size == 0
21          next if !(selection[0].is_a? Sketchup::ComponentInstance)
22          compDef = selection[0].definition
23          defDict = compDef.attribute_dictionary(ObjectDataKey)
24          objDict = selection[0].attribute_dictionary(ObjectDataKey)
25          data = {
26              :typeId => defDict ? defDict[TypeIdKey] : '',
27              :name => objDict ? objDict[NameKey] : '',
28              :floor => objDict ? objDict[FloorKey] : '',
29              :room => objDict ? objDict[RoomKey] : '',
30              :desc => objDict ? objDict[DescKey] : '',
31              :area => objDict ? objDict[AreaKey] : '',
32              :workDetail => objDict ? objDict[WorkDetailKey] : '',
33              :updating => objDict ? objDict[UpdatingKey] : '',
34              :managerId => objDict ? objDict[ManagerIdKey] : '',
35              :managerName => objDict ? objDict[ManagerNameKey] : '',
36              :managerRoles => objDict ? objDict[ManagerRolesKey] : '',
37              :managerContact => objDict ? objDict[ManagerContactKey] : '',
38              :jurisdiction => objDict ? objDict[JurisdictionKey] : '',
39          }
40          s = JSON.generate(data)
41          code = "loadObjectData(#{s})"
42          dlg.execute_script(code) }
43      webDialog.set_on_close {
44          $dataToolShown = false
45      }
46      addJsActions(webDialog)
47      webDialog.set_file(html)
48      webDialog.show()
49      $dataToolShown = true
50  end
```


Figure 6. System code: Object data input.

Figure 7. BFIMS implementation: (**a**) Object visualization; (**b**) single object data output; and (**c**) multiple grouped objects data output.

The "information" module is used to display the defined object information in a tabular form in a single unit (Figure 7b). Moreover, a method exists to display multiple grouped object information by outputting the attribute data pertaining to the objects dependent on the selected types using the list dialog in the "management" module (Figure 7c). In the "report" module, attribute data of all objects stored in the "object" module can be generated in a spreadsheet format. They can be exported to an Excel^{TM} file and saved for additional analysis in the future.

5.2. Application of the BFIMS

To verify the proposed system, an imaginary case-study project, a five-story university building, was applied using a scenario-based framework, as depicted in Figure 8.

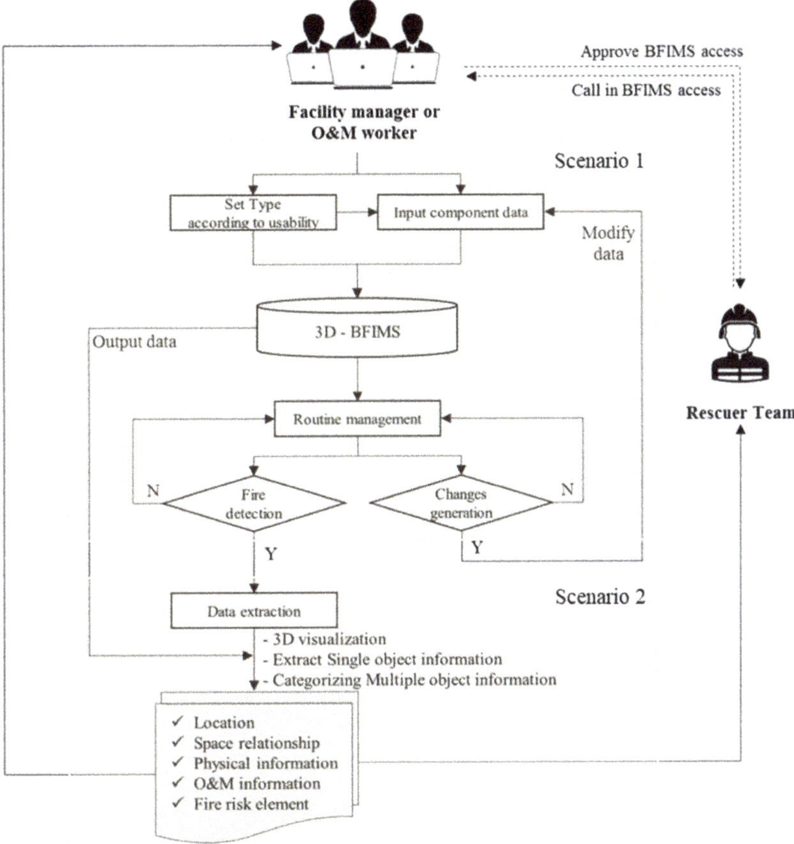

Figure 8. Scenario-based BFIMS application framework.

- Scenario 1: Routine monitoring for facility management

Facility managers or employees regularly inspect the firefighting facilities for fire prevention and preparation. They also manage hazardous materials that are brought into or taken outside the building. The system users can obtain the results in a spreadsheet format as a list of all objects. This information needs to be managed based on the object information stored in the BFIMS database. BFIMS allows them to create checklists to support inspection work items and increase maintenance work productivity. In addition, by managing the detailed information and inspection data of each facility, the manager can use this system to predict the obsolescence of the facility, determine the timing of replacements, and/or enhance the inspection quality.

During the course of routine monitoring, should changes occur during the life cycle of the building, they can be updated with the BFIMS modification function to ensure up-to-date building information. The visualization of this information is based on 3D objects, which provide intuitive location information to the manager, thereby making inspection and maintenance work more convenient.

- Scenario 2: Emergency control in the case of fire occurrence

When a building fire occurs, facility managers move to the site to determine whether or not the fire is real and take actions according to the emergency manual. They also seek the location of the nearest fire hydrant and activate the fire hydrant transmitter to trigger a fire alarm throughout the building. Then, they guide occupants to escape safely through the evacuation stairs, and remove important documents or items from the building. Furthermore, any disastrous area that may affect the entire building owing to fire should be addressed in advance.

However, this fire control process is solely dependent on the manager's experience and memory, which can lead to delays in response time owing to degraded cognitive abilities during challenging events, such as emergencies. Using the proposed BFIMS, emergency responders can quickly identify the visible locations of fire extinguishing facilities, evacuation facilities, and fire risk element information that need to be used. In other words, detailed information, including location data, can be collected, after which the priority of the initial response can be determined based on this reliable objective information. For example, the manager can identify the location of a fire extinguisher or fire hydrant near the ignition point, raise an alarm, broadcast the situation based on the available resource elements, and attempt to control the fire early.

To improve the effectiveness of the proposed system, this study assumes that emergency responders are given the approval to use the BFIMS so as to share the information. In general, firefighting teams are not familiar with a particular building layout. Therefore, it requires time and effort to analyze the spatial relationship inside the building before arriving at the fire site. With this system, prompt decision making can be achieved even if the visible distance inside the building is limited owing to the spread of flames and smoke. In addition, the model supports optimum route calculations and improves the utilization of the fire prevention resources (e.g., fire hydrants, fire extinguishers, sprinklers, standpipe systems, and fire department connections). Because these resources contain certain information required for the fire response activities, the emergency responders can collect the information quickly and assist strategic decision-making based on the linkages of the information with the surrounding environment. Moreover, the firefighting teams can obtain the information on firefighting water facilities and firefighting activity facilities prior to dispatch so that the amount of equipment and water to be transported can be estimated appropriately. This aspect can improve responsiveness to building fires as the rescuers can utilize the resources of the building itself or request additional support in advance to ensure the speed with which the necessary resources need to be procured.

5.3. Expert Feedback on the BFIMS Prototype

The system validation was conducted with experts in building fire disaster. The group of experts included two facility managers (senior manager and practitioner) and two firefighting crew members (crew commander and leading firefighter) with more than 10 years of experience. This focus group interview was conducted to determine how they can effectively identify the location data and relevant information at the case study application project, using the BFIMS. They were also asked about how the accuracy level of the fire-related information can be improved and how they can save time taken to acquire information about the fire site.

In terms of the locations of the building structures and internal facilities, all interviewees replied that 3D models are more powerful than 2D drawings, as the former can easily provide detailed information based on each location. They also responded that the information management of the proposed system can reduce the time and errors in documenting and reporting the building status, compared to the conventional (manual) approach. From the perspective of fire fighters, they responded that an effective and efficient fire response strategy can be obtained with the BFIMS applications.

However, the facility managers who maintain and update the system database should input the raw data in a manual basis, thus generating additional work. The feedback indicates that the system should be sophisticated enough to allow automation of data input. Furthermore, they revealed that public sector buildings are less benefited compared with private sector buildings because the facility managers, who usually reside in the building, are already familiar with the building information, given that they regularly and systematically conduct facility inspections with their own respective systems. However, the firefighting crew asserted that BFIMS would be advantageous in reducing the time required for information collection, and prioritizing the rescue activities under the assumption that the system can be shared among the emergency rescuers. In particular, the firefighting crew responded that they would appreciate if the internal location data could be instantly gathered by the proposed system. Additional feedback indicated that the system effectiveness would be improved if the spreading of flames and smoke during a building fire could be indicated in real time. In summary, it is concluded that the new system demonstrates a potential benefit to emergency responders who are currently managing information conventionally (using paper-based documents alone).

5.4. Limitations of the System Validation and Future Research

The expert feedback on the proposed system has some limitations in validating the effectiveness of the system. First of all, the number of focus group interviews should be expanded. At the time of writing, the BFIMS has only been pilot-tested in terms of potential benefits of information exchange via the 3D object-based information storage and retrieval process. Second, the two groups (facility manger and firefighters) may require different types of information. As such, the validity should be examined using different usage scenarios. Thirdly, the real case project should be applied to fully verify the impact of the new system. As a pioneering study, the validation results from two scenario-based applications reveal that the current fire disaster management has much to be improved. Rather than scenario-based applications, a real case study project may provide a robust validation in a building fire disaster. In future research, this novel system should be further developed in terms of technological aspects. Mobile devices (tablet PC or smart devices) are commonly used in the industry. So, the system would be much benefited when it is operable in a web-based environment. Additionally, a cloud-server programming can be useful to ensure transmission of BIM data at different stages and connectivity between the 3D model and the data because simultaneous information sharing is critical in building fire control. Finally, the sensor information (e.g., fire, smoke, and flame) should be linked with the system in an automatic way. The real-time collection and sharing of information is critical under the fire situation of a building.

6. Conclusions

The urgency and accuracy of information required at an initial fire disaster site are important factors to consider to effectively cope with building fires. First emergency responders should be able to acquire and analyze accurate and detailed information about the indoors of a building to make optimal decisions. This study proposed a system prototype for 3D object visualization-based building fire information management using the Sketchup software. Previous works were reviewed, and interviews were conducted with experts to better understand the sets of key information needed by emergency responders. The main features of the information requirements are the spatial relation and attribute data of the object. The key information for an effective building fire response comprises spatial data and location data. By focusing on these aspects, this study developed a system prototype that first visualizes 3D buildings and objects according to each type using designated colors to utilize such object information and then outputs the object information.

The proposed information breakdown structure also recognizes that appropriate information should be shared among emergency responders in order for them to cope effectively with fires. It is expected that this system can be used as a fundamental source to identify all factors relevant to the integration of information-based facility management and disaster management. The authors contributed to the extension of BIM to fire safety and disaster management by developing an information system based on space and object data. This necessitates digitizing the inspection history and various previously recorded materials (paper-based documents). These improvements in the usability of the information will also translate to more reliable information. In particular, the proposed system improves accessibility to indoor information that is not currently communicated to emergency responders by providing location data and attribute data of the firefighting facilities. Therefore, it is hoped that decision making based on objective information, rather than subjective experiences, would improve building fire safety.

However, the proposed system has some limitations. The scope of the information management system in this study is limited to an after-the-fact approach. In reality, a building fire disaster can be controlled using a preventive maintenance approach. For example, not only the fire disaster facility information but also other various factors (e.g., energy, space layout, and occupants' use) should be incorporated to enhance the fire safety of a building. In addition, it is important to integrate the static and dynamic real-time information generated at fire sites because the information required to construct the database in this study was developed as part of an information system that can gather and store data in advance.

In the future, this system should be extended to serve as an integral disaster management system that is capable of acquiring and analyzing various aspects of both internal and external environments of a building. This would be accomplished by not only obtaining real-time dynamic information through sensing technology integration but also integrating the system with other simulation programs to analyze the diffusion paths of flames and smoke. Moreover, despite the use of a relatively easy-to-use program, the access speed decreases as the number of objects (components) in the 3D model increases. Therefore, future work will investigate possible improvements to the system access speed when using larger amounts of BIM data.

Author Contributions: Conceptualization, S.J. (Suhyun Jung) and H.S.C.; methodology, S.J. (Suhyun Jung) and H.S.C.; software, S.J. (Suhyun Jung); validation, S.J. (Suhyun Jung) and S.J. (Shaohua Jiang); formal analysis, H.S.C.; investigation, S.J. (Shaohua Jiang); resources, S.J. (Suhyun Jung); data curation, S.J. (Suhyun Jung); writing—original draft preparation, S.J. (Suhyun Jung); writing—review and editing, H.S.C. and S.J. (Shaohua Jiang); visualization, S.J. (Suhyun Jung) and H.S.C.; supervision, H.S.C.; project administration, S.J. (Suhyun Jung); funding acquisition, H.S.C. All authors have read and agreed to the published version of the manuscript.

Funding: This research was supported by National Research Foundation of Korea (NRF), grant number 2017R1E1A1A01075266.

Conflicts of Interest: The authors declare no conflict of interest.

References

1. Kim, J.; Hong, C. A study on the Application Service of 3D BIM-based Disaster Integrated Information System Management for Effective Disaster Response. *J. Korea Acad. Ind. Coop. Soc.* **2018**, *19*, 143–150.
2. Nunavath, V.; Prinz, A.; Comes, T. Identifying First Responders Information Needs: Supporting search and rescue operations for fire emergency response. *Int. J. Inf. Syst. Crisis Response Manag. (IJISCRAM)* **2016**, *8*, 25–46. [CrossRef]
3. Kiyomoto, S.; Fukushima, K.; Miyake, Y. Design of Categorization Mechanism for Disaster-Information-Gathering System. *J. Wirel. Mob. Netw. Ubiquitous Comput. Dependable Appl.* **2012**, *3*, 21–34.
4. Tashakkori, H.; Rajabifard, A.; Kalantari, M. A new 3D indoor outdoor GIS model for indoor emergency response facilitation. *Build. Environ.* **2015**, *89*, 170–182. [CrossRef]
5. Li, N.; Yang, Z.; Ghahramani, A.; Becerik-Gerber, B.; Soibelman, L. Situational awareness for supporting building fire emergency response: Information needs, information sources, and implementation requirements. *Fire Saf. J.* **2014**, *63*, 17–28. [CrossRef]
6. Dilo, A.; Zlatanova, S. A data model for operational and situational information in emergency response. *Appl. Geomat.* **2011**, *3*, 207–218. [CrossRef]

7. Wang, S.H.; Wang, W.C.; Wang, K.C.; Shih, S.Y. Applying building information modeling to support fire safety management. *Autom. Constr.* **2015**, *59*, 158–167. [CrossRef]

8. Davtalab, O. Benefits of Real-Time Data Driven BIM for FM Departments in Operations Control and Maintenance. In Proceedings of the ASCE International Workshop on Computing in Civil Engineering, Seattle, WA, USA, 25–27 June 2017.

9. Ramachandran, G. Fire safety management and risk assessment. *Facilities* **1999**, *17*, 363–377. [CrossRef]

10. Chen, H.M.; Wang, Y.H. A 3-dimensional visualized approach for maintenance and management of facilities. *Proc. ISARC* **2009**, 468–475. [CrossRef]

11. Nutt, B. Four competing futures for facility management. *Facilities* **2000**, *18*, 124–132. [CrossRef]

12. Pintelon, L.; Preez, N.D.; Puyvelde, F.V. Information technology: Opportunities for maintenance management. *J. Qual. Maint. Eng.* **1999**, *5*, 9–24. [CrossRef]

13. Elmualim, A.; Pelumi-Johnson, A. Application of computer-aided facilities management (CAFM) for intelligent buildings operation. *Facilities* **2009**, *27*, 421–428. [CrossRef]

14. Becerik-Gerber, B.; Jazizadeh, F.; Li, N.; Calis, G. Application areas and Data Requirements for BIM-enabled Facilities Management. *J. Constr. Eng. Manag.* **2012**, *138*, 431–442. [CrossRef]

15. Seppanen, H.; Virrantaus, K. Shared situational awareness and information quality in disaster management. *Saf. Sci.* **2015**, *77*, 112–122. [CrossRef]

16. Davies, D.K.; Ilavajhala, S.; Wong, M.M.; Justice, C.O. Fire information for resource management system: Archiving and distributing MODIS active fire data. *IEEE Trans. Geosci. Remote Sens.* **2008**, *47*, 72–79. [CrossRef]

17. Shiau, Y.C.; Tsai, Y.Y.; Hsiao, J.Y.; Chang, C.T. Development of Building Fire Control and Management System in BIM Environment. *Stud. Inform. Control* **2013**, *22*, 15–24. [CrossRef]

18. Amaro, G.G.; Raimondo, A.; Erba, D.; Ugliotti, F.M. A BIM-based approach supporting fire engineering. In Proceedings of the International Fire Safety Symposium, Naples, Italy, 7–9 June 2017.

19. Abolghasemzadeh, P. A comprehensive method for environmentally sensitive and behavioral microscopic egress analysis in case of fire in buildings. *Saf. Sci.* **2013**, *59*, 1–9. [CrossRef]

20. Isikdag, U.; Underwood, J.; Aoudad, G.; Trodd, N. Investigating the Role of Building Information Models as a Part of an Integrated Data Layer: A Fire Response Management Case. *Archit. Eng. Des. Manag.* **2007**, *3*, 124–142. [CrossRef]

21. Cheng, M.Y.; Chiu, K.C.; Hsieh, Y.M.; Yang, I.T.; Chou, J.S.; Wu, Y.W. BIM integrated smart monitoring technique for building fire prevention and disaster relief. *Autom. Constr.* **2017**, *84*, 14–30. [CrossRef]

22. Chen, R.; Sharman, R.; Rao, H.R.; Upadhyaya, S.J. Data Model Development for Fire Related Extreme Events: An Activity Theory Approach. *MIS Q.* **2013**, *37*, 125–147. [CrossRef]

23. Lee, J.W.; Jeong, Y.W.; Oh, Y.S.; Lee, J.C.; Ahn, N.S.; Lee, J.H.; Yoon, S.H. An integrated approach to intelligent urban facilities management for real-time emergency response. *Autom. Constr.* **2013**, *30*, 256–264. [CrossRef]

24. Foltz, S.D.; Brauer, B. Communication, Data Sharing, and Collaboration at the Disaster Site. In Proceedings of the ASCE International Conference on Computing in Civil Engineering, Cancun, Mexico, 12–15 July 2005.

25. Xu, W.; Dilo, A.; Zlatanova, S.; Oosterom, P.V. Modelling emergency response processes: Comparative study on OWL and UML. In Proceedings of the Joint ISCRAM-China, GI4DM Conference, Harbin, China, 4–6 August 2008; pp. 493–504.

26. Fan, H.; Meng, L. A three-step approach of simplifying 3D buildings modeled by CityGML. *Int. J. Geogr. Inf. Sci.* **2012**, *26*, 1091–1107. [CrossRef]

27. Schreyer, A.C. *Architectural Design with SketchUp: 3D Modeling, Extensions, BIM, Rendering, Making, and Scripting*, 2nd ed.; John Wiley & Sons: Hoboken, NJ, USA, 2015.

28. Choi, B. A Study of the Use of Computer Softwares in Architectural Design Field. *Proc. Reg. Assoc. Archit. Inst. Korea* **2012**, *8*, 7–8.

Article

Visual Language-Aided Construction Fire Safety Planning Approach in Building Information Modeling

Numan Khan, Ahmed Khairadeen Ali, Si Van-Tien Tran, Doyeop Lee * and Chansik Park *

School of Architecture and Building Science, Chung Ang University, Seoul 06974, Korea;
numanpe@gmail.com (N.K.); ahmedshingaly@gmail.com (A.K.A.); Tranvantiensi1994@gmail.com (S.V.-T.T.)
* Correspondence: doyeop@cau.ac.kr (D.L.); cpark@cau.ac.kr (C.P.)

Received: 31 December 2019; Accepted: 26 February 2020; Published: 2 March 2020

Abstract: Fires pose an enormous threat to human safety and many spectacular fires in under-construction buildings were reported over the past few years. Many construction sites only rely on fire extinguishers, as under-construction buildings do not contain a permanent fire protection system. Traditional safety planning lacks a justified approach for the firefighting equipment installation planning in the construction job site. Even though many government agencies made safety regulations for firefighting equipment installations, it is still a challenge to translate and execute these rules at the job site. Currently, the construction industry is devoted to discovering all the possible applications of Building Information Modelling (BIM) technology in the entire phases of the project life cycle. BIM technology enables the presentation of facilities in 3-D and offers rule-based modeling through visual programming tools. Therefore, this paper focuses on a visual language approach for rule translation and a multi-agent-based construction fire safety planning simulation in BIM. The proposed approach includes three core modules, namely: (a) Rule Extraction and Logic Development (RELD) Module, (b) Design for Construction Fire Safety (DCFS) Module, and (c) Con-fire Safety Plan Simulation (CSPS) Module. In addition, the DCFS module further includes three submodules, named as (1) Firefighting Equipment Installation (FEI) Module, (2) Bill of Quantities (BoQs) for firefighting Equipment (BFE) Module, and (3) Escape Route Plan (ERP) Module. The RELD module converts the OSHA fire safety rule into mathematical logic, and the DCFS module presents the development of the Con-fire Safety Planning approach by translating the rules from mathematical logic into computer-readable language. The three sub-modules of the DCFS module visualize the outputs of this research work. The CSPS module uses a multi-agent simulation to verify the safety rule compliance of the portable firefighting equipment installation plan the system in a BIM environment. A sample project case study has been implemented to validate the proof of concept. It is anticipated that the proposed approach has the potential to helps the designers through its effectiveness and convenience while it could be helpful in the field for practical use.

Keywords: fire safety rule; visual language; building information modeling; portable firefighting equipment

---

1. Introduction

The construction sector includes many unhealthy and unsafe activities, which lead to discouraging workers, delay project progress, affect the cost, reduce productivity, damage reputation and eventually cause human fatalities and injuries [1]. Despite much efforts, construction job sites are still known to be one of the hazardous worksites due to high accidents rate, and thus construction safety remains a vital issue in many countries. Among the construction site accidents, fires pose a significant threat to human safety. Fire safety management must be a concern for every business, but it is particularly important in construction since sites under construction are often at high risk of fire due for several reasons. First, workers usually are exposed to combustible substances, and the presence of wind around the unfinished buildings can immediately cause a fire [2]. Secondly, construction sites only rely on fire extinguishers or sometimes water tanks, as under-construction buildings do not possess permanent and adequate fire protection systems [3]. Thirdly, the unique nature of the construction industry is the reasons for high risks, such as a complex working environment, a certain amount of wastes, unskilled workers, and many other hazards itself [4].

Many people become seriously injured or die due to fire accidents each year [5]. Fire incidents came into existence with the use of fire after its discovery and are intimately related to the progression of human civilization [6]. Regarding the general statistics of fire accidents, the World Health Organization (WHO) reported deaths of more than 300,000 people annually by fire-induced burns. Unfortunately, disturbing statistics are that 95 percent or more of these deaths happen in low-income and middle-income countries [5]. According to the recent report published in Bureau of Labor statistics in 2019, fire and explosion were responsible for 115 workplace deaths in 2018 and 123 deaths in 2017, which is slightly higher than the 88 fatalities recorded in 2016 [7]. The data published in 2018 by the U.S Bureau of Labor Statistics stated that every year, 66 construction workers are killed due to fires and explosion [8]. The National Fire Protection Association (NFPA) carried out a five-year (2010–2014) study on under construction or renovation residential projects (discounting one- and two-unit projects) and found $280 million of direct damage to the property each year [4].

The growing challenges for fire safety concerning economic progress have encouraged the evolution of fire science and technology [6]. The primary aim of fire science and technology in the early time was to protect properties (to save large factories mainly) and avoid sweeping fires conflagration in cities [9]. Ample research has been performed considering fire safety monitoring and early detection. Many studies have been carried out for detecting smoke and by using different technologies such as very early smoke detection apparatus (VESDA), Dual infrared (IR/IR) spectral band flam detection, fiber optic attached to distributed temperature sensing (DTS) and linear infrared flammable gas detection. [6,10,11]. Apart from fire safety monitoring and fire detection, several studies have also focused on fire safety and evacuation planning for buildings and tunnels [12–15]. To overcome the issue of fire safety in an existing building, fire safety evaluation systems for fire prevention, evacuation, and mitigation strategies were studied [13]. Advancements in fire safety science concerning modern-day buildings have gradually urged for the integration of fire safety in the design, yet, very few studies investigated fire safety during the planning and design stage. Wang et al. developed a BIM-based safety management model for escape route planning, fire safety education, and maintenance records of fire equipment are in buildings [16]. Another study integrated BIM with fire dynamics simulator for a personal safety evaluation and escape route plan [17].

Currently, it is evident that many designers do not consider fire safety in the iterative design process of buildings, thus, the merging of fire safety science with the design process has the enormous potential [18]. Also, recent research has depicted that many construction accidents are associated with the design that could be eliminated with appropriate design considerations [19,20].

Equipment operators, electric technicians, carpentry trades, HVAC mechanics, and many other trades are all prone to construction burn injuries and explosion accidents, which could be prevented. The probability of overcoming fire accidents could be increased if fire safety planning is incorporated in the design phase and implement during construction. The quantity of industrial and civil construction projects is rapidly increasing due to socioeconomic development policies and urbanization in many countries [21]. To make sure the fire and rescue operation is practical and useful as possible is a significant task for any building contractor [22]. Very few researches studied fire safety management planning in the design phase, but the firefighting equipment safety plan is not yet discovered. Thus, there is a gap identified in the literature for the portable firefighting equipment safety plan consideration in the design phase.

Likewise, many construction job sites, unfortunately, do not apply the safety rule-based fire safety planning for portable firefighting equipment installation. The Occupational Safety and Health Administration (OSHA) stipulates that the site-specific safety plan should include a fire protection plan for every construction project [23]. In case of fire during the construction phase, the workers in the job site must be aware of escape routes and necessary relevant information about the fire extinguisher. The vital factor is to ensure the equipment is perfectly located and in good order. The scope of this paper is limited to the appropriate placement of the firefighting equipment, while the perfect order issue will be addressed in another research paper. In order to enhance the construction fire safety planning, this paper proposed a safety rule-based visual programming approach for the appropriate portable firefighting equipment installation plan. Figure 1 explains the conceptual framework for the fire safety planning in construction, which is named as the Con-fire Safety Planning (CSP) approach. Related safety rules for portable firefighting equipment's installation are extracted from the construction best practices and OSHA database. As depicted in Figure 1, visual programming is employed to translate the safety rule into machine-readable language, which provides the four outputs: (1) portable firefighting equipment installation plan, (2) bill of quantities for the required portable firefighting equipment, (3) the multi-agent simulation to evaluates and validates the compliance of installation plan with the related OSHA safety rule. This research also integrates the escape route planning for the construction workers as well as the required quantity estimation for the portable firefighting equipment.

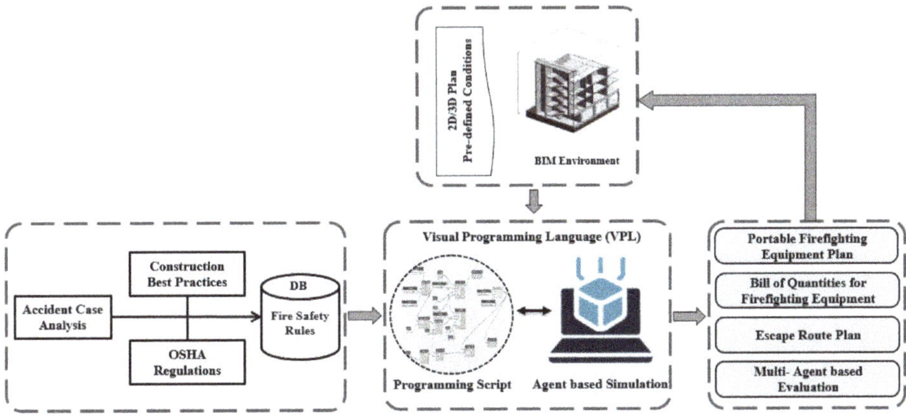

Figure 1. A conceptual framework for fire safety planning in construction.

The significance of the fire safety management, fire accident statistics, current prevention methods, gap in the current fire safety management process, and objectives (solution to the gap) of the research has already been revealed in the introduction section (see Section 1). Next, the literature review (Section 2) presents the current fire safety management status, construction safety planning and fire safety in the design phase, and advance techniques already been applied to the construction safety planning. The system framework constituent of three main modules and three submodules for prototype development. The applicability of the approach is validated through a case study detailed in Section 3. The results and limitations of the research are discussed in Section 4. Finally, this study is ended with the conclusion and future recommendations (Section 5).

2. Fire Safety Management, Construction Planning and BIM Applications

In this regard, to understand about the gap in the literature, fire accident and fatalities reports are reviewed with respect to the current preventive measures developed by different researchers and are summarized as current fire safety management status in construction. Insufficiencies in contemporary fire safety planning were contemplated herein with advanced techniques such as BIM and visual language-based design for safety concepts. Previous efforts on rule-based safety planning are thoroughly studied, and the inevitability of the proposed safety rule-based firefighting equipment installation plan is established.

2.1. Current Fire Safety Management Status in Construction

Occupational Safety and Health (OSHA) reported that one out of five workplace fatalities is the death of a worker from construction [24]. According to the data published by the Federal Emergency Management Agency (FEMA) of the United States (US), approximately 4800 fire accidents annually happened on construction sites [25]. The fire safety engineering emerged in the early 20th century as a response to the fire problem caused by the industrial revolution [6]. The growing urbanization and socioeconomic development reflected an increase in the construction of industrial and civil projects. In recent years, the fire in construction projects was frequently witnessed, and some of them have brought a considerable loss of casualties and property damage [21].

In order to tackle the fire safety problem after the number of severe accidents, several countries around the globe have adopted a series of activities such as fire provisions, fire safety inspections, and up-gradation of fire safety codes [26]. Many early warning systems with different technological approach has been developed for the detection of heat and smoke such as Very Early Smoke Detection Apparatus (VESDA) [10], and Cygnus wireless alarm system [27]. A safe evacuation plan for human is very crucial during fire accidents. BIM-based disaster prevention management system was developed by offering built-in functions such as escape route planning, training, and maintenance record keeping feature for fire safety equipment [17]. In order to understand human behavior during burning tragedies and evacuation, a BIM-based serious human rescue game simulation method was proposed [28]. Generally, construction projects include a vast number of complicated tasks with a tight schedule and fire safety provisions [29]. To ensure the safety from fire in complex buildings, escape routes planning and fire extinguishers are just used to save oneself [15]. Computational Fluid Dynamics (CFD) has been applied to assess the internal condition of the construction workplace during the fire situation [30]. To meet the fire safety requirement, the CFD method is currently common in the design process of construction work. Liu et al. proposed an index system based on-site specification for fire hazard assessment in construction using the fuzzy mathematical method [21].

2.2. Construction Safety Planning and Fire Safety

Fire safety planning holds a key position in the domain of construction safety planning. However, fire safety planning is carried out separately from the project design phase. In the construction industry, the provision of portable firefighting equipment is considered the sole obligation of the contractor [31]. Currently, the construction industry follows site-specific fire safety planning practices [32,33], which need to be reviewed and updated with intervals [34]. Consequently, comprehensive fire safety planning that includes escape route plans, firefighting equipment installation plans, and mandatory education for workers can be an enormously labour-intensive job if done manually. Therefore, updated and practical fire safety plans are not provided in many construction projects [35].

Emerging technologies such as big data, BIM and other computer-aided simulations offer new ways to improve construction safety planning [36]. In order to enhance traditional construction safety management, many researchers have contributed the elementary research studies towards the design for safety concepts [19,36–39]. For instance, a tool to support the hazard identification inherited in the construction process and component, named as design for safety process (DFSP), has been developed by Hadikusmo and Rowinson [40]. Zhang et al. developed a BIM-based safety rule checking system to identify fall hazard through automated checking of 3D models in the design phase. To enhance construction safety planning in the design phase, another study from the same author proposed ontology-based semantic modeling for safety knowledge [41]. A safety rule-based modeling approach has been applied to deal with the excavation related risks, such as cave-in, fall, prohibited zone identification, egress, and ingress [19].

On the contrary, some studies have focused on the BIM application in the design for fire evacuation assessment [17], equipment maintenance, escape route planning, and safety education [16]. However, there is a lake of integration for firefighting equipment installation planning and its integration with fire safety management. Hence, the OSHA rule-based visual language approach is developed to enrich construction fire safety planning.

2.3. BIM and Visual Languages

Generally, the safety management process comprises two aspects, namely safety planning and safety monitoring. Surprisingly, safety planning is generally considered the contractor's liability and thus usually gets ignored in the design phase. However, this concept is being changed due to extensive studies recently considered new technologies such as Building Information Modelling. BIM is actively applied to develop rule-based checking systems for building permits [42]. Many studies have currently developed various algorithms to improve safety planning using BIM-based automatic safety rule checking for unsafe design [43], excavation safety modeling [19], BIM-based scaffolding planning [44], and schedule integrated limited access zone identification and visualization [41]. However, fire safety planning has yet covered education and escape route planning so far.

To translate Korean natural language into computer-readable language, the Korean government has developed KBimCode rule interpretation authoring plug-in, as a part of research [45]. Instead of a hard-coded approach for text-based rule translation into computer understandable language, visual language approaches and parametric input tables have been adopted in different studies recently, such as KBimcode [42], Auto-Exca safety modeling [19], and automated scaffolding risk analysis using BIM [46]. On that account, this research work also utilized the visual language for the development of fire safety management system for construction.

2.4. Need for Rule-Based Firefighting Equipment Installation Plan

The previous efforts on fire safety management and have been reviewed to understand about the accidents and its prevention methods. With this regard, it is concluded that the construction industry currently is focusing on disaster prevention management, which includes escape route planning, fire safety equipment, and educational training intended to ensure the personal safety of individuals [17]. However, conventional disaster prevention methods still depend on manual operating procedures, which is cumbersome. For instance, as a permanent fire safety prevention system that is usually available in build buildings do not exist in many construction job sites. So, they utterly rely on fire extinguishers or water tanks. Average people do not easily understand the correct position for those fire extinguishers in a job site. In the last decade, extensive studies have been carried out by various researchers which reveal the absence of reactive tools to support designers regarding safety in construction [38]. Traditionally, the designer remains surprisingly not aware about the impact of fire safety considerations in the design stage [18]. However, recent advances in fire safety management have disclosed the potential value of fire safety in the design stage. Therefore, a more creative and user-friendly solution is required to sort out the issue. To do so, considering fire safety equipment planning and appropriate allocation in the design stage is inevitable.

3. Structure of Con-fire Safety Planning Approach

The propose system framework for Con-fire Safety Planning Approach consists of three main modules, as revealed in Figure 2, namely: (a) Rule Extraction and Logic Development (RELD) Module, (b) Design for Construction Fire Safety (DCFS) Module, and (c) Con-fire Safety Plan Simulation (CSPS) Module. As depicted by Figure 2, the RELD module is intended to extract the fire safety rules and offer the mathematical structure to the DCFS module. The DCFS module converts mathematical information to computer-readable data by using visual programming tools. This module further includes three sub-modules, named as (1) Firefighting Equipment Installation (FEI), (2) BoQs for firefighting Equipment (BFE), and (3) Escape Route Plan (ERP). The three main modules depict the entire process overview of the Con-fire Safety Planning approach, while the three sub-modules profoundly illustrate the DCFS module. The functions and systematic process of the proposed approach in each module are described in the subsequent sections.

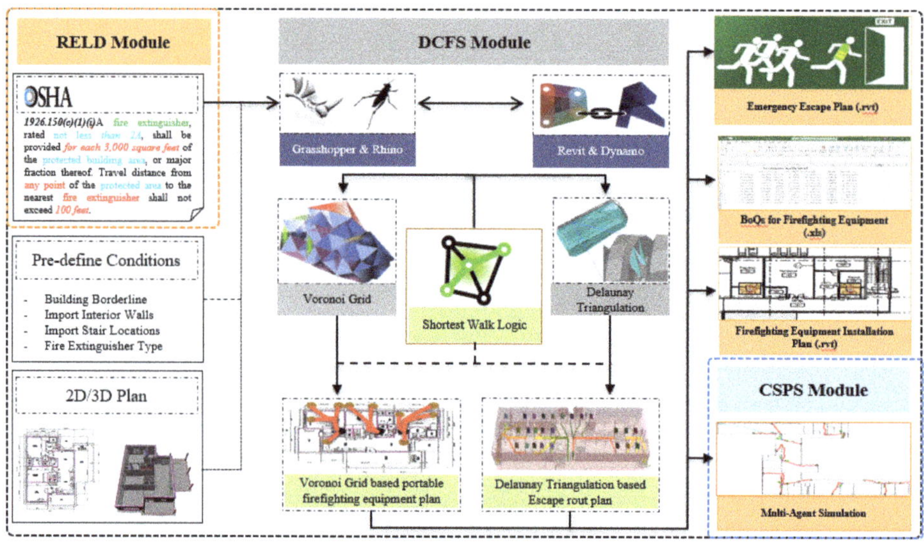

Figure 2. System architecture for Con-fire safety planning approach.

3.1. Rule Extraction and Logic Development (RELD) Module

Accident reports offer vital information such as root causes, responsible roles, and recommended preventive techniques for the avoidance of future accidents. To understand the fire hazards and its preventive methods in construction, this research work was initiated with the assessment of accident reports. Deployment of portable firefighting equipment was found the crucial preventive measure in case of typical under-construction works, while installation of the sprinkler system was noticed vital for fire-sensitive projects such as steel-related projects and tunnels.

In order to ensure a safe environment in terms of fire mishaps, OSHA regulations for fire protection and prevention are investigated in the second step. The safety-related fire regulations have been extracted from the OSHA database. These regulations provide lessons learned from the past and current best practices to minimize the probability of accidents in the construction workplace. Table 1. revealed the OSHA standard number 1926, which describes the related safety and health regulations to the construction industry. Fire protection and prevention rules are depicted under the subpart-F of construction safety and health standard (1926). The standard number 1926.150 demonstrates the fire protection rules, which include general requirements, water supply, portable firefighting equipment, fixed firefighting equipment, and fire alarm devices, as listed in Table 1.

Portable firefighting equipment plays a significant role to deal with the construction fire, as many job sites do not contain permanent firefighting systems [3]. Since the distance between the portable firefighting equipment with the persons working in the area also influence the reactive time required to control the fire when it is in the initial stage, OSHA construction fire standards specify the limit of the distance between the persons and firefighting equipment. According to OSHA, "A fire extinguisher, rated not less than 2A, shall be provided for every 3000 square feet of the protected building area, or major fraction thereof. Travel distance from any point of the protected area to the nearest fire extinguisher shall not exceed 100 feet." To limit the scope of the work, this paper merely considers portable firefighting equipment planning for further research. Apart from that, this research also generates an escape plan as an employee emergency route plan. The manually extracted regulations relevant to fire protection are converted into mathematical logics from text-based information, which is then employed for visual programming in the Design for Construction Fire Safety (DCFS) Module. The following conditions are extracted from the OSHA fire safety rules for construction.

(1)   Placement of at least one fire extinguisher of type, rated not less than 2A, for the area of 3000 square feet.

$$2A = \frac{Total\ Protected\ Area}{3000}$$

(2)   The distance from any point of the protected area to the nearest fire extinguisher should be less than or equal to 100 feet while using the fire extinguisher type, rated not less than 2A.

$$Distance\ from\ any\ point\ \leq\ 100\ feet\ for\ rated\ \geq\ 2A$$

(3)   In the case of using a fire extinguisher, rated not less than 10B, the distance from any point of the protected area to the closest fire extinguisher should be less than or equal to 50 feet.

$$Distance\ from\ any\ point\ \leq\ 50\ feet\ for\ rated\ \geq 10B$$

Table 1. Fire Protection and Prevention related to OSHA regulations.

1926-Safety and Health Regulations for Construction		
1926 Subpart F—Fire Protection and Prevention		
No.	**Standards**	**Explanation**
1926.150(a)	General requirements	This standard explains the employer responsibility of fire protection plan, conspicuously location of firefighting equipment's and its periodic check to assure protection of life
1926.150(b)	Water supply	This standard focuses on the requirements of the fixed or temporary water supply of enough volume and pressure or either completely installed underground water mains.
1926.150(c)	Portable firefighting equipment	This article explains the vital part that determines the location criteria for firefighting equipment based on its types.
1926.150(d)	Fixed firefighting equipment	It focuses on rules related to the installation of fixed sprinkler protection.
1926.150(e)	Fire alarm devices	This part considers the establishment of an alarm system and communication systems with the local fire department.

3.2. Design for Construction Fire Safety (DCFS) Module

The Design for Construction Fire Safety (DCFS) module is a significant module of this research work. This section is devised to convert mathematical logics into computer-readable data, which is obtained from the Rule Extraction and Logic Development (RELD) module. Figure 2 illustrates the system architecture for the Con-fire Safety Planning approach. A commercially available Visual programming language (VPL) tools, named as Grasshopper (BIM authoring platform and a plug-in for Rhinoceros) and Dynamo (BIM authoring platform and a plug-in for Autodesk Revit) were availed to achieve the task of rule conversion. Visual programming is employed in this research study due to several reasons. It is relatively convenient to use as compared with other programming languages [19] such as python, java, and many more. This language effectively represents the information flow through visual symbols (nodes and connections) [42] with precise inputs and outputs. Therefore, this paper proposed a visual language approach for the translation of fire safety rules, which is established from the subpart F (Fire Protection and Prevention) of the OSHA-1926. This module further contains three submodules of the proposed approach.

3.2.1. Design for Firefighting Equipment Installation (FEI) Module

This subsection focuses on fire protection and leverages the visual language approach to acquire BIM-based portable firefighting equipment safety planning in construction. The script is initiated with the extraction of geometric data as input from the 2D plan. As depicted in Figure 3, the method allows the user to define the following four (U1, U2, U3 and U4) conditions: U1 is importing building borderlines (BL) as a 2D plan geometry, U2 is to determine the plan interior walls (IW) as a 2D line geometry, U3 is to define stair locations (SL) as pick points, and U4 is intended to choose fire extinguisher type from the dropdown list that includes type 2A and 10B. According to the OSHA rules, 1926.150(c)(1)(i) A fire extinguisher, rated not less than 2A, shall be provided for every 3,000 square feet of the protected building area, or significant fraction thereof. Travel distance from any point of the protected area to the nearest fire extinguisher shall not exceed 100 feet. Where the Dividing Distance of 2A (DDA) fire extinguisher type is 100ft, and the area is 3000 square feet. Similarly, The OSHA article 1926.150(c)(1)(vi) revealed the Dividing Distance of rated not less than 10B (DDB) fire extinguisher type is 50 feet.

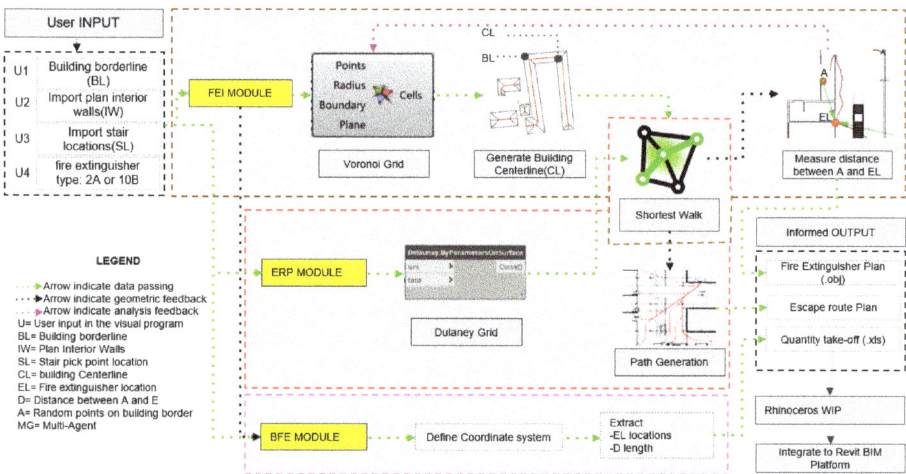

Figure 3. System architecture for Con-fire safety planning approach.

In the FEI Module, the algorithm uses the Voronoi Grid to generate a building centerline (CL), as shown in Figure 3. The building centerline (CL) used in this research refers to a 2D line located in the center of a closed curve of the building borderline (BL). To generate the fire extinguisher's initial locations, the program divides the building centerline by the number of the fire extinguisher obtained from the division of the surface area of the building and the advised required area by OSHA, which is 300 square feet. Thereafter, these initial locations are moved to their closest interior walls (IW) in order to create a fire extinguisher location (EL). The algorithm divides building borderline (BL) into a random number of points (A) to test the shortest walk distance between BL and EL. The algorithm verifies the appropriate location of the fire extinguishers by using the shortest walk logic. The shortest walk logic is a pre-designed component in the visual programming tool (grasshopper). It populates the building surface geometry with points then creates connection lines between points using proximity two-dimensional logic. After that, the algorithm calculates the shortest walk between the locations of random points A and their nearest fire extinguisher cluster locations (EL) and calculates a network of paths between them. The system then measures the closest route curve length (D) and evaluate it in an if statement: as, If D is less or equal to distance advised by the rule that could be DDA or DDB (based on the user selection U4), then do nothing, If D is higher than U4 then divide D over U4 and round up as illustrated in the below equation.

$$\text{if}(D \leq U4),\ \textit{Return Null}$$
$$\textit{else}$$
$$\textit{Roundup}\left(\tfrac{D}{U4}\right)$$

Based on the result of the if statement, the algorithm considers the addition of new fire extinguisher to the existing fire extinguisher cluster. Then, the system automatically moves the fire extinguisher cluster to the closest wall (interior walls imported by the user) and moves in the z-direction by 3 ft. The algorithm creates this portable firefighting equipment plan inside the rhinoceros environment as an (.obj) file. Then this plan is integrated into the BIM model inside a commercially developed Revit program using Rhinoceros WIP add-on.

3.2.2. BoQs for Firefighting Equipment (BFE) Module

This submodule is envisioned to develop a creative solution for estimating the Bill of Quantities (BoQs) for portable firefighting equipment using the visual programming approach. Currently, tools for digital quantity take-off revolutionized the cost estimation aspect in the design stage by saving the time, increased accuracy, and are sophisticated enough to be used. However, to take the cost estimation process to the next level, additional concentration is needed to propose a more creative solution. To do this, a rule-based visual algorithm supported safety information model, which has a fire extinguishers placement plan obtained from the firefighting equipment installation (FEI) module.

Based on the fire extinguisher's placement plan, the system will automatically calculate the quantity take-off for the required portable firefighting equipment. The quantity take-offs list is exported to Microsoft Excel (Ms. Excel) for the cost estimation. Cost estimator and quantity surveyors with this powerful built-in function can quickly develop the required cost plan of the fire protection equipment for the project. In order to generate the quantity, take off, required data was acquired from the Firefighting Equipment Installation (FEI) Module.

The categoric process of the algorithm development for this module can be seen in Figure 3. The visual algorithm is designed to automatically extract the list of geolocations for each fire extinguisher (x-coordinates, y-coordinates, and z-coordinates) from the previous FEI module. The system will calculate the number of locations and generate quantity take-off for the fire extinguishers. The code will then merge the quantity take-off with the predefined fire extinguisher specification. The designed approach also establishes the tag codes of each fire extinguisher (see the details in the case study). The program generates the choices for the users by using the default database, such as relevant data associated with the fire extinguishers chosen by the user, such as manufacturer, family, expiry, and inspection date. As a byproduct of the proposed construction fire safety planning, the program automatically generates a bill of quantities report in excel format (.xls) and saves it on the local machine.

3.2.3. Escape Route Plan (ERP) Module

The purpose of this submodule is to develop the escape route plan for humans in case of an emergency evacuation. The concept of an escape route plan is not new and has been extensively studied by many researchers. An automated escape route plan establishing tools for emergency evacuation has also been developed recently in some commercially available software. The proposed approach in this paper also offers the integration of the escape route plan module. The devised script computes the escape route from the doors of the rooms towards the designated emergency exits.

The visual script developed in Dynamo, available for generation of the fire escape plan with an open application programming interface (API) is integrated with the designed approach for portable firefighting equipment plan. The dynamo script automatically extracts the level of the building as an input. The designed algorithms need few predefined conditions as inputs such as specifying circulation area, regular doors location, and emergency exits. The program deals with the corridor/lobby/stairs as a space family (Stairs are added separately as the category). Once the user defines and highlights the main exit, emergency exit, and staircases as a destination and click the evacuation path button, then the designed algorithm generates a grid on the interest region using the Dulaney triangulation grid method. The Dulaney triangulation is the alternate of the Voronoi grid logic and is used in the dynamo (a plug-in to the Revit). Eventually, the system applies the shortest walk logic from the regular doors to the nearest exit doors, which generate an appropriate fire escape plan and visualize back it in the Revit. The generated escape route plan can be seen in the case study.

3.3. Con-fire Safety Plan Simulation (CSPS) Module

The con-fire safety plan simulation module further verifies and evaluates the OSHA rule-based Con-fire Safety Planning (CSP) approach. This algorithm uses a minimalistic and lightweight crowd simulation library named as PedSim, which is based on the social forces model algorithm [47] and offers a real-time multiple pedestrian simulation in Grasshopper. In PedSim, people move from Start Gate (SG) to Destination Gate (DG), following the best route, avoiding obstacles, and other people. In this research, the Start Gate (SG) means the starting point of an agent and Destination Gate (DG) is the nearest fire extinguisher location (EL) point. The algorithm once again divides the borderline of the building (building envelope) (BL) into several points (in our case) within 100ft distance from each other and offsets them inside the building to be used as a Start Gate (SG). These points will act as start gates for Human Agent (HA), which could be increased or decreased based on user choice and the level of accuracy required for simulation. However, increasing these points (start gates) on the borderlines will reflect an increase in the computational process. The interior walls (IW) are used as obstacles for the agents to avoid in the simulation. The target destination (DG) of the agents is the closest fire extinguisher location (EL). The simulation shows the live indicators: (1)agent position, (2) the body radius, (3)velocity, (4)acceleration, (5) current path, (6) behaviors as list (6.1) maintain closeness to goal, 6.2) obstacle collision avoidance force, 6.3) anticipatory collision avoidance force, 6.4) passive collision avoidance force. The simulation generates a heat map of the agents' trace movement showing the most used corridors and paths by the agents to reach their closest fire extinguisher by using GridPersonCounter component in the grasshopper. The travel distance of any agent A (human) towards the closest portable firefighting equipment is updated in the table. In this way, we can see the traveled path of the agent, and our system shows that every generated agent is in the range of 100 ft (1200 inches).

4. Case Study

This section validates a case study carried out to evaluate the approach by using a sample project from the Autodesk Revit file. The aim was to apply the Con-fire Safety Planning approach developed for portable firefighting equipment installation, bill of quantities estimation, and escape plan. Figure 4 demonstrates the Isometric view and plan of the sample project. The 2D blueprints made in Revit with the entire property's information was used for the experiment. In order to conduct the experiment in a common environment, an add-in to the Revit, Rhino WIP package was installed on the local machine. The designed visual algorithm was imported from the rhinoceros-Grasshopper to the Revit and Rhino-Grasshopper environment using the WIP package. The sample project from the Revit was opened, and exterior walls were converted into a closed curve. The stair landing locations were highlighted by a point object. The interior wall partitions were grouped as curve objects. The user chose the desired family type of the fire-extinguishers from the top-down list, such as 2A or 10 B; in our case study, we assumed 2A. Then the program generates the optimized building centerline using VORONOI grid logic. The program finds the best locations for the fire extinguishers using shortest walk logic and calculates the distance between the building envelope (border), and the fire extinguishers then add additional fire extinguishers in the area longer, outside the range of the selected fire extinguishers type. By pressing the run button for the fire extinguishers plan, a portable firefighting equipment installation plan is generated.

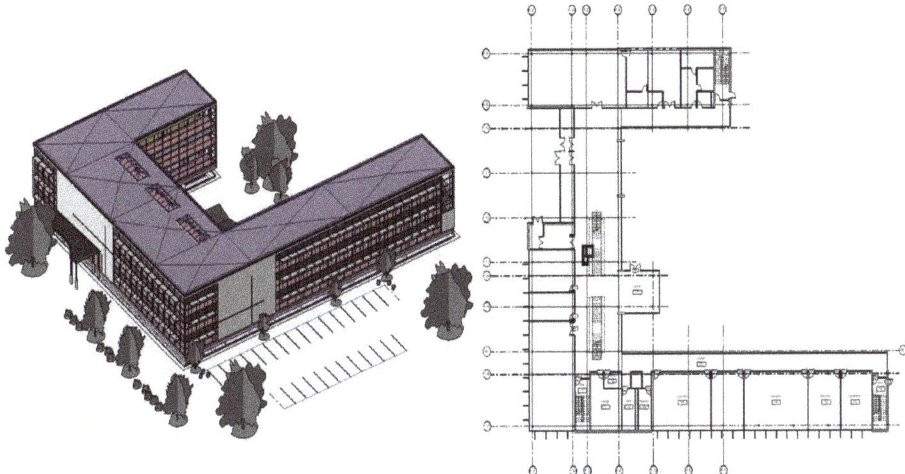

Figure 4. Isometric view and plan of Sample Project.

In order to change the shape of the building to obtain a more verified evaluation from the case study, we added one big room to the building shown in Figure 4. To further explains the process flow of the developed algorithm, the graphical illustration of the process workflow is depicted in Figure 5. This graphical illustration includes five sections. In Section 1 (S-1), the user needs to define the algorithm inputs: building outline, interior walls, stair locations, and fire extinguisher type. As illustrated in Figure 5, the intent of Section 2 (S-2) is to generate a centerline (CL) of the given building 2D plan using a Voronoi grid after evaluating the continuous line clusters among the grid collection. Subsequently, in order to locate the initial position of portable firefighting equipment, the algorithm divide centerline (CL) in several segments to generate points based on user preferences (U4 = fire extinguisher type). Afterward, the program moves the fire extinguishers to the nearest interior walls. In Section 2 of the Figure 5, the Voronoi Grid demonstrates that how the program generates proximity 2d grid in the interior space surface. After that, the algorithm divides the building borderline (BL) into specific segments for the generation of random points (as mentioned in Section 3.2.1). Thereafter, the program uses the shortest walk path logic to generate the shortest path from the A points to the fire extinguisher location (EL); in this case study, we named it as point B (see Figure 5 of Section 3). Moreover, Section 3 also visualize the shortest walk from the collection of path solutions and measures the distance between the building's borderline to the fire extinguishers for further evaluation. The intention of the fourth section (S-4) is to highlight the outreach (longer) paths by using the lightweight crowd simulation library called PedSim, as an add-on in Grasshopper, a multi-agent simulation to verify the compliance of portable firefighting equipment plan with OSHA rule. In this case, the travel distance length is greater than the advised length, so the program creates an additional fire extinguisher on the longer path and moves the new fire extinguisher to the nearest wall. In the last section (S-5), the algorithm uses PedSim and GridPersonCounter component to create a heat map from the multi-agent trace for further analysis.

Figure 6 shows the multi-agent simulation results for Con-fire Safety Planning. The multi-agent simulation is adopted using the lightweight crowd simulation library named as PedSim, which is based on the social forces model algorithm. In PedSim, people move from Start Gate (A) to Destination Gate (B), following the best route, avoiding obstacles and other people, as revealed in Figure 6. The building's envelope (border) is divided into several segments with an interval of 100 ft and offsets them inside the building for the purpose as a start gate (A) using the developed algorithm. The program considered interior walls (IW) as obstacles for the agents and avoided in the simulation while traveling to the destination (B) considering the closest fire extinguisher location (EL). The traveled path of the agent and our system shows that every generated agent is in the range of 100 ft (1200 inches) except number 1, which is highlighted in the table of Figure 6.

The table highlighted the agent, which is not in compliance with the OSHA regulations. Thus, new firefighting equipment (fire extinguisher) is added to the fire extinguishers cluster, as shown in the Figures 6–8. The portable firefighting equipment plan depicts the appropriate location of the fire-extinguishers. Figure 7 shows the 3D visualization of the Con-fire Safety Planning (CSP), while Figure 8 depicts the heatmap generated after the CSP simulation.

In addition to the portable firefighting equipment installation plan, the system also generated the bill of quantities for the required equipment. A layer of firefighting equipment family is generated, and the location coordinates of the fire extinguisher are extracted from the 3D-model using visual programming. The program then automatically counts the location and export it to the excel environment. The generated excel report for the quantity take-off can be seen in Figure 9. Similarly, the algorithm automatically saves that report on the local machine of the user.

Figure 5. Graphic illustration of the process steps followed in the case study.

Correspondingly, the visual algorithm already developed for the escape route planning is applied for the purpose of finding the evacuation path. The visual code is downloaded from the GitHub and imported to the dynamo environment. The required packages were installed, and the code was simulated on the 2D plan. The system automatically generated the escape route from the regular doors to the defined emergency doors, as shown in Figure 10. In summary, the portable firefighting equipment installation plan and escape plan is saved as a single .obj file, while the machine automatically stored the bill of quantities as a .xls file.

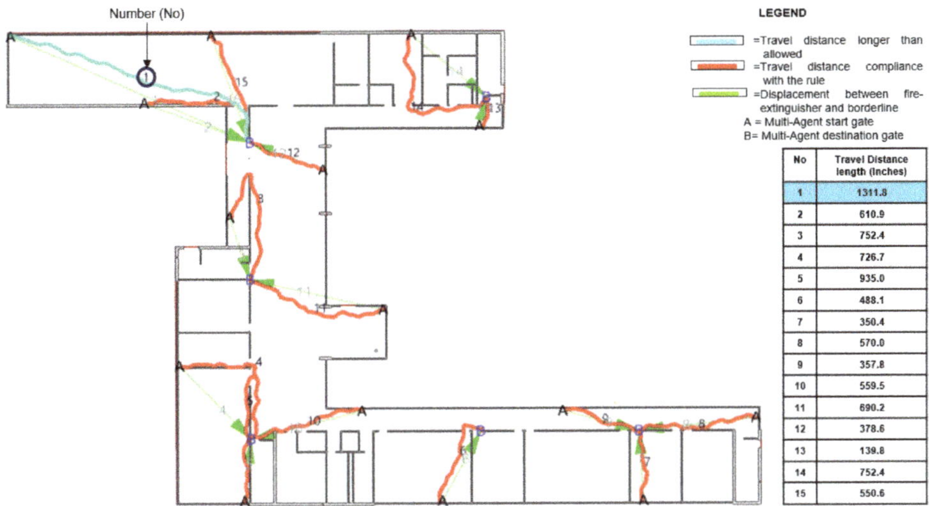

No	Travel Distance length (Inches)
1	1311.8
2	610.9
3	752.4
4	726.7
5	935.0
6	488.1
7	350.4
8	570.0
9	357.8
10	559.5
11	690.2
12	378.6
13	139.8
14	752.4
15	550.6

Figure 6. Multi-Agent Simulation for the Con-fire Safety Planning.

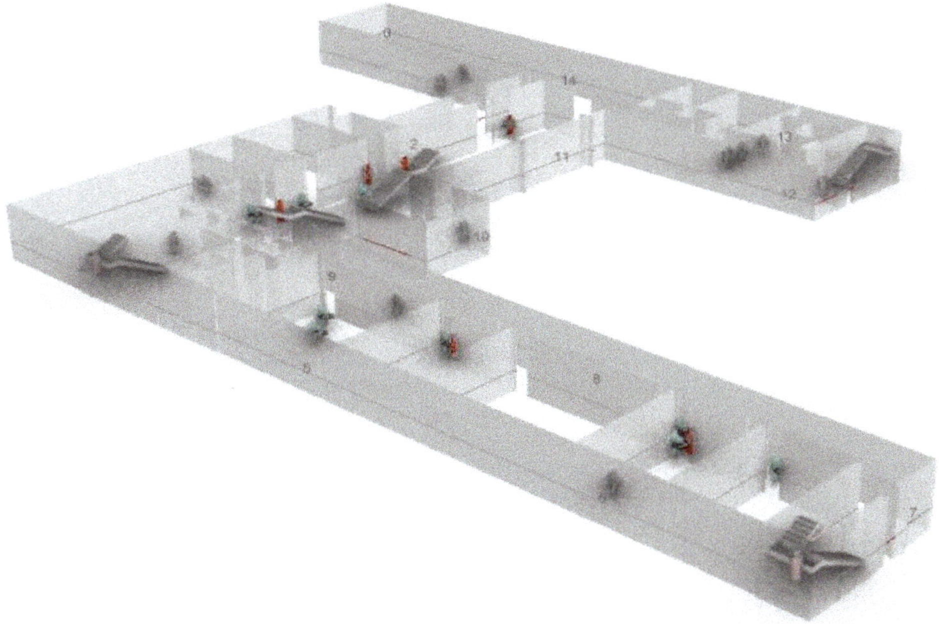

Figure 7. 3D-BIM Model of the con-fire safety planning simulation.

Figure 8. Heatmap of the con-fire safety planning Simulation.

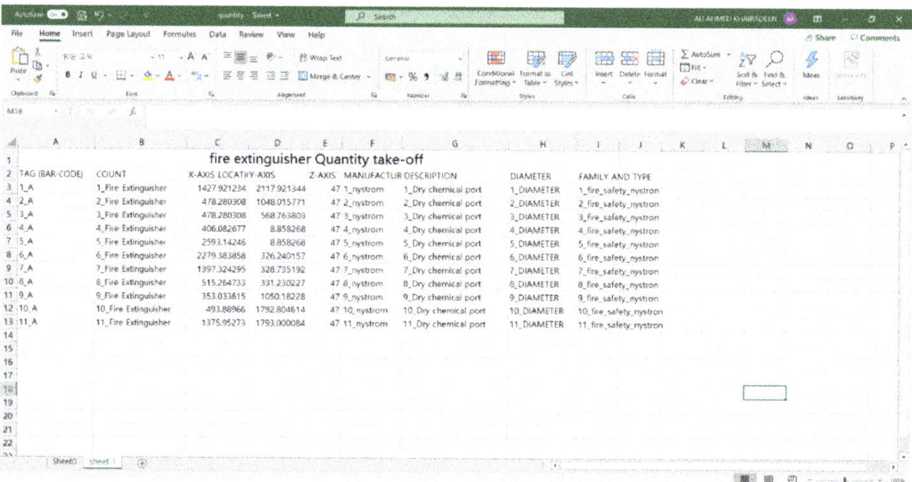

Figure 9. Bill of quantities (BoQs) for portable firefighting Equipment.

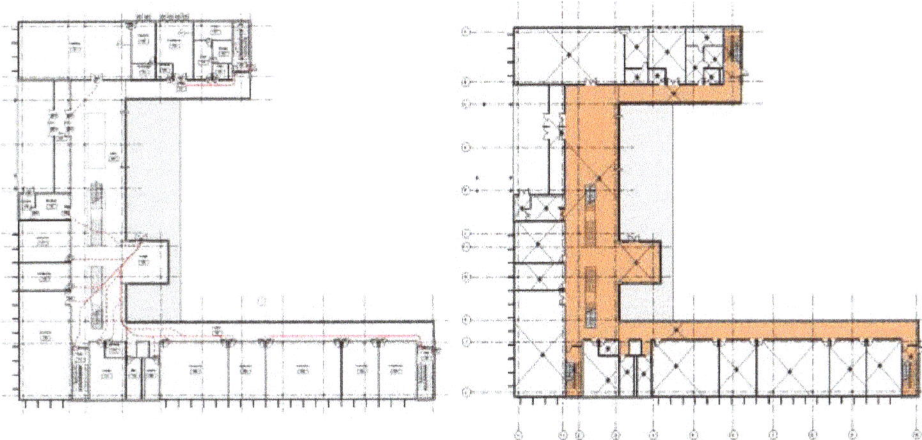

Figure 10. Escape route planning in Revit environment using an open API visual program.

5. Discussion

This research work presented a framework for construction fire safety planning using a visual algorithmic modeling approach to integrate firefighting equipment installation plans with construction fire safety planning. While previous studies in automated fire safety planning did not consider the rule-based equipment installation plan. The developed visual language and BIM-based con-fire safety planning (CSP) approach integrated firefighting equipment plan as an essential part of the construction fire safety. The case study test of a sample Revit project disclosed that the proposed approach leveraging visual language and BIM technology could provide more advanced and inclusive assessment for the rule-based construction fire safety planning. In order to prevent the fire risks in the construction job site, the designed approach has successfully developed the portable firefighting equipment installation plan semiautomatically, and in compliance with the OSHA guidelines.

Construction safety is not only limited to controlling and managing the safety behavior of the workers, but it also includes the design, procurement, and installation of the safety equipment such as guardrails, safety nets, scaffolding, and firefighting equipment. It is vital to be modeled in BIM for quantification and visualization purposes [36]. As the current fire safety planning practices in the construction industry follows the site-specific fire safety plan [33], which is time-consuming and labour-intensive if it is done with contemporary methods [35]. Based on the complexity and the scale of the project, the tedious modeling process generally needs days or even weeks [36]. However, as mentioned in the literature section, there are many software packages and approaches that can create sophisticated solutions to reduce the workload of the safety planner. This work also proposed a convenient approach to visualize the portable firefighting equipment and determine the actual cost estimation of the portable firefighting equipment with respect to the just on time approach. Hence, it is anticipated that the development of cost estimate plans could be cycle down from weeks to days or from days to even hours in some cases, depending on the size and complexity of the project. Moreover, the ambiguity of overestimating or underestimate can be eliminated, and the obtain cost information will be shared to others with great certainty. In this research work, the 2-D plan stair location, outline curve, interior walls need to be defined manually, automatic detection of 2D plan elements could be developed using space family tool in BIM environment, where the user does not need to define 2D plan elements and hence could be defined automatically. Moreover, to measure the travel distance trace length, the multi-agent simulation from the building corners towards the fire-extinguishers was out of the scope of this paper, and will be integrated with the future work.

6. Conclusions

The growing challenges for fire safety concerning economic progress have encouraged the evolution of fire science and technology. Many researchers have currently developed building information modeling (BIM) technologies for construction safety planning to enhance safety in the pre-construction phase, such as checking the BIM model to prevent fall risks, BIM 4D-supported limited access zone identification and visualization, and excavation safety modeling. However, fire safety planning—mainly firefighting equipment planning—still relies on conventional methods and has not yet been appropriately explored. To address the issue, this research work offers rule-based construction fire safety planning using a visual language approach in BIM. The proposed approach includes three core modules, such as the RELD module, DCFS module, and CSPS module. The RELD module converts the OSHA fire safety rules into mathematical logic, and the DCFS module presents the development of the proposed approach, such as the conversion of rules from mathematical logic into computer language. The three sub-modules of the DCFS module visualize the results of the research work, for instance, generation of appropriate portable firefighting equipment planning, bill of quantities of firefighting equipment, and escape route plan. The CSPS module simulates the system into a common BIM environment. The developed fire safety rule-based approach has been successfully implemented in a sample project case study, and the vital benefits are summarized as follows.

a. It is found that the con-fire safety system semi-automatically generates and visualizes the portable firefighting equipment installation plan based on OSHA safety rules, which will minimize the workload and reduce the time of the safety planner. This research depicted that the proposed approach has significant potentials to enhance construction fire safety planning, which is inevitable to deal with the recently reported fatalities and property damages.

b. Integrating firefighting equipment installation plan with the escape route planning is a significant additional contribution of this study. It is anticipated that this approach would help decision-makers in developing practical fire safety plans. This full package of the construction fire safety planning and multi-agent simulation could also be used for the training of the workers and can be extended to the virtual contents by employing Virtual Reality (VR) technology.

c. Portable firefighting equipment installation was witnessed along with their quantity take-off and appropriate locations in the case study. Hence, the con-fire safety system can also predict the location coordinates and required fire preventive resource quantities in advance. Likewise, integrating this fire safety planning with BIM-4D will provide the cost based on the just-in-time approach.

Furthermore, the developed approach could open a new direction of research in the digital twin domain, such as integrating the multi-agent simulation with real humans and updating. In this way, the workers will understand the shortest path towards the fire extinguishers or outside, whatever the case may be. It is expected that the proposed approach has the potential to help designers in practical job sites.

Author Contributions: Conceptualization, N.K. and D.L.; methodology, N.K. and A.K.A.; software, A.K.A. and N.K.; validation, A.K.A., S.V.-T.T. and D.L.; formal analysis, C.P.; writing—original draft preparation, N.K.; writing—review and editing, N.K. and C.P.; visualization, X.X.; supervision, C.P.; project administration, D.L. All authors have read and agreed to the published version of the manuscript.

Funding: This research was supported by the Chung-Ang University research grant in 2018 and National Research Foundation of Korea (NRF) grant funded by the Korea government (MSIP) (No. NRF- 2019R1A2B5B02070721).

Conflicts of Interest: The authors declare no conflicts of interest.

References

1. Kumar, S.; Bansal, V.K. Construction safety knowledge for practitioners in the construction industry. *J. Front. Constr. Eng.* **2013**, *2*, 34–42.

2. Mangement, L. The Importance Of Fire Safety on Construction Sites. Available online: https://lifesafetymanagement.com/importance-fire-safety-construction-sites/ (accessed on 30 January 2020).

3. Rentals, U. Fire Extinguishers on Construction Sites: What's Required | United Rentals. Available online: https://www.unitedrentals.com/project-uptime/safety/fire-extinguishers-construction-sites-whats-required#/ (accessed on 30 January 2020).

4. Campbell, R. Fires in structures under construction, undergoing major renovation or being demolished fact sheet. *NFPA Res. Data Anal.* **2017**, *4*, 1–11.

5. Shokouhi, M.; Nasiriani, K.; Cheraghi, Z.; Ardalan, A.; Khankeh, H.; Fallahzadeh, H. Preventive measures for fire-related injuries and their risk factors in residential buildings: A systematic review. *J. Inj. Violence Res.* **2019**, *11*, 1–14.

6. Guo, T.N.; Fu, Z.M. The fire situation and progress in fire safety science and technology in China. *Fire Saf. J.* **2007**, *42*, 171–182. [CrossRef]

7. Bureau of Labor Statistics National Census of Fatal Occupational Injuries in 2018—2019. Available online: https://www.bls.gov/news.release/pdf/cfoi.pdf (accessed on 3 February 2020).

8. Bureau of Labor Statistics Census of Fatal Occupational Injuries (CFOI)—Current and Revised Data. Available online: https://www.bls.gov/iif/oshcfoi1.htm (accessed on 3 October 2018).

9. Gehandler, J. The theoretical framework of fire safety design: Reflections and alternatives. *Fire Saf. J.* **2017**, *91*, 973–981. [CrossRef]

10. Johnson, P.; Beyler, C.; Croce, P.; Dubay, C.; McNamee, M. Very early smoke detection apparatus (VESDA), David Packham, John Petersen, Martin Cole: 2017 dinenno prize. *Fire Sci. Rev.* **2017**, *6*, 1–12. [CrossRef]

11. Cram, D.; Hatch, C.E.; Tyler, S.; Ochoa, C. Use of distributed temperature sensing technology to characterize fire behavior. *Sensors (Switzerland)* **2016**, *16*, 1712. [CrossRef]

12. Kumm, M.; Bergqvist, A. Fire and rescue operations during construction of tunnels. Mälardalen University, Västerås, Sweden, Greater Stockholm Fire Brigade & Karlstad University, Karlstad, Sweden. 2010. Available online: https://pdfs.semanticscholar.org/7638/c3d0c37ce0c79d54755964524ffda62b71aa.pdf (accessed on 3 February 2020).

13. Chen, Y.Y.; Chuang, Y.J.; Huang, C.H.; Lin, C.Y.; Chien, S.W. The adoption of fire safety management for upgrading the fire safety level of existing hotel buildings. *Build. Environ.* **2012**, *51*, 311–319. [CrossRef]

14. Yang, P.; Shi, C.; Gong, Z.; Tan, X. Numerical study on water curtain system for fire evacuation in a long and narrow tunnel under construction. *Tunn. Undergr. Sp. Technol.* **2019**, *83*, 195–219. [CrossRef]

15. Cheng, M.Y.; Chiu, K.C.; Hsieh, Y.M.; Yang, I.T.; Chou, J.S. Development of BIM-based real-time evacuation and rescue system for complex buildings. In Proceedings of the ISARC 2016—33rd International Symposium on Automation and Robotics in Construction, Auburn, AL, USA, 18–21 July 2016; pp. 999–1008.

16. Wang, S.H.; Wang, W.C.; Wang, K.C.; Shih, S.Y. Applying building information modeling to support fire safety management. *Autom. Constr.* **2015**, *59*, 158–167. [CrossRef]

17. Wang, K.C.; Shih, S.Y.; Chan, W.S.; Wang, W.C.; Wang, S.H.; Gansonre, A.A.; Liu, J.J.; Lee, M.T.; Cheng, Y.Y.; Yeh, M.F. Application of building information modeling in designing fire evacuation-a case study. In Proceedings of the ISARC 2014—31st International Symposium on Automation and Robotics in Construction and Mining, Sydney, Australia, 15 January 2014; pp. 593–601.

18. Maluk, C.; Woodrow, M.; Torero, J.L. The potential of integrating fire safety in modern building design. *Fire Saf. J.* **2017**, *88*, 104–112. [CrossRef]

19. Khan, N.; Ali, A.K.; Skibniewski, M.J.; Lee, D.Y.; Park, C. Excavation safety modeling approach using BIM and VPL. *Adv. Civ. Eng.* **2019**, *2019*, 1515808. [CrossRef]

20. Hossain, M.A.; Abbott, E.L.S.; Chua, D.K.H.; Nguyen, T.Q.; Goh, Y.M. Design for Safety knowledge library for BIM-integrated safety risk reviews. *Autom. Constr.* **2018**, *94*, 290–302. [CrossRef]

21. Liu, H.; Wang, Y.; Sun, S.; Sun, B. Study on safety assessment of fire hazard for the construction site. *Procedia Eng.* **2012**, *43*, 369–373.

22. Kumm, M.; Bergqvist, A. Fire and rescue operations during construction of tunnels. In Proceedings of the ISTSS—Fourth International Symposium on Tunnel Safety and Security, Frankfurt am Main, Germany, 17–19 March 2010; pp. 383–394.

23. OSHA 1926.150—Fire protection. | Occupational Safety and Health Administration. Available online: https://www.osha.gov/laws-regs/regulations/standardnumber/1926/1926.150 (accessed on 30 January 2020).

24. OSHA Commonly Used Statistics | Occupational Safety and Health Administration. Available online: https://www.osha.gov/data/commonstats (accessed on 31 January 2020).

25. Walnut Creek Personal Injury Lawyers Construction Site Fires—The Appel Law Firm. Available online: https://www.appellawyer.com/practice-areas/construction-accidents/construction-site-fires/ (accessed on 9 December 2019).

26. Chow, W.K. Fire safety in green or sustainable buildings: Application of the fire engineering approach in hong kong. *Archit. Sci. Rev.* **2003**, *46*, 297–303. [CrossRef]

27. Cygnus wireless alarm system Cygnus Wireless Fire Alarms | Construction Site Alarms. Available online: https://www.cygnusalarms.com/ (accessed on 12 December 2019).

28. Rüppel, U.; Schatz, K. Designing a BIM-based serious game for fire safety evacuation simulations. *Adv. Eng. Inform.* **2011**, *25*, 600–611. [CrossRef]

29. Chow, W.K. Performance-based approach to determining fire safety provisions for buildings in the Asia-Oceania regions. *Build. Environ.* **2015**, *91*, 127–137. [CrossRef]

30. Sztarbala, G. An estimation of conditions inside construction works during a fire with the use of computational fluid dynamics. *Bull. Polish Acad. Sci. Tech. Sci.* **2013**, *61*, 155–160. [CrossRef]

31. Fire emergency services Alaska. *Contractor's Guide Safeguarding Construction/Demolition Operations from Fire Contractor's Guide*; Fire emergency services Alaska: Anchorage, AK, USA, 2017; Volume 8, pp. 1–7.

32. Fire Safety Plan for Construction, Renovation and Demolition Sites Background Information. *Edmont. Fire Rescue Serv.* **2019**, *11*, 1–11.

33. Building, B.C.; Requirements, F.C. Fire safety plan guidelines for construction and demolition sites. *City Colwood Fire Dep.-Fire Prev. Div.* **2012**, 1–4.

34. Choe, S.; Leite, F. Construction safety planning: Site-specific temporal and spatial information integration. *Autom. Constr.* **2017**, *84*, 335–344. [CrossRef]

35. Kim, K.; Lee, Y.C. Automated generation of daily evacuation paths in 4D BIM. *Appl. Sci.* **2019**, *9*, 1789. [CrossRef]

36. Zhang, S.; Sulankivi, K.; Kiviniemi, M.; Romo, I.; Eastman, C.M.; Teizer, J. BIM-based fall hazard identification and prevention in construction safety planning. *Saf. Sci.* **2015**, *72*, 31–45. [CrossRef]

37. Gambatese, J.A.; Hinze, J.W.; Haas, C.T. Tool to design for construction worker safety. *J. Archit. Eng.* **1997**, *3*, 32–41. [CrossRef]

38. Zhang, S.; Teizer, J.; Lee, J.K.; Eastman, C.M.; Venugopal, M. Building information modeling (BIM) and safety: Automatic safety checking of construction models and schedules. *Autom. Constr.* **2013**, *29*, 183–195. [CrossRef]

39. Gangolells, M.; Casals, M.; Forcada, N.; Roca, X.; Fuertes, A. Mitigating construction safety risks using prevention through design. *J. Safety Res.* **2010**, *41*, 107–122. [CrossRef]

40. Hadikusumo, B.H.W.; Rowlinson, S. Capturing safety knowledge using design-for-safety-process tool. *J. Constr. Eng. Manag.* **2004**, *130*, 281–289. [CrossRef]

41. Zhang, S.; Frank, B.; Jochen, T. Ontology-based semantic modeling of construction safety knowledge: Towards automated safety planning for job hazard analysis (JHA). *Autom. Constr.* **2015**, *52*, 29–41. [CrossRef]

42. Kim, H.; Lee, J.K.; Shin, J.; Choi, J. Visual language approach to representing KBimCode-based Korea building code sentences for automated rule checking. *J. Comput. Des. Eng.* **2019**, *6*, 143–148. [CrossRef]

43. Hongling, G.; Yantao, Y.; Weisheng, Z.; Yan, L. BIM and safety rules based automated identification of unsafe design factors in construction. *Procedia Eng.* **2016**, *164*, 467–472. [CrossRef]

44. Kim, K.; Teizer, J. Automatic design and planning of scaffolding systems using building information modeling. *Adv. Eng. Inform.* **2014**, *28*, 66–80. [CrossRef]

45. Lee, J. The automated design rule checking system. In Proceedings of the 24th International Conference on Computer-Aided Architectural Design Research in Asia: Intelligent and Informed, CAADRIA 2019, Wellington, New Zealand, 15–18 April 2019; Volume 1, pp. 795–804.

46. Feng, C.W.; Lu, S.W. Using BIM to automate scaffolding planning for risk analysis at construction sites. In Proceedings of the ISARC 2017—34th International Symposium on Automation and Robotics in Construction, Taipei, Taiwan, 28 June–1 July 2017; pp. 610–617.

47. Payet, G. Developing a Massive Real-Time Crowd Simulation Framework on the Gpu. Bachelor's Thesis, University of Canterbury, Canterbury, UK, 2016.

 *applied sciences*

Article

A Web-Based BIM–AR Quality Management System for Structural Elements

Mehrdad Mirshokraei, Carlo Iapige De Gaetani * and Federica Migliaccio

Politecnico di Milano, Dept. of Civil and Environmental Engineering, 20133 Milan, Italy; mehrdad.mirshokraei@mail.polimi.it (M.M.); federica.migliaccio@polimi.it (F.M.)
* Correspondence: carloiapige.degaetani@polimi.it

Received: 28 August 2019; Accepted: 18 September 2019; Published: 23 September 2019

Abstract: This paper investigates quality management (QM) during the execution phase of structural elements by proposing, developing, and testing a complete framework by integrating building information modeling (BIM) and augmented reality (AR) technology. QM during execution is boosted by BIM–AR integration through a dedicated web-based system aimed at reducing the occurrence of omissions and negligence. With such a system, efficiency is improved by allowing the entering of inspection data directly in a shared digital environment, where people involved in QM have permanent access to updated information and inspection results, clearly organized, and entered in real time. The system has been developed in the asp.net framework using C# language where, by generating a web-based checklist and establishing its link to AR, it can enhance the process of information extraction from industry foundation class (IFC) 4D BIM models and the recording of inspection data. A test has been performed on a real case study in Budapest, to assess the effectiveness of the system in the field. Results demonstrate the following benefits brought by such a type of QM system: improved understanding of the design, access to information, and overview of the quality status of the project, leading to reductions in defects and reworking, as well as improved and quicker response and decision-making.

Keywords: Building Information Modeling; process improvement; construction management; information and communication technologies; Augmented Reality

---

1. Introduction

Presently, due to customer demand and high competition in the market, there is pressure on construction enterprises to improve quality in their projects. The main way to achieve this goal and achieve a competitive edge in the market is by adopting a sound quality management (QM) system [1]. Although QM must be applied to all phases of the building process, from conceptual design to demolition, the main challenge of projects and construction managers is controlling the quality during the execution phase, which calls for more resources and time. Based on surveys conducted by Gottfried et al. [2] and Alpsten [3], it has been found that failures ascribed to the execution phase are more prevalent than errors in the design process, and the construction phase appears particularly prone to errors. On the other hand, structural elements are considered to be the most fundamental components of a building to be controlled for their quality, since they are directly responsible for structural integrity, strength, and safety, and any defects in them will cause fatal accidents, severe additional costs, or delay [4,5]. In many projects, structural elements have also proved to have defects after execution but directly related to the execution phase, such as incorrect positioning of the frame in relation to the foundation, or insufficient length of the reinforcement bars [6].

Since there may be differences in the perception of the quality of an object, quality must be defined in a clear way. The ISO standard [7] defines quality as "degree to which a set of inherent characteristics of an object fulfils requirements". Quality in construction projects refers not only to the quality of products and equipment used in the construction of a building or facility, but also to the adopted management approach. As Chung states [8], both the construction cost and time of delivery are also important quality characteristics. Construction project quality is managed through quality assurance (QA) and quality control (QC). Turner interprets QA as "preventive medicine" [9], which consists of steps taken to increase the likelihood of obtaining a good-quality product and management processes. The aim of QA is to ensure that the project scope, cost, and time functions are fully integrated. QC as part of QA is "curative medicine", which recognizes human fallibility and takes steps to ensure that any (hopefully small) variations from standards that do occur are eliminated. As such, QC is the specific implementation of the QA program and related activities. Effective QC reduces the possibility of changes, mistakes, and omissions, which in turn result in fewer conflicts and disputes, and reduced waste of project resources. Although the procedure of quality check during the execution phase is consolidated and seems to be well organized, it does not work out properly in practice, for instance, due to intensive manual data collection entailing frequent transcription or data entry errors. Supporting this statement, examples can be found in reports and surveys presented by Glagola et al. [10] and Meijer & Visscher [11]. However, the complexity of properly controlling the quality of execution of structural elements can be easily recognized by having in mind, e.g., the number of processes and stakeholders involved in the execution of the concrete structure.

Recently, information technology (IT) has gained much attention as a key driver of change in the architecture, engineering and construction (AEC) industry. Developments in IT have provided numerous opportunities for the AEC industry, one of which is building information modeling (BIM) [12]. BIM serves as a central data repository that can store information about a facility and is currently regarded as an essential tool in managing the lifecycle of a construction project from the initial design to its maintenance. BIM is not just a technology change, but also a process change. In fact, unlike the traditional approach, the BIM approach allows the project team and the stakeholders to share information and to be constantly aware about the project. BIM is considered to be a multi-dimensional digital representation of the physical and functional characteristics of a project. Every time a specific type of information is added to the model, a different dimension is set, and, for this reason, various dimensions have been defined. Three dimensions are generally sufficient for geometric purposes, but new descriptive modalities and quantities, such as time or costs, introduce a different type of information. According to BIM fundamentals there are seven recognized "dimensions". 3D (three-dimensional rendering of the artefact), 4D (time and duration analysis), 5D (cost), 6D (sustainability assessment), and 7D (management phase). Taking advantage of its potential, the BIM methodology can be exploited to manage all the QC data and the complex relationships between them, establishing an effective approach to realize improvements in construction quality management. Various researchers have already proposed to implement BIM concepts into a quality management [13]. As an example, the QC framework by Chen & Luo [14] consisted of a 4D model combined with a specific company's process, organization and product (POP) model. According to Turk [15]: "BIM refers to a combination or a set of technologies and organizational solutions that are expected to increase inter-organizational and disciplinary collaboration in the construction industry and to improve the productivity and quality of the design, construction, and maintenance of buildings". In this sense, several authors proposed the combination of BIM with other technologies aiming at exploiting their potential in the framework of quality management. There are many examples, involving different techniques and technologies, such as personal digital assistants [16], mobile devices to access design information and to capture work progress [17], radio frequency identification [18–20], laser-scan point clouds [21–25], and indoor positioning through magnetic fields and wi-fi signals [26].

In the last decade, augmented reality (AR) has received a considerable amount of attention from researchers in the AEC community [27]. According to Wang et al. [28] AR and BIM are complementary technologies. AR could represent the site extension of the BIM concept and approach, and maximize the potentials of BIM in the construction site. AR allows the overlaying of a virtual object into the real world and can present information on site where it is needed. Rankohi and Waugh [29] classified AR application areas in the AEC industry as follows: (1) visualization or simulation; (2) communication or collaboration; (3) information modeling; (4) information access or evaluation; (5) progress monitoring; (6) education or training; and (7) safety or inspection. Therefore, the benefits of bringing AR to the job site could be truly remarkable. In the framework of quality management, remarkable works can be found in the literature. Golparvar-Fard et al. [30] implemented the D4AR system for visualizing the deviation from the construction schedule by registering new daily site images and using a traffic light metaphor as feedback to represent discrepancies between the as-planned and the as-built. Wang et al. [28,31,32] developed a conceptual framework to investigate how BIM can be extended to the site via AR and investigated the use of BIM and AR for project control, procurement monitoring, visualization of design during construction, and linking virtual to physical objects. Park et al. [33] presented a conceptual system framework for construction defect management using AR and BIM technologies to enable the storage and retrieval of defect data visually. Following that study, Kwon et al. [34] proposed a defect management system for reinforced concrete work by integrating BIM, image-matching, and AR.

The aim of this paper is to propose a complete framework to integrate BIM and AR to improve the quality management of the execution of structural elements on site. Such a framework has been verified by implementing a prototype BIM–AR QM web-based system that is platform-independent and fully customizable, as well as able to be modified depending on the needs of the users. This prototype has been tested on a real-life test case to assess benefits, issues, and key points requiring further investigation.

2. The Proposed Framework for BIM–AR Integration for Quality Control

In this section, the developed framework integrating BIM and AR into the two pillars of quality management (QA and QC) will be described. The proposed procedure is illustrated in Figure 1. The QA starts by the customer defining his/her requirements, which are the basis for the design team to define the specifications. To be able to realize the constructed elements with a quality consistent with the specification, quality parameters and QC activities must be identified along with a schedule with the time when they need to be controlled. All this information will result in a QC plan which is the base for QC. After integrating quality information into a BIM, a quality model will be obtained to be shared between project participants, which will be the basis for inspection on site by AR technology. For the sake of simplicity, in this study the integration with a 4D BIM has been considered (i.e., including the project time schedule). Of course, with a 5D BIM model (hence including also cost information), a 5D quality model would be obtained, providing additional information, e.g., about the cost of possible required interventions.

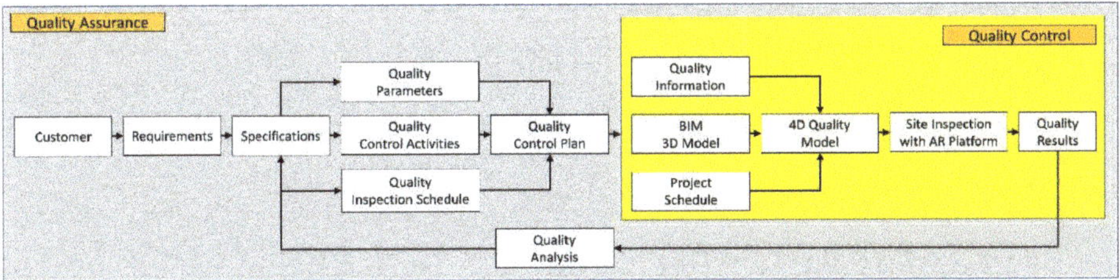

Figure 1. Framework of the proposed quality management (QM) procedure.

2.1. Quality Control Workflow

The QC workflow is represented in more detail in Figure 2. Using the 4D quality model as a reference, the supervisor can determine beforehand which parts need to be controlled; then, when entering the construction site, the position of the elements to be controlled can be identified with respect to the environment in the AR mobile application. On site, the required information can be extracted and visualized from the updated BIM model and, using a web-based checklist of all the quality parameters to be checked, a decision can be made on the quality conformance of each specific element. Once the evaluation is completed, a notification will be sent to the responsible contractor. The evaluation could result in a corrective action which needs to be taken—in this case, the BIM model will be modified based on the change. Further inspections could be required to determine if the corrective actions have been successfully performed. The project manager can also use the output from the inspection to have both an overall view of the project quality and an insight on current quality control processes and their effectiveness in limiting defects. This can help the manager to decide if any adjustment needs to be done in quality management, e.g., updating project scheduling techniques, adding/removing quality requirements, or modifying current checks/inspections.

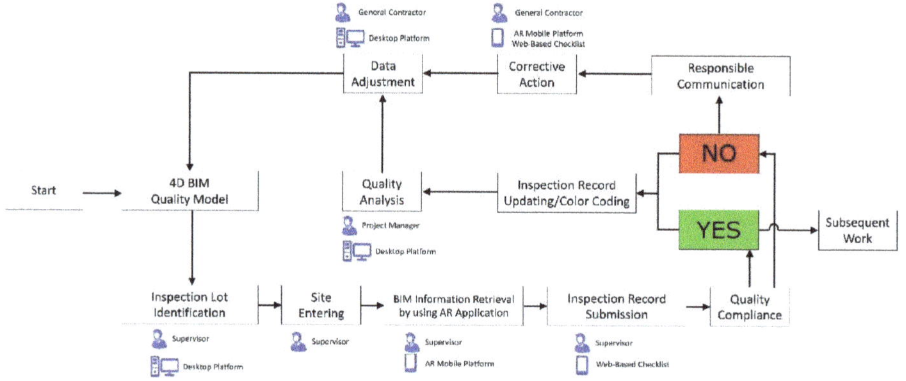

Figure 2. The proposed BIM–AR QC workflow.

2.2. Data Needed for Quality Control

In the proposed BIM–AR QC procedure, the information needed for the quality management of structural elements is represented by three types of data that must be collected, exchanged, and synchronized to have a 4D BIM quality model:

- model of physical objects involved in the construction processes such as column and beam components and equipment employed, including the geometric data and the data on materials and other specifications; an example of generic physical objects that can be modeled is presented in Figure 3, in the case of a steel and cast-on-site reinforced concrete structure;
- work schedule data based on the project tasks, their relationship, and their time schedule, where inspection lots and their parameters can be related to these tasks to give them a time dimension;
- QC data, including definition of inspection lots and relevant quality parameters serving as a checklist to be controlled for each element, acceptance criteria, decisions, and instructions to the persons in charge in case of rework.

All these data can be divided into three categories, based on when they need to be used, namely: (a) before, (b) during, and (c) after execution. The aforementioned categorization of the data for QC purposes is quite basic to apply in real-life projects, but structured enough to comprise the most relevant aspects and verify the proposed framework.

According to European standard 13670:2009 [35] on the execution of concrete structures, quality parameters must be controlled for material, process, and geometry. The used material must be consistent with the required specification as well as the compliance of the finished element. The related activities and preventive measures must follow specific guidelines, and defined tolerances are admitted in terms of geometry of both simple and composite structural elements. As an example, Figure 4 shows how activities for the execution of a concrete structure are divided into the three phases; in each phase, quality parameters are controlled for material, process, and geometry. In Table A1 of

Appendix A, all the quality parameters for the control of the material of a concrete structure are listed, with the corresponding inspection lots, related activity, relevant standards, and execution phase.

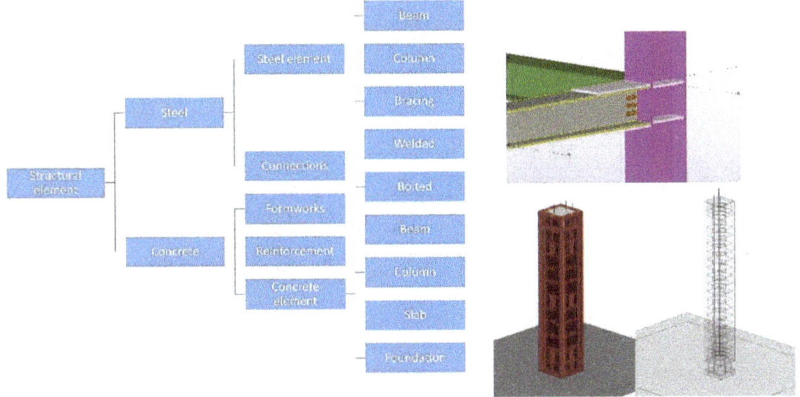

Figure 3. Example of generic physical object elements.

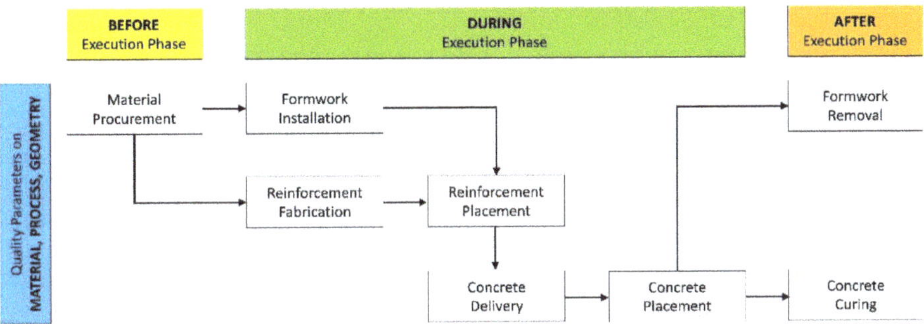

Figure 4. Quality control phases for a concrete structure execution.

3. Quality Control System Development

The proposed workflow needs a suitable system for it to be realized. Such a system has been developed based on a stepwise procedure. Ideally, the starting point is the 3D BIM model of the structure in industry foundation class (IFC) format. IFC is one of the open standards in the buildingSMART portfolio; it is a neutral data format to describe, exchange, and share all building information including geometry, spatial relationships, attributes, and quantity [36]. With such a format, the model can be then imported in suitable software to prepare the time schedule of the project, define the actors involved, and assign the physical elements to the work schedule, obtaining the 4D BIM model. The developed QC system synchronizes the 4D model and the quality information, generating a web-based checklist, storing all the inspection results and generating the link in the 4D IFC model to the AR application. By importing the synchronized 4D IFC model into the AR application, it is possible to visualize the BIM model on site, while the checklist is also accessible and can be updated.

The QC system was developed using the asp.net framework based on C# programming language to synchronize all information needed. Since the system is platform-independent and it can be accessed through any type of web browser, it can be accessed through either desktop or mobile devices and there are no requirements on the type of operating system. The input/output scheme of the system is represented in Figure 5.

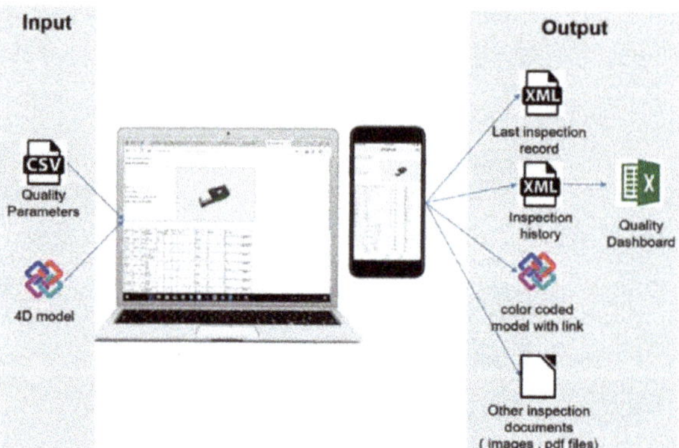

Figure 5. Input/output scheme of the proposed QC system.

The main packages/modules used to generate the quality system are the xBIM Toolkit which is a .NET open-source software development BIM toolkit that supports IFC: it allows reading, creating, and viewing BIM Models in the IFC format. Two core libraries of the xBIM toolkit are xBIM Essentials and xBIM Geometry which are written in C# and C++. To allow addition of the BIM model to the checklist, WeXplorer was used; it is the visualization part of xBIM toolkit and uses WebGL technology giving 3D viewing control.

There are no specific entities for quality information in the IFC standard, therefore new entities describing the quality management process and their relationships with other information have been created [37]. The entity IfcQuality has been defined as an entity connected to other different entities describing the quality of an element. One of them is IfcInspectionLot and its subset IfcQualityParameters. The latter is categorized with respect to the epoch (IfcQualityPhase) and the type (IfcQualityCategory) of the control. IfcQualityAcceptance and its subset IfcQualityDoc describe the information regarding the inspection results and the related uploaded documentation. IfcInspectionPlan describes the assignments of activities and their schedule information in the framework of QC. These new entities are related to other entities already defined in the IFC standard Version 4 such as IfcProduct and its inherited entities describing the designed products, or IfcWorkSchedule describing the timing of the activities. Figure 6 shows the IFC-based process model and the relevant entities.

As an example, in the case of a wall that should be under inspection after concreting, the geometrical data and its properties are represented by IfcWall as a type of IfcProduct and it has one assigned IfcQuality entity. This entity is defined by three IfcInspectionLot, specifically cross-sectional geometry, surface control, and hardened concrete quality. Regarding the cross-sectional geometry of IfcInspectionLot, three parameters (dimensions, skewness, orthogonality) are related to geometry (IfcQualityCategory) and execution phase (IfcQualityPhase) with their own upper and lower deviation limits. In IfcQuality, the related quality parameter activities (the concrete placement in the case of wall geometrical quality parameters) are recorded in IfcInspectionplan, while the results of the inspection are recorded in IfcQualityAcceptance.

The 4D BIM model in IFC format, inspection lots, their corresponding quality parameters and related activities in comma-separated values format (CSV) are prerequisites as input for the system. The application initially parses the CSV file and stores the data in a dictionary which will be used to determine the value of quality classes, their subtypes, and their attributes. Loading the IFC model, the system parses all the IfcBuildingElements that are in the model and retrieves their globally unique identifier (GUID) and type. For concrete structures, the type of IfcBuildingElement could be IfcColumn, IfcBeam, IfcWall, IfcFooting, IfcSlab, IfcReinforcementRebar, and IfcBuidlingElementProxy. Based on such types, the associated quality classes will be dynamically loaded, also containing the GUID of the specific elements. Since each element has its own unique GUID, for each element a specific webpage address will be generated and this web address will be added as a user-defined property (IfcPropertySet) in the 4D IFC model, creating the link between this system and the web-based QC checklist. Opening this webpage will direct to the checklist of the related quality parameters, where the inspector can enter and record all the requested information. The results will be saved in IfcQualityAcceptance class and stored as two XML files. One of them is used to save the results of the latest inspection performed and the other to save all the inspection results and the documents related to each quality parameter. Whenever the checklist opens, the results of the latest inspection are shown. The other file can be imported into a spreadsheet for data queries, filtering, and quantification. Results can then be summarized to generate a quality dashboard of the project. Another output of the system is a color-coded IFC model, which provides the construction team with direct visual feedback on the contents of the inspection results. This reduces the time needed to analyze data and allows for a faster corrective action of quality defects to take place on site. The color coding is based on the legend reported in Figure 7.

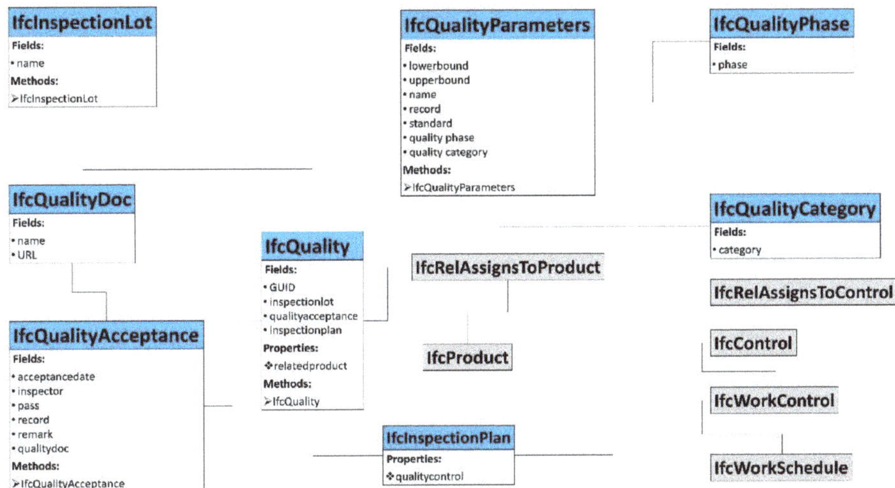

Figure 6. IFC-based process model. New entities are highlighted in blue.

The on-site visualization of the color-coded model with a mobile device exploiting the integration with AR can be useful for both the inspector and the people responsible for rework. In the former case it will make up the basis for subsequent inspections, while in the latter case it can be used to access the inspection checklist to identify the element, its problem, and the inspector's comments. Through a central interface, projects are managed and IFC models are uploaded and visualized on site. Figure 8 shows the interface of the AR application and the steps needed to visualize the model on site and to access the web-based checklist; as an example, the figure reports the inspection of a column element.

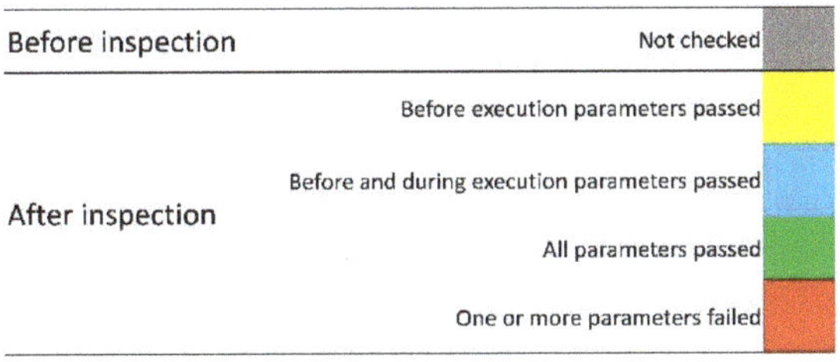

Figure 7. Color coding legend for the visual feedback on the inspection results.

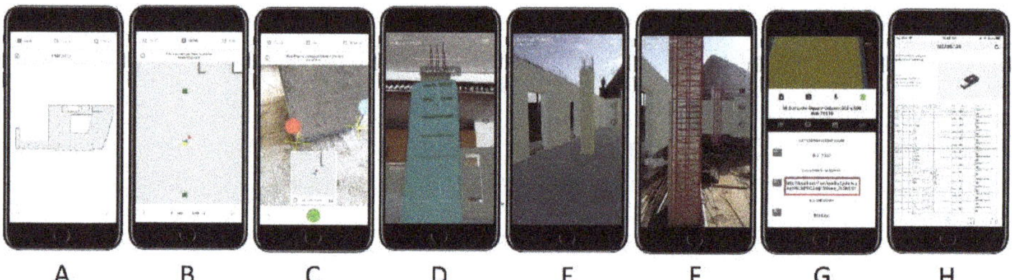

Figure 8. AR application user interface. (**A**) Opening the file and choosing the level. (**B**) Choosing an element for positioning. (**C**) Positioning by superimposing two corners of the element. (**D**) Augmented column with formwork. (**E**) Hiding formwork. (**F**) Column reinforcement. (**G**) Retrieving element properties and linking the checklist web address. (**H**) Accessing the selected element quality checklist.

4. On-Site Test of the Proposed BIM–AR QM System and Results

The proposed system has been tested in the case study of a reinforced concrete building under construction in Budapest, Hungary. Trial inspections have been performed on one level of the building to check the quality of structures and test the developed system. The test followed a stepwise procedure. First, the 3D model of the building was created in Autodesk Revit®and exported to IFC format. Secondly, the model was imported in Synchro Pro®establishing the connection between schedule information and the building elements and then exported again in IFC format. Such an IFC 4D model was finally imported into the developed system to synchronize the quality information with the physical and schedule information of the building and to generate the web-based checklist and the link for accessing the AR application. The Gamma AR Pro®application was used as the AR platform to visualize the BIM model on site. Such application directly uses the IFC format and gives the possibility of accessing information regarding the elements by just clicking on the object. This application positioning system is marker-less, and uses depth-sensing as the tracking system to overlay the BIM model to the reality on site. For positioning, the inspector needs to choose the floor and room and then select two corners of a wall or column in the model to superimpose it to the reality. An Apple iPhone 7®was used as the mobile device for the AR application.

The project used for the test is an office building with reinforced concrete structure which is being reconstructed after demolishing the old one, to match the architecture of the building next to it. Considering the building is being used as offices, based on EN1990-2002 Annex B [38], the consequence class CC2 has been chosen. Moreover, based on EN13670 [35], Execution Class 2 has been considered for assigning the range of admissible tolerances and the severity of the inspection. A total of 36 elements were inspected and the results were recorded. At the time that the test was done, most of the structural elements had been constructed and the activities related to the architectural part of the project were ongoing. Therefore, the inspection lots related to the "after-execution" phase of QC and were added as an input to the web-based checklist. In this phase there are no inspection lots related to formwork and rebar elements; however they were modeled in Revit®and then exported in IFC to examine how the inspection could have been performed on site using AR. In this way, it was also possible to assign and then retrieve their corresponding quality parameters. In Figure 9, the model of a column is shown, together with its on-site visualization using the AR application.

Inspection records were saved as XML and visualized using the color-coded model of the quality system; in this way it is also possible to find the location of the defects (see Figure 10). Furthermore, the dashboard of the quality status of the projects shows the results of the inspection (see Figure 11). During the inspection test, the response was negative in three cases (8% of the items checked); all of them concerned the concrete placement, two were related to walls, and one was related to a slab. Two of the failed inspection lots concerned the geometry and one concerned the material. These cases are represented in Figure 12.

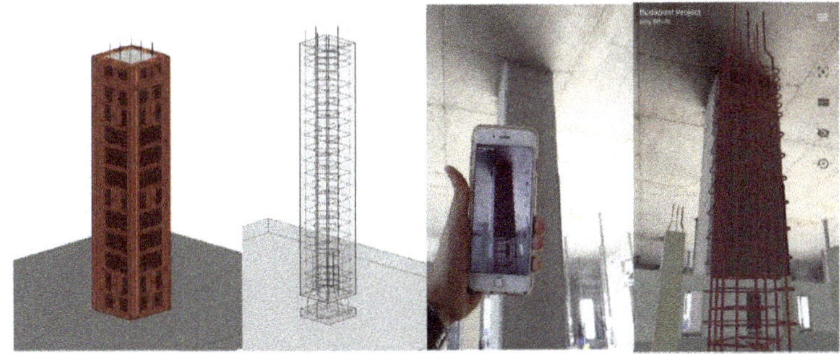

Figure 9. Model of a reinforced concrete column with its formwork, and how it is visualized on site.

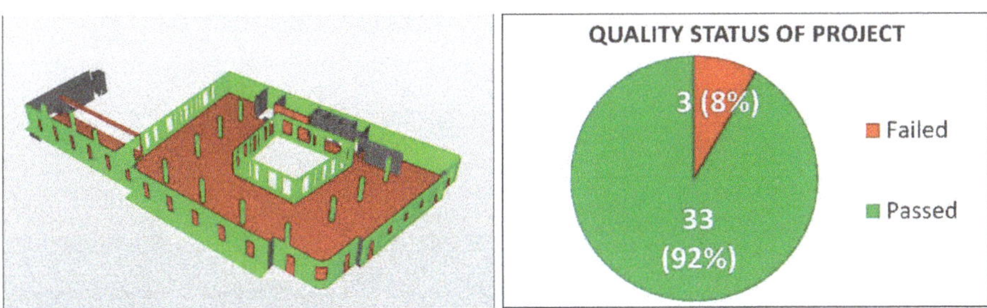

Figure 10. Case study, screenshots of the produced quality dashboard: visualization of the color-coded model (**left**) and report on the quality status of the project (**right**).

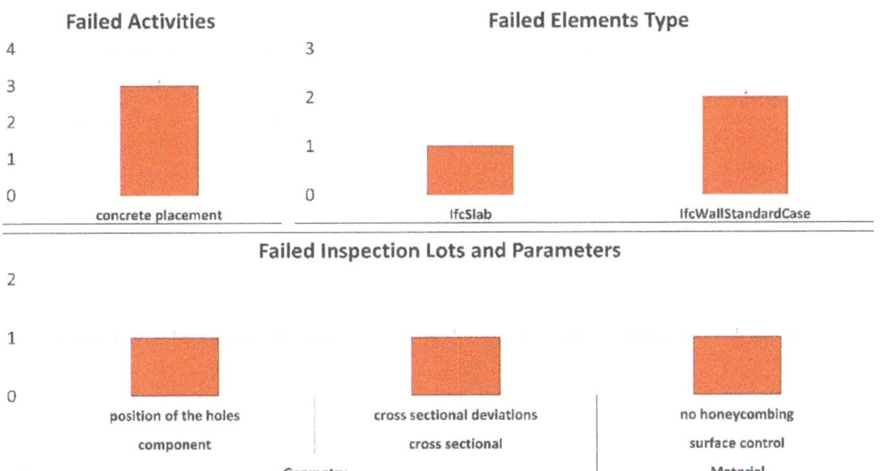

Figure 11. Case study, screenshots of the produced quality dashboard and summary of the defects: three defects in the concrete placement have been reported (**top-left**), a slab and two walls are involved (**top-right**), two defects regard the geometry of the objects, the remaining regard the material (**bottom**).

Figure 12. Case study, shots of the failed inspection lots: wall concrete surface with honeycombing (**left**), wall cross-sectional geometry (**center**), hole in a slab (**right**).

5. Discussion

The results showed that the main weaknesses of the current QM practice (namely: poor management, poor understanding of the scope of quality, and poor communication between stakeholders) can be overcome, while the practice itself can be made more efficient and effective by exploiting the integrated use of BIM and AR technologies in a web-based collaborative system. The case study allowed identification of benefits in adopting the proposed framework. Figure 13 maps the possible outcomes of the proposed system while pursuing the objectives of budget, time, and quality of the construction project.

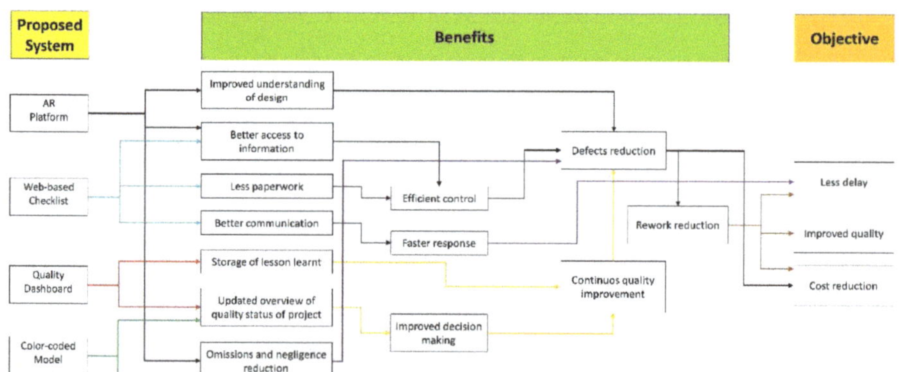

Figure 13. Benefit map of the proposed system.

On the other hand, during the development and testing of the proposed system, some limitations and barriers have been identified. One of the most important is the problem related to the correct positioning of the virtual model on the real environment. In fact, the accuracy of the positioning is not sufficient to rely on this system as a tool for the measurement of geometry, and therefore geometric measurements should still be done using traditional approaches. Furthermore, due to the nature of the construction sites and the dependency of the AR system on depth sensors, it could be difficult to superimpose the pins in the correct positions to achieve an AR experience with acceptable accuracy during the first tries. A possible solution could be the integration of both marker-based and marker-less tracking systems in the AR platform. With a proper enhancement of the software, the indoor positioning could exploit, e.g., near-real-time photogrammetric techniques to guide the procedure.

Occlusions create difficulties, as well. In AR, an object closer to the viewer obscures the view of objects further away along the line of sight, affecting the interpretation of the environment by the AR application in the case of complex models with lots of elements and details. Again, it the opinion of the authors that this could be managed by the AR platform at software level so that, on the basis of a more accurate positioning of the device, it could recognize the relative distance of objects and manage the overlaps between superimposed elements. Examples of these problems are shown in Figure 14.

Figure 14. Examples of problematic behaviors: occlusion of the real objects by the wall model (**left and middle**) and positioning accuracy problems (**right**).

From this point of view, several different techniques aiming at improving the accuracy of the positioning during the AR experience are under investigation by the scientific community [39], due to the huge potential that such technology has in various fields of application.

Another shortcoming of the proposed system for quality inspection is represented by the dependency on Internet access, which may cause problems when the signal is poor. Finally, there is the need to train people to use the system properly for safety reasons: in fact, working with mobile AR applications may cause the user to lose attention to the surrounding area, which could be very dangerous in a construction site and cause safety issues [40].

6. Conclusions and Future Work

In this paper, a comprehensive approach for QM has been proposed by integrating BIM and AR technology. Exploiting advantages and help from the potentials of BIM and AR and their integration, a system has been developed for acquiring inspection data, processing the results, and facilitating collaboration among the actors involved in QM, through the web. The approach has been implemented and a preliminary test has been carried out on a real building project to study the efficiency and effectiveness of the system.

The results confirm that objectives such as delay reduction, quality improvement, and cost optimization can be pursued through the proposed quality management system. The produced web-based checklist makes the access to updated information easier, by improving the communication between the involved parties. Moreover, the quality dashboard summarizing the results, combined with the color-coded model, gives an overview of the current quality status of the project both in qualitative and quantitative terms.

In the frame of the experimental approach followed for the proposed methodology check, difficulties also encountered during the testing phase can be considered to be positive outcomes, and the seeds for further considerations and research work.

Regarding possible future developments to optimize the proposed approach for integrating BIM and AR for QM purposes, a more stable and accurate positioning of the AR platform would improve the experience, offering the possibility to add tools to measure and record the as-built situation for real-time comparisons with the designed plan and make the control of geometric quality parameters more efficient. Furthermore, other improvements could be represented by a priority index for each defect, ranking its influence on the others, and by a feature allowing superimposition of the model onto the appropriate LOD (Level Of Detail), to avoid visualizing too many details, and to make the AR experience better.

Author Contributions: Conceptualization, M.M. and C.I.D.G.; Methodology, M.M. and C.I.D.G.; Software, M.M.; Validation, M.M. and C.I.D.G.; Formal analysis, M.M. and C.I.D.G.; Investigation, M.M.; Writing—original draft preparation, M.M., C.I.D.G. and F.M.; Writing—review and editing, C.I.D.G. and F.M.; Visualization, M.M. and C.I.D.G.; Supervision, C.I.D.G. and F.M.

Funding: This research received no external funding.

Conflicts of Interest: The authors declare no conflict of interest.

Appendix A

Table A1. List of quality parameters for the control of the material of a concrete structure.

Project Item	Related Activity	Inspection Lot	Quality Parameter	Standard	Execution Phase
formwork	material procurement	mechanical properties	adequate stiffness	EN12812 EN12813	Before
		surface condition	clearness of surface		
		release agents	no unintended effect on the color and surface quality no detrimental effect on permanent structures		
reinforcing steel		mechanical properties	steel class tensile test	EN10080	
		surface condition	free from loose rust and deleterious substances no cracks and other damage		
		spacers	protection against corrosion		
concrete	concrete delivery	concrete ingredients	cement	EN197-1	
			aggregates	EN1260 EN13055	
			mixing water	EN1008	
			admixtures	EN934-2	
			mixed design		
concrete	concrete delivery	visual inspection of delivered concrete	no aggregates segregation no concrete bleeding no paste loss		During
		fresh concrete tests	cube/cylinder strength test	EN12350-1	
			slump test	EN12350-2	
			temperature test		
			air content test	EN12350-7	
concrete	concrete delivery	hard concrete test	non-destructive tests		After
	concrete placement	surface control	crack formation no honeycombing no damage or disfiguration		

References

1. Building in Quality Working Group. Building in Quality Initiative: A Guide to Achieve Quality and Transparency in Design and Construction. Available online: https://www.architecture.com/working-with-an-architect/building-in-quality-tracker (accessed on 3 September 2019).

2. Gottfried, A.; Di Giuda, G.; Villa, V.; Piantanida, P. Controls on Structure execution: Acceptance condition and Types of inspection for cast on site reinforced concrete. In *New Developments in Structural Engineering and Construction, Proceedings of the 7th International Structural Engineering and Construction Conference, Honolulu, HI, USA, 18–23 June 2013*; Research Publishing Services: Singapore, 2013; pp. 461–466. [CrossRef]

3. Alpsten, G. Causes of Structural Failures with Steel Structures. In *IABSE Symposium Report*; International Association for Bridge and Structural Engineering: Zurich, Switzerland, 2017; Volume 107, pp. 1–9.

4. Gorse, C.; Sturges, J. Not what anyone wanted: Observations on regulations, standards, quality and experience in the wake of Grenfell. *Constr. Res. Innov.* **2017**, *8*, 72–75. [CrossRef]

5. Knyziak, P. The impact of construction quality on the safety of prefabricated multi-family dwellings. *Eng. Fail. Anal.* **2019**, *100*, 37–48. [CrossRef]

6. Forcada, N.; Macarulla, M.; Gangolells, M.; Casals, M. Assessment of construction defects in residential buildings in Spain. *Build. Res. Inf.* **2014**, *42*, 629–640. [CrossRef]

7. International Organization for Standardization. ISO Standard No. 9000: Principles of Quality Management. Available online: https://www.iso.org/publication/PUB100080.html (accessed on 27 August 2019).

8. Chung, H.W. *Understanding Quality Assurance in Construction: A Practical Guide to ISO 9000 for Contractors*, 1st ed.; Routledge: London, UK, 2002.

9. Turner, J.R. *The Handbook of Project-Based Management: Leading Strategic Change in Organizations*, 3rd ed.; McGraw-Hill: New York, NY, USA, 2008.

10. Glagola, C.R.; Ledbetter, W.B.; Stevens, J.D. *Quality Performance Measurements of the EPC Process: Current Practices*; Construction Industry Institute, University of Texas at Austin: Austin, TX, USA, 1992.

11. Meijer, F.; Visscher, H. Quality control of constructions: European trends and developments. *Int. J. Law Built Environ.* **2017**, *9*, 143–161. [CrossRef]

12. Morgan, B. Organizing for digitalization through mutual constitution: The case of a design firm. *Constr. Manag. Econ.* **2019**, *37*, 400–417. [CrossRef]

13. Chen, K.; Lu, W. Bridging BIM and building (BBB) for information management in construction: The underlying mechanism and implementation. *Eng. Constr. Archit. Manag.* **2019**, *26*, 1518–1532. [CrossRef]

14. Chen, L.; Luo, H. A BIM-based construction quality management model and its applications. *Autom. Constr.* **2014**, *46*, 64–73. [CrossRef]

15. Turk, Ž. Ten questions concerning building information modelling. *Build. Environ.* **2016**, *107*, 274–284. [CrossRef]

16. Kimoto, K.; Endo, K.; Iwashita, S.; Fujiwara, M. The application of PDA as mobile computing system on construction management. *Autom. Constr.* **2005**, *14*, 500–511. [CrossRef]

17. Davies, R.; Harty, C. Implementing 'Site BIM': A case study of ICT innovation on a large hospital project. *Autom. Constr.* **2013**, *30*, 15–24. [CrossRef]

18. Jaselskis, E.J.; Anderson, M.R.; Jahren, C.T.; Rodriguez, Y.; Njos, S. Radio-frequency identification applications in construction industry. *J. Constr. Eng. Manag.* **1995**, *121*, 189–196. [CrossRef]

19. Wang, L.C. Enhancing construction quality inspection and management using RFID technology. *Autom. Constr.* **2008**, *17*, 467–479. [CrossRef]

20. Moon, S.; Yang, B. Effective monitoring of the concrete pouring operation in an RFID-based environment. *J. Comput. Civ. Eng.* **2009**, *24*, 108–116. [CrossRef]

21. Akinci, B.; Boukamp, F.; Gordon, C.; Huber, D.; Lyons, C.; Park, K. A formalism for utilization of sensor systems and integrated project models for active construction quality control. *Autom. Constr.* **2006**, *15*, 124–138. [CrossRef]

22. Tang, P.; Akinci, B.; Huber, D. Quantification of edge loss of laser scanned data at spatial discontinuities. *Autom. Constr.* **2009**, *18*, 1070–1083. [CrossRef]

23. Tang, P.; Anil, E.B.; Akinci, B.; Huber, D. Efficient and effective quality assessment of as-is building information models and 3D laser-scanned data. In Proceedings of the International Workshop on Computing in Civil Engineering 2011, Miami, FL, USA, 19–22 June 2011; pp. 486–493. [CrossRef]

24. Anil, E.B.; Tang, P.; Akinci, B.; Huber, D. Deviation analysis method for the assessment of the quality of the as-is Building Information Models generated from point cloud data. *Autom. Constr.* **2013**, *35*, 507–516. [CrossRef]

25. Bosché, F.; Ahmed, M.; Turkan, Y.; Haas, C.T.; Haas, R. The value of integrating Scan-to-BIM and Scan-vs-BIM techniques for construction monitoring using laser scanning and BIM: The case of cylindrical MEP components. *Autom. Constr.* **2015**, *49*, 201–213. [CrossRef]

26. Ma, Z.; Cai, S.; Mao, N.; Yang, Q.; Feng, J.; Wang, P. Construction quality management based on a collaborative system using BIM and indoor positioning. *Autom. Constr.* **2018**, *92*, 35–45. [CrossRef]

27. Wang, K.C.; Wang, S.H.; Kung, C.J.; Weng, S.W.; Wang, W.C. Applying BIM and visualization techniques to support construction quality management for soil and water conservation construction projects. In Proceedings of the 35th International Symposium on Automation and Robotics in Construction (ISARC 2018), Berlin, Germany, 20–25 July 2018; Volume 35, pp. 1–8. [CrossRef]

28. Wang, X.; Love, P.E.; Davis, P.R. BIM+AR: A framework of bringing BIM to construction site. In Proceedings of the Construction Challenges in a Flat World 2012, West Lafayette, IN, USA, 21–23 May 2012; Cai, H., Kandil, A., Hastak, M., Dunston, P.S., Eds.; ASCE: Reston, VA, USA, 2012; pp. 1175–1181. [CrossRef]

29. Rankohi, S.; Waugh, L. Review and analysis of augmented reality literature for construction industry. *Vis. Eng.* **2013**, *1*, 9. [CrossRef]

30. Golparvar-Fard, M.; Peña-Mora, F.; Savarese, S. Integrated sequential as-built and as-planned representation with D4AR tools in support of decision-making tasks in the AEC/FM industry. *J. Constr. Eng. Manag.* **2011**, *137*, 1099–1116. [CrossRef]

31. Wang, X.; Kim, M.J.; Love, P.E.; Kang, S.C. Augmented Reality in built environment: Classification and implications for future research. *Autom. Constr.* **2013**, *32*, 1–13. [CrossRef]

32. Wang, X.; Truijens, M.; Hou, L.; Wang, Y.; Zhou, Y. Integrating Augmented Reality with Building Information Modeling: Onsite construction process controlling for liquefied natural gas industry. *Autom. Constr.* **2014**, *40*, 96–105. [CrossRef]

33. Park, C.S.; Lee, D.Y.; Kwon, O.S.; Wang, X. A framework for proactive construction defect management using BIM, augmented reality and ontology-based data collection template. *Autom. Constr.* **2013**, *33*, 61–71. [CrossRef]

34. Kwon, O.S.; Park, C.S.; Lim, C.R. A defect management system for reinforced concrete work utilizing BIM, image-matching and augmented reality. *Autom. Constr.* **2014**, *46*, 74–81. [CrossRef]

35. European Committee for Standardization. EN Standard No. 13670:2009. Eurocode—Execution of Concrete Structures. Available online: https://standards.cen.eu (accessed on 27 August 2019).

36. BuildingSmart. Industry Foundation Classes (IFC) for Data Sharing in the Construction and Facility Management Industries. Available online: https://www.buildingsmart.org/standards/bsi-standards/industry-foundation-classes/eu (accessed on 27 August 2019).

37. Ding, L.; Li, K.; Zhou, Y.; Love, P.E. An IFC-inspection process model for infrastructure projects: Enabling real-time quality monitoring and control. *Autom. Constr.* **2017**, *84*, 96–110. [CrossRef]

38. European Committee for Standardization. EN Standard No. 1990(2002). Eurocode—Basis of Structural Design. Available online: https://eurocodes.jrc.ec.europa.eu (accessed on 27 August 2019).

39. Joshi, R.; Hiwale, A.; Birajdar, S.; Gound, R. Indoor Navigation with Augmented Reality. In *Lecture Notes in Electrical Engineering*; Kumar, A., Mozar, S., Eds.; Springer: Singapore, 2019; Volume 570.

40. Sabelman, E.; Lam, R. The real-life dangers of augmented reality. *IEEE Spectr.* **2015**, *52*, 48–53. [CrossRef]

Article

Extending BIM Interoperability for Real-Time Concrete Formwork Process Monitoring

Morteza Hamooni [1], Mojtaba Maghrebi [2], Javad Majrouhi Sardroud [1] and Sungjin Kim [3],*

[1]   Department of Civil Engineering, Central Tehran Branch, Islamic Azad University, Tehran 1584743311, Iran; hamooni.morteza@gmail.com (M.H.); j.majrouhi@iauctb.ac.ir (J.M.S.)

[2]   Department of Civil Engineering, Ferdowsi University of Mashhad, Mashhad 9177948974, Iran; mojtabamaghrebi@um.ac.ir

[3]   Department of Civil, Construction, and Environmental Engineering, University of Alabama, 401 7th Avenue, 264 Hardaway Hall, Box 870205, Tuscaloosa, AL 35487, USA

*   Correspondence: sungjin.kim@eng.ua.edu; Tel.: +1-205-348-0369; Fax: +1-205-348-0783

Received: 10 January 2020; Accepted: 3 February 2020; Published: 6 February 2020

Abstract: The concrete formwork process is a critical component of construction project control because failing to gain the necessary concrete strength can lead to reworks and, consequently, project delays and cost overruns during the project's execution. The goal of this study is to develop a novel method of monitoring the maturity of concrete and providing reduced formwork removal time with the strength ensured in real-time. This method addresses the wireless sensors and building information modeling (BIM) needed to help project management personnel monitor the concrete's status and efficiently decide on the appropriate formwork removal timing. Previous studies have focused only on the monitoring of concrete's status using sensor data or planning the formwork layout by integrating the BIM environment into the design process. This study contributes to extending BIM's interoperability for monitoring concrete's maturity in real-time during construction, as well as determining the formwork removal time for project control. A case study was conducted at a building construction project to validate the developed framework. It was concluded that BIM can interoperate with the data collected from sensors embedded in concrete, and that this system can reduce formwork removal time while retaining sufficient strength in the concrete, rather than adhering to the removal time given in building code standards.

Keywords: concrete formwork; concrete maturity; building information modeling (BIM); interoperability; real-time monitoring

---

1. Introduction

Concrete is widely used in the construction industry. The material requires sufficient strength and durability to be used in building construction for different types of structural members, such as foundations, beams, walls, slabs, or columns. During the concrete's placement, it is important to monitor its status throughout the entire concrete formwork process, until the material cures. The formwork must remain in place to support a concrete structure during the construction. This process must consider factors such as the class of concrete, cement type or grade, humidity, temperature, and size/type of the structural components [1–3]. The formwork removal time is critical to the concrete gaining sufficient strength, as well as to the cost and time management of the entire construction project; this is because early removal leads to rework, which leads to project delays [4,5]. The current state of practice has established a standard time to remove concrete formworks during a construction project [6–8]. Tailored time determination involves employing a concrete maturity method that estimates the mechanical properties of the concrete, based on the temperature during curing concrete [9,10]. This approach involves monitoring the concrete's temperature and using the data acquired to estimate its strength and the time needed for the formwork [11].

To monitor the temperature and strength of concrete, the current state of practice uses a mobile-based or wireless sensor-embedded monitoring system [9,12,13]. Such studies have focused only on the schematic design for monitoring and collecting sensor data. The concrete formwork is placed for all concrete structural elements such as columns, walls, and slabs. Therefore, sensor data regarding the concrete's maturity obtained from different elements can be stored and managed in the building information modeling (BIM) environment because BIM can store geometric and semantic information of the construction project. Previous systems have focused only on collecting data and monitoring the condition of the concrete, and not providing real-time information about the appropriate time to remove the formwork. To address this concern, the present research expanded BIM's interoperability to monitor concrete's maturity and determine the formwork removal time by using wireless sensor data.

BIM is capable of integrating all stages of construction into a single digital model, achieving high-quality construction, accurate performance and cost estimates, and continuous monitoring and control, as well as maintaining an up-to-date information source that facilitates project stakeholders achieving a clear overview of the project for use in informed decision-making [14,15]. This study incorporates BIM and a sensor-based concrete maturity method to monitor and manage data related to various concrete-based structural components (e.g., slabs, walls, columns) and their strengths. This research used a module and wireless sensors embedded in fresh concrete, measured the concrete's temperature, and then automatically calculated its maturity so that its strength would be predictable. To expand BIM's interoperability, a BIM application programming interface (API) was applied that can transfer data related to the concrete's maturity to the BIM environment from sensors in real-time. All information about the temperature and maturity of the concrete can be monitored through planted sensors to the BIM environment, and stored in an industry foundation class (IFC) format. To test the interoperability, this research conducted a field case study in a building construction project in Iran. The results showed that BIM was able to interoperate with the data collected from the sensors, receive data related to the concrete's maturity, and calculate the appropriate concrete formwork removal time. The time to remove the formwork was less than what would have resulted from using the standardized regulation alone, and still clearly ensured the necessary level of strength. Thus, this interoperable BIM system contributes to monitoring the concrete curing conditions, receiving maturity data from the sensors, reducing concrete formwork removal times, and positively preventing wasted time and cost from reworks resulting from concrete failing to gain the necessary strength.

2. Literature Review

2.1. Concrete Maturity Method

The concrete maturity approach generally assumes that concrete's strength and durability depend on its temperature and moisture [16,17]. These parameters significantly affect how concrete's properties (such as strength, elastic modulus, deformation, and shrinkage) develop during the early stages of a formwork [18]. Since concrete maturity is calculated based on the temperature of the concrete between its place and form, the in situ concrete's strength and time of removal can be estimated in real-time if it is possible to collect temperature data at the construction site [11,19,20]. Among the various properties of concrete, strength is particularly critical to measuring and evaluating its quality and performance [21]. To measure the performance, a sample is tested to see if the concrete can bear the maximum load it should be able to resist. However, this direct measurement method consumes both time and money because it requires a number of samples and the test must be repeated until an allowable strength is acquired [22]. The process may also ignore the time and temperature of mature concrete. Moreover, results based on this type of manual test can be affected by external or environmental factors such as human mistakes, errors in sample specifications, and problems with loading configurations and settings [23]. To overcome this challenge, a maturity index for calculating in situ concrete's strength during construction can be used to test field-cured concrete cylinders. It is also important to predict and document the concrete's strength from the early work stages, in order to reduce the time for removing the concrete formwork [24,25].

2.2. Wireless Concrete Monitoring Systems and Formwork Removal

Over the years, studies have tried to implement indirect methods of evaluating the strength performance of concrete, developing sensor-based systems to monitor environmental conditions and identifying the parameters affecting concrete's performance [23,26–30]. For example, temperature data can be collected while the concrete is forming by using temperature sensors and radio frequency identification (RFID) tags or resistance rigidity thermometers [31]. Kumbhar and Chaudhari [32] proposed a wireless sensor for measuring temperatures in early-age concrete structures in order to determine the concrete's strength. That study found that this sensor-based method could be used to predict concrete's maturity. Cabezas et al. also presented a new embedded sensor system for monitoring the status of concrete [33]. Such sensor-based maturity methods are very effective at estimating the properties of concrete, based on data collected via temperature monitoring [33,34]. Dong et al. [35] conducted a case study measuring the temperature of concrete on a highway construction project by using a sensor network system. The temperature data obtained allowed researchers to continuously estimate the concrete's strength and maturity throughout the project's lifecycle, showing that the sensor-based maturity method could also be used to estimate the appropriate time to detach the concrete mold, based on the temperature [35]. Such interdisciplinary research has shown new techniques based on wireless sensor systems that allow for the continuous measurement and real-time monitoring of both the temperature and humidity level of concrete [24,36]. As a result, these sensor systems have underscored the potential to compute more accurate and consistent outputs based on concrete's various properties. However, it should be noted that sensor-based systems embedded in concrete are still not reliable in terms of the concrete's stability, lifespan, signal transmission requirements, and protection against various types of damage [37].

Several standards are widely used to concrete formworks suggest removal at different structural ranges [6–8]. Vertical structural elements require a shorter amount of time to form than do horizontal components with large spans (greater than six meters) that require at least 14 to 28 days. The standards dictate that large structures (greater than six meters) must acquire about 85% of the recommended strength level before formwork removal [3]. Several studies have reported that for vertical elements the formwork can be detached earlier, before what is recommended in the standard [20,38,39]. However, the manual maturity method does not provide sufficient information about the ideal time to remove the formwork. In addition, it requires workers to follow the standard even if they could remove the formwork earlier without any damage. This can cause significant delays in a panel's removal because there is still the possibility of serious damage during the early formwork stages [20].

2.3. Concrete Formwork Applications in BIM

In the construction industry, BIM has recently been used to facilitate the integration of the design and construction processes, saving both cost and time [14,40]. BIM can store and manage geometric and semantic information for design components and integrate all descriptions of visualized objects and related data during a construction project. Many BIM applications have focused on pre-construction planning activities [41]. A few studies have tried to develop BIM applications for decision support systems facilitating construction project control and operation [42,43]. Particularly for concrete formwork, Romanovskyi et al. [44] recently proposed a BIM-based decision support system for concrete formwork design that managed all information needed and collected during the concrete formwork construction process. BIM has been implemented to help the design and planning process, as well as simulate the designed formwork. Meadati et al. [45] provided a conceptual overview of a knowledge repository for concrete formworks via BIM. The initial concept was to connect 3D models and associated information, saving on parameters related to cost and concrete formwork materials such as unit price, specifications, and quantity [45]. Kannan and Santhi [46] developed a BIM-based automated concrete formwork layout and simulation system that could predict construction schedules based on a BIM formwork simulation. Another study in this domain developed a BIM-based concrete formwork management system for improving the efficiency of formwork use, such as by optimizing reuse and maximizing return-on-investment [47]. The current state of knowledge has contributed significantly to developing BIM-based management systems, especially in terms of concrete formwork design and planning during the pre-construction step [44,46,47]. However, the end-users of such systems employ static data instead of real-time information collected by sensors or other information technologies during the construction process [48]. Real-time data collected in BIM can support project stakeholders' decision-making by controlling and interpreting data pattern changes and trends [44,47–49]. Current research has focused on using BIM for concrete formwork planning and design rather than monitoring and calculating the maturity of concrete during the formwork process [44,47–50].

Therefore, the present study expanded the interoperability of BIM for real-time monitoring of the maturity of concrete by using sensor data and calculations conducted through the API interface, in addition to conveying the possible formwork opening time that does not risk damage to the concrete's structural elements. The interoperability of exchanging BIM models and sensor data can also extract all computed outcomes to an IFC data format in the BIM environment.

3. Research Methodology

This study consisted of three steps: (1) creating a sensor-based monitoring system (both module and application design), (2) developing a BIM-integrated data exchange system through a Revit API add-in, and (3) applying and validating the BIM-integrated system in a building construction project (i.e., a case study). This work employed three sensors (i.e., SHT21, Ds18b20, and LM3) to develop the sensor-based monitoring system. The first sensor, SHT21, collected temperature and humidity data, while the other two were only capable of measuring the temperature of the concrete. In the first step, the sensors' hardware and software modules were designed, allowing for the retrieval of two types of data that could be stored on an SD card. The software module was designed based on Java and .NET frameworks in order to allow for visualization of the temperature and humidity data collected by the sensors. Through Wi-Fi, the data on the SD card could be forwarded to the application developed for this step. The information in this application could then be used to calculate the concrete's maturity and estimate its strength. The second step involved the development of a Revit add-in to extend the BIM interoperability for calculating the concrete's maturity and storing all information in a database as an IFC format (i.e., Maturity 1, Maturity 2, and Maturity 3). In the final step, a case study was conducted at a construction project in Iran to apply and validate the interoperability and system. Figure 1 shows an overview of the research methodology.

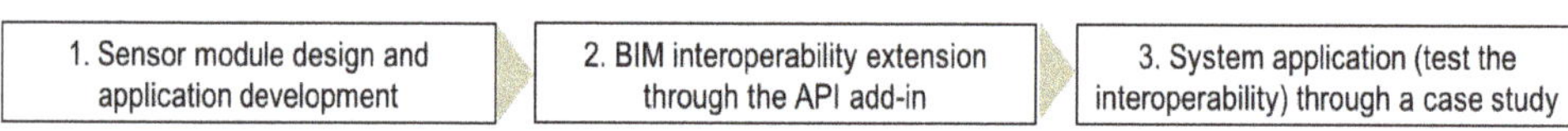

Figure 1. Overview of research methodology.

3.1. Sensor Module Design and Application Development

This study used three different sensors to collect humidity and temperature data from the tested concrete. The first sensor was an SHT21, which can collect both temperature and humidity information and produce digital output according to an I2C communication protocol. This sensor has a 14-bit resolution for temperature and 12-bit resolution for humidity. Its operating range is −40 °C to 125 °C for temperature and 0% to 100% for humidity. The Ds18b20 sensor can measure temperatures between −55 °C and 125 °C. It operates based on a one-wire communication protocol connected to a board via three wires (i.e., VCC, GND, and Date). This sensor has a nominal error of ±0.5 °C in the temperature range of −10 °C to 85 °C and an error rate of up to ±2 °C in the rest of its operating range [51]. The third sensor was an LM35, which can produce the analog output. The voltage output increases linearly by 10mV with each 10 °C increase in temperature. It has an error of ±0.5 °C at 25 °C. After connecting the LM35 to a micro-analog/digital converter, the voltage outputs can be turned into temperature data with 12-bit resolution [29]. The combination of these three sensors made the proposed system capable of monitoring temperatures ranging between 0 °C and 80 °C while the concrete cured. High-performance sensors capable of indicating high or low temperatures between −100 °C and 300 °C were not required for this research. The sensors used in this study have a relatively high rate of precision with regards to temperature (±0.5 °C); the level of accuracy required is at least ±1 °C to determine the concrete's maturity. Also, these sensors resist changes to the properties of the concrete that occur during hardening; previous studies have also used these sensors for similar purposes.

The hardware module consisted of data acquisition, data storage, and wireless transmission circuits. The module stored the retrieved information on an SD card mounted on the board. The data were processed by software designed with a .NET framework, using Android for visualization of the server or mobile device. This small module was equipped with a self-designed ad-hoc board appropriate for concrete formwork on construction sites. The module program was developed using Cube software (see Figure 2a). The microcontroller employed on the board was a STM32F130RET6, which is a 32-bit microcontroller from the Cortex M3 family. The microcontroller had an operating frequency of 72MHz and enjoyed very high processing speed and precision. The IEEE 802.11 (Wi-Fi) topology and, specifically, an HLK-M35 module were used to establish a connection with the computer or smartphone. The module used the TCP/IP operating protocol and universal asynchronous receiver/transmitter communication protocol to link with the microcontroller. Figure 2 shows the HLK-M35 communication module setup (see Figure 2b) and the configuration of the microcontroller connection (see Figure 2c). The Wi-Fi module was connected to the microcontroller through the serial port, which was specified on the PA0 to PA3 bases. Two analog inputs were inserted into PA5 and PA6 to connect the sensors. A one-wire protocol was used on the PA9 base to connect the DS18b20 sensor. The SDIO protocol developed by STMicroelectronics was connected through the PC to PC12 bases for connection with the SD card. Pull-up 47K resistors were used to reduce noise.

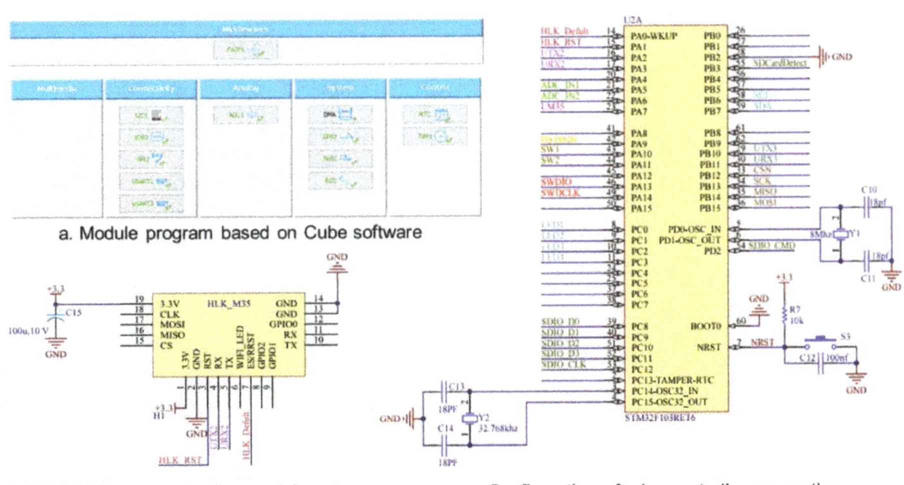

a. Module program based on Cube software

b. HLK-M35 communication module setup   c. Configuration of microcontroller connection

Figure 2. Sensor module program and design.

The data logger system was designed in Visual Studio with C#, and the DevExpress software suite was used alongside to improve the user interface (UI) design. The developed program could not be directly executed on the operating system and needed the .NET framework to function. This framework is preinstalled on all newer versions of Windows. Communication with the board was designed using .NET Framework sockets, which is a package for network-managed communications. Software performance was improved for multithread processing, which while making the coding more difficult, leading to easier event management and execution. In the next stage, a mobile application was developed to monitor the output temperature via a smartphone. The UIs of data logger and mobile application are shown in Figure 3a,b, respectively. The application was designed in Android Studio based on the Navigation Drawer platform. The design was accomplished with through Java, using the model-view-controller pattern and consisting entirely of object-oriented components. The model-view-controller is an architectural pattern commonly used in UI design; it partitions a program into three parts: the model, view, and controller. After connecting through Wi-Fi (by clicking on the Connect button), the program automatically connected to the electronic circuit and relayed the ReqData command to initiate information retrieval from the data logger. The information received from the data logger was then uploaded to the software dashboard. Fragment design was used to optimize RAM usage and increase program performance. Hence, the program contained three fragments managed by the program once the Main Activity was executed. Keeping every page in memory, this structure prevented unnecessary opening and closing and improved the software efficiency. The Dashboard component was built using the well-known MPAndroidChart library, which allows developers to create a variety of different charts. The Android app consisted of four main components. The first, Main Activity, was the main UI for the program and where views such as the bottom navigation bar and fragment holders were maintained. The second component, Connection Fragment, involved wireless communication and related codes. The third component was the Dashboard Fragment, which displayed the data charts. The fourth was the Log Fragment, which logged the program events for later review and analysis.

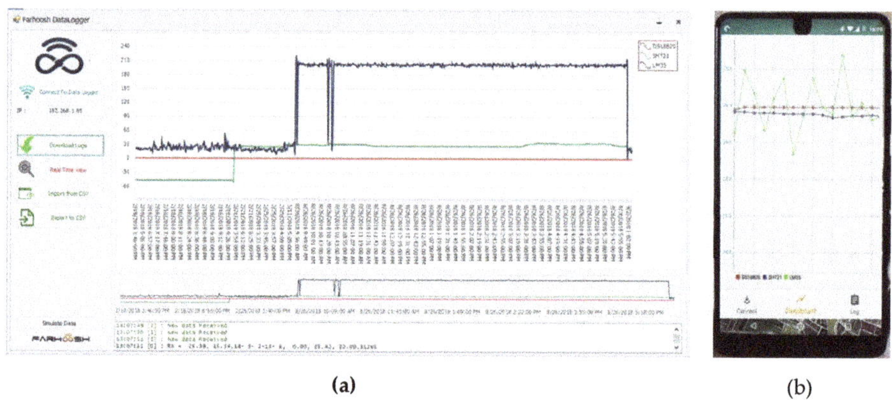

(a)   (b)

Figure 3. User interface ((**a**) data logger; (**b**) mobile application).

3.2. Extending BIM Interoperability for Concrete Maturity Calculation

This study developed a BIM API add-in to bring the temperature data from the developed module and calculate the concrete formwork removal time. Figure 4 shows the process to link the BIM environment and the sensors. The process consists of sending the concrete temperature data from the sensors to the BIM model over Wi-Fi. The sensors planted in fresh concrete can collect the temperature data and convey the data to the module system (see Figure 4a,b). All data can be sent to the server via Wi-Fi (see Figure 4c). Once the sensor module is connected to the BIM, two extra buttons appear in the Add-Ins pane in Revit (see Figure 4d). The BIM can receive the sensor data from the module, and those data can then be analyzed to compute the concrete maturity on the associated sections in the 3D model. Since the sensor data are constantly transmitted to the BIM, this system monitors the concrete's strength in real-time. To create a link between the module and the BIM-based system (see Figure 4d), an object was defined as the Socket during the coding process. Once the systems were connected, two queues were also defined to command the system: DataRXQueue for incoming and DataTXQueue for outgoing commands. The information was placed in DataTXQueue to automatically transmit the data. To receive data, the information was simply read from the queue. The data were processed and listed as a record that could be accessed at any time. Since the commands received were added to the end of the buffer, short interruptions occurred and caused premature connection termination, as well as an #Enddata message to be displayed before the data were completely received. This meant that the system would start to send partial data before collecting the entire dataset. Therefore, a timer was set up to process the data during the designated time. The time interval was computed to be two seconds, so a multi-second delay was applied after reading each data line. Data processing began when no new data were received during this period. The system could calculate the concrete's maturity by using the data collected. This conveyed the maturity as determined by each sensor. The Revit add-in system was designed to log the data hourly in an IFC format through a computed command. Then, each value from each sensor could be computed:

#data:(The number of data): time:(Time of logging)

ObjectStrength ((Third sensor value), (second sensor value), (first sensor value)).

#enddata:(Data number): time:(Time of logging)

Based on the data received, it was possible to compute the concrete's maturity based on the temperature in the maturity index, according to ASTM C1074 [11]. The result was displayed on the associated structural members (e.g., slab, column, wall) of the 3D BIM model. Finally, the system could determine the time for formwork removal based on comparing the standard strength and computed values. The information regarding the status of the concrete could then be exported to an IFC format. Based on this, BIM's interoperability through IFC can be extended to communicate temperature data for concrete collected from embedded sensors to an objective 3D model in the BIM.

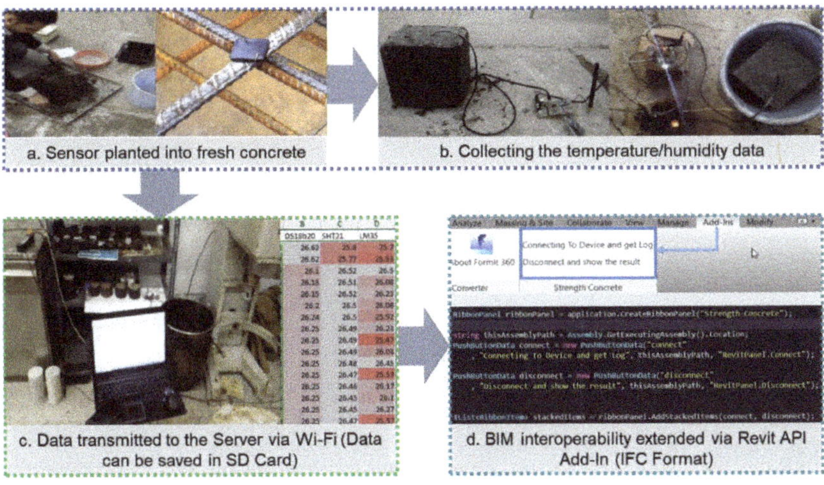

Figure 4. Process to link building information modeling (BIM) and sensors planted in concrete.

4. System Validation and Case Study

To apply and validate the developed BIM-based concrete formwork control system, a case study was conducted on a beam with a span of seven meters in a building construction project (the Baran building) in Mashhad, Iran. A geometric 3D model of the Baran building was developed in BIM. The module was simultaneously connected to three sensors. During the construction project execution, the sensors were embedded in the concrete, and the temperature data was collected during the concrete placement. The average of the temperatures measured by these three sensors was used in the calculations. Figure 5 shows the core temperature of the concrete automatically measured by the SHT21, Lm35, and Ds18b20 sensors at 1-minute intervals. Considering the size of the dataset collected, the figure presented only covers 44 min.

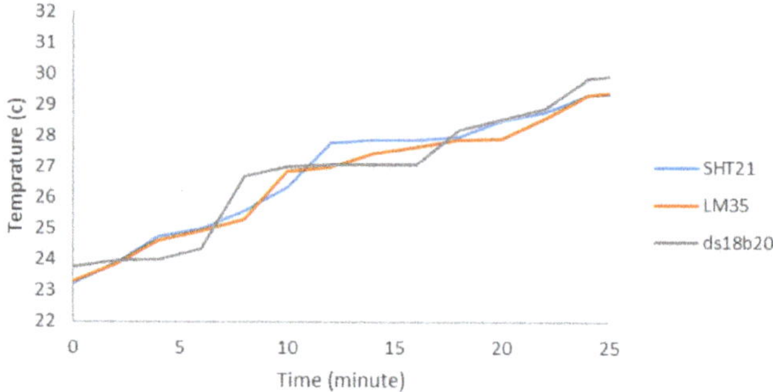

Figure 5. Temperature data collected from the SHT21, Lm35, and Ds18b20 sensors.

The compressive strength of concrete can be formulated based on the maturity concept, as defined by Carino [20,52]. The equivalent age was computed using Equation (1):

$$t_e = \sum_0^t exp\left[\frac{-E_a}{R}\left(\frac{1}{T} - \frac{1}{T_r}\right)\right]\Delta t \tag{1}$$

(where, t_e: equivalent age at the temperature; E_a: activation energy; R: universal gas constant; T: average absolute temperature during interval time, Δt)

The activation energy was considered to be constant for all cement types (in accordance with the recommendation in Carino and ASTM C1074). Many studies have argued that activation energy can depend on time and temperature [10,52–55]. However, it is customary to assume constant activation energy. The concrete's strength was estimated using parabolic Equation (2).

$$S = S_\infty \frac{\sqrt{K_T(t_e - t_0)}}{1 + \sqrt{K_T(t_e - t_0)}} \tag{2}$$

(where, S: strength; K: constant rate)

The cement used in the case study was ordinary Portland cement (OPC). As suggested by Carino and Tank [56], the activation energy for OPC was assumed to be 41 kJ. It should be noted that some researchers have recommended slightly different values for this parameter [57,58]. Table 1 shows the performance coefficient and activation energy values used in this case study. By employing Equation (1), the equivalent time t_{e24} was calculated for a normal day of the month. The times $t_{50\%}$, $t_{70\%}$, and $t_{85\%}$ were required for the concrete to gain 50%, 70%, and 85% of its designed compressive strength, respectively, at the reference temperature (i.e., 20 °C); these were calculated using Equation (2). The times $t_{e50\%}$, $t_{e70\%}$, and $t_{e85\%}$ were required for the concrete to gain 50%, 70%, and 85% of its designed compressive strength, respectively, at the site temperature.

Table 1. Constant coefficient and activation energy used.

Concrete Class	Type of Cement	E_a ⟦*kj*⟧	k ⟦−⟧	t_0 ⟦*h*⟧	S_u ⟦*MPa*⟧	$t_{50\%}$ ⟦*h*⟧	$t_{50\%}$ ⟦*days*⟧	$t_{70\%}$ ⟦*h*⟧	$t_{70\%}$ ⟦*days*⟧	$t_{85\%}$ ⟦*h*⟧	$t_{85\%}$ ⟦*days*⟧
C30	OPC	41	0/018	11/30	23	61/78	2/23	130/7	5/92	301/69	12/02

The t_e value obtained from Equation (1) depended on the cement type (i.e., activation energy) and daily temperature variations. In this case study, particularly for a beam with a span of seven meters were used to test the BIM interoperability developed. For the selected beam element (spanning over six meters), 14 days were indicated in the standard building code as the minimum time required before formwork removal (BS8110:1997) and 28 days are required as the maximum time to strike the formwork (PN-B-06251:1963). The results of the t_e calculations for the selected beam are presented in Table 2. Temperature data have been collected through three sensors during two tests. It is assumed that the concrete should gain at least 85% of its design strength before the formwork can be removed [8]. As a result, the formwork removal time was computed as 9.6 days, ensuring 85% of the designed concrete strength at the second test.

Table 2. Computed concrete formwork removal times and temperature.

Factors	1st Test (days)	2nd Test (days)
Average minimum temperature	19.3	19.8
Average maximum temperature	35.8	37.9
t_e, 85% [days]	9.8	9.6
$T_{85\%}$ [days] (cured at 20 °C)	11.6	11.5
t_e, 70% [days]	4.65	4.08
$T_{70\%}$ [days] (cured at 20 °C)	5.6	5.6
t_e, 50% [days]	2.01	1.98
$T_{50\%}$ [days]	3.65	3.65
t_e, 24 [h]	28.25	29.75

The developed BIM-based system predicted that the formwork could be removed 4.4 days earlier than the requirement in the BS 8110:1997 (31.4% time-saving), 11.4 days earlier than the standard of IS 456:2000 (54.3% time-saving), and 18.4 days earlier than the PN-B-06251:1963's formwork strike standard time (65.8% time-saving). In addition to the beam, the case study also conducted at the slab with a span of less than 4.5 m. The calculated time for removing the formwork was 6.9 days, which can reduce the time about 50.7% in maximum and minimum 1.4%. In the case of the slab more than 4.5 m, the computed time was indicated as 11.3 days. This reduced the time about 18.6% based on both IS456:2000 and BS8110:1997 standards. The results indicate that the proposed BIM interoperability in this study can aim in monitoring the concrete curing status based on the temperature data received from the sensors and computing the removal strike time based on the maturity method. Also, the BIM model could receive and compute the concrete maturity data through Wi-Fi and the information was exported into the IFC format. Figure 6 shows the schematic of the BIM's interoperability to compute the concrete maturity data from the three sensors.

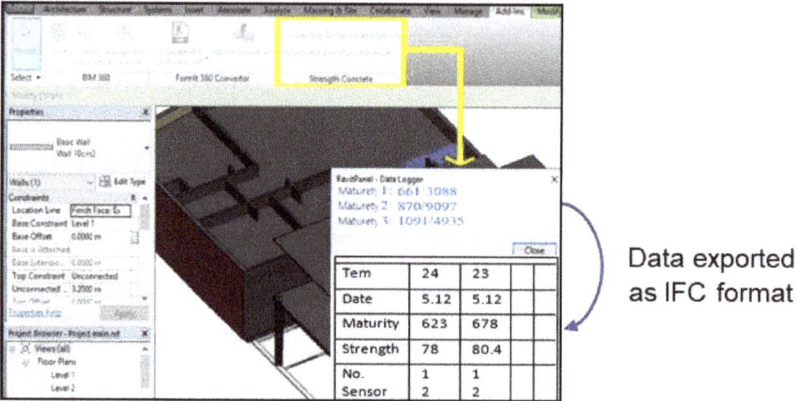

Figure 6. Schematic of BIM interoperability for concrete maturity monitoring.

5. Discussion

This study proposed a method for expanding the interoperability of BIM with wireless sensors in order to monitor the concrete maturity and control the concrete formwork process. While previous efforts to create BIM-assisted systems focused on concrete formwork design and the planning process, the system developed in this research targeted a concrete formwork control that potentially affects construction management and project controls. Particularly, the extended capability of BIM allows for the calculation of formwork removal time based on the maturity and strength data collected from planted sensors. In the case study, the overall approach, including wireless sensors, concrete maturity, and formwork removal control, was tested at a real construction project. As a result, this study identified the amount of time reduction in the removal of concrete formwork, as compared to the time given in the standard. Based on the case study, this study has four noteworthy implications.

First, sensor data can be transmitted to the BIM environment via a Wi-Fi signal with a TCP-IP protocol and API add-in interface. Data related to the concrete's maturity can then be connected to the associated 3D objective model in the BIM environment. Therefore, this system is capable of monitoring the status of concrete and providing maturity data for users of the BIM environment. Therefore, the sensor module and BIM add-in developed in this study contribute to extending BIM's interoperability in terms of real-time concrete monitoring and formwork process control.

Second, this interoperable BIM-based system can calculate the removal times of concrete formworks according to different parts of a structure (i.e., slab, beam, or column). This system will help practitioners to continuously monitor and control the concrete placement process and deliver an earlier time to remove the formwork. The current standards manually regulates the recommended time for removing the formwork to maintain the concrete's performance. However, the results of the case study imply that the time for removing the formwork was reduced beyond what was recommended in the current standards. Removing the formwork earlier while still ensuring strength will help to avoid project delays and unnecessary formwork rental expenses. This will lead to reducing the time needed for the concrete formwork process and, consequently, project completion.

Third, it can be noticed that the use of BIM can be expanded from the concrete formwork design and layout to the concrete placement control. The previous studies contributed to documenting the design process of the concrete formwork by using BIM. This research described a proof-of-concept to exchange real-time sensor data in the BIM environment. By integrating two concepts in the BIM environment, it can provide more efficiency to design and control the concrete formwork.

Lastly, the developed BIM interoperability will help to prevent concrete rework from failing to gain the necessary strength. Based on the case study, the system demonstrated less than a 13%

difference in concrete strength between the manual and newly developed system-based methods. Also, this implies that using the average temperature data collected from three sensors to calculate the strength and formwork removal time is of low tolerance. Since concrete work can be as high as 40%~60% of the total project cost and concrete formation usually represents over 50% of the entire set of concrete-based tasks in a construction project [59], minimizing the reworking of concrete jobs is very significant for preventing total cost overruns. Consequently, this system will save time and cost on construction projects.

In this sense, this new BIM interoperability is beneficial for project control and management of construction. However, this study did not consider: (1) that environmental factors may affect concrete formation; (2) that the number of planted sensors may affect the performance of fresh concrete; (3) natural sensor errors, and (4) total cost and time impacts. Based on the limitation and implications documented in this research, it is recommended that this system potentially be integrated with the BIM-integrated concrete formwork layout and planning system proposed in previous studies. In addition, the sensor-embedded BIM interoperability could be also extended to the operation and maintenance steps of the constructed facilities. The sensors will continuously collect the data in terms of the concrete temperature and strength during the life cycle of the project. This will provide a more visible platform for implementation into the construction industry. Also, a deep investigation of the potential cost and time savings resulting from early formwork removal is needed. This would allow for an investigation of how this extended BIM interoperability impacts the total cost and time of a construction project.

6. Conclusions

Removing concrete formwork after a sufficient level of strength has been reached is critical to the quality of the concrete and the consequent safety and health of the structural components. The current study has documented various sensor-based concrete monitoring systems for use during the formwork process. Studies integrating BIM into the concrete formwork process have to date focused on design and planning rather than controlling the project and formwork use. Therefore, this study extended the interoperability of BIM for monitoring the status of curing concrete and computing its maturity and the timing of the formwork during the concrete placement. A sensor module was designed to receive data from three sensors collecting concrete temperature data. The data were conveyed to the associated parts of an objective 3D model via Wi-Fi. A database was then created via the BIM API add-in and based on the concrete's maturity as computed and the formwork removal time, thus assisting the user's decision-making with regards to the concrete work. A 3D model of an active construction project in Iran was established to validate the system. During the case study, sensors were planted in various structural components, columns, beams, floors, and slabs. The data collected by the sensors and module were then transmitted to the designed software. The BIM-based system presented here was capable of receiving and documenting data in real-time. The data computed in BIM were automatically stored in an IFC format. These data were capable of describing the strength of the concrete, building codes, and associated criteria, as well as opportunities to remove the formwork. The results of the case study showed that the concrete formwork could be removed approximately 40% earlier than the documented standard time without resulting in any structural damage.

In summary, this study proposed a new form of BIM interoperability for concrete formwork control and monitoring. This work ignored natural errors from the sensors, as well as other external factors affecting the concrete formwork. However, it should be underscored that this research provides extended interoperability for managing project activities, particularly concrete formwork, and the developed system can reduce the time spent before removing concrete formworks while still ensuring the concrete's strength. In the future, this BIM-based system could be incorporated with BIM-based concrete formwork design and planning processes described in previous studies to improve the efficiency and productivity of concrete formwork use, as well as reduce redundancy. Other future efforts should investigate the economics of this new system's use, as well as other issues affecting management feasibility. Such efforts will lead to an overall reduction in the time and money consumed by construction projects.

Author Contributions: Conceptualization, M.H. and J.M.S.; methodology, M.H., J.M.S., and S.K.; software, M.M.; validation, M.H., M.M., and J.M.S.; formal analysis, M.H.; investigation, J.M.S. and S.K.; resources, M.M. and J.M.S.; data curation, M.H., M.M., and J.M.S.; writing—original draft preparation, M.H.; writing—review and editing, S.K.; visualization, M.H. and S.K.; supervision, J.M.S. and S.K.; project administration, J.M.S.; All authors have read and agreed to the published version of the manuscript.

Funding: This research received no external funding.

Conflicts of Interest: The authors declare no conflict of interest.

References

1. Johnson, R.P. *Composite Structures of Steel and Concrete*; John Wiley & Sons: Hoboken, NJ, USA, 2008; ISBN 978-1-405-16839-7.
2. Hyun, C.; Jin, C.; Shen, Z.; Kim, H. Automated optimization of formwork design through spatial analysis in building information modeling. *Autom. Constr.* **2018**, *95*, 193–205. [CrossRef]
3. Skibicki, S. Optimization of Cost of Building with Concrete Slabs Based on the Maturity Method. In *IOP Conference Series: Materials Science and Engineering*; IOP Publishing: Bristol, UK, 2017; p. 022061.
4. Hanna, A.S.; Senouci, A.B. Material Cost Minimization of Concrete Wall Forms. *Build. Environ.* **1997**, *32*, 57–67. [CrossRef]
5. Okafor, F.; Ewa, D. Estimating Formwork Striking Time for Concrete Mixes. *Niger. J. Technol.* **2015**, *35*, 1–7. [CrossRef]
6. PN-B-06251:1963. *Concrete and Reinforced Concrete Work—Technical Specification (Roboty Betonowe i Żelbetowe—Wymagania Techniczne)*; Polish Committee for Standardization: Warsaw, Poland, 1963. (In Polish)
7. Indian Standard, I.S. *IS 456 2000: Plain and Reinforced Concrete. Code of Practice*, 4th ed.; Bureau of Indian Standards: New Delhi, India, 2000.
8. British Standards Institution. *Structural Use of Concrete—Part 1: Code of Practice for Design and Construction*; BSI Press: London, UK, 1998.
9. Yikici, T.A.; Chen, H.L. Use of maturity method to estimate compressive strength of mass concrete. *Constr. Build. Mater.* **2015**, *95*, 802–812. [CrossRef]
10. Topçu, İ.B.; Karakurt, C. Strength-Maturity Relations of Concrete for Different Cement Types. In *International Sustainable Buildings Symposium*; Springer: Cham, Switzerland, 2017; pp. 435–443.
11. American Society for Testing and Materials (ASTM). *ASTM C1074-17 Standard Practice for Estimating Concrete Strength by the Maturity Method*; ASTM Internationa: West Conshohocken, PA, USA, 2017.
12. Badawi, Y.; Shamselfalah, A.; Elbadri, A.; Badawi, D.; Abdelazim, M.; Wahid, R.A.; Ali, A.M.; Ahmed, Y. Use of Maturity Sensors to Predict Concrete Compressive Strength in Different Curing and Compaction Regimes. In Proceedings of the 2nd Conference on Civil Engineering, Khartoum, Sudan, 3–5 December 2018; pp. 203–210.
13. Moon, S. Application of Mobile Devices in Remotely Monitoring Temporary Structures during Concrete Placement. *Procedia Eng.* **2017**, *196*, 128–134. [CrossRef]
14. Akula, M.; Lipman, R.R.; Franaszek, M.; Saidi, K.S.; Cheok, G.S.; Kamat, V.R. Real-time drill monitoring and control using building information models augmented with 3D imaging data. *Autom. Constr.* **2013**, *36*, 1–15. [CrossRef]
15. Joblot, L.; Paviot, T.; Deneux, D.; Lamouri, S. Building Information Maturity Model specific to the renovation sector. *Autom. Constr.* **2019**, *101*, 140–159. [CrossRef]
16. Paulsson, J.; Farhang, A. Measurements on the moisture state in a heavily trafficked concrete flat slab repaired with a bonded conrete overlay. In Proceedings of the Nordic Mini-seminar of the Nordic Concrete Federation, Espoo, Finlad, 22 August 1997; Volume 174, pp. 5–19.
17. Paroll, H.; Nyknen, E. Measurement of relative humidity and temperature in a new concrete bridge vs. laboratory samples. *Nord. Concr. Res.* **1999**, *23*, 116–118.
18. Kim, J.K.; Lee, C.S. Moisture diffusion of concrete considering self-desiccation at early ages. *Cem. Concr. Res.* **1999**, *29*, 1921–1927. [CrossRef]

19. Klieger, P.; Lamond, J.F. *STP 169C Significance of Tests and Properties of Concrete and Concrete-Making Materials*; Klieger, P., Lamond, J.F., Eds.; ASTM International: West Conshohocken, PA, USA, 1994; ISBN 0-8031-2053-2.

20. Malhotra, V.M.; Carino, N.J. *Handbook on Nondestructive Testing of Concrete*; CRC Press: Boca Raton, FL, USA, 2003.

21. Sadrossadat, E.; Ghorbani, B.; Hamooni, M.; Moradpoor Sheikhkanloo, M.H. Numerical formulation of confined compressive strength and strain of circular reinforced concrete columns using gene expression programming approach. *Struct. Concr.* **2018**, *19*, 783–794. [CrossRef]

22. Morel, J.C.; Pkla, A.; Walker, P. Compressive strength testing of compressed earth blocks. *Constr. Build. Mater.* **2007**, *21*, 303–309. [CrossRef]

23. Lee, S.-C. Prediction of concrete strength using artificial neural networks. *Eng. Struct.* **2003**, *25*, 849–857. [CrossRef]

24. Norris, A.; Saafi, M.; Romine, P. Temperature and moisture monitoring in concrete structures using embedded nanotechnology/microelectromechanical systems (MEMS) sensors. *Constr. Build. Mater.* **2008**, *22*, 111–120. [CrossRef]

25. Benaicha, M.; Burtschell, Y.; Alaoui, A.H. Prediction of compressive strength at early age of concrete—Application of maturity. *J. Build. Eng.* **2016**, *6*, 119–125. [CrossRef]

26. Schlicke, D.; Kanavaris, F.; Lameiras, R.; Azenha, M. On-site Monitoring of Mass Concrete. In *RILEM State-of-the-Art Reports*; Springer International Publishing: Berlin/Heidelberg, Germany, 2019; pp. 307–355. ISBN 978-3-319-76616-4.

27. Providakis, C.; Liarakos, E. T-WiEYE: An early-age concrete strength development monitoring and miniaturized wireless impedance sensing system. *Procedia Eng.* **2011**, *10*, 484–489. [CrossRef]

28. Hue, F.; Serrano, G.; Bolaño, J.A. Öresund Bridge. Temperature and cracking control of the deck slab concrete at early ages. *Autom. Constr.* **2000**, *9*, 437–445. [CrossRef]

29. Lee, H.-S.; Cho, M.-W.; Yang, H.-M.; Lee, S.-B.; Park, W.-J. Curing Management of Early-age Concrete at Construction Site using Integrated Wireless Sensors. *J. Adv. Concr. Technol.* **2016**, *12*, 91–100. [CrossRef]

30. Zekavat, P.R.; Moon, S.; Bernold, L.E. Performance of short and long range wireless communication technologies in construction. *Autom. Constr.* **2014**, *47*, 50–61. [CrossRef]

31. Kang, J.H.; Gandhi, J. Readability test of RFID temperature sensor embedded in fresh concrete. *J. Civ. Eng. Manag.* **2010**, *16*, 412–417. [CrossRef]

32. Kumbhar, D.S.; Chaudhari, H.C. Development of Wireless Sensor Network for Monitoring and Analysis of Concrete Compressive Strength Development in Early Age Concrete Structure. In Proceedings of the International Conference on Functional Materials and Microwaves ICFMM—2015, Aurangabad, India, 28–30 December 2015.

33. Cabezas, J.; Sánchez-Rodríguez, T.; Gómez-Galán, J.A.; Cifuentes, H.; Carvajal, R.G. Compact embedded wireless sensor-based monitoring of concrete curing. *Sensors* **2018**, *18*, 876. [CrossRef]

34. Lee, S.; Park, S. Development of Integrated Wireless Sensor Network Device with Mold for Measurement of Concrete Temperature. *J. Korea Inst. Struct. Maint. Insp.* **2012**, *16*, 129–136.

35. Dong, Y.; Luke, A.; Vitillo, N.; Ansari, F. In-Place estimation of concrete strength during the construction of a highway bridge by the maturity method. *Concr. Int.* **2000**, *24*, 61–66.

36. Ilc, A.; Turk, G.; Kavčič, F.; Trtnik, G. New numerical procedure for the prediction of temperature development in early age concrete structures. *Autom. Constr.* **2009**, *18*, 849–855. [CrossRef]

37. Chang, C.Y.; Hung, S.S. Implementing RFID and sensor technology to measure temperature and humidity inside concrete structures. *Constr. Build. Mater.* **2012**, *26*, 628–637. [CrossRef]

38. Rudeli, N.; Santilli, A.; Arrambide, F. Striking of vertical concrete elements: An analysis using the maturity method. *Eng. Struct.* **2015**, *95*, 40–48. [CrossRef]

39. Sofi, M.; Mendis, P.A.; Baweja, D. Estimating early-age in situ strength development of concrete slabs. *Constr. Build. Mater.* **2012**, *29*, 659–666. [CrossRef]

40. Charles, M.E.; Teicholz, P.; Sacks, R.; Liston, K. *BIM Handbook: A Guide to Building Information Modeling for Owners, Managers, Designers, Engineers and Contractors*; John Wiley & Sons: Hoboken, NJ, USA, 2008; ISBN 978-0-470-18528-5.

41. Karan, E.P.; Irizarry, J. Extending BIM interoperability to preconstruction operations using geospatial analyses and semantic web services. *Autom. Constr.* **2015**, *53*, 1–12. [CrossRef]

42. Kyungki, K.; Yong, C.; Kinam, K. BIM-Driven Automated Decision Support System for Safety Planning of Temporary Structures. *J. Constr. Eng. Manag.* **2018**, *144*, 4018072.
43. Jeong, W.; Chang, S.; Son, J.; Yi, J.-S. BIM-Integrated Construction Operation Simulation for Just-In-Time Production Management. *Sustainability* **2016**, *8*, 1106. [CrossRef]
44. Romanovskyi, R.; Sanabria Nejia, L.; Rezazadeh Azar, E. BIM-based Decision Support System for Concrete Formwork Design. In Proceedings of the 36th International Association for Automation and Robotics in Construction, Banff, AB, Canada, 21–24 May 2019; pp. 1129–1135.
45. Meadati, P.; Irizary, J.; Aknoukh, A. BIM and Concrete Formwork Repository. In Proceedings of the 47th Associated School of Construction, Omaha, NE, USA, 6–9 April 2011.
46. Ramesh, K.M.; Santhi, M.H. Automated Construction Layout and Simulation of Concrete Formwork Systems Using Building Information Modeling. In Proceedings of the 4th International Conference of European Asian Civil Engineering Forum (EACEF), Singapore, 26–28 June 2013; pp. C7–C12.
47. Mansuri, D.; Chakraborty, D.; Elzarka, H.; Deshpande, A.; Gronseth, T. Building Information Modeling enabled Cascading Formwork Management Tool. *Autom. Constr.* **2017**, *83*, 259–272. [CrossRef]
48. Chen, J.; Bulbul, T.; Taylor, J.E.; Olgun, G. A Case Study of Embedding Real-time Infrastructure Sensor Data to BIM. In Proceedings of the Construction Research Congress (CRC) 2014: Construction in a Global Network, Atlanta, GA, USA, 19–21 May 2014; pp. 269–278.
49. Kannan, M.R.; Santhi, M.H. Constructability Assessment of Climbing Formwork Systems Using Building Information Modeling. *Procedia Eng.* **2013**, *64*, 1129–1138. [CrossRef]
50. Singh, M.M.; Sawhney, A.; Sharma, V. Utilising Building Component Data from BIM for Formwork Planning. *Constr. Econ. Build.* **2017**, *17*, 20–36. [CrossRef]
51. Runjing, Z.; Hongwei, X.; Guanzhong, R. Design of Temperature Measurement System Consisted of FPGA and DS18B20. In Proceedings of the 2011 International Symposium on Computer Science and Society, Kota Kinabalu, Malaysia, 16–17 July 2011; pp. 90–93.
52. Xuan, D.; Zhan, B.; Poon, C.S. A maturity approach to estimate compressive strength development of CO2-cured concrete blocks. *Cem. Concr. Compos.* **2018**, *85*, 153–160. [CrossRef]
53. Zhang, J.; Cusson, D.; Monteiro, P.; Harvey, J. New perspectives on maturity method and approach for high performance concrete applications. *Cem. Concr. Res.* **2008**, *38*, 1438–1446. [CrossRef]
54. Yang, K.-H.; Mun, J.-S.; Cho, M.-S. Effect of Curing Temperature Histories on the Compressive Strength Development of High-Strength Concrete. *Adv. Mater. Sci. Eng.* **2015**, *2015*, 12. [CrossRef]
55. Kamkar, S.; Eren, Ö. Evaluation of maturity method for steel fiber reinforced concrete. *KSCE J. Civ. Eng.* **2018**, *22*, 213–221. [CrossRef]
56. Tank, R.C.; Carino, N.J. Rate Constant Functions for Strength Development of Concrete. *Mater. J.* **1991**, *88*, 74–83.
57. Kaszyńska, M. Early age properties of high-strength/high-performance concrete. *Cem. Concr. Compos.* **2002**, *24*, 253–261. [CrossRef]
58. Saul, A.G.A. Principles underlying the steam curing of concrete at atmospheric pressure. *Mag. Concr. Res.* **1951**, *2*, 127–140. [CrossRef]
59. Robert, L. Think Formwork - Reduce Costs. *Struct. Mag.* **2007**, *4*, 14–16.

Article

A Deep Learning-Based Method to Detect Components from Scanned Structural Drawings for Reconstructing 3D Models

Yunfan Zhao [1], Xueyuan Deng [1,*] and Huahui Lai [2]

[1] Department of Civil Engineering, School of Naval Architecture, Ocean and Civil Engineering, Shanghai Jiao Tong University, Shanghai 200240, China; zhaoyunfan@sjtu.edu.cn

[2] Shenzhen Municipal Design & Research Institute Co., Ltd, Shenzhen 518029, China; laihuahui81665@alumni.sjtu.edu.cn

* Correspondence: dengxy@sjtu.edu.cn

Received: 17 February 2020; Accepted: 11 March 2020; Published: 19 March 2020

Abstract: Among various building information model (BIM) reconstruction methods for existing building, image-based method can identify building components from scanned as-built drawings and has won great attention due to its lower cost, less professional operators and better reconstruction performance. However, this kind of method will cost a great deal of time to design and extract features. Moreover, the manually extracted features have poor robustness and contain less non-geometric information. In order to solve this problem, this paper proposes a deep learning-based method to detect building components from scanned 2D drawings. Taking structural drawings as an example, in this article, 1500 images of structural drawings were firstly collected and preprocessed to guarantee the quality of data. After that, the neural network model—You Only Look Once (YOLO) was trained, verified and tested. In addition, a series of metrics were utilized to evaluate the performance of recognition. The results of test experiments show that the components in structural drawings (e.g., grid reference, column and beam) can be successfully detected, while the average detection accuracy of the whole image is over 80% and the average detection time for each image is 0.71 s. The experimental results demonstrate that the proposed method is robust and timesaving, which provides a good basis for the reconstruction of BIM from 2D drawings.

Keywords: 3D Reconstruction; BIM; 2D structural drawing; object detection; deep learning; YOLO

---

1. Introduction

In recent years, more and more practical cases [1–3] have verified the promotion of BIM technology in operation and maintenance (O&M) management of buildings. The application of BIM technology can guarantee the information passing from construction phase to O&M phase and help owners or managers to make better decisions on O&M activities, such as fault diagnosis, facilities management, space management, data monitoring, energy consumption analysis, and emergency evacuation, based on more accurate and comprehensive information [4]. However, most of the existing buildings are currently designed and built based on 2D drawings due to their long history [5], which results in the lack of building information models. Therefore, how to quickly, accurately and cost-effectively reconstruct 3D models of existing buildings has become the focus of research work in this field.

Numerous techniques and methods have been proposed to build effective and reliable building information models for existing buildings. For example, photogrammetry [6] can obtain building information (e.g., shape, size and position of building) and build 3D models by processing the building image, which is acquired by optical camera, but this type of model only contains geometric and physical information, while the precision for single building is lower compared to other methods. Laser scanning [7] can efficiently create 3D models of existing buildings through point clouds, however, it is invalid when dealing with the components hidden above ceilings or behind walls [8]. In addition, the cost of using laser scanning devices is relatively high [1]. Other methods like reconstructing 3D models based on 2D vector drawings [9] can extract not only geometrical information but also topological and semantic information of buildings. The downside of this method is that there are a number of drafting errors [10,11] in CAD drawings, which will definitely have a bad effect on the precision of reconstruction models. Moreover, a large proportion of existing buildings have been built for more than 20 years, so instead of CAD vector drawings, only paper-based drawings or hand-drawn blueprints may remain [12].

Different from the methods mentioned above, the method that scans CAD drawings or hand-drawn blueprints into digital formats and extracts information from those raster images using image processing technique [5] has received more attention from researchers due to its high efficiency, low cost and widespread applicability. As the first step of an image-based method, component recognition plays a significant role in reconstructing BIM for existing buildings. Traditionally, geometric primitives (lines, polylines, arcs, etc.) obtained by feature extraction are combined into building components based on predefined definitions about geometric constraints and topological relations [13]. However, components may be represented in different ways due to the diversity of drawing standards and drawing conventions in different countries [9]. As a result, the traditional component recognition approach, which predefines the rules, cannot satisfy all kinds of expressions. Therefore, a smarter method that can be widely applied to various conditions is needed to identify building components from raster images of CAD drawings [14].

In this paper, a novel method that automatically detects structural components from scanned CAD drawings was proposed based on a convolutional neural network called You Only Look Once (YOLO). The reasons for choosing structural components as research objects are listed as follows: (1) the continuous occurrence of decoration, refurbishment and reconstruction project during O&M phase may lead to a marked change of existing buildings, especially for the components in the architecture and mechanical system, and (2) structural framings (e.g., columns, beams and shear walls) are hardly changed since they are the most fundamental and significant part of the entire building. Hence, the actual condition of structural components will be consistent with those shown in structural CAD drawings.

2. Literature Review

2.1. Traditional Methods for Components Recognition

The general process to identify building components based on the raster image of the CAD drawing can be divided into two steps: primitive recognition and building element recognition [5].

The primitive recognition identifies basic geometric primitives (e.g., lines, arcs and circles) from the images of CAD drawings using image processing techniques and computer vision techniques. The essence of this step is feature extraction, which determines whether each pixel in the image belongs to a feature and extracts core information (e.g., shape, color, texture, spatial relations or text) based on these image features. Numerous feature extraction algorithms, like Hough Transform [15], scale invariant feature transform (SIFT) [16], maximally stable extremal regions (MSER) [17], and speeded-up robust features (SURF) [18], have been proposed in the past few decades for different tasks of object recognition. Among these algorithms, Hough Transform is one of the most common and significant methods to recognize geometric primitives from raster images of CAD drawing. More than 2500 papers focus on its improved algorithms and applications [19]. Some improved algorithms have been developed based on classical Hough Transform, such as Generalized Hough Transform [20], Progressive Probabilistic Hough Transform [21], Random Hough Transform [22], Digital Hough Transform [23], and Fuzzy Hough Transform [24].

After identifying geometric primitives, the next step, namely building element recognition, mainly defines several classification rules based on geometric constraints and topological relations, and categorizes these geometric primitives according to pre-established rules. Many researchers have been working on the recognition of building components in architectural floor plan for a long time [25]. Macé et al. [26] proposed a method to detect walls based on the distance and texture between the parallel lines. Ahmed et al. [27] distinguished walls from other components by dividing the lines into three levels: thick, medium, and thin. Riedinger et al. [28] took the binarization process for the floor plan first, then the main walls and the dividing walls were detected and located based on the thickness of dark sketches, which represent wall seams. Grimenez et al. [9] identified building components like walls, doors, windows, and rooms from floor plans using the method of pattern recognition. Meanwhile, others focus on the recognition of structural components in 2D drawings. Lu et al. [29] put forward a shape-based method to detect parallel pairs (PPs) of structural elements like shear walls and bearing beams. Moreover, Lu et al. [30] divided the whole image of structural drawing into several segments through detecting the symbol of grids, and then the location information of columns, beams and slabs were extracted with the help of Optical Character Recognition (OCR) technology. Instead of dealing with structural components, Cho et al. [8] payed more attention to the 2D mechanical drawings and developed an algorithm with a relatively high accuracy to extract the geometrical information of mechanical entities such as ducts, elbows, branches and pipes as well as their corresponding semantic information.

Although existing component recognition methods can be used to identify specific components from CAD drawings, a major disadvantage of these methods is that it spends a lot of time to manually design and extract features, and the robustness of the features extracted in these conventional methods is poor. In addition, since the components obtained from existing methods are composed of several geometric primitives or symbols defined by the fixed pre-established regulations, the generalization ability is weak when dealing with the same components represented in different standards or drawn with different design conventions. Therefore, it is necessary and valuable to propose a more intelligent method to identify structural components in different types of structural CAD drawings. Due to the higher accuracy, strong generalization ability, simple operation, and lower cost (unnecessary to buy expensive measuring instruments), the deep learning technology was performed to detect components in this paper. Moreover, deep learning-based object detection method has been successfully applied in other fields of civil engineering [31–33], but no researches using this method were found to identify components in CAD drawings, especially for structural components. The development of deep learning methods in object detection will be reviewed in following section.

2.2. Deep Learning Methods for Object Detection

In recent years, deep learning is one of the research hotspots in the field of artificial intelligence. The rationale of deep learning is to make a computer learn to simulate the way of thinking in the human brain as well as the transmission mode of signal in nervous systems. At present, deep learning has made breakthroughs in object detection such as gesture detection [34], iris detection [35], license plate detection [36], face detection [37], and human action detection [38]. According to the process of detection, deep learning algorithms can be divided into two categories: two-stage object detection algorithm and one-stage object detection algorithm.

The two-stage object detection algorithm converts the object detection problem into a classification problem. The overall process can be divided into two stages. First, the region proposal is generated, and then the classifier is utilized to classify and amend the region proposal. Ross Girshick et al. firstly proposed Region-Based Convolutional Neural Network (R-CNN) [39] by using a selective search [40]. However, this algorithm requires a large amount of time to calculate and detection speed is slow. To improve R-CNN, Girshick developed Fast Region-Based Convolutional Neural Network (Fast R-CNN) [41] based on the idea of Spatial Pyramid Pooling Layer (SPP) [42]. Fast R-CNN greatly improves the speed of detection since it only performs convolution calculation once for the whole image. However, the process of a selective search for generating a region proposal in Fast R-CNN run on the CPU, still spends so much time on convolution calculation. Based on Fast R-CNN, Ren et al. put forward a new algorithm, namely Faster Region-Based Convolutional Neural Network (Faster R-CNN) [43], to merge the generation of region proposal and the classification of CNN together and take all the computation with the help of GPU. As a result, there is a significant increase in speed as well as accuracy. He et al. [44] proposed Mask R-CNN based on Faster R-CNN for object detection and instance segmentation. A limitation of this algorithm is that the cost of labeling when segmenting the instance is too expensive. Moreover, its detection speed still cannot reach the real-time level.

The essence of one-stage object detection approach is to transform the object detection problem into a regression problem. Different from the process in the two-stage algorithm that generates the region proposal first, one-stage object detection algorithm can directly create the class probability and coordinate information of the target object through the CNN, which immensely improves the efficiency of objection detection and meets the requirements of real-time detection in computing speed. Redmon et al. [45] proposed an algorithm named YOLO through dividing an image into N×N grids and predicting the two bounding boxes and their corresponding category information for each grid. On the basis of YOLO, Liu et al. [46] came up with the SSD based on the anchor mechanism of Faster R-CNN, which guarantees the high accuracy as well as the fast speed of the detection. However, the effect for recognizing a small target is not particularly desirable. To address these problems, researchers proposed some improved algorithms based on YOLO. The YOLOv2 [47] was developed to improve the detection accuracy and speed by adding batch normalization, multi-scale training and anchor box after each convolutional layer. The YOLO9000 [47] combined the ImageNet dataset [48] and the COCO dataset [49] together and achieved the detection of 9418 kinds of objects using the method of WordTree hierarchical classification.

2.3. Selection for Structural Component Detection

As shown in Section 2.2, when the accuracy meets certain baselines, deep learning approaches are continuously developed for the speed of detection. Besides, in some application scenarios of object detection, it is noted that researchers prefer to use deep learning algorithms with a simple structure and faster detection speed. In particular, YOLO and improved algorithms based on YOLO are widely applied. YOLO is an end-to-end model that directly predicts the location of bounding boxes and the class probabilities of objects from the original image. Due to this concise and straightforward detection process, the detection speed of YOLO is extremely fast and the object in video can be detected in real-time. Meanwhile, comparing the two-stage object detection approaches like R-CNN, Fast R-CNN and Faster R-CNN, YOLO has less background errors since it trains on the whole image, which effectively helps to acquire contexture information about the target object. In addition, YOLO has characteristics of quick convergence and strong generalization ability.

Furthermore, structural components such as beams and columns in structural drawings have several characteristics like small size, high similarity and less features. When detecting these components using a deep learning-based method, a deep convolutional neural network is needed to form more abstract features to represent the location and category information. Moreover, a structural drawing always contains hundreds of components, and there are dozens of such drawings in a construction project. As a result, the detection method is needed to meet the precision for object detection as well as the speed nearly up to real-time level.

Therefore, according to the analysis mentioned above, YOLO is recommended to use for detecting structural components in scanned CAD drawings.

3. Methodology

Figure 1 illustrates the overall process of proposed YOLO-based component detection method. First of all, the images of 2D structural drawings are collected and classified into five classes: grid reference, column, horizontal beam, vertical beam, and sloped beam. Then a series of image processing operations are performed to decrease the noises and improve the quality of images for the following detection step. After image preprocessing, dataset management is carried out to label and augment image data. Next, images in an augmented dataset are utilized to train YOLO, and the test set is applied to evaluate the detection performance of structural components. Finally, the detection results are output to a file in TXT format, which includes information about the classification and localization of components. The following provides a detailed explanation for the main steps in the proposed method.

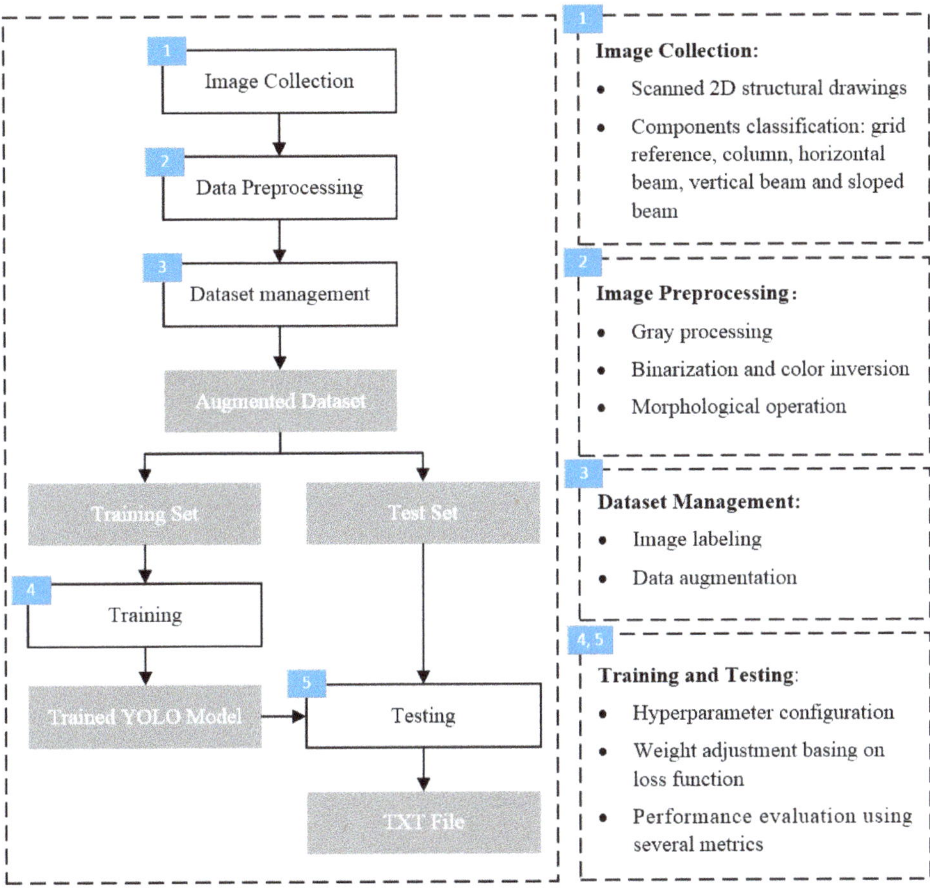

Figure 1. Overall process of YOLO-based component detection method.

3.1. Image Preprocessing

It can be seen in Figure 2a that components in structural drawing images mainly have two characteristics: (1) a single component only occupies a small proportion of the entire image, and (2) these structural components, which are generally composed of lines or arcs, do not differ significantly from each other. As a result, most of structural components in 2D drawings lack strong and distinctive features. Besides, since the images utilized in this study are generated by scanning paper drawings, noises would be produced during this process, which may have a bad effect on the performance of detection. In addition, the images of scanned structural drawings normally contain three image channels of red, green and blue. Training YOLO through these raw images directly without any processing will result in a large amount of computational consumption. To overcome these problems, a series of image preprocessing steps are carried out to reduce the data amount and improve images quality used in subsequent detection steps. The main processes include: gray processing, binarization and color inversion, and morphological operation.

3.1.1. Gray Processing

Gray processing is used to convert color images into grayscale images. In RGB color mode, the color of each pixel is co-determined by the R, G, and B components and each component may take a value between 0 and 255. When R = G = B, the resulting color is a gray color, and the value of R = B = G is called grayscale value. At this time, only 1 byte is needed for each pixel to store the grayscale value. Same as the value range of the three components, the grayscale value of each pixel varies from 0 to 255 and represents the different intensity of light. In general, image gray processing includes three common-used methods [50]: average method, weighted average method and maximum method. In this paper, the weighted average method is utilized to calculate the grayscale value for each pixel according to Equation (1). The effect of gray processing for the structural drawing image is shown in Figure 2b.

$$F = 0.2989R + 0.5870G + 0.1140B, \tag{1}$$

3.1.2. Binarization and Color Inversion

Binarization is the process of setting the grayscale value of pixels to 0 or 255, so that the image only shows black or white color. The mathematical expression of this process is shown as follows:

$$G(x,y) = \begin{cases} 255 & f(x,y) \geq T \\ 0 & f(x,y) < T \end{cases}, \tag{2}$$

where f (x, y) represents the original grayscale value of a pixel in structural drawing image, T is the threshold value, and G (x, y) denotes the grayscale value of a pixel after binarization. If the grayscale value f (x, y) is smaller than a given threshold T, the grayscale value of this pixel is set to 0; otherwise, the value changes to 255.

Furthermore, Color inversion is performed once the grayscale image has been converted into a binary image. If the original grayscale value of a pixel is 255, its grayscale value is reset to 0. Otherwise, the grayscale value is reset to 255.

The processes of image binarization and color inversion further reduce the amount of data in the image as well as the interference of background. Consequently, features of structural components will be highlighted. The effect of binarization and color inversion for a structural drawing image is shown in Figure 2c.

3.1.3. Morphological Operation

The aforementioned steps have reduced the data size and alleviate the influence of image background on final detection performance. However, features of components themselves are still weak since lines or arcs that normally make up a structural component in drawings are thin. This has become the main question in this step.

To address this issue, a morphological processing technology called dilation was applied to deal with the semi-finished images of structural drawings. Dilation expands the white part of the image so that the new image has larger bright regions. In other words, structural components in images will be thicker than before. It is assumed that A is an image of structural drawings to be processed, and B is the core used to dilate A. The dilation equation is defined as follows [51]:

$$A \oplus B = \cup\{A + b : b \in B\}, \tag{3}$$

where $\oplus$ represents dilation operation.

The final image after dilation is shown in Figure 2d. Dilation makes it easier for YOLO to extract features from scanned structural drawings and thus improves the performance of component detection. Furthermore, semantic information corresponding to the structural components in the images is also dilated, which promotes the robustness of subsequent extraction of semantic information from 2D structural drawings by using Optical Character Recognition (OCR) technique.

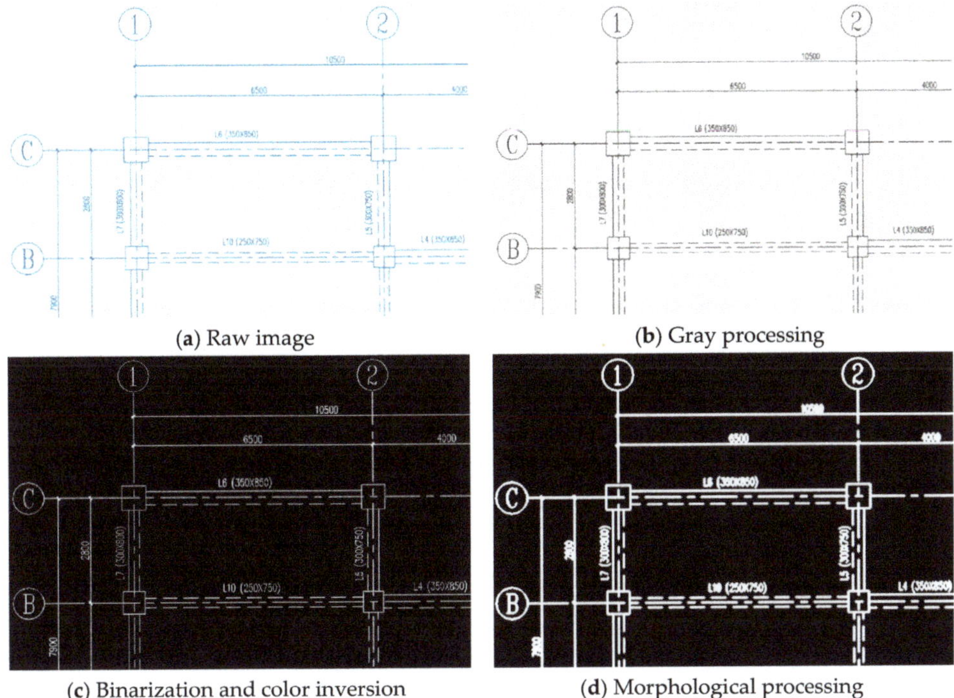

(**a**) Raw image   (**b**) Gray processing

(**c**) Binarization and color inversion   (**d**) Morphological processing

Figure 2. Image preprocessing results in different phases.

3.2. YOLO for Structural Components Detection

When YOLO is utilized to detect components in 2D structural drawings, images of these drawings are input and divided into N × N grids firstly, and then each grid cell is responsible for three types of predictions (as shown in Figure 3): (1) pixel coordinates and size of two bounding boxes, (2) confidence of two bounding boxes, and (3) the probabilities that a cell contained five classes of components. Finally, these predictions are output as a N × N × (5 × 2 + 5) tensor.

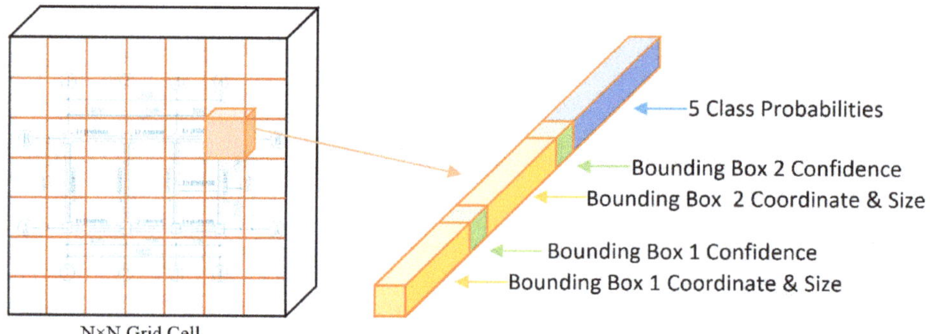

Figure 3. The prediction information contained in each grid cell.

3.2.1. Architecture of YOLO

As shown in Table 1, the neural network of YOLO used in this paper has 24 convolution layers with different core sizes and two fully connected layers, and the max-pooling technique is applied to extract the main characteristics as well as to reduce the computational complexity [45].

Table 1. Architecture of YOLO.

Layer	Name	Filter	Kernel Size/ Stride	Layer	Name	Filter	Kernel Size/ Stride
0	Covn.1	64	$7 \times 7 / 2$	15	Conv.13	256	$1 \times 1 / 1$
1	Maxp.1	/	$2 \times 2 / 2$	16	Covn.14	512	$3 \times 3 / 1$
2	Covn.2	192	$3 \times 3 / 1$	17	Covn.15	512	$1 \times 1 / 1$
3	Maxp.2	/	$2 \times 2 / 2$	18	Covn.16	1024	$3 \times 3 / 1$
4	Covn.3	128	$1 \times 1 / 1$	19	Maxp.4	/	$2 \times 2 / 2$
5	Covn.4	256	$3 \times 3 / 1$	20	Covn.17	512	$1 \times 1 / 1$
6	Covn.5	256	$1 \times 1 / 1$	21	Covn.18	1024	$3 \times 3 / 1$
7	Covn.6	512	$3 \times 3 / 1$	22	Covn.19	512	$1 \times 1 / 1$
8	Maxp.3	/	$2 \times 2 / 2$	23	Covn.20	1024	$3 \times 3 / 1$
9	Covn.7	256	$1 \times 1 / 1$	24	Covn.21	1024	$3 \times 3 / 1$
10	Covn.8	512	$3 \times 3 / 1$	25	Covn.22	1024	$3 \times 3 / 2$
11	Covn.9	256	$1 \times 1 / 1$	26	Covn.23	1024	$3 \times 3 / 1$
12	Covn.10	512	$3 \times 3 / 1$	27	Covn.24	1024	$3 \times 3 / 1$
13	Covn.11	256	$1 \times 1 / 1$	28	Conn.1	/	/
14	Covn.12	512	$3 \times 3 / 1$	29	Conn.2	/	/

Convolution layer: The purpose of convolutional layers is to extract features from structural drawing images by convolving the input data with convolutional cores of fixed size and to produce feature maps as an output. According to Table 1, the convolutional layers of YOLO in this paper include 3×3 and 1×1 convolutional cores. The 3×3 convolutional cores are used for convolution, while the 1×1 cores are used to reduce the number of input channels, decrease the parameters generated in the neural network, and lighten the computation burden when training YOLO.

Max-pooling layer: The max-pooling layer segments the feature map according to the spatial position of the feature matrix in the feature map. The maximum value obtained from each segment area is then used as a new eigenvalue. Table 1 reveals that the filter of max pooling in this paper is 2×2 with stride 2. It means that the maximum values, taken respectively in each segment area, with 2 strides, will be as the new eigenvalue, and the new feature map is generated based on these eigenvalues. After max-pooling operation, the new feature map is reduced by four times while the important information is remained in the former feature map. In addition, the image is abstracted to a higher level, which allows the following convolutional layers to extract features from it at multiple scales.

Fully connected layer: The fully connected layer mainly deals with the dimensionality reduction problem, which turns the input 2D feature matrices into feature vectors, to facilitate the prediction and classification of structural components in the output layer. In the neural network of YOLO, two fully connected layers are connected to the feature map, which has undergone the process of convolution and max pooling to generate an output tensor.

Moreover, in order to increase the nonlinearity of each layer of the neural network in YOLO, a linear activation function is applied for the final fully connected layer, while the leaky rectified linear activation function [45] is utilized for the rest of the layers and is shown as:

$$\varnothing(x) = \begin{cases} x, & x > 0 \\ 0.1x & x \le 0 \end{cases},\tag{4}$$

3.2.2. Loss Function

In order to keep a balance among coordinate predictions, category predictions and confidence predictions when training YOLO, an improved sum-squared error (SSE) method [45] is applied in this study to optimize the loss function. The specific loss function is shown as follows:

$$\begin{aligned}
&\lambda_{coord} \sum_{i=0}^{S^2} \sum_{j=0}^{B} \mathbb{1}_{ij}^{obj}\left[(x_i - \hat{x}_i)^2 + \left(y_i - \hat{y}_i\right)^2\right] \\
&+\lambda_{coord} \sum_{i=0}^{S^2} \sum_{j=0}^{B} \mathbb{1}_{ij}^{obj}\left[(\sqrt{w_i} - \sqrt{\hat{w}_i})^2 + (\sqrt{h_i} - \sqrt{\hat{h}_i})^2\right] \\
&+ \sum_{i=0}^{S^2} \sum_{j=0}^{B} \mathbb{1}_{ij}^{obj}(C_i - \hat{C}_i)^2 \\
&+\lambda_{noobj} \sum_{i=0}^{S^2} \sum_{j=0}^{B} \mathbb{1}_{ij}^{noobj}(C_i - \hat{C}_i)^2 \\
&+ \sum_{i=0}^{S^2} \mathbb{1}_{i}^{obj} \sum_{c \in classes} (p_i(c) - \hat{p}_i(c))^2,
\end{aligned}\tag{5}$$

where λ_{coord} denotes the weight of the localization error while λ_{noobj} is the weight of the classification error when the bounding box contain no objects. (x, y, w, h) reveals pixel coordinate information and size information of the target's bounding box appearing in grid i, and $(\hat{x}, \hat{y}, \hat{w}, \hat{h})$ represents the same information for the actual object in grid i. $\mathbb{1}_{i}^{obj}$ indicates that there is an object in the jth bounding box of the grid i, while $\mathbb{1}_{ij}^{noobj}$ means no object exists in the same bounding box. B denotes the number of bounding boxes predicted in each grid, and S^2 grid denotes the number of grids partitioned by the input image. C_i is a component class predicted by grid i, and $\hat{C}_i$ is the actual class of the component in the image.

There are five parts in Equation (5), the first two parts denote the localization error about the center coordinate and the width and height of the bounding box. The third part calculates the loss of confidence of the bounding box containing objects, while the fourth part obtains the loss without objects. The final part represents the conditional probability error for the class of objects.

3.3. Detection Result Export

According to Section 3.1, it can be seen that each grid cell predicts five probabilities and two candidate bounding boxes. To acquire the most possible class and its corresponding bounding box from the grid cell, non-maximal suppression (NMS) algorithm [52] is applied. NMS algorithm suppresses non-maximal elements and acts as a local maximum search. As a result, the bounding box with the highest class-specific confidence score will be extracted.

However, since the detection results obtained above are generated in the code, they cannot be utilized directly following researches of semantic information matching and BIM model reconstruction of existing buildings. Moreover, some of the data in detection results like the width and height of bounding box are useless in the subsequent study, applying these results without any change will result in an increase of output. Therefore, to optimize the output data, a file in txt format is exported in this study to record the detection results of structural components. As shown in Figure 4, four types of information are required to contain in this txt file: component ID, class of detection, pixel coordinate, and confidence of bounding box.

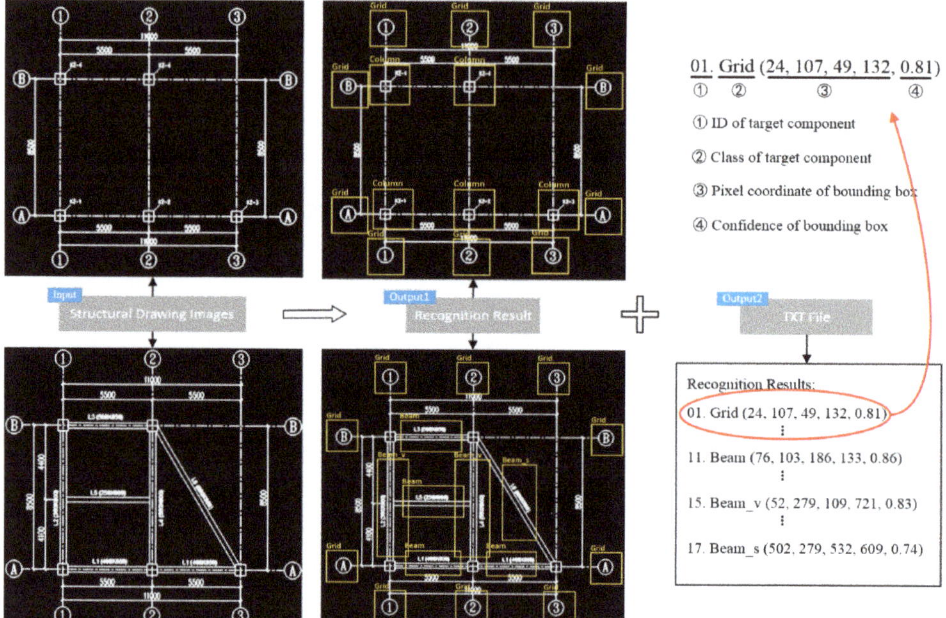

Figure 4. Example of output for component detection result.

4. Experiment and Results

4.1. Data Preparation

To identify structural components from 2D drawings, a large number of images with class labels are necessary to train and test the YOLO model. However, there is no public and general-purpose image dataset of structural drawings with multiple labeled classes in the field of object detection so far. Therefore, this study collected and established a dataset consisting of 500 images of 2D structural drawing with a fixed resolution of 850 × 750. Two types of structural drawing images were included in this dataset: the column layout plan image (CLPI) and the framing plan image (FPI). Each image was obtained by scanning a paper-based structural drawing. The components to be recognized in the image were divided into five classes: grid reference, column, horizontal beam, vertical beam, and sloped beam. Here grid reference refers to the corresponding numbers of grids, and the reason for regarding the grid reference as one of the structural components is that the successful recognition of the grid reference will be beneficial to the establishment of component coordinate systems in the future, which plays a significant role in matching the topological and semantic information of structural components. Moreover, the names of three types of beams are used to represent the different placement directions of them on 2D structural drawings. For example, if a beam spans horizontally, it will be regarded as a horizontal beam. Otherwise it will be considered as a vertical beam or sloped beam.

4.1.1. Image Labeling

Once images in the dataset had been completed all the operations of image preprocessing mentioned above, the structural components in the images were labeled based on their classes. In this paper, LabelImg [53] was used to label the scanned 2D drawings. The components to be annotated were selected with a box respectively, and then tags were assigned to the corresponding boxes. Finally, the labeled images were output as XML format. To increase the features of structural components, which have a significant influence on the final results of detection, when boxing the components, the semantic information corresponding to the components was also selected. Taking beam as an example, correct and incorrect labeling on the image of the structural drawing are shown in Figure 5.

(**a**) Incorrect labeling   (**b**) Correct labeling

Figure 5. Labeling components in the image of structural drawings.

4.1.2. Data Augmentation

Based on Section 3.2.1, it can be inferred that the neural network of YOLO used in this study has a multi-layer structure that contains millions of parameters. To make these parameters work well, a great deal of data is needed to be trained. Otherwise, overfitting may occur during the training process and then there will be a poor performance of component detection on the new image. Therefore, to avoid the overfitting problem when training on a small dataset and improving the generalization ability of the YOLO model, image augmentation technique was implemented in this study to increase the size of the dataset by randomly cropping and translating images in the original dataset. A total of 1000 new images were generated in this phase, and all original and newly generated images were mixed together to form an augmented dataset. After that, this augmented dataset was used for training and testing to prevent classification bias in the process of structural component detection.

4.2. Experiment Design

4.2.1. Experiment Environment and Strategy

The experiment was carried out on a PC with a processor of Intel Core (TM) i7-6700, RAM of 32 GB and GPU of NVIDIA GeForce GTX 1080 Ti. In addition, Python 2.7.15 was utilized as programming language on the operating system of Ubuntu 16.04, and structural component detection was performed using the deep-learning framework of Tensorflow-gpu 1.1.0. Moreover, CUDA 8.0 was applied to make GPU acceleration, and OpenCV [54] was used for image pre-processing and image visualization during the process of training and testing.

When training YOLO in this article, a weight decay of 0.0005 was applied, and 0.9 for the momentum. Moreover, epoch and batch size were manually set to 160 and 32, respectively. The learning rate was initially set to 0.01 since a high initial learning rate made the neural network of YOLO converge faster. This learning rate maintained for 60 epochs and was adjusted to 0.001, and then 0.0001 after 90 epochs. The reason to decrease the learning rate with the process of training is that it is beneficial for fine-tuning of parameters in YOLO.

4.2.2. Dataset Division

In this paper, the augmented dataset was selected to train and test the YOLO model, and the detection results were analyzed (Section 4.3). Meanwhile, the same operations were performed on the original dataset as a comparison to study the effect of dataset size on the detection accuracy (Section 5.1).

To prevent overfitting, the original and augmented datasets were divided into two parts, respectively. Eighty percent of each dataset were used to train the YOLO model while the remaining 20% was applied as a test set to evaluate the performance of the trained neural network of YOLO. The numbers of training and testing images in each dataset are shown in Table 2.

Table 2. The number of training and testing images in original and augmented dataset.

Dataset	Total	Training	Testing
Original	500	400	100
Augmented	1500	1200	300

4.2.3. Metrics for Performance Evaluation

In order to quantify the detection results of structural components in scanned 2D drawings based on YOLO, several commonly used metrics in the field of object detection such as precision (Pre.), recall (Rec.), missing rate (MR), F1 score (FS), overall accuracy (OA), and actual accuracy (AA) were utilized in this article. For one class of structural components, Pre. represents the ratio of the number of correct predictions to the total number of predicted components belonging to this class, while Rec. reveals the proportion of the number of correct predictions to the actual number of detected components in this class no matter whether the classification is right or not. These two metrics represent the discrimination ability of YOLO to negative samples and identification ability to positive samples, respectively. Normally, the detection results are expected with high Pre. and Rec. However, in fact, the components in scanned structural drawings in this work are not classified equally. For instance, the number of columns in CLPI is much more than other components, and the same thing happens for beams in FPI. This inequality may lead to the situation where one of these two metrics is high and another is low based on their definitions. Hence, FS was also adopted to evaluate detection performance of structural components. FS can be regarded as a weighted average of Pre. and Rec. It simultaneously accounts for both metrics and ranges between 0 and 1. The higher the FS is, the better the detection performance of YOLO will be. OA is the number of correct predictions to all predictions for this certain class of components. In contrast, AA stands for the number of correct predictions to the actual number of components that belong to this class. The definitions of these metrics are as follows:

$$Precision\ (Pre.) = \frac{TP}{TP + FP}, \tag{6}$$

$$Recall\ (Rec.) = \frac{TP}{TP + FN}, \tag{7}$$

$$Missing\ Rate\ (MR) = \frac{FN}{TP + FN}, \tag{8}$$

$$F1\ Score\ (FS) = 2 \cdot \frac{Precision \times Recall}{Precision + Recall} = \frac{2TP}{2TP + FP + FN}, \tag{9}$$

$$Overall\ Accuracy\ (OA) = \frac{TP}{TD} = \frac{TP}{TP + FP + FN}, \tag{10}$$

$$Actural\ Accuracy\ (AA) = \frac{TP}{TA}, \tag{11}$$

where TP, FP and FN represent the number of true-positive, false-positive and false-negative detections for each class of components, respectively. Taking columns in CLPI as an example, according to the definitions of TP, FN and FP (shown in Table 3), TP is the number of columns that identifies correctly, FN is the number of columns that misidentifies as other types of components, and FP is the number of non-column components that misidentifies as columns. The reason why TN is not defined in this article is that for the detection process of a certain class of structural component, all regions that do not contain this type of component were negative, so it is meaningless to measure TN. TD is the total number of detections for this kind of component, while TA represents the actual number of components belonging to this class.

Table 3. Definitions of TP, FP and FN for component detection.

Type	Predicted Objects	Actual Objects
TP	√	√
FP	√	×
FN	×	√

4.3. Experiment Results

The detection performance for each type of component is shown in Table 4. It can be seen that grid references have the best results with Pre. of 86.41%, Rec. of 99.16% and FS of 92.35%. The reason for this situation is that grid references appear most frequently in both CLPIs and FPIs, and their features are more distinctive than other components. On the other hand, horizontal beams are identified with the lowest Pre. of 82.67%, Rec. of 91.18% and FS of 86.71% as well as the highest MR of 8.82%. It demonstrates that beams, especially referring to those placing horizontally on 2D drawings, cannot be accurately recognized by YOLO in some cases. For example, when the aspect ratio of the beam is too large, the horizontal beam is often neglected to detect. When the aspect ratio is small, this type of beam is easily identified as a vertical beam. A horizontal beam tends to be identified as a column if its aspect ratio is close to 1. In addition, it also can be observed from Table 4 that the Pre. of all components is lower than their Rec. According to the definitions of precision and recall (Section 4.2.3), this result can be attributed to the fact that FP is always greater than FN for each type of component. In other words, other components are frequently misidentified as this kind of component during the process of object detection.

Table 4. Detection performance for different objects in test set.

	TP	FP	FN	Pre.	Rec.	MR	FS
Grid	2967	467	25	86.41%	99.16%	0.84%	92.35%
Column	833	158	17	84.03%	98.04%	1.96%	90.50%
Beam	517	108	50	82.67%	91.18%	8.82%	86.71%
Beam_v	500	100	42	83.33%	92.31%	7.69%	87.59%
Beam_s	125	17	8	88.24%	93.75%	6.25%	90.91%

Note: Grid, Column, Beam, Beam_v and Beam_s in the table denote grid references, columns, horizontal beams, vertical beams and sloped beams, respectively.

In addition to the aforementioned metrics, OA, AA and speed of component detection by YOLO are also evaluated in this experiment. It can be seen from Table 5 that grid references and vertical beams have the highest OA and AA, while horizontal beams perform the worst in these two metrics. Moreover, the total averages of OA and AA for the whole image rather than for one type of component are calculated as comprehensive performance indexes to measure the detection results of the image when using YOLO. Since large differences between the total amount of each class exist (for instance, the number of grid references is about 20 times than that of sloped beams), weighted average method instead of arithmetic average method is applied in this study to obtain these two image-level accuracies. The equation for calculating these two metrics is $(x_1 N_1 + x_2 N_2 + \ldots + x_k N_k)/k$, where x_i ($i = 1, 2 \ldots k$) represents the OA or AA for each class of component, and N_j ($i = 1, 2 \ldots k$) denotes the actual total number component of each class. Referring to Table 5, the total average of OA reaches 83.31%, with 81.25% of the components in image being correctly recognized. Also, the average computing time for an image is 0.71 s. These results demonstrate that the proposed YOLO-based method has the ability to detect multiple structural components from structural 2D drawings with comparatively high accuracy and a short detection time.

Table 5. TD, TA, OA, AA, and speed for different objects in test set.

	TD	TA	OA	AA	Speed
Grid	3459	3125	85.78%	82.67%	-
Column	1008	925	82.64%	80.18%	-
Beam	675	583	76.54%	77.14%	-
Beam_v	642	583	77.92%	78.57%	-
Beam_s	150	142	83.33%	85.00%	-
Average	2338	2113	83.31%	81.25%	0.71 s

5. Discussion

During the process of component detection with YOLO, many factors, such as the size of dataset and the number of components in images, may have a significant impact on final detection results. Therefore, several accuracy-influencing factors are discussed in this section. Moreover, the errors generated in this research are also analyzed.

5.1. Accuracy-Influencing Factors Analysis

5.1.1. The Size of Dataset

Generally, in the field of object detection, with the increase of the training dataset, the effect of detection will be improved. This is because the neural network will constantly adjust the weights in the hidden layer during the training process. The larger the dataset is, the smaller the error of the result will be as well as the risk of overfitting, and the closer the detection result will be to the truth. In this paper, the detection effect of structural components using YOLO is tested under the original dataset and the augmented dataset. Since each image in each different dataset contains a different number of structural components, for a better understanding, the weighted averages of several metrics (like Pre., Rec., MR, FS, OA and AA) are calculated to measure the detection performance of YOLO in a different dataset.

According to Figure 6, it is obviously observed that the YOLO-based method trained by the augmented dataset receives much better results with Pre. of 85.30%, Rec. of 97.21% and FS of 90.86%, and OA and AA are improved by 5.79% and 5.32%, respectively. Meanwhile, MR decreases from 5.89% to 2.79%. Therefore, the thorough improvements using the augmented dataset rather than the original dataset demonstrate that the size of the dataset does have a marked impact on the performance of object detection. When using more data for training YOLO, the better the result of structural object detection will be obtained.

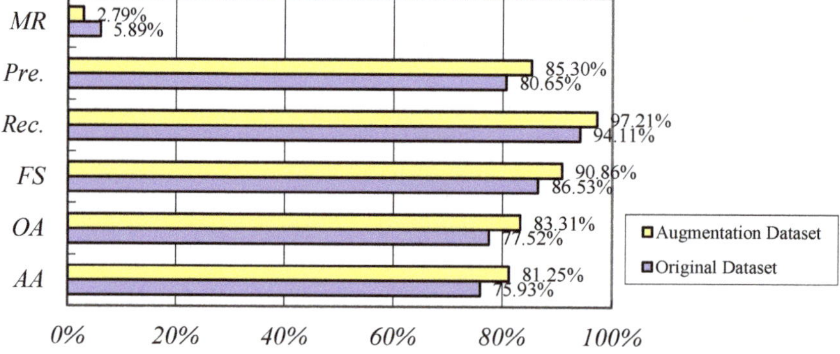

Figure 6. Comparison of the performance in different sizes of datasets via YOLO.

5.1.2. The Number of Components in the Image

Another factor that affects the accuracy of the structural component detection is the number of components to be detected in the image of 2D drawings. With the fixed resolution of the input image, if more components are contained in an image, each component will occupy a smaller pixel size and have less features. Consequently, the probability of missing or false detection will increase when classifying the structural components based on YOLO. To verify the relationship expounded above, in this paper, 300 images of structural drawings used for testing are classified into CLPIs and FPIs, and the total number of components in each image is counted. After that, the average of OA and AA for each image are calculated based on FP, TP, FN, and total number of components. The relationships between the number of structural objects and the detection accuracy in CLPI and FPI are shown in Figure 7.

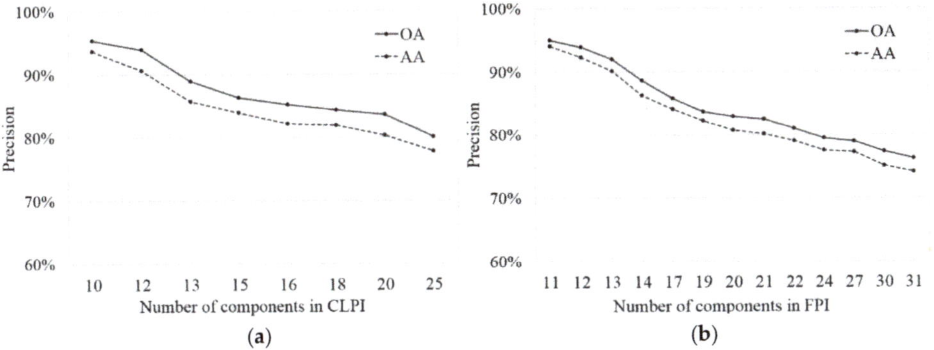

Figure 7. The relation between the number of structural components and the detection accuracy: (**a**) In the column layout plan image; (**b**) In the framing plan image.

In Figure 7a, the total numbers of structural components of each CLPI in the test set range from 10 to 25. Meanwhile, the average OA and AA of CLPI decrease with the number of components increasing. The maximum average OA of 95.42% and AA of 93.67% are obtained in CLPI with 10 components, while the minimum average OA of 80.07% and AA of 77.81% are acquired in the image with 25 components. The same trend also happens with FPI, as shown in Figure 7b. Moreover, it is distinctly noticed that the value of average OA (95.01%) is nearly equal to the value of average AA (94.06%) when the number of structural objects in FPI is small. As the number of components increases, the declining trend of AA is bigger than that of OA. The main reason for this inconsistency, based on the definitions of OA and AA explained in Section 4.2.3, is that more missing errors are produced during the detection process rather than classification errors. Therefore, based on the aforementioned analysis, it can be concluded that the number of structural components does have a significant impact on the accuracy of structural component detection.

5.1.3. Other Factors

Other factors, such as the hyperparameter of YOLO and the size of image resolution, also have effects on the detection result of structural components. Hyperparameter refers to the parameter that is preset based on the experience before training rather than obtained through training. The selection of the hyperparameter has a significant influence on the detection results. For instance, the batch size, which is one of the most common hyperparameters, represents the number of samples sent into the model of the neural network in each training. Generally, the smaller batch size is prone to longer training time and non-convergence of the model, while the larger batch size makes the model converge faster, but easily results in the out-of-memory error or program crash. Therefore, choosing the appropriate batch size within a reasonable range can achieve the best balance between time and accuracy when training the YOLO model. Moreover, raising the size of the input image may also affect the accuracy of the structural component detection based on YOLO. This is because when the size of image resolution increases, the number of pixels occupied by a single component will also increase. In consequence, the image features of each component will be richer than before. A disadvantage of raising the image size is that it greatly increases the computing workload as well as the time for training.

5.2. Error Analysis

In order to further analyze the recognition results of the structural components, the errors resulted from the proposed YOLO-based method are divided into five categories based on their types: localization error, background error, similarity error, classification error, and missing error [55]. Localization error indicates that the classification result is correct but the location of the bounding box is biased and not rightly enclosing the structural component. On the contrary, classification error happens when the bounding box locates at the right place while the target object is detected in other categories. Background error means that a bounding box appears in background areas without any component. Similarity error occurs when two or more bounding box exist for one target object, and missing error refers to the situation where the structural components in the image of 2D drawings are not detected. The causes of different types of errors vary from each other, but most of them are closely linked to the selected neural network. For instance, in YOLO used in this paper, each grid predicts two bounding boxes to detect the structural components whose coordinates of central position fall within the grid area. This may lead to missing error if the centers of multiple components fall in the same grid.

Similar to the method mentioned in Section 4.3, the weighted average method was adopted to calculate the average proportion of all kinds of errors across five classes. As shown in Table 6, when using YOLO for structural component detection, 69.54% of the images are correctly recognized, while 30.46% of the detections have errors. It is also obviously noticed that localization error is the largest of all errors and is about five times the total amount of classification error, which indicates that YOLO has problems with locating the identified components correctly. In addition, 8.03% of detections have background error, and 5.65% and 4.43% of detections have the problems of multi-detection and omission of bounding boxes, respectively.

Table 6. Proportion of different types of errors across five classes of components.

	Correct	Localization	Background	Similarity	Classification	Missing
Average	69.54%	10.09%	8.03%	5.65%	2.26%	4.43%

6. Conclusions

In recent years, BIM has been continuously regarded as a core tool to operate and manage the as-built and as-is buildings. However, the lack of BIM models in existing buildings limits BIM applications in O&M phase. Building 3D models based on 2D engineering drawings is regarded as a feasible way to reconstruct BIM models. As the first step of this kind of method, the accuracy of component detection from 2D drawings and the quality of extraction of corresponding geometric information have a significant impact on the final results of reconstruction. Therefore, in this paper, a novel component detection method based on YOLO was proposed to quickly and accurately identify structural components in scanned structural drawings. The experimental results show that Pre. and Rec. for all five classes are above 80% as well as the average of OA and AA for the entire image. Moreover, the average time for identifying an image is only 0.71 s. All these results strongly prove the feasibility and potential of the proposed YOLO-based method for recognizing components from 2D structural drawings.

The main contribution of this study is that it opens up a new way to detect structural components for reconstructing a BIM model of existing buildings from paper-based 2D drawings. Also, it is of important referential value for the application of the deep learning-based component detection method in other professions like architecture and mechanical systems. The second contribution is that it clearly certifies the advantages of the proposed method over traditional component recognition methods. Compared with the existing approaches mentioned in Section 2.1, YOLO-based method not only guarantees the accuracy of components detection, but also has more advantages in detection speed and cost. What is more, the method proposed in this paper has the trait of universality due to its powerful learning ability, which means it can detect components in scanned 2D drawings generated in different countries and design conventions if the training data is sufficient. The final contribution of this study is that several optimal experiment settings have been found for better performance of the proposed YOLO-based method, though there is still room for improvement. It offers some references for the following researchers to improve the work mentioned in this paper.

Additionally, the method proposed in this study can be further improved. The future research work is concluded as follows:

- The experiment of component detection in this paper is focused on limited structural components (e.g., beam and column). To further improve the proposed method, more structural elements (such as floor, shear wall and pile) need to be tested in future research.
- More advanced YOLOv2 with batch normalization, high resolution classifier, anchor box, and multi-scale training will be tried. In addition, more sensitivity analysis will be conducted to study the influence of various accuracy-influencing factors on the final recognition results.
- Last but not least, as the first step to reconstruct BIM models for existing buildings, this study only solves the problem of extracting geometric information of structural components. How to extract topological and semantic information and match with the corresponding components will be the key research direction in the future.

Author Contributions: Conceptualization, Y.Z. and X.D.; methodology, Y.Z.; validation, Y.Z. and H.L.; writing—original draft preparation, Y.Z.; writing—review and editing, Y.Z.; visualization, Y.Z.; supervision, X.D. and H.L.; project administration, X.D.; funding acquisition, X.D. All authors have read and agreed to the published version of the manuscript.

Funding: The research work was supported by the National Key Research and Development Program of China during the 13th Five-year Plan (No. 2016YFC0702001) and Project Funded by China Postdoctoral Science Foundation (No. 2019M663115).

Conflicts of Interest: The authors declare no conflict of interest.

References

1. Volk, J.; Stengel, J.; Schultmann, F. Building information modeling (BIM) for existing buildings—Literature review and future needs. *Autom. Constr.* **2014**, *38*, 109–127. [CrossRef]
2. Becerik-Gerber, B.; Jazizadeh, F.; Li, N.; Calis, G. Application areas and data requirements for BIM-enabled facilities management. *J. Constr. Eng. Manag.* **2012**, *138*, 431–442. [CrossRef]
3. Akcamete, A.; Akinci, B.; Garrett, J.H. Potential utilization of building information models for planning maintenance activities. In Proceedings of the 13th International Conference on Computing in Civil and Building Engineering, Nottingham, UK, 30 June–2 July 2010; pp. 8–16.
4. Hu, Z.Z.; Peng, Y.; Tian, P.L. A review for researches and applications of BIM-based operation and maintenance management. *J. Graph.* **2015**, *36*, 802–810.
5. Gimenez, L.; Hippolyte, J.; Robert, S.; Suard, F.; Zreik, K. Review: Reconstruction of 3d building in-formation models from 2d scanned plans. *J. Build. Eng.* **2015**, *2*, 24–35. [CrossRef]
6. Li, J.; Huang, W.; Shao, L.; Allinson, N. Building recognition in urban environments: A survey of state-of-the-art and future challenges. *Inf. Sci.* **2014**, *277*, 406–420. [CrossRef]
7. Tang, P.B.; Huber, D.; Akinci, B.; Lipman, R.; Lytle, A. Automatic reconstruction of as-built building information models from laser-scanned point clouds: A review of related techniques. *Autom. Constr.* **2010**, *19*, 829–843. [CrossRef]
8. Cho, C.Y.; Liu, X. An automated reconstruction approach of mechanical systems in building infor-mation modeling (BIM) using 2d drawings. In Proceedings of the ASCE International Workshop on Computing in Civil Engineering 2017, Seattle, WA, USA, 25–27 June 2017; pp. 236–244.

9. Gimenez, L.; Robert, S.; Suard, F.; Zreik, K. Automatic reconstruction of 3d building models from scanned 2D floor plans. *Autom. Constr.* **2016**, *63*, 48–56. [CrossRef]

10. Huang, H.C.; Lo, S.M.; Zhi, G.S.; Yuen, R.K.K. Graph theory-based approach for automatic recognition of cad data. *Eng. Appl. Artif. Intell.* **2008**, *21*, 1073–1079. [CrossRef]

11. Yin, X.; Wonka, P.; Razdan, A. Generating 3d building models from architectural drawings: A survey. *IEEE Comput. Graph. Appl.* **2008**, *29*, 20–30. [CrossRef]

12. Klein, L.; Li, N.; Becerik-Gerber, B. Imaged-based verification of as-built documentation of operational buildings. *Autom. Constr.* **2012**, *21*, 161–171. [CrossRef]

13. Dominguez, B.; Garcia, A.L.; Feito, F.R. Semiautomatic detection of floor topology from CAD architectural drawings. *Comput. Aided Des.* **2012**, *44*, 367–378. [CrossRef]

14. Lu, Q.C.; Lee, S.; Chen, L. Image-driven fuzzy-based system to construct as-is IFC BIM objects. *Autom. Constr.* **2018**, *92*, 68–87. [CrossRef]

15. Hough, P.V.C. Method and Means for Recognizing Complex Patterns. U.S. Patent 3,069,654, 18 December 1962.

16. Lowe, D.G. Object recognition from local scale-Invariant features. In Proceedings of the Seventh I-EEE International Conference on Computer Vision, Kerkyra, Greece, 20–27 September 1999; pp. 1150–1157.

17. Matas, J.; Chum, O.; Urban, M.; Pajdla, T. Robust wide baseline stereo from maximally stable extremal regions. *Image Vis. Comput.* **2004**, *22*, 761–767. [CrossRef]

18. Bay, H.; Ess, A.; Tuytelaars, T.; Gool, L.V. Speeded-up robust features (SURF). *Comput. Vis. Image Underst.* **2008**, *110*, 346–359. [CrossRef]

19. Mukhopadhyay, P.; Chaudhuri, B.B. A survey of Hough Transform. *Pattern Recognit.* **2015**, *48*, 993–1010. [CrossRef]

20. Ballard, D.H. Generalizing the Hough Transform to detect arbitrary shapes. *Pattern Recognit.* **1981**, *13*, 111–122. [CrossRef]

21. Galambos, C.; Kittler, J.; Matas, J. Gradient based progressive probabilistic Hough Transform. *IEE Proc. Vis. Image Signal Process.* **2001**, *148*, 158. [CrossRef]

22. Xu, L.; Oja, E.; Kultanen, P. A new curve detection method: Randomized Hough transform (RHT). *Pattern Recognit. Lett.* **1990**, *11*, 331–338. [CrossRef]

23. Kiryati, N.; Lindenbaum, M.; Bruckstein, A.M. Digital or analog hough transform? *Pattern Recognit. Lett.* **1991**, *12*, 291–297. [CrossRef]

24. Izadinia, H.; Sadeghi, F.; Ebadzadeh, M.M. Fuzzy generalized hough transform invariant to rotation and scale in noisy environment. In Proceedings of the IEEE International Conference on Fuzzy Systems, Jeju Island, Korea, 20–24 August 2009; pp. 153–158.

25. Dosch, P.; Tombre, K.; Ah-Soon, C.; Masini, G. A complete system for the analysis of architectural drawings. *Int. J. Doc. Anal. Recognit.* **2000**, *3*, 102–116. [CrossRef]

26. Mace, S.; Locteau, H.; Valveny, E.; Tabbone, S. A system to detect rooms in architectural floor plan images. In Proceedings of the Ninth IAPR International Workshop on Document Analysis Systems, Boston, MA, USA, 9–11 June 2010; pp. 167–174.

27. Ahmed, S.; Liwicki, M.; Weber, M.; Dengel, A. Improved automatic analysis of architectural floor plans. In Proceedings of the 2011 International Conference on Document Analysis and Recognition, Beijing, China, 18–21 September 2011; pp. 864–869.

28. Riedinger, C.; Jordan, M.; Tabia, H. 3D models over the centuries: From old floor plans to 3D representation. In Proceedings of the 2014 International Conference on 3D Imaging, Liege, Belgium, 9–10 December 2014; pp. 1–8.

29. Lu, T.; Yang, H.F.; Yang, R.Y.; Cai, S.J. Automatic analysis and integration of architectural drawings. *Int. J. Doc. Anal. Recognit.* **2007**, *9*, 31–47. [CrossRef]

30. Lu, Q.C.; Lee, S. A semi-automatic approach to detect structural components from CAD drawings for constructing as-Is BIM objects. In Proceedings of the ASCE International Workshop on Computing in Civil Engineering 2017, Seattle, WA, USA, 25–27 June 2017; pp. 84–91.

31. Fang, Q.; Li, H.; Luo, X.C.; Ding, L.Y.; Luo, H.B.; Rose, T.M.; An, W.P. Detecting non-hardhat-use by a deep learning method from far-field surveillance videos. *Autom. Constr.* **2018**, *85*, 1–9. [CrossRef]

32. Fang, Q.; Li, H.; Luo, X.C.; Ding, L.Y.; Luo, H.B.; Li, C. Computer vision aided inspection on falling prevention measures for steeplejacks in an aerial environment. *Autom. Constr.* **2018**, *93*, 148–164. [CrossRef]

33. Zhang, A.; Wang, K.C.P.; Li, B.X.; Yang, E.; Dai, X.X.; Yi, P.; Fei, Y.; Liu, Y.; Li, J.Q.; Chen, C. Automated pixel-level pavement crack detection on 3d asphalt surfaces using a deep-learning network. *Comput. Aided Civ. Infrastruct. Eng.* **2017**, *32*, 805–819. [CrossRef]

34. Wu, D.; Pigou, L.; Kindermans, P.J.; Le, N.D.H.; Shao, L.; Dambre, J.; Odobez, J.M. Deep dynamic neural networks for multimodal gesture segmentation and recognition. *IEEE Trans. Pattern Anal. Mach. Intell.* **2016**, *38*, 1583–1597. [CrossRef] [PubMed]

35. Gangwar, A.; Joshi, A. DeepIrisNet: Deep iris representation with applications in iris recognition and cross-sensor iris recognition. In Proceedings of the IEEE International Conference on Image Processing, Phoenix, AZ, USA, 25–28 September 2016.

36. Hu, C.; Bai, X.; Qi, L.; Chen, P.; Xue, G.; Mei, L. Vehicle color recognition with spatial pyramid deep learning. *IEEE Trans. Intell. Transp. Syst.* **2015**, *16*, 2925–2934. [CrossRef]

37. Saxena, S.; Verbeek, J. Heterogeneous face recognition with CNNs. In Proceedings of the 14th European Conference on Computer Vision, Amsterdam, The Netherlands, 11–14 October 2016; pp. 483–491.

38. Kong, Y.; Ding, Z.; Li, J.; Fu, Y. Deeply learned view-invariant features for cross-view action recognition. *IEEE Trans. Image Process.* **2017**, *26*, 3028–3037. [CrossRef] [PubMed]

39. Girshick, R.; Donahue, J.; Darrell, T.; Malik, J. Rich feature hierarchies for accurate object detection and semantic segmentation. In Proceedings of the 2014 IEEE Conference on Computer Vision and Pattern Recognition, Columbus, OH, USA, 23–28 June 2014; pp. 580–587.

40. Uijlings, J.R.R.; Sande, K.E.A.; Gevers, T. Selective search for object recognition. *Int. J. Comput. Vis.* **2013**, *104*, 154–171. [CrossRef]

41. Girshick, R. Fast R-CNN. In Proceedings of the 2015 IEEE International Conference on Computer Vision, Santiago, Chile, 7–13 December 2015; pp. 1440–1448.

42. He, K.M.; Zhang, X.Y.; Ren, S.Q.; Sun, J. Spatial pyramid pooling in deep convolutional networks for visual recognition. In Proceedings of the 13th European Conference on Computer Vision, Zurich, Switzerland, 6–12 September 2014; pp. 1904–1916.

43. Ren, S.Q.; He, K.M.; Girshick, R.; Sun, J. Faster R-CNN: Towards real-time object detection with re-gion proposal networks. *IEEE Trans. Pattern Anal. Mach. Intell.* **2017**, *39*, 1137–1149. [CrossRef]

44. He, K.M.; Gkioxari, G.; Dollár, P.; Girshick, R. Mask R-CNN. In Proceedings of the 2017 IEEE International Conference on Computer Vision, Venice, Italy, 22–29 October 2017.

45. Redmon, J.; Divvala, S.; Girshick, R.; Farhadi, A. You only look once: Unified, real-time object detection. In Proceedings of the 2016 IEEE Conference on Computer Vision and Pattern Recognition, Las Vegas, NV, USA, 27–30 June 2016; pp. 779–788.

46. Liu, W.; Anguelov, D.; Erhan, D.; Szegedy, C.; Reed, S.; Fu, C.Y.; Berg, A.C. SSD: Single shot multibox detector. In Proceedings of the 14th European Conference on Computer Vision, Amsterdam, The Netherlands, 11–14 October 2016; pp. 21–37.

47. Redmon, J.; Farhadi, A. YOLO9000: Better, faster, stronger. In Proceedings of the IEEE Conference on Computer Vision and Pattern Recognition, Honolulu, HI, USA, 21–26 July 2017.

48. Russakovsky, O.; Deng, J.; Su, H.; Krause, J.; Satheesh, S.; Ma, S.; Huang, Z.; Karpathy, A.; Khosla, A.; Bernstein, M.; et al. ImageNet large scale visual recognition challenge. *Int. J. Comput. Vis.* **2014**, *115*, 211–252. [CrossRef]

49. Lin, T.Y.; Maire, M.; Belongie, S.; Hays, J.; Perona, P.; Ramanan, D.; Dollár, P.; Zitnick, C.L. Microsoft COCO: Common objects in context. In Proceedings of the 13th European Conference on Computer Vision, Zurich, Switzerland, 6–12 September 2014; pp. 740–755.

50. Yang, H.; Zhou, X. Deep learning-based ID card recognition method. *China Comput. Commun.* **2016**, *21*, 83–85.

51. Gonzalez, R.C.; Woods, R.E. *Digital Image Processing*, 3rd ed.; Pearson Prentice Hall: Upper Saddle River, NJ, USA, 2008.

52. Neubeck, A.; Gool, L.V. Efficient non-maximum suppression. In Proceedings of the 18th International Conference on Pattern Recognition, Hong Kong, China, 20–24 August 2006; pp. 850–855.

53. LabelImg. Available online: https://github.com/tzutalin/labelImg (accessed on 15 October 2019).

54. Opencv. Available online: https://opencv.org (accessed on 20 December 2019).

55. Hoiem, D.; Chodpathumwan, Y.; Dai, Q. Diagnosing error in object detectors. In Proceedings of the 12th European Conference on Computer Vision, Florence, Italy, 7–13 October 2012.

Article

Industry 4.0 for the Construction Industry—How Ready Is the Industry?

Raihan Maskuriy [1,2], **Ali Selamat** [1,3,4,5,*], **Kherun Nita Ali** [6], **Petra Maresova** [5] and **Ondrej Krejcar** [5]

[1]   Malaysia Japan International Institute of Technology (MJIIT), Universiti Teknologi Malaysia Kuala Lumpur, Jalan Sultan Yahya Petra, Kuala Lumpur 54100, Malaysia
[2]   Department of Architecture, Faculty of Design and Architecture, Universiti Putra Malaysia (UPM), Serdang 43400, Malaysia
[3]   Media and Games Center of Excellence (MagicX), Universiti Teknologi Malaysia, Skudai 81310, Malaysia
[4]   School of Computing, Faculty of Engineering, Universiti Teknologi Malaysia (UTM), Skudai 81310, Malaysia
[5]   Faculty of Informatics and Management, University of Hradec Kralove, Rokitanskeho 62, 500 03 Hradec Kralove, Czech Republic
[6]   Department of Quantity Surveying, Faculty of Built Environment and Surveying, Universiti Teknologi Malaysia (UTM), Skudai 81310, Malaysia
*   Correspondence: aselamat@utm.my

Received: 23 April 2019; Accepted: 6 July 2019; Published: 15 July 2019

Abstract: Technology and innovations have fueled the evolution of Industry 4.0, the fourth industrial revolution. Industry 4.0 encourages growth and development through its efficiency capacity, as documented in the literature. The growth of the construction industry is a subset of the universal set of the gross domestic product value; thus, Industry 4.0 has a spillover effect on the engineering and construction industry. In this study, we aimed to map the state of Industry 4.0 in the construction industry, to identify its key areas, and evaluate and interpret the available evidence. We focused our literature search on Web of Science and Scopus between January 2015 and May 2019. The search was dependent on the following keywords: "Industry 4.0" OR "Industrial revolution 4.0" AND TOPIC: "construction" OR "building". From the 82 papers found, 20 full-length papers were included in this review. Results from the targeted papers were split into three clusters: technology, security, and management. With building information modelling (BIM) as the core in the cyber-physical system, the cyber-planning-physical system is able to accommodate BIM functionalities to improve construction lifecycle. This collaboration and autonomous synchronization system are able to automate the design and construction processes, and improve the ability of handling substantial amounts of heterogeneity-laden data. Industry 4.0 is expected to augment both the quality and productivity of construction and attract domestic and foreign investors.

Keywords: Industry 4.0; construction industry; building information modeling; cyber-planning-physical system

---

1. Introduction

The world has never evolved as fast as in the last couple of decades. To provide context, the global construction industry has been affected by the world's urban population rising by 200,000 people per day [1]. The demand for affordable housing has never been higher [2], affected by a concomitant need for social, utility, and transportation infrastructure. Such challenges have ensured that the construction industry continues to review and revamp itself. The changes that occur here impact on society as a whole—the construction costs will fall, and the environment will benefit. This is achieved by efficiently using scarce resources and/or ensuring that buildings are being constructed with eco-efficiency in mind [3]. This positively impacts the economy by ensuring that the global infrastructure gap is narrowed, and economic development is boosted overall. During the last couple of decades, most industries have undergone an evolution and have instilled product and process innovations into the core of their operations. The engineering and construction sector has not kept pace in terms of technological opportunities that can help improve production and productivity, resulting in a stagnation of labor productivity as well [4]. Several internal and external challenges are responsible for this situation, including the industry dealing with consistent fragmentation, trouble recruiting a workforce with the right talent, insufficient links to contractors and suppliers, and inadequate transfer of knowledge from one project to another [5]. Despite the industry's vast potential, increasing efficacy and productivity can only result from digitalization, new techniques for construction, and innovations. Tools such as three-dimensional (3D) scanning, building information modelling (BIM), drones, and augmented reality have all reached market maturity [6]. By incorporating these innovations, firms can exploit them to increase productivity level, safety, and quality, and improve project management. To use this potential, a strategy must be devised for concerted and committed efforts across many different areas, including operations, technology, personnel, regulation, and more.

The fourth industrial revolution 4.0 or Industry 4.0 has introduced digital technologies, sensor systems, intelligent machines, and smart materials to the construction industry where BIM has become the central repository for collating digital information about a project [5,7]. BIM is an ideal stage for the development of powerful and innovative applications for the construction industry by providing an additional layer of data that are able to interact and collaborate in real time throughout the project life cycle [8]. The innovation of BIM manages computational data for improving construction efficiency and economy [9]. With open BIM, existing construction management tools can be integrated with BIM to extend its capabilities in the construction ecosystem [10]. BIM has been widely accepted in the construction industry [11] though not many firms have taken full advantage of its potential despite the investment in BIM being proven to be well worth the cost of implementation to organizations [12]. Therefore, we decided to examine the present status of BIM in the Industry 4.0 in the context of the construction industry to identify evidence on the integration of BIM with digital technologies such as intelligent machines, sensor systems, and smart materials. To achieve our goal, we conducted a bibliometric mapping study using a scoping review technique to discuss the current trend of Industry 4.0 in the construction industry to identify the patterns of existing research in the aforementioned context, identify gaps in the research, and provide suggestions for future research directions.

2. Background

Humanity has had a significant impact on the industrial landscape, which includes the construction industry [13]. The basic technological revolution elements can be divided into three aspects, material, energy power, and control technology, which determine the operation and power modes of the things designed, thereby determining the ways people feel, the mode and scale of human perception, and the method through which people cognize knowledge, so that a new living method can be constructed [14]. Prior to the 19th century, clear engineering limits defined the building weight, height, and strength due to the limited types of material such as humanmade materials along with those available in nature: timber, stone, lime mortar, and concrete [13]. During the first industrial revolution (from 1760 to about 1830), the mechanical heavy industry experienced exponential growth leading to the creation of a whole host of new building materials. These included glass, cast iron, and even steel—all of which were created by engineers and architects to create buildings no one had ever imagined could exist in terms of form, frame, and functionality [15]. The second industrial revolution (1870–1914) was characterized by dense innovation based on useful knowledge being mapped onto technologies that drove the industry with cheap and more efficient mass production of steel, electricity, telegraphs, and railroads [16]. This revolution drove the construction industry in terms of innovation in architectural design and lightening vertical space [13] alongside new prefabrication technology and the beginning of computer-aided design (CAD) [17], which provided numerous unforeseen opportunities [9].

In the late 1950s, the Internet, information technology (IT), and the availability of personal computers enabled a new digital revolution where mechanical and analogue were digitized and mass production shifted to mass customization [9,18]. The third industrial revolution created a new relationship between architecture and technology that challenged the production industry [19]. Architects started using diffused 3D computer-aided design software as a representational tool to improve precision and expand the limits of their creations [20]. Digital architecture creations require architects to use their perceptual and cognitive abilities to construct digital geometric forms in the computer program using certain underlying computational foundations with action and reaction rules [19]. The complex constructed geometries were able to be constructed using computer numerically controlled (CNC) fabrication processes, which perform the basic controlling functions over the movement of a machine tool using a set of coded instructions [19]. CNC became the connector between 3D-CAD models and computer aided manufacturing (CAM) to produce mass-customization and computerized production of building elements [21]. Complex construction became more economical and seemed unnecessarily difficult to build as the digital fabrication and digital architecture tools evolved and matured [20].

Research on Industry 4.0 is relatively a new topic. The term is well known not only in academia but also in industry [22]. By definition, Industry 4.0 or the forth industrial revolution refers to the Internet of Things (IoT) and the Internet of Services integrated with the manufacturing environment where all industrial businesses around the globe connect and control their machinery, factories, and warehousing facilities intelligently through cyber-physical systems by sharing information that triggers actions [23]. This revolution has challenged the industry by demonstrating the construction digitization potential with the availability of digital data and online digital access that can be used to automatically gather and process electronic data discrete tasks into the value chain [24]. Robotic technologies have been merged with the construction industry, known as construction automation technologies, to create elements of buildings, building components, and building furniture [25]. The integration of BIM into the IT environment enables transitioning the current 'react to event' practice to 'predict the event' practice [26]. The integration of BIM into cloud computing allows project stakeholders to collaborate in real-time from different locations to enhance decision-making and ensure project deliverability [5]. The IoT, together with BIM, is able to maximize productivity [27], enhance the information flow during a project life cycle [28], optimize energy efficiency [29], and improve security and safety [30,31], as well as the planning, managing, and monitoring of resources [31].

The number of research publications covering topics related to Industry 4.0 and construction have grown tremendously. For example, in the Scopus database, for "Industry 4.0" OR "Industrial revolution 4.0" AND TOPIC "construction" OR "building" search keywords, the number of publications have increased every year. In year 2015, eight papers were recorded. This increased to 27 papers in 2016, 59 in 2017, 163 in 2018, and the numbers is growing in 2019 with 364 published papers as of May 2019. Of these, 196 papers were conference papers, 105 were journal articles, 10 papers were review papers, and the rest were other types of publications. Most of the papers were published in *Advances in Intelligent Systems and Computing* (17 papers), *Procedia Manufacturing* (13 papers), *IFIP Advances in Information and Communication Technology* (11 papers), and *IOP Conference Series Materials Science and Engineering* (11 papers).

From all the papers found, only three review papers provided significant impact related to Industry 4.0 in construction: "Understanding the implications of digitization and automation in the context of Industry 4.0: A triangulation approach and elements of a research agenda for the construction industry" [32], "Industry 4.0 as an enabler of proximity for construction supply chains: A systematic literature review" [33], and "Digital construction: From point solutions to IoT ecosystem" [26]. These three publications provided an overview of the potential of the industrial revolution in the construction industry in term of practice and supply chain. Much room exists for exploration related to this topic in the construction industry as the pattern and structure in this area of study are still in their infancy and conceptual.

A mapping study was used to provide a systematic and objective procedure to identify the pattern and structure of an existing study [34] to discover little evidence likely exists in a broad topic [35] such as construction within Industry 4.0. Our underlying research question was to determine how extensive the studies related to the topic are, with a secondary question to discuss the methodological approaches used in the studies to understand the movement of the topic. The objectives of this systematic review were (1) to identify, assess, and analyze the published studies to understand the prevalent patterns and structures in existing work related to Industry 4.0; (2) to identify the gaps in this topic; and (3) to propose a framework that links the current research topics to future research directions. With these objectives, we identified the most relevant authors, the topics that have already been covered, and, finally, potential future research directions. This overview will help researchers understand the current state of the construction industry within the Industry 4.0 era and reduce the digital gap throughout a project life-cycle.

This paper is structured as follows. Section 3 explains the selection of the information resources and information mapping. The next section provides reports and discussion of the results in some fields, knowledge area, authors, and keywords. Finally, Section 5 provides our conclusions and suggestions for future research.

3. Methodology

3.1. Literature Search

In this review-based study, we adopted the bibliometric mapping study method [36] and scoping review technique [37] to provide a systematic and holistic review of Industry 4.0 and the construction industry and the two are linked. We collected bibliometric studies of papers published from 2015 to May 2019 in Web of Science and Scopus. The following keywords were used for the search: "Industry 4.0" OR "Industrial revolution 4.0" AND TOPIC "construction" OR "building" the result obtained as listed in Tables 1 and 2.

Table 1. Numbers of papers in Web of Science (WoS) database based on the query "Industry 4.0" OR "Industrial revolution 4.0" AND TOPIC "construction" OR "building".

Type	2015	2016	2017	2018	2019
Article	1	7	13	51	10
Review	-	-	1	3	1
Proceeding	7	12	35	40	3
Total	**8**	**19**	**49**	**94**	**14**

Table 2. Numbers of papers in Scopus database based on the query "Industry 4.0" OR "Industrial revolution 4.0" AND TOPIC "construction" OR "building".

Type	2015	2016	2017	2018	2019
Article	3	9	23	56	14
Review	1	2	1	2	4
Proceeding	4	16	35	105	36
Total	**8**	**27**	**59**	**163**	**54**

The methodological framework introduced by Arksey and O'Malley involves five stages of research flow. First, the research questions are outlined, then appropriate research work is found, and then the process of selection occurs. Subsequently, the data are charted, after which information is collated and summarized, and the results are reported [37]. The flow diagram in Figure 1 demonstrates the Preferred Reporting Items for Systematic Reviews and Meta-Analyses (PRISMA 2009) flow of articles from search to final selection.

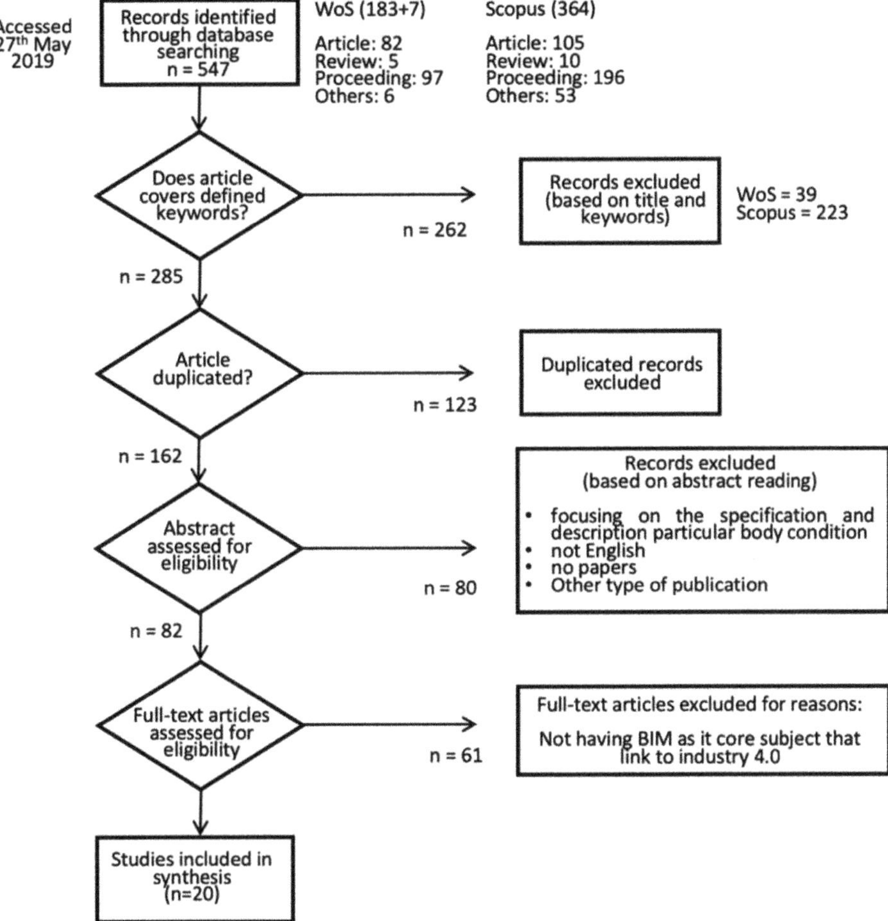

Figure 1. Preferred Reporting Items for Systematic Reviews and Meta-Analyses (PRISMA) flow diagram of search process.

A total of 547 papers related to Industry 4.0 studies on the construction industry were identified, with 183 papers and 364 papers on Web of Science and Scopus, respectively, with keyword search "Industry 4.0" OR "Industrial revolution 4.0" AND TOPIC "construction" OR "building. An additional 7 papers were found in Web of Science using the snowballing method. Manual scoping was required to finalize the desirable papers due to the broad keyword search and inability to refine search by specific subject areas and keywords in the database. Almost all the works identified were false positives: they contained the right properties; however, the content therein had no relevance to the topic under discussion.

Once the unrelated works were eliminated through the process of keyword screening, a total of 285 papers remained for further processing. The removal of duplicated works eliminated 123 papers. After the final manual screening of abstracts, 82 papers were eligible for full paper assessment and only 20 full-length papers were assessed and synthesized.

3.2. Data Extraction and Study Quality Evaluation

Eligible work was sorted by researchers who had independently reviewed the studies. Every paper was examined in terms of the following elements: the country of publication, title, the author(s), and publication type. For a study to be chosen for further review, it had to meet the following specific set of criteria:

(1) Published until May 2019,
(2) Focusing on the construction industry,
(3) Dealing with questions concerning technology adoption,
(4) Discussing the Industry 4.0 trends, opportunities, and challenges for the construction industry,
(5) Discussing the innovative approaches for managing operational processes of Industry 4.0 for the construction industry, and
(6) Written in English.

Publications were excluded if the following criteria applied:

(1) Focusing on the specification and description of a particular condition, for example, "Industry 4.0 for construction industry in circular economy" or "smart manufacturing";
(2) Focusing on specific technology, for example, "smart metering", "measuring application", and "human behavior sensing";
(3) Discussing a specific solution;
(4) Written in a language other than English;
(5) Inaccessible papers; or
(6) Not having BIM as its core subject that link to Industry 4.0.

4. Results

We started with a broader theme to identify, assess, and analyze the published papers to discover the structure and patterns for the query "Industry 4.0" OR "Industrial revolution 4.0" AND TOPIC: "construction" OR "building". At this stage, the identified papers indicated that these were the most popular keywords used for Industry 4.0 that could be clustered into at minimum three clusters, as illustrated in Figures 2 and 3.

Figure 2 shows the most co-occurring keyword search related to "Industry 4.0" OR "Industrial revolution 4.0" AND TOPIC "construction" OR building" in Web of Science with the minimum number of occurrence of keywords being at least five occurrences. From 490 keywords, 13 met the threshold. For each of the 13 keywords, the total strength of the co-occurrence links with other keywords was generated. Figure 2 shows the most co-occurring keyword search related to "Industry 4.0" OR "Industrial revolution 4.0" AND TOPIC "construction" OR "building" in Scopus with the minimum number of occurrence of keywords being at least five occurrences. From 3633 keywords, 197 met the threshold. Only the 30 strongest keywords were selected.

For each of the selected keywords, the total strength of the co-occurrence links with other keywords was generated. Only the keywords with the greatest total link strength were selected. Co-occurrence is the term used to describe the proximity of keywords in the title, abstract, or keyword list in publication [38] to find connections so that the research topic can be identified [39]. The link indicates the strength of their occurrences.

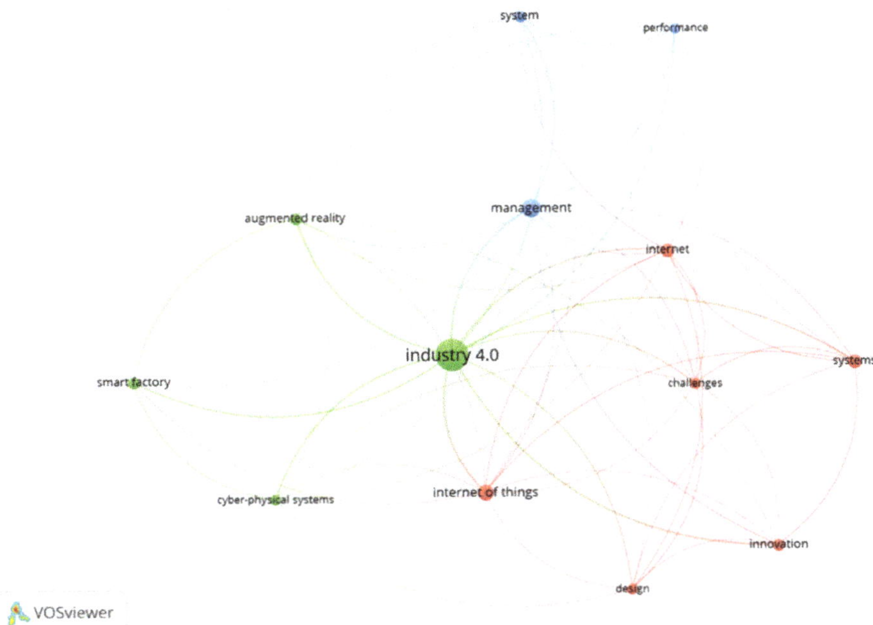

Figure 2. The most commonly co-occurring keyword search related to "Industry 4.0" OR "Industrial revolution 4.0" AND TOPIC "construction" OR "building" in Web of Science, with 183 publications.

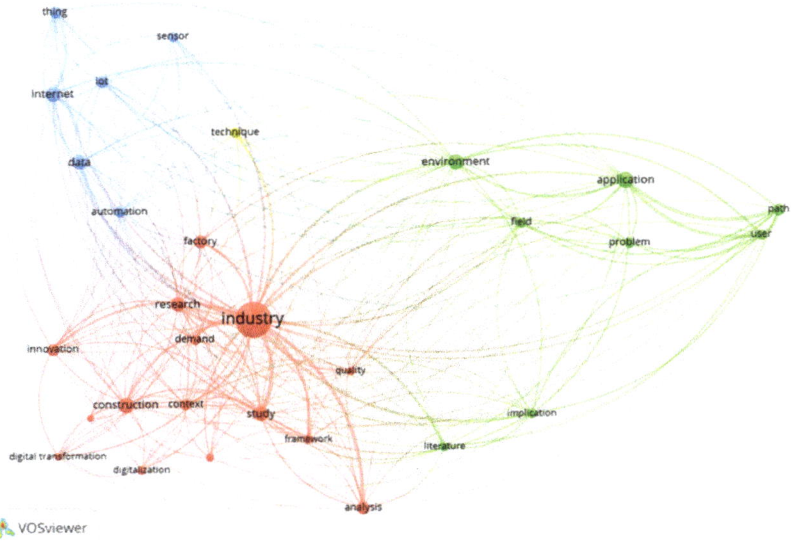

Figure 3. The most commonly co-occurring keyword search related to "Industry 4.0" OR "Industrial revolution 4.0" AND TOPIC "construction" OR "building" in Scopus with 364 publications.

Tables 3 and 4 show the available cluster of the co-occurrence keywords and strongest occurrences from the selected keyword search in Web of Science and Scopus, respectively. Tables 5 and 6 show the total link strength of the linked keywords to the keyword search. Industry 4.0, Management, Internet,

Internet of Things, and System were the major keywords in Web of Science; Industry, Application, Internet, Construction, Environment, and Data were the major keywords in Scopus. These tables show that using the keyword search "Industry 4.0" OR "Industrial revolution 4.0" AND TOPIC "construction" OR "building", the keyword Industry 4.0 co-occurred 43 times, Management co-occurred 14 times, Internet co-occurred 8 times, Internet of Things co-occurred 10 times, and System co-occurred 8 times in Web of Science. Industry 4.0 had 11 links that held the strongest links of all keywords. It was linked to all the keywords, forming the core of every available keyword. In Scopus, Industry co-occurred with the keywords search for 250 times, Application co-occurred 51 times, Internet 49 times, Construction 48 times, Environment 46 times, and Data 41 times.

Table 3. Cluster of keywords that co-occurred together in Web of Science under keyword search related to "Industry 4.0" OR "Industrial revolution 4.0" AND TOPIC "construction" OR "building".

Cluster	Keywords
1	Design, Innovation, Internet, Internet of Things (IoT), Systems
2	Augmented reality, Cyber-Physical System, Industry 4.0
3	Management, Performance, System

Table 4. Cluster of keywords that co-occurred together in Scopus under keyword search related to "Industry 4.0" OR "Industrial revolution 4.0" AND TOPIC: "construction" OR "building".

Cluster	Keywords
1	China, Construction, Construction industry, Digital transformation, Digitalization, Factory, Innovation, Quality
2	Application, Environment
3	Automation, Data, Internet, IoT, Sensor, Thing

Table 5. Most common co-occurring keywords of "Industry 4.0" OR "Industrial revolution 4.0" AND TOPIC "construction" OR "building" in Web of Science.

Keyword	Cluster	Links	Total Link Strength	Occurrence
Industry 4.0	2	11	53	43
Management	3	11	28	14
Internet	1	9	17	8
Internet of Things	1	10	18	10
Systems	1	7	15	8
Augmented Reality	2	9	15	6
Innovation	1	6	14	6
System	3	6	13	6
Cyber-Physical System	2	7	12	5
Design	1	7	10	5
Performance	3	6	10	5
Small Factory	2	5	11	6

Tables 7 and 8 show the total link strength of the keyword Industry 4.0 with its linked keywords. In Web of Science, keyword Industry 4.0 was linked to Management, Internet, IoT, System, Augmented Reality, Innovation, System, Cyber-Physical System, Design Performance, and Small Factory. Construction Industry, Construction, Digital Transformation, Innovation, Digitalization, Factory, Automation, Data, Internet, Sensor, IoT, Thing, Environment, and Application were the keywords linked to the Industry 4.0 keyword in Scopus. Keywords that hold higher strength with the linked keywords show how tightly these keywords are networked together. For example, keywords Data and Internet have the strongest link to keyword Industry 4.0. This means that keywords Data and Internet are significantly related to Industry 4.0 as data are the proverbial new oil in the digital era and are only useful used in real time and are exchangeable with other applications using an Internet connection.

Table 6. Most commonly co-occurring keywords of "Industry 4.0" OR "Industrial revolution 4.0" AND TOPIC "construction" OR "building" in Scopus.

Keywords	Cluster	Links	Total Link Strength	Occurrence
Industry	1	29	2023	250
Application	2	26	901	51
Internet	3	25	618	49
Construction	1	29	350	48
Environment	2	28	672	46
Data	3	28	503	41
Factory	1	26	234	32
IoT	3	25	385	30
Thing	3	24	445	29
Innovation	1	24	294	28
Automation	3	26	313	26
Sensor	3	22	264	23
Quality	1	27	157	16
Digitalization	1	14	94	15
Digital transformation	1	15	135	13
Construction industry	1	23	119	12

Table 7. Total link strength of keyword Industry 4.0 with its linked keywords in Web of Science.

Keyword	Total Link Strength
Management	8
Internet	4
Internet of Things	6
System	5
Augmented Reality	5
Innovation	6
System	3
Cyber-Physical System	5
Design	3
Performance	3
Small Factory	5

Table 8. Total link strength of keyword Industry 4.0 with its linked keywords in Scopus.

Keyword	Total Link Strength
Construction industry	29
Construction	78
Digital transformation	31
Innovation	79
Digitalization	25
Factory	75
Automation	61
Data	109
Internet	104
Sensor	59
IoT	69
Thing	65
Environment	98
Application	83

From all the keywords search findings, we conclude that all the keywords fall under one of three clusters: technology, security, and management cluster (Figure 4). Note that these keywords have highest total link strength, and other keywords that related to Industry 4.0 and the construction industry were ignored. The first cluster includes a wide range of technologies available in Industry 4.0; the second cluster deals with cybersecurity used to protect systems, networks, and programs from digital attacks, unauthorized access, changing, or destroying sensitive information, and others; and the third cluster focuses on the management issues in Industry 4.0 in the construction industry. These clusters are further explained in the discussion section. This clustering helped us to find patterns in the literature from the concept matrix (Tables 9 and 10).

Cluster Technology

Internet of Things	Augmented Reality
Internet	Digitization
Automation	Construction
Application	Sensor

Cluster Security

Cyber Physical System

Environment

Legislative Aspects — Data

Cluster Management

Digital Transformation

Performance	Innovation
Industry 4.0	Design
Management	Quality

Figure 4. Clustering of the keywords (Source Author's illustration).

Table 9. Clustered topics discussed in 20 targeted papers (articles).

Authors	Clusters																			
	Technology									Security					Management					
Keywords	A	B	C	D	E	F	G	H	I	J	K	L	M	N	O	P	Q	R	S	T
Oesterreich and Teuteberg (2016)	x	x	x	x	x	x	x	x	x	x	x	x	x	x	x	x	x	x	x	x
Dallasega et al. (2018)	x	x	x	x	x	x	x	x	x			x		x	x	x	x	x	x	x
Woodhead et al. (2018)	x	x	x	x	x	x	x	x	x		x			x	x			x	x	x
Theiler and Smarsly (2018)						x	x	x	x		x			x			x	x		
Pasetti Monizza et al. (2018)	x					x		x	x	x	x	x		x	x	x	x		x	x
Bianconi et al. (2019)								x	x		x	x				x	x	x	x	x

A. Internet; B. Automation; C. Internet of Things; D. Augmented Reality; E. Construction; F. Application; G. Sensor; H. Digitization; I. System; J. Cyber-Physical System; K. Data; L. Environment; M. Legislative Aspects; N. Industry 4.0; O. Management; P. Innovation; Q. Design; R. Performance; S. Quality; T. Digital Transformation.

Concept matrix in Tables 9 and 10 elaborate upon the topic discussed by the targeted papers. The pattern in the table shows that the topics focused more on the possible application of Industry 4.0 for the construction industry from a managerial aspect and less on the application of technology and security. The pattern also shows a lack of studies from the legislative perspective and cyber physical systems under cluster security for Industry 4.0 within the construction industry.

Table 10. Clustered topics discussed in 20 targeted papers (conference papers).

Authors	Technology							Security						Management						
Keywords	A	B	C	D	E	F	G	H	I	J	K	L	M	N	O	P	Q	R	S	T
Scheffer et al. (2018)					X		X	X			X									
Li and Yang (2017)	X		X	X	X	X		X	X		X		X	X	X		X	X	X	X
Dallasega et al. (2016)	X		X			X		X	X	X	X			X	X	X	X			X
Axelsson et al. (2018)		X	X		X			X	X		X			X	X			X	X	
Ding et al. (2018)	X		X		X	X	X	X	X	X	X			X	X	X	X	X	X	X
Delbrügger et al. (2018)				X	X	X		X	X		X			X		X	X			X
Trapp and Richter (2018)						X	X	X			X			X	X	X		X	X	X
Pruskova (2019)					X			X	X		X		X	X	X		X		X	X
Correa (2018)		X			X	X	X	X	X	X	X	X		X		X	X		X	X
Hotový (2018)						X		X	X		X	X		X	X	X		X	X	X
Schweigko et al. (2018)				X	X	X		X	X		X			X	X	X		X	X	X
King et al. (n.d.)					X			X			X	X		X	X					X
De Lange et al. (2017)	X		X		X	X		X	X		X		X	X	X	X	X	X	X	X

A. Internet; B. Automation; C. Internet of Things; D. Augmented Reality; E. Construction; F. Application; G. Sensor; H. Digitization; I. System; J. Cyber-Physical System; K. Data; L. Environment; M. Legislative Aspects; N. Industry 4.0; O. Management; P. Innovation; Q. Design; R. Performance; S. Quality; T. Digital Transformation.

The contents of the keyword search "Industry 4.0" OR "Industrial revolution 4.0" AND TOPIC "construction" OR "building" were analyzed after the first keyword and abstract inspection. Then, we examined the selected papers and their citations to further evaluate a particular author's and publication's contributions. The number of citations increased drastically in 2018 both in Web of Science and Scopus, which indicates the increase in the interest in and the value of this research topic (Figures 5 and 6). The papers most frequently cited are listed in Table 11.

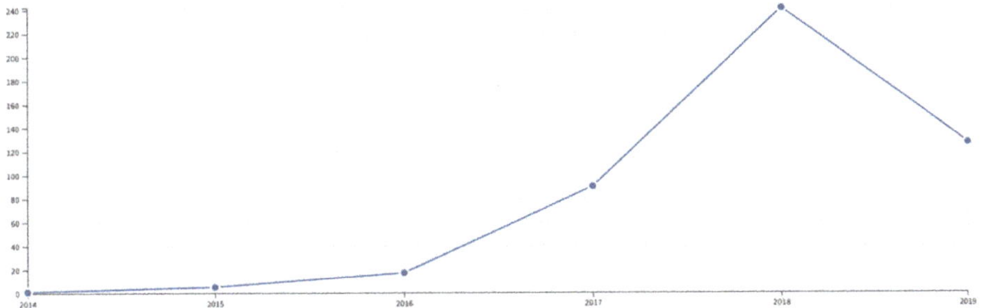

Figure 5. Sum of times cited per year of "Industry 4.0" OR "Industrial revolution 4.0" AND TOPIC "construction" OR "building" keyword from 2014 to 2019 in Web of Science.

Table 11. Most frequently cited papers.

Author	Paper Title	No. of Times Cited	Source of Publication
Oesterreich and Teuteberg (2016)	Understanding the Implications of Digitization and Automation in the Context of Industry 4.0: A Triangulation Approach and Elements of a Research Agenda for the Construction Industry [32]	176	*Computers in Industry*
Baccarelli et al. (2017)	Fog of Everything: Energy-Efficient Networked Computing Architectures, Research Challenges, and a Case Study [40]	98	*IEEE Access*
Shafiq et al. (2015)	Virtual Engineering Object/Virtual Engineering Process: A Specialized Form of Cyber-Physical System for Industry 4.0 [41]	68	*Knowledge-Based and Intelligent Information & Engineering Systems 19th Annual Conference, Kes-2015*
Dallasega et al. (2018)	Industry 4.0 As an Enabler of Proximity for Construction Supply Chains: A Systematic Literature Review [33]	15	*Computers in Industry*
Kleineidam et al. (2016)	The Cellular Approach: Smart Energy Region Wunsiedel. Testbed for Smart Grid, Smart Metering and Smart Home Solutions [42]	9	*Electrical Engineering*
Theiler and Smarsly (2018)	IFC Monitor – An IFC Schema Extension for Modelling Structural Health Monitoring Systems [43]	9	*Advanced Engineering Informatics*
Nguyen and Lo Iacono (2016)	RESTful IoT Authentication Protocols [44]	7	*Mobile Security and Privacy: Advances, Challenges and Future Research Directions*
Li and Yang (2017)	A Research on Development of Construction Industrialization Based on BIM Technology under the Background of Industry 4.0 [45]	7	*MATEC Web of Conferences*
Dallasega et al. (2016)	A Decentralized and Pull-based Control Loop for On-demand Delivery in ETO Construction Supply Chains [46]	7	*IGLC 2016 - 24th Annual Conference of the International Group for Lean Construction*
Delbrügger et al. (2018)	A Navigation Framework for Digital Twins of Factories Based on Building Information Modeling [47]	7	*IEEE International Conference on Emerging Technologies and Factory Automation, ETFA*

A detailed description of the targeted papers that concentrated on BIM as its core subject that link to Industry 4.0 is provided in Table 12, which lists the main objective, methodology used, findings, and the limitation of each paper.

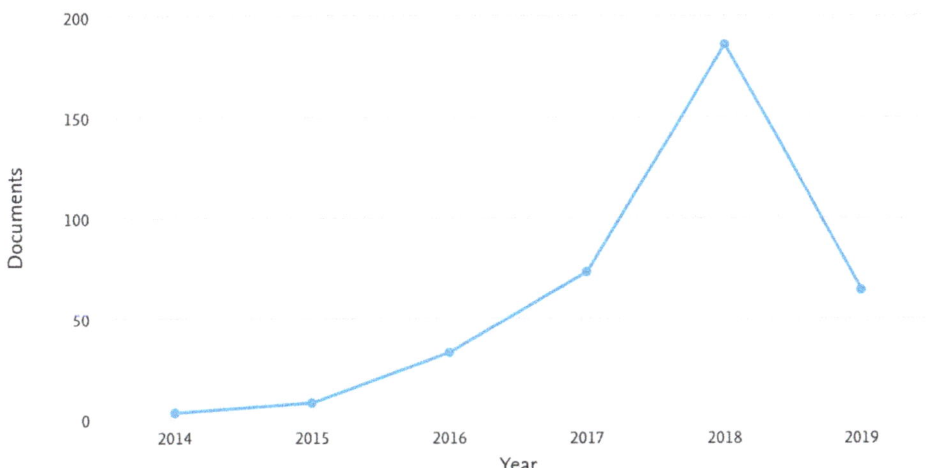

Figure 6. Sum of times cited per year of "Industry 4.0" OR "Industrial revolution 4.0" AND TOPIC "construction" OR "building" keyword from 2014–2019 in Scopus.

Table 12. Summary of 20 targeted papers.

Title	Document Type	Objective	Methodology	Main Finding	Limit of the Study
Understanding the implications of digitization and automation in the context of Industry 4.0: A triangulation approach and elements of a research agenda for the construction industry [32]	Review	Explore the state of the art as well as the state of practice of Industry 4.0	Triangulation approach: a comprehensive systematic literature review and case study research	Industry 4.0 technologies are far-reaching despite the maturity and availability of technologies. To adopt, many political, economic, social, technological, environmental and legal challenges have to be embraced	Did not cover all available published papers and included non-peer-reviewed publications in the review due to the novelty of the topic
Industry 4.0 as an enabler of proximity for construction supply chains: A systematic literature review [33]	Review	Explain Industry 4.0 concepts that increase or reduce proximity of construction supply chain	Systematic literature review and analysis of applicability through practical publications and examples from industrial case studies	Industry 4.0 technologies mainly influence technological, organizational, geographical, and cognitive proximity dimensions by de-territorializing closeness of major processes, routines, and procedures using Industry 4.0 concept	Used standard application and did not cover social, cultural, and economic norms
Digital construction: From point solutions to IoT ecosystem [26]	Review	Explain the construction industry in a transformational stage of a larger evolutionary process	Used longitudinal view of literature to explain the current period as disruptive technology driving an evolutionary adaptation of the construction industry in a historical socio-technological process	BIM is similar to PIM (Project Information Model)except BIM does not have the ability to integrate other data types from across the whole lifecycle because the construction industry has fragmented data compared to manufacturers who possess end to end lifecycle data	The similar pattern in every revolution system was not clearly elaborated

Table 12. *Cont.*

Title	Document Type	Objective	Methodology	Main Finding	Limit of the Study
Evaluation of Open Data Models for the Exchange of Sensor Data in Cognitive Building [48]	Conference Paper	Present the linkage between sensor and BIM using IFC (Industry Foundation Classes)	Describe the state of art of ODM (Open Data Models)	More dynamic concept of IFC connection has been investigated and further testing was required	Too conceptual
IFC Monitor—An IFC schema extension for modelling structural health monitoring systems [43]	Article	Propose SHM (Structural Health Monitoring) to monitor information using IFC, and open BIM standard is used to facilitate interoperability	Describe the possibilities and constraints, extend IFC schema to enable interoperability, verify IFC schema monitoring, then validate prototype	IFC monitor schema proposed in this study for SHM	Need to enhance communication protocol and system dynamics inherent in SHM systems
A research on development of construction industrialization based on BIM technology under the background of Industry 4.0 [45]	Conference Paper	Introduce a new production pattern of construction industrialization that is BIM-based	Comparative analysis	Countermeasures and suggestions to promote BIM: (1) Seize opportunities, and take own advantages; (2) attach importance to research and development of construction industrialization; (3) improve laws, regulations, and construction standards	The proposed pattern needs to have support from the government in terms of legalization and standardization
An approach supporting real-time project management in plant building and the construction industry [49]	Conference Paper	Proposal allowing scheduling and monitoring a building project in real time	Process planning-pitching-synchronization. Process templates as "planning configurator" for new projects	The first prototypical application of a scheduling and monitoring prototype that supports project management in real time and showed great potential.	Need to enhance the capability to plan several orders/projects in parallel
Towards a system-of-systems for improved road construction efficiency using lean and Industry 4.0 [50]	Conference Paper	Identify efficiency attributes and wastes in current practices, which led to a conceptual solution that focuses on improved coordination of working machines	Identifying similarities and differences between the construction and other industries	Outlined SoS (System of Systems) concept for improving productivity in road construction using lean principles for waste reduction and applying Industry 4.0 concepts to address different architectural concerns	Detailed applied studies and prototypes are needed
Smart steel bridge construction enabled by BIM and Internet of Things in industry 4.0: A framework [51]	Conference Paper	Discuss smart steel bridge construction enabled by BIM and IoT to deal with the uncontrollability and inefficiency problem of construction progress, quality, and cost in traditional steel bridge construction projects	Proposed framework, pointed out research directions and roadmap, applied data-driven methods and model-based analytics to realize real-time collaborative management and closed-loop control of steel bridge lifecycle activities	Position the BIM- and IoT-enabled steel bridge construction mode in the context of Industry 4.0	The integration of BIM and IoT is still under conceptualization and need to be validated
Parametric and Generative Design techniques in mass-production environments as effective enablers of Industry 4.0 approaches in the Building Industry [52]	Article	Investigate potential and criticisms of parametric and generative design techniques in mass-production environments of the BI though a pilot-case-study analysis in the GLT (Glue Laminated Timber) industry	Program a parametric algorithm for GLT engineering-measured manufacturing effectiveness and manufacturing efficiency through a value-stream map of an ordinary GLT supply chain system	Parametric and generative design techniques improved manufacturing effectiveness and manufacturing efficiency	Absence of Industry 4.0 approaches and technologies in an ordinary GLT supply chain system

Table 12. *Cont.*

Title	Document Type	Objective	Methodology	Main Finding	Limit of the Study
A navigation framework for digital twins of factories based on building information modelling [47]	Conference Paper	Propose a framework that allows for effortless inclusion of BIM and factory equipment, as well as a plugin system for a variety of spatial representations	Used BIM and interrelated technologies, then applied the framework to simulate digital twin factory	Developed framework that combines building, agent, and machine information into one dynamic factory model to support pathfinding and path following	Extension is required to the use of sensors in the navigation framework to complete construction life cycle
Towards the generation of digital twins for facility management based on 3D point clouds [53]	Conference Paper	Present the current research and development progress of a service-oriented platform for generation of semantically rich 3D point cloud representations of indoor environments	Described O&M stages within the FM operations, particularly on space management	The preliminary results of a prototypical web-based application demonstrate the feasibility of service-oriented platform for FM using a service-oriented paradigm	The generation of as-is BIM data from point clouds; the selection and labelling of segmented areas to be converted to IFC components required manual user input
Automated design and modeling for mass-customized housing. A web-based design space catalog for timber structures [54]	Article	To develop a cross-laminated timber (CLT) model for AEC (Architecture, Engineering and Construction) industry	Using generative models and evolutionary principles to inform the customization process in the early stage of design to explore different design solutions	The presented collaborative strategy and web-based catalogue represent a first step in developing a comprehensive methodology for wooden architecture to enable mass-customized housing	The strategy is still in its infancy
Beginning of Real Wide us of BIM Technology in Czech Republic [55]	Conference Paper	Reveals the reason and method to achieve BIM adoption in Czech Republic	Review	Users see more barriers and complications than benefits	Proper investigation into BIM adoption is needed
Cyber-physical systems for construction industry [56]	Conference Paper	Present a framework in which cyber-physical systems (CPS) for construction based on virtual models of construction processes	CPS for construction based on virtual models of construction processes, implemented via Petri Nets, and connected to both BIM models and hardware (sensors and actuators) working in on-site production or assembly. Then, the proposed framework was 'relaxed' and extended to be applied to a scenario where CPS are only a bi-directional link between virtual models and its real counterparts on-site, without hardware requiring control, but with observations based on data acquired via sensing	To reduce or eliminate manual data gathering in the field, about construction progression, and reducing the time to produce construction progress reports.	Further elaboration is required to have practical value on the field, especially for the Petri Nets simulations and CPS
Dynamic model of implementation efficiency of BIM in relation to the complexity of buildings and the level of their safety [57]	Conference Paper	Present the efficiency of BIM on the created dynamic model for the complexity of buildings	Basic dynamic BIM implementation model (without implementation of the subset employees)	BIM achieved the efficiency of the construction and management throughout the life cycle despite of the complexity of the construction process and the corresponding cost of the BIM	The dynamic BIM model and employment model, training should be further elaborated

Table 12. *Cont.*

Title	Document Type	Objective	Methodology	Main Finding	Limit of the Study
Development of a digital platform based on the integration of augmented reality and BIM for the management of information in construction processes [58]	Conference Paper	Describe the development process of a digital platform that uses AR combined with BIM to provide workers with relevant information in real-time based on their current position on the construction site	The location system uses sensory information collected by mobile devices to provide location awareness to the application; the integration of 3D BIM model metadata to contextualize tasks and instructions and provide building components information	A platform for the integration of augmented reality and BIM	The benefit of having this digital platform in construction processes can be incorporated with economic gain for the implementation to be rationalized
How industry 4.0 and BIM are shaping the future of the construction environment [7]	Conference Paper	Discuss on the potential of Industry 4.0	Not stated	Industry 4.0 promotes valuable data exchange as well as trust in collaboration	Limited to the overview of Industry 4.0 only, the superficial beauty of Industry 4.0
Socio-Technical challenges in the digital gap between building information modelling and industry 4.0 [59]	Conference Paper	Envision a socio-technical solution strategy based on the common understanding that communication and cooperation are mission-critical for the overall success of the deployed information system, the design process, and the final result of the mission, which is the building.	Sketched the challenges and discussed a running construction project as a real application scenario including the use of serious gaming strategies, near real-time collaboration, and mixed reality.	The results show that despite the cost and time restrictions, innovative and relevant research in interdisciplinary research and development teams is feasible	The 90% sociology on the 90/10 rule of people adoption can be elaborated as well as suggestions for future solution

5. Discussion

5.1. State-Of-The-Art Industry 4.0 in the Construction Industry

Whereas the first industrial revolution introduced mechanical power, the second industrial revolution lightened up the industry, and the third industrial revolution digitized the information and production [15,16,19], Industry 4.0 has amalgamated the physical world with the information era directed by the cyber-physical system approach [24]. Industry 4.0 is also known as "smart manufacturing", "industrial internet", or "integrated industry", which has shifted the value chain of organization and management across the lifecycle of products by integrating complex devices, machines, and networked sensors and software, deployed to predict, control, and plan for better business and societal outcomes [60]. The combination of these digital innovations is collectively called the Internet of Things (IoT) in the cyber-physical system, which are able to meet new emergent needs and provide capabilities that instigate the next evolution of society and its organization or institution [26] and transform how products are designed, fabricated, used, operated, maintained, and serviced [61]. Robotic technologies have been merged with the construction industry, known as construction automation technologies, to create elements of buildings, building components, and building furniture [25].

Industry 4.0 has challenged the construction industry by providing a glimpse of the construction digitization potential with the availability of digital data and online digital access that automatically gather and process electronic data into the value chain on discrete tasks [24]. BIM (within the planning domain), as the center of construction digitization together with Industry 4.0 (production domain), is able to close the digital gap that still exists and sustain the impact on future building processes [59]. The innovation in and approaches to construction automation are still in their infancy [62] and not fully employed as the technical aspects of the available technologies are still being investigated though some technologies have reached maturity, such as BIM, cloud computing, mobile computing, and modularization [32]. The fact that construction projects are becoming increasingly complex despite construction being a flat market for the previous five decades requires Industry 4.0 as a solution for a new business model [24,25]. This has occurred because, currently, the construction industry has one of the lowest capital investments as well as low capital intensity [25] with the lowest R&D intensity [32] compared to other sectors despite being a major contributor to the employment and economy of many countries [63]. The fragmented supply chain of the construction industry, which includes several small- and medium-sized enterprises (SMEs), limits the ability to invest in innovative technologies [33]. Another reason behind the slow adoption is because the gap between construction and manufacturing is relatively huge though both industries are categorized into the same group and work together [24]. The unavailability of a dedicated process change strategy, dedicated implementation plan, and business strategy alignment have also contributed to the slow adoption. Since Industry 4.0 creates value that transforms the overall business strategy in the construction industry, there is a need to propose some strategies for implementation. With the capability to automate both design and manufacturing processes and the possibility of handling a heterogeneous and significant amount of information, Industry 4.0 is expected to be able to improve the quality and productivity of construction and attract domestic and foreign investors.

This review, though we are not claiming it to be exhaustive, has provided an overview of Industry 4.0 concept in construction industry in the last five years. The selected papers revealed the active collaboration between BIM with technologies from Industry 4.0 (Figure 7), such as the use of BIM to support design decisions for mass customization production [54], structural health monitoring (SHM) using open BIM [43], allowing schedule monitoring in real time [50], smart steel bridge construction enabled by BIM and IoT [51], and a digital platform that uses augmented reality (AR) combined with BIM to provide workers with relevant information in real-time [58]. However, the pattern of topics discussed are broad and conceptual. The targeted papers only mention the benefit of Industry 4.0 to the construction industry conceptually. A detailed study should be completed to understand the benefits brought by Industry 4.0 to the construction industry. In addition, studies are lacking on management processes for overall project life cycle as well as the operation, and tactical and strategic planning in this collaborative and autonomous synchronization system. Studies in this area are needed in order to transform the construction network and construction economy and to integrate BIM with Industry 4.0 technology. The relationship of Industry 4.0 as the production domain with BIM as the planning domain acts as the core structure of the cyber-planning-physical system, influenced by the benefits and challenges of Industry 4.0 for the construction industry is illustrated in Figure 8 [32,59,64]. Figure 7 shows how the physical and cyber domains are controlled by the planning domain.

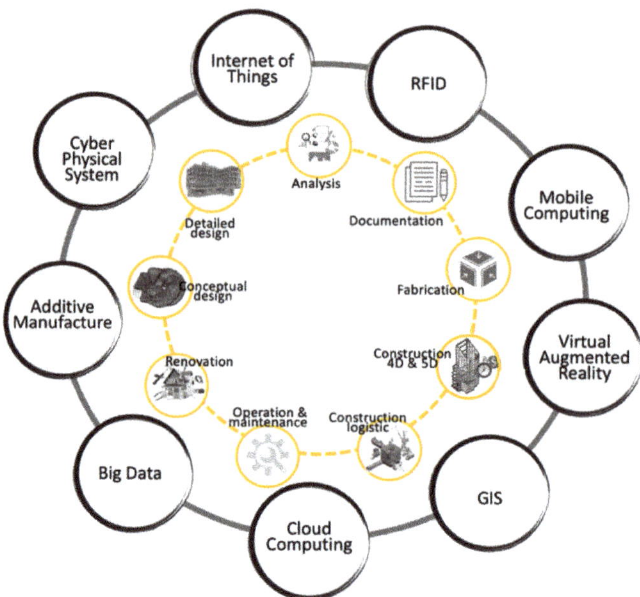

Figure 7. Concept of technologies in Industry 4.0 with BIM as its core structure with collaboration and an autonomous synchronization system.

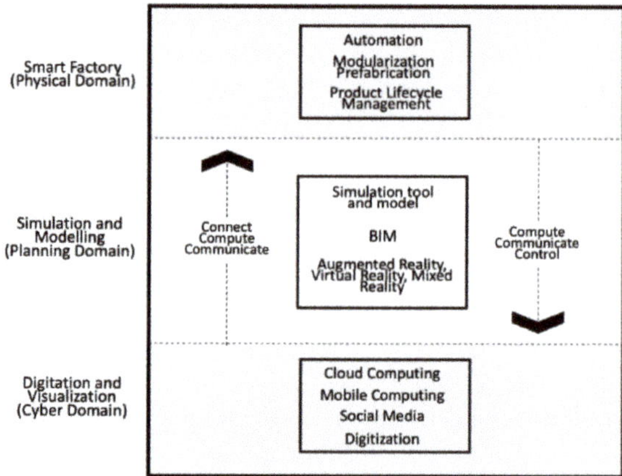

Figure 8. The relationships in the cyber-planning-physical system with BIM as its core. Adapted from previous studies [32,59,64].

The bi-directional coordination between the physical domain and cyber domain has the potential to improve real-time progress monitoring and control the construction process, track changes, model updates, and exchange information between the design and operational stages [65]. This is a solution to the infamous construction practice epitomized by the management inefficiencies that result in delays, unforeseen costs, and poor work quality [66]. Since BIM is the core of this bi-directional coordination, its role is to digitize and control the overall process of the construction life cycle. However, for this to be realized, the construction industry needs to accommodate their activities with BIM functionalities, as BIM tools have the potential to be used for managing different activities [67]. BIM functionalities include six components [68]:

(1)   Team communication and integration,
(2)   Parametric modelling and visualization,
(3)   Building performance analysis and simulation,
(4)   Automatic document generation,
(5)   Improved building lifecycle management, and
(6)   Software interoperability with other applications.

The relationship between the cyber-planning-physical system, BIM functionalities, and construction phases is illustrated in Figure 9. However, this improvement requires data transparency, concurrent viewing and editing of a single federated model, and controlled coordination of information access [69]. For this to be implemented, the three main components clustered in the findings need to be highlighted.

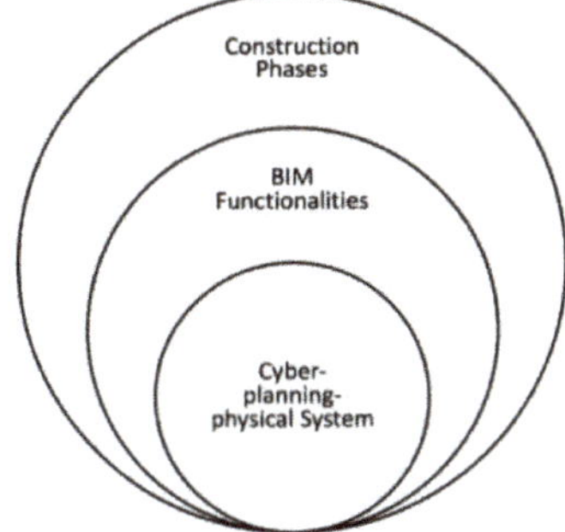

Figure 9.   Relationship between cyber-planning-physical system, BIM functionalities, and construction phases.

5.1.1. Cluster Technology

The first cluster includes a wide range of technologies available in Industry 4.0: internet, automatic equipment, Internet of Things, augmented reality technology, and sensors. This cluster completes the cyber-planning-physical ecosystem by integrating physical machineries and devices, non-physical technologies, and BIM to accommodate BIM functionalities to improve construction activities during different phases. The application of these technologies in the construction industry could not possibly be realized without the digitization of data from BIM as the collaboration medium. For example, design automation combining the BIM model with advanced simulation tools and genetic algorithms was able to mass-customize housing construction [54]; the use of BIM in the navigation core provided an augmented reality navigation system to navigate pathways in building [47]; the use of Bluetooth Low Energy (BLE) Beacons, Extended Markup Language (XML) formatted informative BIM, and BIM model on the Unity platform provided workers with relevant information in real-time based on their current position on the construction site through augmented reality [59]; and BIM, IoT, sensors, data computing, and advanced analytics tools were used to simulate, re-simulate, and map the simulation for steel bridge performance monitoring as the core enabling the technology system [51]. Automated real-time construction technologies would help the construction industry to improve the productivity and quality of a project throughout its lifecycle. As such, the availability of the Internet and Internet of Things enables the creation of a cyber domain to support these smart technologies in the physical domain with the support of BIM in the planning domain.

5.1.2. Cluster Security

Cluster two includes system, cyber-physical system, data, environment, and legislative aspects that can be categorized as a security cluster in Industry 4.0 for construction. Since Industry 4.0

involves data and systems in a virtual environment, it is crucial to be concerned about the security issues as any wrong information has the potential to have negative consequences [70]. BIM, as the planning domain, possesses rich information and data about the construction life cycle that can easily be extracted and reused [71]. BIM open standards were developed to represent information in a building information model and openly exchange this information [68]. The BIM open standard has been recognized a standardization to exchange information and documents with other partners that previously could not be executed automatically [69]. These standards include the Industry Foundation Classes (IFC), Green Building XML, and the newer Construction Operations Building Information Exchange [68]. Green Building XML is an open schema created to facilitate the transfer of building data stored in BIM to engineering analysis tools. However, most construction professionals are still unaware of the legal implications arising from BIM adoption, although several BIM protocols and contracts have been developed [11]. This requires protection of the ownership of the model through copyright laws [72] to avoid data loss, theft of intellectual property, or misuse during data exchanges within a common data environment [73]. However, security issues faced by Industry 4.0 in the construction industry are not only limited to the ownership of the data in BIM, but also include the system security and data security associated with the cyber-planning-physical system security, software, and hardware. The main objectives of cyber-physical system security are confidentiality, integrity, availability, and authenticity [74]. Without a properly designed cyber-planning-physical system security, the whole system might be at risk of cyber-attacks such DDOS (Distributed Denial of Service), data theft, eavesdropping, and malicious software. The attacker can delete, modify, steal, or exploit the information and resources for inappropriate reasons [75]. Most published papers did not properly cover providing security involved with BIM. Further studies on the cyber-planning-physical system security are required.

5.1.3. Cluster Management

Cluster three included articles related to management related to Industry 4.0, including management, innovation, design, performance, quality, and digital transformation, which are related to BIM functionalities. Management in the Industry 4.0 era with BIM as its core has slowly been revolutionized, as shown by higher performance and good quality in construction practices. The targeted papers have demonstrated the successful implementation of Industry 4.0 technologies by achieving real-time project management [49], smart technology management [51], smart indoor navigation management [47], creating a digital twin for facility management [53], road construction management [50], as well as mass-customization of design management [54]. With the clear distribution of managerial framework for the integration of BIM and Industry 4.0, the construction industry is able to capture the benefits from BIM and Industry 4.0 from a management perspective and is expected to develop and deploy more technologies to enhance productivity. As the core of the project with a collaborative and autonomous synchronization system, BIM provides a new means to predict, manage, and monitor the quality and performance of the project throughout the whole project life cycle.

5.2. Methodological Concerns

Concept papers have flourished in the Industry 4.0 for construction research topic. This review provided a summary of the current state of the research related to Industry 4.0 in the construction industry, both conceptually and providing an in-depth systematic discussion. From 20 targeted papers, 9 papers were concept papers. Concept papers generally contain a clear description of the research topic, including a summary of what is already known about that topic and the importance of the studies without documenting statistical evidence of the sources for the purpose of attracting readers to understand what the researcher is currently investigating, usually published during earlier stages of research [76]. Four papers were categorized as systematic literature reviews (SLRs), reviewing the trend toward digitization and automation of the construction industry and identifying and classifying the pattern of the research themes. SLRs are used to comprehensively locate and synthesize related research, using organized, transparent, and replicable procedures at each step in the process [77]. Another method of review used in the summarized papers was Longitudinal Literature Review (LLR). LLR is the science of tracing changes by repeatedly measuring the same phenomenon under the same circumstances over a long period of time [78]. This procedure is often used in standard SLR, but is differentiated by the extensive length of the observation period to identify the changing trend and determine how the trend influences the surroundings. This targeted review of these specific types of papers is able to help other researchers to overview what is currently happening to the construction industry due to the dramatic change due to the Industry 4.0 era. The suggestions for future research and the overview of the benefits and challenges of the research topic were rigorously documented using the evidence from the implementation in other industries.

Original research (primary sources) of the targeted papers was limited. Only seven original papers were found, all of which used quantitative research on specific research topics that were not related to each other: augmented reality [58], facility management [53], structural monitoring [43], data management [48], design management [52], and real-time project management [49]. All selected papers discussed experiments and the result obtained. The articles included detailed descriptions of the methods used to produce the results for future verification or knowledge transfer by other researchers.

To obtain an in-depth understanding of the theories underpinning Industry 4.0 for construction, qualitative research is needed as it focuses on "why" rather than the "what" to examine the natural phenomena of the research topic. However, the selected papers contained no qualitative original research, as the research topic is still in its infancy. Qualitative research papers are able to dive deeper into the problem of the research topic by reporting the phenomena from multiple perspectives, identifying many factors involved in the situation, and generally sketch the bigger picture that emerges. This limitation provides an opportunity to further explore this research topic qualitatively as the phenomena is immature due to a lack of theory and a limited number of available studies.

6. Conclusions

Findings show a clear, active, and unfinished discussion about Industry 4.0 in the construction industry. This review demonstrates the lack of a complete understanding on what Industry 4.0 entails for the construction industry as the number of original papers are limited. The available studies focused on the concept of the possibility of adopting Industry 4.0 in the construction industry rather than providing a solid theoretical development to realizing the adoption. Most of the papers elaborated upon the existing technologies and how creative innovation can be used to adopt these technologies into different sectors or industries. The selected papers revealed the active collaboration between BIM and technologies from Industry 4.0—how BIM became the agent of collaboration between cyber systems and physical system to complete a cyber-planning-physical ecosystem to accommodate BIM functionalities throughout construction project lifecycles. However, the pattern of topics discussed are broad and conceptual. The targeted papers only conceptually mentioned the benefit of Industry 4.0 in the construction industry and how BIM can possibly be deployed to enhance productivity. Detailed studies on the implementation should be completed to evaluate the potential of Industry 4.0 in the construction industry. In additional, studies are lacking on the management processes of the overall project life cycle as well as the operation, tactical, and strategic planning in this collaborative and autonomous synchronization system. Studies in this area are needed in order to transform the construction network and construction economy and to integrate BIM with Industry 4.0 technology. We are currently moving in the right direction as BIM together with Industry 4.0 are being introduced to the industry, transforming how the industry has been operating for decades, improving the quality and performance of the overall construction life cycle. More qualitative research is needed to explore and understand the issue, challenges, and future direction of these new technologies to allow more experimental design research for realizing BIM and Industry 4.0 in the construction industry.

We discussed the eligibility of the selected papers in our research approach in Section 3.2. However, there is a limitation in this search. We reviewed all publications in Web of Science and Scopus for the past five years. We focused on a wider view of the overall Industry 4.0 component to depict overall mapping. Since the topic Industry 4.0 is relatively new, the extracted data were unable to justify the relatedness of every component. A new search with different keywords from different databases needs to be completed to extend the findings. It may be helpful to perform the same type of analysis with different keywords. This could provide a different perspective and a different aspect from which to understand the trends in BIM in Industry 4.0 in the construction industry.

Future research on collaborative and autonomous synchronization systems with BIM as the core of the structure in the cyber-planning-physical system could fill the gap that we highlighted. It would be interesting to gain further insights into the management scope, especially on the role and responsibility of the stakeholders as well as the operation, tactical, and strategic management in construction companies during different construction phases throughout the construction lifecycle. More analyses and experimentation are required about the changes in working culture in the new construction network. Finally, we think that our contribution provides a resourceful foundation about the collaborative and autonomous synchronization in cyber-planning-physical systems. With the capability to automate both the design and construction processes and the possibility of handling large amounts of heterogeneous data, BIM in Industry 4.0 for the construction industry is expected to improve the quality and productivity of construction and attract domestic and foreign investors.

Author Contributions: Conceptualization, R.M., A.S. and K.N.A.; Methodology, R.M., A.S., P.M. and O.K.; Analysis, R.M., A.S., P.M. and O.K.; Writing-Original Draft Preparation, R.M., A.S. and P.M.; Writing-Review & Editing, R.M., A.S., O.K., and P.M.; Visualization, R.M.; Supervision, A.S. and K.N.A.; Project Administration, R.M.; Funding Acquisition, A.S., O.K., and P.M.

Funding: This research was funded by Universiti Teknologi Malaysia (UTM) under Research University Grant Vot-20H04, Malaysia Research University Network (MRUN) Vot 4L876 and the Fundamental Research Grant Scheme (FRGS) Vot 5F073 supported under Ministry of Education Malaysia for the completion of the research. The work was partially funded by the SPEV project, University of Hradec Kralove, FIM, Czech Republic (ID: 2103–2019).

Acknowledgments: We are also grateful for the support of student Jan Hruska and Sebastien Mambou in consultations regarding application aspects.

Conflicts of Interest: The authors declare no conflict of interest.

References

1. UN. 68% of the world population projected to live in urban areas by 2050, says UN | UN DESA | United Nations Department of Economic and Social Affairs. *United Nations News*, 16 May 2018; pp. 2–5.
2. Adabre, M.A.; Chan, A.P.C. Critical success factors (CSFs) for sustainable affordable housing. *Build. Environ.* **2019**, *156*, 203–214. [CrossRef]
3. De Wilde, P. Ten questions concerning building performance analysis. *Build. Environ.* **2019**, *153*, 110–117. [CrossRef]
4. Livotov, P.; Sekaran, A.P.C.; Law, R.; Reay, D.; Sarsenova, A.; Sayyareh, S. *Eco-Innovation in Process Engineering: Contradictions, Inventive Principles and Methods*; Elsevier Ltd.: Amsterdam, The Netherlands, 2019; Volume 9.
5. Craveiro, F.; Duarte, J.P.; Bartolo, H.; Bartolo, P.J. Additive manufacturing as an enabling technology for digital construction: A perspective on Construction 4.0. *Autom. Constr.* **2019**, *103*, 251–267. [CrossRef]
6. Gao, H.; Koch, C.; Wu, Y. Building information modelling based building energy modelling: A review. *Appl. Energy* **2019**, *238*, 320–343. [CrossRef]
7. King, M. How Industry 4.0 and BIM are Shaping the Future of the Construction Environment. *GIM Int. Worldw. Mag. GEOMATICS* **2018**, *31*, 24–25.
8. Bilal, M.; Oyedele, L.O.; Qadir, J.; Munir, K.; Akinade, O.O.; Ajayi, S.O.; Alaka, H.A.; Owolabi, H.A. Analysis of critical features and evaluation of BIM software: Towards a plug-in for construction waste minimization using big data. *Int. J. Sustain. Build. Technol. Urban Dev.* **2015**, *6*, 211–228. [CrossRef]
9. Yanagawa, K. reIndustrializing Architecture. *Int. J. Archit. Comput.* **2016**, *14*, 158–166. [CrossRef]
10. Autodesk, BIM 360 API. Available online: https://forge.autodesk.com/en/docs/bim360/v1/overview/introduction/ (accessed on 1 May 2019).
11. Fan, S.L.; Chong, H.Y.; Liao, P.C.; Lee, C.Y. Latent Provisions for Building Information Modeling (BIM) Contracts: A Social Network Analysis Approach. *KSCE J. Civ. Eng.* **2019**, *23*, 1427–1435. [CrossRef]

12. Porter, S.; Tan, T.; Tan, T.; West, G. Breaking into BIM: Performing static and dynamic security analysis with the aid of BIM. *Autom. Constr.* **2014**, *40*, 84–95. [CrossRef]

13. Gildow, C. M9—Architecture and the Industrial Revolu on. Module 9. Architecture, 2012. Available online: https://learn.canvas.net/courses/24/pages/m9-architecture-and-the-industrial-revolution?module_item_id=44477 (accessed on 30 December 2018).

14. Wang, Y.; Li, X.; Li, X. Eco-Topia: Design 4.0 and the Construction of E-Image City. In Proceedings of the 1st International Conference on Arts, Design and Contemporary Education, Moscow, Russia, 22–24 April 2015; Volume 23, pp. 414–417.

15. Sreekanth, P.S. Impact of Industrial revolution on architecture. *The Archi Blog Not Just another Architecture Blog.* 2011. Available online: https://thearchiblog.wordpress.com/2011/06/02/impact-of-industrial-revolution-on-architecture/ (accessed on 20 April 2019).

16. Mokyr, J. The Second Industrial Revolution, 1870–1914. In *Storia dell'economia Mondiale*; Valerio, C., Ed.; Laterza: Rome, Italy, 1998.

17. Bethany, How CAD Has Evolved Since 1982. 2017. Available online: https://www.scan2cad.com/cad/cad-evolved-since-1982/#history-of-cad (accessed on 30 December 2018).

18. Fisher, T. Welcome to the Third Industrial Revolution: The Mass-Customisation of Architecture, Practice and Education. In *Architectural Design*; John Wiley & Sons Ltd.: Hoboken, NJ, USA, 2015; Vol 85, pp. 40–45. Available online: https://onlinelibrary.wiley.com/doi/abs/10.1002/ad.1923 (accessed on 11 May 2019).

19. Kolarevic, B. Designing and Manufacturing Architecture in the digital age. In *Architectural Information Management, Proceedings of the 19th eCAADe Conference*; Helsinki University of Technology: Helsinki, Finland, 2001; pp. 117–123.

20. Naboni, R.; Paoletti, I. *Advanced Customization in Architectural Design and Construction*; Springer: Milano, Italy, 2015.

21. Howard, R. Describing the Changes in Architectural Information Technology to Understand Design Complexity and Free-Form Architectural Expression. *J. Inf. Technol. Constr. ITcon* **2006**, *11*, 395–408.

22. Oztemel, E.; Gursev, S. Literature review of Industry 4. 0 and related technologies. *J. Intell. Manuf.* **2018**, 1–56. [CrossRef]

23. Gilchrist, A. *Industry 4.0: The Industrial Internet of Things*; Apress: New York, NY, USA, 2016.

24. Alaloul, W.S.; Liew, M.S.; Amila, N.; Abdullah, W.; Mohammed, B.S. Industry Revolution IR 4.0: Future Challenges in Construction Industry Opportunities and. *MATEC Web Conf.* **2018**, *02010*, 1–7. [CrossRef]

25. Bock, T. The future of construction automation: Technological disruption and the upcoming ubiquity of robotics. *Autom. Constr.* **2015**, *59*, 113–121. [CrossRef]

26. Woodhead, R.; Stephenson, P.; Morrey, D. Digital construction: From point solutions to IoT ecosystem. *Autom. Constr.* **2018**, *93*, 35–46. [CrossRef]

27. Li, C.Z.; Hong, J.; Xue, F.; Shen, G.Q.; Xu, X.; Luo, L. SWOT analysis and Internet of Things-enabled platform for prefabrication housing production in Hong Kong. *Habitat Int.* **2016**, *57*, 74–87. [CrossRef]

28. Dave, B.; Kubler, S.; Pikas, E.; Holmström, J.; Singh, V.; Främling, K.; Koskela, L. Intelligent Products: Shifting the Production Control Logic in Construction (With Lean and BIM). In Proceedings of the 23rd Annual Conference of the International Group for Lean Construction, Perth, Australia, 29–31 July 2015; pp. 341–350.

29. Bottaccioli, L.; Aliberti, A.; Ugliotti, F.; Patti, E.; Osello, A.; Macii, E.; Acquaviva, A. Building Energy Modelling and Monitoring by Integration of IoT Devices and Building Information Models. In Proceedings of the 2017 IEEE 41st Annual Computer Software and Applications Conference (COMPSAC), Turin, Italy, 4–8 July 2017; pp. 914–922.

30. Li, C.Z.; Hong, J.; Xue, F.; Shen, G.Q.; Xu, X.; Mok, M.K. Schedule risks in prefabrication housing production in Hong Kong: A social network analysis. *J. Clean. Prod.* **2016**, *134*, 482–494. [CrossRef]

31. Fang, Y.; Cho, K.Y.; Zhang, S.; Perez, E. Case Study of BIM and Cloud–Enabled Real-Time RFID Indoor Localization for Construction Management Applications. *J. Constr. Eng. Manag.* **2016**, *142*, 1–12. [CrossRef]

32. Oesterreich, T.D.; Teuteberg, F. Understanding the implications of digitisation and automation in the context of Industry 4.0: A triangulation approach and elements of a research agenda for the construction industry. *Comput. Ind.* **2016**, *83*, 121–139. [CrossRef]

33. Dallasega, P.; Rauch, E.; Linder, C. Industry 4.0 as an enabler of proximity for construction supply chains: A systematic literature review. *Comput. Ind.* **2018**, *99*, 205–225. [CrossRef]

34. Budgen, D.; Turner, M.; Brereton, P.; Kitchenham, B. Using mapping studies in software engineering. *PPIG* **2008**, *8*, 195–204.

35. Kitchenham, B.; Charters, S. Guidelines for performing Systematic Literature reviews in Software Engineering Version 2.3. *Engineering* **2007**, *45*, 1051.

36. Leung, X.Y.; Sun, J.; Bai, B. Bibliometrics of social media research: A co-citation and co-word analysis. *Int. J. Hosp. Manag.* **2017**, *66*, 35–45. [CrossRef]

37. Arksey, H.; O'Malley, L. Scoping studies: Towards a methodological framework. *Int. J. Soc. Res. Methodol. Theory Pract.* **2005**, *8*, 19–32. [CrossRef]

38. Van Eck, N.J.; Waltman, L. *Visualizing Bibliometric Networks*; Springer: Berlin/Heidelberg, Germany, 2014.

39. Wang, W.; Laengle, S.; Merigó, J.M.; Yu, D.; Herrera-Viedma, E.; Cobo, M.J.; Bouchon-Meunier, B. A Bibliometric Analysis of the First Twenty-Five Years of the International Journal of Uncertainty, Fuzziness and Knowledge-Based Systems. *Int. J. Uncertainty Fuzziness Knowl.-Based Syst.* **2018**, *26*, 169–193. [CrossRef]

40. Baccarelli, E.; Naranjo, P.G.V.; Scarpiniti, M.; Shojafar, M.; Abawajy, J.H. Fog of Everything: Energy-Efficient Networked Computing Architectures, Research Challenges, and a Case Study. *IEEE Access* **2017**, *5*, 9882–9910. [CrossRef]

41. Shafiq, S.I.; Sanin, C.; Szczerbicki, E.; Toro, C. Virtual engineering object/virtual engineering process: A specialized form of cyber physical system for industrie 4.0. *Procedia Comput. Sci.* **2015**, *60*, 1146–1155. [CrossRef]

42. Kleineidam, G.; Krasser, M.; Reischböck, M. The cellular approach: Smart energy region Wunsiedel. Testbed for smart grid, smart metering and smart home solutions. *Electr. Eng.* **2016**, *98*, 335–340. [CrossRef]

43. Theiler, M.; Smarsly, K. IFC Monitor—An IFC schema extension for modeling structural health monitoring systems. *Adv. Eng. Inform.* **2018**, *37*, 54–65. [CrossRef]

44. Nguyen, H.V.; Iacono, L.L. *RESTful IoT Authentication Protocols*; Elsevier Inc.: Amsterdam, The Netherlands, 2016.

45. Li, J.; Yang, H. A Research on Development of Construction Industrialization Based on BIM Technology under the Background of Industry 4.0. *MATEC Web Conf.* **2017**, *100*, 02046. [CrossRef]

46. Dallasega, P.; Marcher, C.; Marengo, E.; Rauch, E.; Matt, D.T.; Nutt, W. A Decentralized and Pull-based Control Loop for On-Demand Delivery in ETO Construction Supply Chains. In Proceedings of the 24th Annual Conference of the International Group for Lean Construction, Boston, MA, USA, 18–24 July 2016; pp. 33–42.

47. Delbrügger, T.; Lenz, L.T.; Losch, D.; Roßmann, J. A navigation framework for digital twins of factories based on building information modeling. In Proceedings of the 2017 22nd IEEE International Conference on Emerging Technologies and Factory Automation (ETFA), Limassol, Cyprus, 12–15 September 2017; pp. 1–4.

48. Scheffer, M.; Konig, M.; Engelmann, T.; Tagliabue, L.C.; Ciribini, A.L.C.; Rinaldi, S.; Pasetti, M. Evaluation of Open Data Models for the Exchange of Sensor Data in Cognitive Building. In Proceedings of the 2018 Workshop on Metrology for Industry 4.0 and IoT, Brescia, Italy, 16–18 April 2018; pp. 151–156.

49. Dallasega, P.; Frosolini, M.; Matt, D.T.T. An approach supporting real-time project management in plant building and the construction industry. In Proceedings of the 21st XXI Summer School Francesco Turco, Naples, Italy, 13–15 September 2016; pp. 247–251.

50. Axelsson, J.; Froberg, J.; Eriksson, P. Towards a system-of-systems for improved road construction efficiency using lean and industry 4.0. In Proceedings of the 2018 13th Annual Conference on System of Systems Engineering (SoSE), Paris, France, 19–22 June 2018; pp. 576–582.

51. Ding, K.; Shi, H.; Hui, J.; Liu, Y.; Zhu, B.; Zhang, F.; Cao, W. Smart steel bridge construction enabled by BIM and Internet of Things in industry 4.0: A framework. In Proceedings of the 2018 IEEE 15th International Conference on Networking, Sensing and Control (ICNSC), Zhuhai, China, 27–29 March 2018; pp. 1–5.

52. Monizza, G.P.; Bendetti, C.; Matt, D.T. Parametric and Generative Design techniques in mass-production environments as effective enablers of Industry 4.0 approaches in the Building Industry. *Autom. Constr.* **2018**, *92*, 270–285. [CrossRef]

53. Trapp, M.; Richter, R. Towards The Generation of Digital Twins for Facility Management Based on 3D Point Clouds Urban Analytics View project Visual Analytics for Sensor Data View project Vladeta Stojanovic Hasso Plattner Institute. In Proceedings of the ARCOM 2018: 34th Annual Conference, Belfast, UK, 3–5 September 2018.

54. Bianconi, F.; Filippucci, M.; Buffi, A. Automated design and modeling for mass-customized housing. A web-based design space catalog for timber structures. *Autom. Constr.* **2019**, *103*, 13–25. [CrossRef]

55. Pruskova, K. Beginning of Real Wide us of BIM Technology in Czech Republic. *IOP Conf. Ser. Mater. Sci. Eng.* **2019**, *471*, 102010. [CrossRef]

56. Correa, F.R. Cyber-physical systems for construction industry. In Proceedings of the 2018 IEEE Industrial Cyber-Physical Systems (ICPS), St. Petersburg, Russia, 15–18 May 2018; pp. 392–397.

57. Hotový, M. Dynamic model of implementation efficiency of Building Information Modelling (BIM) in relation to the complexity of buildings and the level of their safety. *MATEC Web Conf.* **2018**, *146*, 01010. [CrossRef]

58. Schweigko, A.; Monizza, G.P.; Domi, E.; Popescu, A.; Ratajczak, J.; Marcher, C.; Riedl, M.; Matt, D. Development of a digital platform based on the integration of augmented reality and BIM for the management of information in construction processes. In Proceedings of the IFIP International Conference on Product Lifecycle Management, Turin, Italy, 2–4 July 2018; Volume 540, pp. 46–55.

59. De Lange, P.; Bähre, B.; Finetti-Imhof, C.; Klamma, R.; Koch, A.; Oppermann, L. Socio-Technical challenges in the digital gap between building information modeling and industry 4.0. *CEUR Workshop Proc.* **2017**, *1854*, 33–46.

60. Maresova, P.; Soukal, I.; Svobodova, L.; Hedvicakova, M.; Javanmardi, E.; Selamat, A.; Krejcar, O. Consequences of Industry 4.0 in Business and Economics. *Economies* **2018**, *6*, 48. [CrossRef]

61. Rahim, Z.A.; Yusoff, M.R.M. *Standard and Guidelines to Malaysia Fourth Industrial Revolution: Assessment, Readiness and Development Program*, 1st ed.; Siri Sdn Bhd: Kuala Lumpur, Malaysia, 2018.

62. Niu, Y.; Lu, W.; Chen, K.; Huang, G.G.; Anumba, C. Smart Construction Objects. *J. Comput. Civ. Eng.* **2016**, *30*, 04015070. [CrossRef]

63. Hampson, K.; Kraatz, J.A.; Sanchez, A.X. The Global Construction Industry and R & D. In *R&D Investment and Impact in the Global Construction Industry*, 1st ed.; Routledge: Oxon, UK, 2014; p. 364.

64. Boulila, N. *Guidelines for Modeling Cyber-Physical Systems–A Three-Layered Architecture for Cyber Physical Systems*; Siemens AG, Corporate Technology: Munich, Germany, 2017.

65. Akanmu, A.; Anumba, C.J. Cyber-physical systems integration of building information models and the physical construction. *Eng. Constr. Archit. Manag.* **2015**, *22*, 516–535. [CrossRef]

66. Ellis, M. Industry 4.0 and The Future of Construction. 2018. Available online: https://rebim.io/industry4-0-and-the-future-of-construction/ (accessed on 30 May 2019).

67. Latiffi, A.A.; Mohd, S.; Kasim, N.; Fathi, M.S. Building Information Modeling (BIM) application in Malaysian construction industry. *Int. J. Constr. Eng. Manag.* **2013**, *2*, 1–6.

68. Akinade, O.O. *Bim-Based Software for Construction Waste Analytics Using Artificial Intelligence Hybrid Models*; University of the West of England: Bristol, UK, 2017.

69. Grilo, A.; Jardim-Goncalves, R. Value proposition on interoperability of BIM and collaborative working environments. *Autom. Constr.* **2010**, *19*, 522–530. [CrossRef]

70. Carneiro, J.; Rossetti, R.J.F.; Silva, D.C.; Oliveira, C. BIM, GIS, IoT, and AR/VR Integration for Smart Maintenance and Management of Road Networks: A Review. In Proceedings of the 2018 IEEE International Smart Cities Conference, ISC2, Kansas City, MO, USA, 16–19 September 2018.

71. Fan, S.-L. Intellectual Property Rights in Building Information Modeling Application in Taiwan. *J. Constr. Eng. Manag.* **2014**, *139*, 556–563. [CrossRef]

72. Azhar, S. Research Impact Principles and Framework. Australian Research Council. 2013. Available online: https://www.arc.gov.au/policies-strategies/strategy/research-impact-principles-framework (accessed on 1 July 2019).

73. Boyes, H. Security, Privacy, and the Built Environment. *IT Prof.* **2015**, *17*, 14–18. [CrossRef]

74. Ali, S.; Al Balushi, T.; Nadir, Z.; Hussain, O.K. *ICS/SCADA System Security for CPS*; Studies in Computational Intelleigence; Springer-Verlag: Cham, Switzerland, 2018; Volume 768, pp. 89–113.

75. Huda, S.; Miah, S.; Hassan, M.M.; Islam, R.; Yearwood, J.; Alrubaian, M.; Almogren, A. Defending unknown attacks on cyber-physical systems by semi-supervised approach and available unlabeled data. *Inf. Sci.* **2017**, *379*, 211–228. [CrossRef]

76. Spickard, J. Useful Ideas for Doctoral Research. 2008. Available online: http://www.coolsociology.net/Handouts/Spickard---WhatisaConceptPaper(CCLicense).pdf (accessed on 1 January 2019).

77. Böke, J.; Knaack, U.; Hemmerling, M. State-of-the-art of intelligent building envelopes in the context of intelligent technical systems. *Intell. Build. Int.* **2018**, *11*, 27–45. [CrossRef]

78. Kehr, F.; Kowatsch, T. Quantitative Longitudinal Research: A Review of IS Literature, and a Set of Methodological Guidelines. In Proceedings of the ECIS 2015, Münster, Germany, 26–29 May 2015.

Article

Behavior and Performance of BIM Users in a Collaborative Work Environment

Eric Forcael [1], Alejandro Martínez-Rocamora [2,*], Javier Sepúlveda-Morales [3], Rodrigo García-Alvarado [4], Alberto Nope-Bernal [5] and Francisco Leighton [6]

[1] GeDIE, Department of Civil and Environmental Engineering, Faculty of Engineering, Universidad Del Bío-Bío, Av. Collao 1202, Concepción, Chile; eforcael@ubiobio.cl

[2] ArDiTec, Department of Architectural Constructions II, IUACC, Higher Technical School of Building Engineering, Universidad de Sevilla, Av. Reina Mercedes 4-a, 41012 Sevilla, Spain

[3] Integrated Project Design Lab., Department of Civil and Environmental Engineering, Faculty of Engineering, Universidad Del Bío-Bío, Av. Collao 1202, Concepción, Chile; luisepul@egresados.ubiobio.cl

[4] GeDIE, Department of Architectural Design and Theory, Faculty of Architecture, Construction and Design, Universidad Del Bío-Bío, Av. Collao 1202, Concepción, Chile; rgarcia@ubiobio.cl

[5] GeDIE, Faculty of Architecture, Universidad La Gran Colombia, Carrera 6, Cundinamarca 12B-40, Colombia; yuber.nope@ugc.edu.co

[6] Department of Civil and Environmental Engineering, Faculty of Engineering, Universidad Del Bío-Bío, Av. Collao 1202, Concepción, Chile; fleighto@alumnos.ubiobio.cl

* Correspondence: rocamora@us.es; Tel.: +34-675041010

Received: 29 January 2020; Accepted: 3 March 2020; Published: 24 March 2020

Featured Application: This research has specific applications in architecture firms developing Building Information Modeling (BIM) projects where coordinators and managers are willing to analyze behavioral patterns and efficiency of their modelers and to identify further training needs and opportunities to increase their performance.

Abstract: Collaborative work in Building Information Modeling (BIM) projects is frequently understood as the interaction of modelers in an asynchronous way through modification requests or via e-mail/telephone. However, alternative work methodologies based on creating a common and synchronous environment allow solving issues instantaneously during the design process. This study aimed to analyze the behavior and performance of BIM users with different specialties who were subjected to an experimental exercise in a collaborative environment. For this purpose, a process was devised to collect, sort, and select the data from the log files generated by the BIM software. A timeline of the experiment was populated with data on the intensity and types of commands used by each specialist, which allowed determining behavioral patterns, preferred commands, indicators of their experience, further training needs, and possible strategies for improving the team's performance. In the experiment, the mechanical designer's performance was 49% and the rest approximately 64%, with respect to that of the architect. An average rate of 1.66 necessary or auxiliary commands for each contributory command was detected. The average performance was 200–400 commands per hour, which intensified by the end of the experiment. Further training needs were detected for the plumbing designer to reduce the use of backwards commands. Conversely, the electrical designer showed a positive evolution regarding this aspect during the experiment. The analysis methods here described become useful for the aforementioned purposes. Nevertheless, combinations with methods from existing research might improve the outcomes and therefore the specificity of recommendations.

Keywords: building information modeling (BIM); log data mining; modeling performance; collaborative environment; behavioral patterns

---

1. Introduction

With the advance of technology, improvements in industry are generally observed as an increase in productivity. However, the architecture, engineering and construction (AEC) industry has low productivity, with fragmentation of activities, lack of added-value and weak standardization [1]. This is partly due to limited collaboration among the participants in this sector. In this context, Building Information Modeling (BIM) arises as a collaborative methodology to facilitate information transfer between the various specialties involved in the design process of a construction project, thus improving decision-making in the early stages [2]. This methodology enables users to create parametric models based on multidimensional objects, which are a tool for managing construction projects throughout their life cycle [3–6].

In recent years, the construction sector has experienced changes in its traditional work system with the introduction of BIM-based technological solutions. This phenomenon is generating a revolution in building work practices, which both poses a challenge and provides an opportunity to achieve the objectives of a given project [7]. Without forgetting BIM's numerous advantages, it is important to point out that there are important challenges to be faced regarding the many disciplines that comprise construction companies, who must prepare to adopt the concept of BIM [8,9].

Despite the great potential that BIM presents for the improvement of collaboration between the parties involved [9,10], some difficulties have been detected in its successful implementation. For this reason, in recent years numerous studies have analyzed the gap between the construction industry and the introduction of Information and Communication Technologies (ICT). Since this problem affects the performance of the entire process, there is a need to diagnose the cause of current difficulties [11]. From previous studies, Mutai [12] associates the success of BIM with the BIM skills and training demonstrated by a project team, which are determined by their collaboration. In addition, Azhar [13] indicates that collaboration between different disciplines and users is essential for the successful implementation of BIM. Therefore, the need arises to study the collaborative behavior of BIM users in more detail.

When used properly, BIM promises to improve all the processes in the different stages of design and construction, reducing the quantity and severity of problems associated with traditional practice [13], and avoiding nearly all communication difficulties related to information transfer [14]. However, to achieve this transition process, users face various challenges, mainly derived from the collaboration required between a variety of specialists in the early stages: architects, engineers, analysts and others, in conjunction with the client, which is lacking in the current design process [15]. Furthermore, collaboration in BIM projects usually occurs in an asynchronous way, with each of the specialists working in their respective offices and communicating whether via e-mail or modification requests on the central model. This frequently causes problems between the specialists, which eventually translate into increased redesign time and costs [16]. It is important to notice that BIM is directly driven by human activities [17] and software merely generates a platform that makes this encounter possible.

Research on collaborative work in BIM has made it possible to detect the possibilities it has to offer and also the difficulties that arise from it. However, a methodology has not been established yet to evaluate the behaviors and performance of the different participants of a BIM project, which would allow developing more effective group work procedures. In order to successfully implement this methodology, it is necessary to train existing work teams in the development of collaboration skills. To that end, collaborative work environments have emerged as a suitable means to identify those skills that each specialist should improve, and therefore as a way to design personalized training.

In this research, the aim was to analyze the behavior and performance of diverse profiles of BIM modelers during an experimental exercise of BIM collaborative work. To that end, a methodology was designed and its potential application was demonstrated for their analysis based on the data generated by a BIM software in its log files. In that experiment, users worked under an architect's leadership in the same physical environment with the aim of accelerating the design process and improving the quality of results [18,19]. This work methodology is based on the hypothesis that a better understanding of the performance and activities among participants can provide ideas on how to increase their efficiency, effectiveness, and organization [20], as well as the interactions between them that can help to identify bad practices and further training needs.

In the next section, background is provided on collaborative work environments and the study of BIM users' behavior through the analysis of log files generated by the software. In Section 3, the design of the experiment and the case study are presented. The methodology for studying the participants' behavior and performance is organized following a sequence of data collection, processing, and analysis. Results of the identification of relevant actions that contribute to generating a model of parameterized design and the analysis of information exchanges between the project developers are presented and discussed in Section 4, based on a comparison of the data obtained from the work carried out by the different specialists involved. These results are discussed and compared to other findings from similar studies based on BIM log mining in Section 5, where future research directions are also highlighted.

2. Background

2.1. Collaborative Work Environments

A collaborative work environment is referred to in this study as a common physical room with the optimal conditions to develop the project, where a group of various professionals work simultaneously in order to allow direct and synchronous interactions between them [18,19]. In this environment, there are two communication networks: the human network that is based on the interactions between the individuals; and the electronic network that computers use to exchange information. While virtual collaboration experiences based on cloud systems do exist and are more frequent in the development of BIM projects, physical interaction between the different agents makes it possible to increase the efficiency of the design process and results in quicker problem-solving and decision-making.

The organization of a collaborative work environment is a complex process in which diverse participants must adapt their behavior to the peculiarities of this new methodology, selectively monitor information, and continuously give feedback in order to solve problems, improve project design, and finally, decrease the associated costs. In these work environments, work is usually organized into one to three sessions per week; teams consist of a leader and engineers specialized in design sub-areas; each designer is allowed to work on their design and move freely around the room in search of the information needed to solve their problems; and when an difficulty pertains to the whole team, the session is stopped, and the entire group focuses on resolving the problem [21].

Due to its great success, in recent years, the AEC industry has adapted and modified this methodology. In this case, the sessions are divided into cycles, which should not last more than two weeks depending on the phase of the project [21]. The work teams are comprised of a leader, designers, builders, the client, and the suppliers, with some or all of the parties participating, depending on which are necessary for each cycle. Despite high interest in the use of this methodology with BIM projects, there is limited evidence regarding the behavior and performance of BIM users in these work environments.

2.2. BIM Users' Behavior and Performance and BIM Log Files

One key factor in the study of BIM users' behavior and performance in a collaborative work environment is the identification and development of appropriate collaboration models to improve user performance rates [22]. However, this area of research has mainly focused on using data from BIM logs to improve collaborative, design, and non-design practices [23–25].

Autodesk Revit is a BIM software with a set of tools to coordinate the different disciplines in a project, thus minimizing the risk of errors in execution and decreasing production times and associated costs [10]. This software automatically generates log files as a database in which all the relevant information is recorded, from the technical specifications of the computer to the commands executed by the user, including when and for how long the program was used. Due to the huge amount of data they contain, reviewing these log files is a considerably complex and time-consuming task; consequently, they are frequently only used to investigate possible failures in the software or the computer in use. However, lately, new and innovative uses are being given to these sources of valuable information.

For example, Yarmohammadi et al. [26] analyzed the content of log files from an architecture and design firm to investigate the presence of design patterns and to characterize the performance of BIM modelers. To that end, they searched the log files of different modelers, identified patterns of at least three sequential commands, and quantified the time it took these modelers to execute them, thus determining both individual work behavior and performance of those modelers.

Zhang and Ashuri [22] developed a systematic procedure for extracting data from BIM log files in order to monitor and measure productivity in the design process. They used a set of techniques known as 'process mining', which is usually divided into three branches according to the objective: (1) 'process discovery', (2) 'process conformance', and (3) 'process improvement' [27]. Of these, 'process discovery', which involves using saved records as an input to generate a process model, is the most widely employed [28]. In their case, Zhang and Ashuri [22] presented a novel methodology that used these records to capture and model collaboration patterns between designers and generated social networks according to their behavior. To do this, they used the social network analysis (SNA) metric-type 2, which, in a design context, measures the number of times two designers interact with each other to contribute to the same project.

In a second study, Zhang et al. [23] focused on identifying patterns in designer behavior and found that in general there are three commands of greatest use, which vary among the different designers. It was also possible to measure which patterns were the most efficient when working on a project. In this research, the pattern discovery process focused on the description of user operations as represented by executed commands.

In addition, Oraee et al. [24] developed a conceptual model of collaboration in BIM through a literature review in order to identify collective behavior. And Kouhestani and Nik-Bakht [25] studied event logs archived during the design process of building projects using IFC (Industry Foundation Classes) files, discovering different approaches to support BIM management.

Finally, Pan and Zhang [29] applied fuzzy clustering methods to explore a massive amount of BIM log files from an international architecture design firm in order to identify productivity patterns in the modelers' behavior. They organized the process in three main stages: data preparation, clustering, and knowledge discovery. The analysis at individual level revealed specific time periods where the modelers showed better performance. The application of team-level clustering made it possible to evaluate if grouping designers with similar design productivity would provide better results for the firm.

3. Materials and Methods

3.1. Design of the Experiment

To implement this experiment for the study of collaborative behavior and performance of BIM modelers, it was first necessary to determine the technologies the different designers would use, in addition to organizing their participation in the case study. This process was based on an initial technical characterization of the project that identified the design and construction conditions for the various disciplines. This analysis is essential for the development of successful BIM-based projects since it facilitates understanding and coordination of the various roles and their roles in the design process.

The participants in this experiment included a group of three civil engineers, a building engineer, and an architect who also had a coordination role. Each participant had received previous training in Autodesk Revit and collaborative work with different roles. The participants were junior professionals aged between 25 and 35 years old with an average experience of three years, except for the architect and coordinator, who had 10 years of experience in architecture firms. As shown in Figure 1, all the designers interacted in a common work environment, supported by modeling technologies to carry out the work. The work was done using a central file shared through a local network, and a local copy for each specialist. Access was granted to the central file to save or to obtain the data of the other members in real time through Autodesk Revit´s synchronization features.

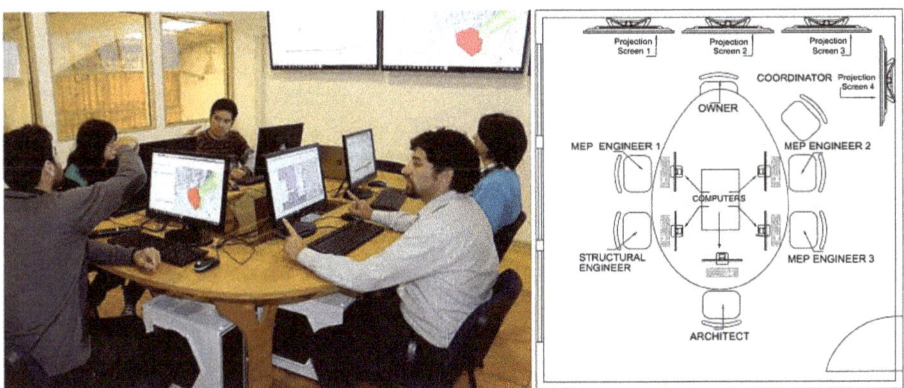

Figure 1. Layout and implementation of the collaborative work environment.

The activity was organized into 18 3-h sessions conducted over six weeks, in which the participants worked together on the modeling of a two-story building for university classrooms (see Figure 2) based on previously existing plans of the project. The participants attended to all the sessions, with the exceptions of the mechanical and the electrical specialists, who were absent in three and two entire sessions, respectively, and took several long pauses throughout the experiment. The parts of the project were first distributed among the disciplines in sub-projects (namely 'worksets' in the selected BIM software) in order to organize permissions to modify objects. The experiment was performed in a collaborative work environment so that the participants were able to directly compare, evaluate, and solve problems due to clashes between models—whether that be the coincidence of two elements in the same physical space or inconsistencies in the design detected by the program or the specialists themselves-by generating a continuous exchange of information between them.

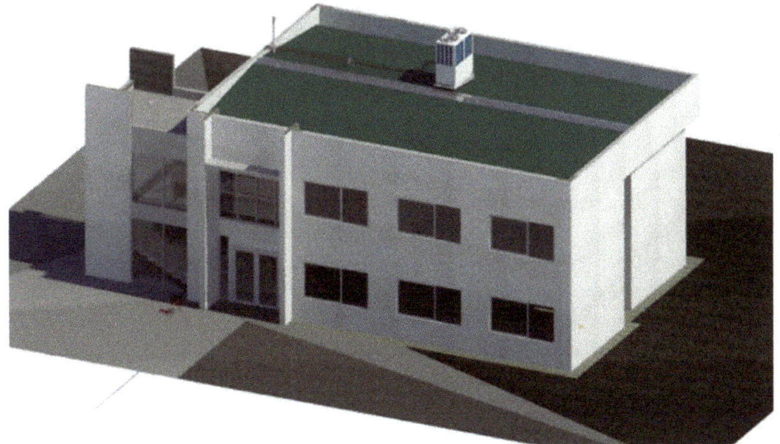

Figure 2. Classroom building model in Autodesk Revit.

3.2. Collection and Processing of Data from Log Files

Subsequently, the activities performed by each specialist were analyzed through the content of the log files, which also allowed identifying certain behavioral patterns regarding pauses in modeling activity or dependencies among the various specialties in the design process. Autodesk Revit records activities in a text file that describes how the designer uses the software [23]. Log files are generated for each session with the corresponding date and time of each action. These files include information about the project name, current user, session, commands used, and computer characteristics (see Figure 3), thus enabling the extraction of information about all the operations performed within the model.

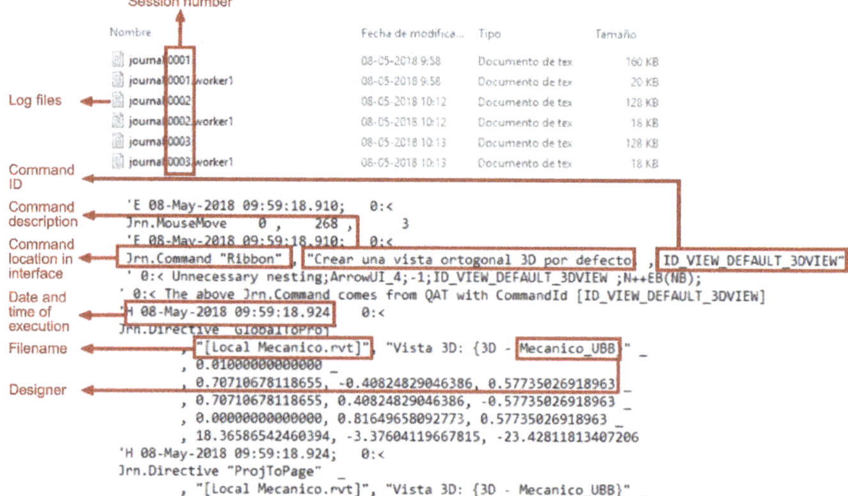

Figure 3. An example of a Building Information Modeling (BIM) log file and its content.

During this experiment, the log files were collected from each computer after the end of each session and subsequently stored and organized into folders per day of work to correctly identify the data corresponding to each session. A session is defined as the set of activities carried out by a user from the moment the BIM software is loaded until the moment it is closed [30].

Data registered in the server generated a large number of files, thereby making it necessary to clean and organize the information to obtain a structured, reliable, and integrated database to study the frequency of execution of each command. These cleaning and organization processes were carried out as follows:

1. *Data cleaning*: In this first step, a simple program was developed to parse the information from the log files and exclude a large number of records related to the configuration and monitoring of operations, which are not relevant for the characterization of tasks. The result was a text file containing all the commands executed by each user and indicating their ID (command identification code), access route, a short description, and the date and time of execution.

2. *Data organization*: After the previous data cleaning process, the information was transferred to a spreadsheet and organized into tables for each designer, which made it possible to count the number of executions of each command for an initial analysis.

Additionally, each session was divided into five periods of equal duration in order to generate more data for the analysis of distributions and to better allocate isolated events that might arise related to users' behavior. Two main categories, contributory and non-contributory commands, with three sub-categories each (described in Table 1), were created to classify the commands executed by each designer, in this way producing a more precise analysis.

Table 1. Classification system for executed commands.

Category	Sub-Category	Description	Examples
Contributory	Geometrical modeling	Modeling commands that translate into a virtual object in the model	Wall, Door, Beam, Finish sketch, Move, Align
	Non-geometrical modeling	Commands to modify object parameters, project configuration, or templates	Edit type, Annotations, Level, Grid, Schedule
	Collaborative	Commands with an implied interaction with the other designers	Save to master file, Relinquish all mine, Editing requirements
Non-contributory	Backwards	Commands that imply a step backwards in the contribution to the final product	Undo, Delete, Cancel sketch
	Necessary	Commands that do not contribute to the modeling process, but are necessary for software design reasons	Cancel, Open file, Temporal hide, Visualization option
	Unnecessary	Commands that should not have been executed, whether because they correspond to a different specialty, or are completely unnecessary for the advance of the project	More than two 'Cancel' in a row

3.3. Methods for Analysis

To analyze the behavior and performance of users, the relevance of the commands used by each specialist was studied. First, a preliminary analysis was conducted to detect certain behavioral patterns for each designer based on the overall quantity of executions of each sub-category of commands and the effect that these had on the development of the project. Additionally, the compliance of each designer's contributions with the Pareto principle (also known as the 80/20 rule) was studied. The Pareto principle states that in any group of elements that contribute to the same effect, a few (~20%) are usually responsible for most of that effect (~80%). Therefore, non-contributory commands were excluded from this analysis.

Pareto charts consisting of a bar graph and a curve were created for each designer. In these charts, the commands are sorted in descending order by their total number of executions and the curve represents the relative frequency of these commands. The Pareto principle, supported by the chart, is a simple visual tool for focusing and analyzing relationships, studying results, and planning for continuous improvement, and has been proven to be valid in numerous situations [31]. Thus, in accordance with the Pareto principle, the number of contributory commands that make up 80% of the total number of executions (relative frequency) should be approximately 20% of the number of different contributory commands executed by a designer.

In order to gain a better understanding of the events occurred during the experiment and their timing, the 18 sessions were divided into five periods each, for a total of 90 periods. This made it possible to detect behavioral patterns, such as pause times, non-modeling activities, and performance metrics, such as modeling intensities-the rate of commands per hour-, as well as training priorities. As a first step, pauses longer than two minutes were quantified, considering that designers frequently stop modeling for a short time to think, to consult other documents, such as standards, books or catalogues, or to solve issues with other designers, given that BIM workflows are known to be hybrid and not solely BIM-based [32], and sometimes they take longer pauses to rest. Therefore, the rate of commands per hour (work intensity) for each specialist in each section was calculated both with respect to real time (the actual duration of the period) and to active time (subtracting pauses longer than two minutes).

Then, the distributions of real-time execution intensities of each command sub-category during the experiment were studied by generating boxplot charts for each designer, which help detecting differences between the types of commands each specialist executed, as well as identifying those modelers with higher collaborative behavior and those applying bad practices and therefore needing further training to improve their performance.

Finally, timelines of the shares of commands of each sub-category executed by the different designers were generated that allow detecting sessions without active modeling participation, excessive shares of non-contributory commands, and evolution of the modeling behavior of each specialist during the experiment, among others.

4. Results

4.1. Preliminary Analysis

In the experiment, each designer worked for approximately 54 h total. The data extracted from the log files were analyzed to evaluate the performance and collaborative behavior of each user, which can be related to the execution of commands from each sub-category in Table 1. In addition, the percentage of use of each contributory command executed was analyzed for all the designers in order to validate the Pareto principle.

The intensity of commands executed by each designer can be considered a confounding indicator in this preliminary analysis since this measurement can alternately demonstrate high efficiency in the use of commands and/or greater experience in the use of the BIM software, or on the contrary, lack of knowledge or lower experience as evidenced by the execution of commands that provide little advance to the project. Therefore, higher performance cannot be inferred from this raw data, and further analysis is required in order to identify those commands actually contributing to the model.

As can be seen in Table 2, while the total number of commands used by the architect was the highest, this designer also executed a great quantity of unnecessary commands, which were nearly all 'Cancel' (ESC key). In Autodesk Revit, a designer must use the ESC key twice to exit the current command, although on occasion users push this key more than twice, perhaps while thinking. Although this does not affect their performance, it must not be taken into account as productive work. Conversely, the mechanical designer executed the lowest number of commands, both total and unnecessary, which apparently could indicate a more efficient use of commands, but also lower performance.

Table 2. Total number of commands executed by each designer.

Designer	Total	Contributory Commands			Non-Contributory Commands			Contributory Vs Backwards
		G	NG	C	N	U	B	
Architectural	9430	1770	169	372	3291	3228	657	2150
Structural	6433	1353	203	228	2470	920	1296	1425
Mechanical	5555	1032	121	141	2496	726	1039	1052
Electrical	6523	1296	165	184	2918	994	966	1401
Plumbing	7360	1190	366	165	3101	776	1762	1309

G: Geometrical modeling; NG: Non-geometrical modeling; C: Collaborative; N: Necessary; U: Unnecessary; B: Backwards.

As was expected, the architectural designer also stands out with the highest number of geometrical modeling commands, which not only reflects the need to define the building's spaces and shape, but also the subsequent modifications to improve the design and resolve clashes between models. On the other hand, the plumbing designer executed the highest quantity of non-geometrical modeling commands. Through a deeper analysis of the specific commands executed by this specialist, it was detected that this was mainly due to the definition of numerous family types, creation of schedules, and inclusion of annotations in the model. The architectural and structural designers executed the highest number of collaborative commands, the former significantly more, which was also expected given that these two specialties provide most of the information that mechanical, electrical and plumbing (MEP) specialties require. In return, the three MEP designers supplied a similar amount of information that made it possible to detect and solve clashes between models.

Regarding non-contributory commands, all the specialists executed a large amount of necessary commands. These auxiliary commands, such as those to modify the view configuration or temporarily hide elements, are frequently needed to advance, but do not provide new information or elements to the model. While it has not still been determined what would be an appropriate rate of necessary to contributory commands, in this experiment an average rate of 1.66 necessary commands for each contributory command was detected. Therefore, the data indicate how the specialists rely on the use of auxiliary commands during their modeling process. As was mentioned previously, the architectural designer pushed the ESC key excessively, while the other specialists' use of unnecessary commands was all similar to one another. Alternately, the architectural designer used the lowest quantity of backwards commands, whilst the plumbing designer executed a large amount of these. This may mean a greater number of corrections, less experience in BIM modeling, and/or bad modeling practices, and therefore a need for further training. However, identifying specific recommendations for improving a modeler's modeling practices would require an analysis of command execution patterns, which is not within the scope of the present study.

Additionally, the balance between backwards and contributory commands was studied since it can be assumed that these cause opposite effects that essentially cancel each other out. Given that 'Undo' commands can affect commands from all the sub-categories including 'Delete' commands in the backwards sub-category, they were subtracted proportionally from the quantity of contributory commands. From this examination, it can be inferred that the architectural designer had the most effective modeling behavior, while the mechanical designer presented the lowest performance. These overall results show that the mechanical designer's performance in the experiment was 49% with respect to that of the architect, while the rest of designers' was around 64%. This means that, even though the architect used a significant number of unnecessary commands, that designer had the highest performance during the experiment by a significant difference. This excessive use of the ESC key, while considered a bad practice, did not affect his/her performance. However, it could be advisable to address it through further training focused on developing a more fluent modeling process.

In this preliminary analysis, absolute and relative frequencies were obtained for all the commands executed by each designer in order to identify those of preferential use and to check compliance with the Pareto principle. As an example, Table 3 shows the contributory commands most frequently used by the architectural designer, which account for 80% of his/her total contributory command executions. It can be seen that most of these commands are related to geometrical modeling, whether direct commands for creating objects or those for defining sketches that eventually will enable the creation of sketch-based objects. Similar tables for the other four designers are shown in Appendix A, where the close relationship between the commands executed and their corresponding specialty can be observed.

Table 3. Contributory commands most frequently used by the architectural designer.

Command ID	Description	Absolute Frequency	Relative Frequency
ID_EDIT_MOVE	Move selected objects or their copies	369	16.0%
ID_FINISH_SKETCH	Finish sketch	339	30.6%
ID_OBJECTS_WALL	Create a wall	237	40.9%
ID_FILE_SAVE_TO_MASTER_SHORTCUT	Save the active project to the central model again	182	48.8%
ID_OBJECTS_CURVE_LINE	Create a line	169	56.1%
ID_FILE_SAVE_TO_MASTER	Save the active project to the central model again	125	61.5%
ID_EDIT_MOVE_COPY	Move copies of selected objects	113	66.4%
ID_OBJECTS_CURVE_RECT	Create a rectangle	56	68.8%
ID_SPLIT	Divide walls and lines	45	70.7%
ID_REVIT_FILE_SAVE	Save the active project	44	72.7%
ID_EDIT_MATCH_TYPE	Copy the object type to other objects	40	74.4%
ID_OBJECTS_DOOR	Create a door	39	76.1%
ID_OBJECTS_CW_GRID	Create a grid line in a curtain wall	38	77.7%
ID_OBJECTS_MULLION	Create a mullion	33	79.1%
ID_LOAD_INTO_PROJECTS	Load document into open projects	33	80.6%

Based on this analysis, the Pareto charts in Figure 4 were produced for each designer, where the horizontal axes show the total number of different contributory commands executed, and the command (in descending order by quantity of executions) with which at least 80% of the contributory command executions is reached. By adopting a simplified assumption whereby each contributory command makes the same contribution to the final product, it can be said that the Pareto principle applies to all the specialists except for the structural designer, who showed a more balanced number of executions of a variety of commands: the 80% threshold was reached with 31.3% of the contributory commands used. Although the architectural and plumbing designers attained 80% with 22% of their contributory commands, this was considered a fair approximation to the Pareto principle. The identification of each designer's preferred commands can be useful for detecting bad modeling practices or which actions should be given focus in further personalized training.

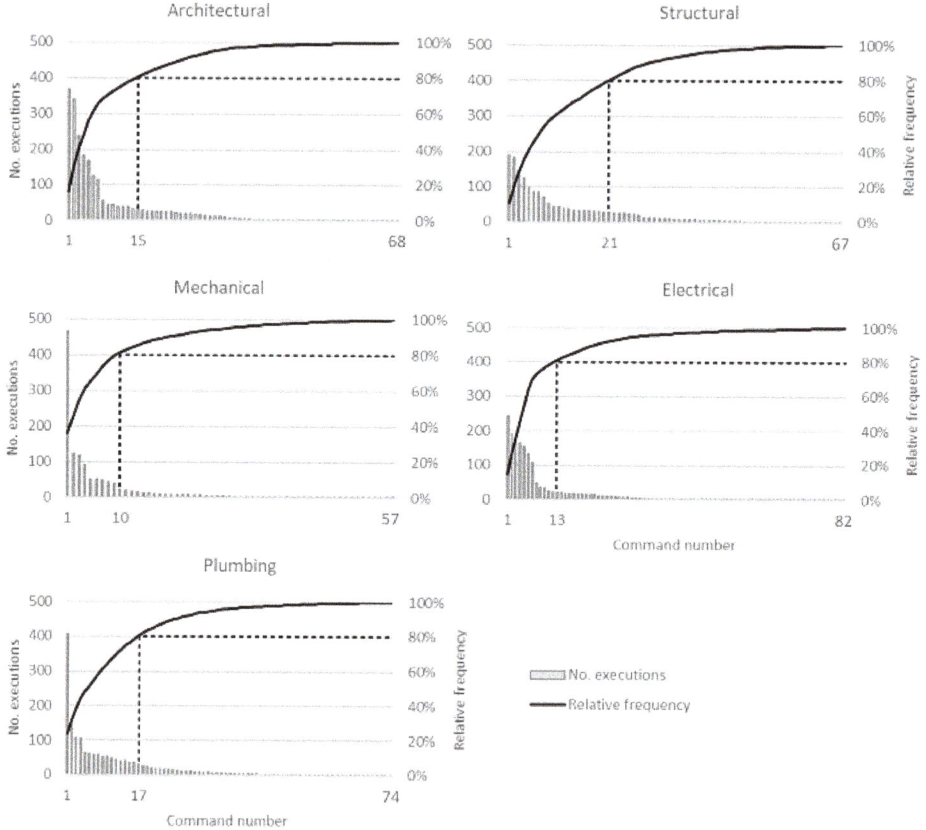

Figure 4. Pareto charts of contributory commands for each specialist.

4.2. Session/Period-Based Analysis

A detailed analysis of the commands executed by each specialist during the different sessions/periods of time of the experiment was carried out according to the methods explained in Section 3.3. Thus, as can be seen in Figure 5, real time intensities are lower than active time intensities, since the latter do not take into account pause times. The real time and active time intensity curves are slightly separate in sections where the designer took a few short pauses, while there is greater separation where the designer took a longer pause (>10 min) or a greater quantity of short pauses. In those periods where the designers did not execute any command, both curves go to zero, which represents a long pause in their work.

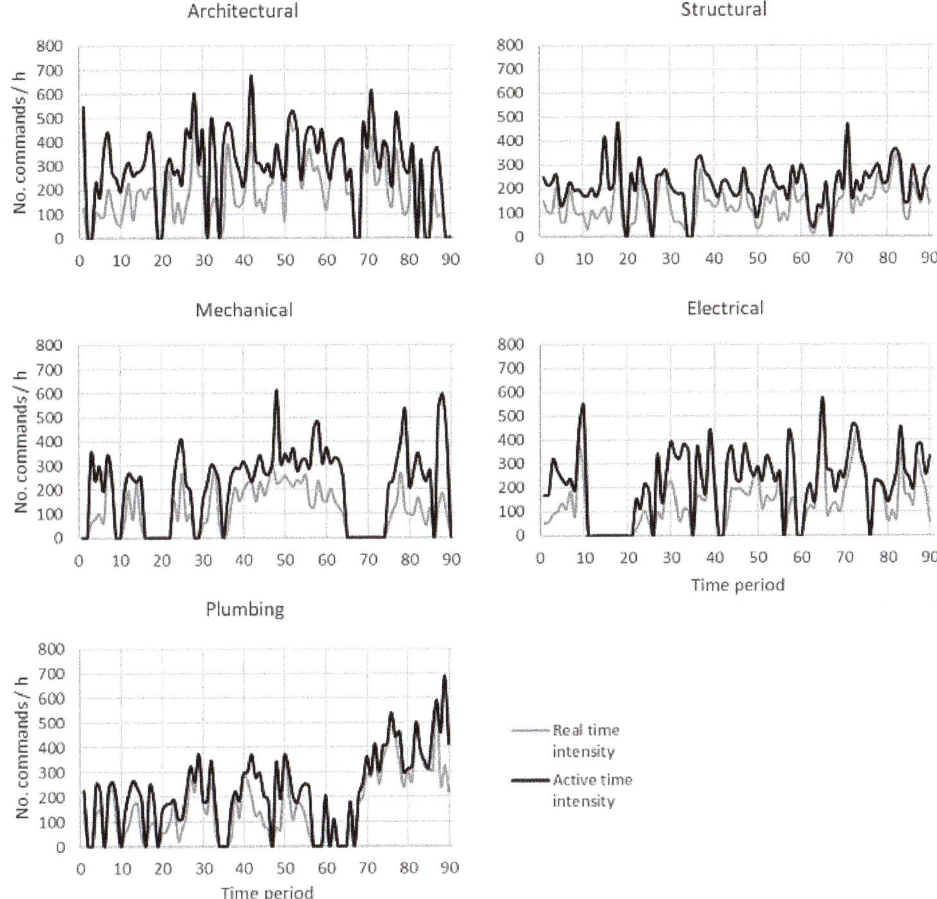

Figure 5. Real time and active time work intensities for each designer.

From this analysis, it can be observed that all the designers frequently took short and long pauses, except for the plumbing specialist, who often took long pauses that bring the curve to zero. However, while working this designer executed commands more uniformly, which is reflected in the frequent coincidence of both curves. The intensity of the architectural designer's commands was clearly the highest, with rates of between 300 and 500 commands per hour. The rest of specialists presented rates of between 200 and 350 commands per hour, with the exception of the plumbing designer, whose work intensity increased to 300 to 600 commands per hour in the last 20 periods (i.e., four sessions). It is worth mentioning that the calculations for these intensities included non-contributory commands and consequently a deeper analysis should be carried out in order to identify anomalies. For example, given the data shown in Table 2, the architectural designer's higher command intensity could have been triggered by the great amount of 'Cancels' he executed during the experiment.

During the experiment, the team worked together, but not everyone worked with the same intensity at all times mainly due to the need to generate a base model for other specialists to begin. Figure 5 shows the evolution of each specialist's work over the 18 sessions (90 time periods), of which the first several were dedicated to preparation and to determining and creating elements and families to be used within the project. In these initial sessions, the architectural and structural designers were of great importance, since their early work determined heights and spaces where the MEP specialists could install their elements. It can also be observed how their command intensities followed a somehow paired curve, with long pauses in periods 20, 34, and 67.

As the experiment progressed, it can be observed how the MEP designers intensified their work, which was initially on standby due to required prior information to be able to advance. In the final phase of the project, all the specialists intensified their activity. This was on account of delays generated during the process that had to be compensated for to finish the project on time. In addition, the need to model construction details, solve clashes between models, and generate quantification tables had a significant influence in this final stage. While the mechanical and electrical designers had a uniform work intensity during the experiment, the plumbing designer showed a remarkable increment from period 70 until the end, whose cause would be worth studying in detail.

In Figure 6, the distribution of the real-time execution intensities for each sub-category of commands is presented. The architectural designer provided geometrical content for the model with the highest uniform intensity of all the specialties (~10 to ~60 commands per hour in 50% of the time periods). Alternately, the plumbing designer stands out with the highest rate of non-geometrical modeling commands (0 to ~11 commands per hour), while the mechanical designer hardly ever provided this kind of information to the model. It is worth mentioning that all the designers present numerous non-frequent values (points) in this chart, indicating frequent peaks in the provision of non-geometrical information, which is a common behavior in BIM modeling. The architectural and structural designers had a higher rate of collaborative actions, through which they updated the central model and received new information from it 0 to ~8–10 times per hour in 50% of the time periods. This was an expected behavior from these specialists, since information was frequently requested from them during the experiment. These events affected their focus, thereby generating errors in their work that involved redoing elements and delivering the information to the central file once again. A possible improvement could consist in the more precise timing of information delivery in order to achieve a better flow of activities and more equitable work. This should be organized by the coordinator previously to the design process to alleviate the pressure on the main designers to deliver information.

Regarding non-contributory commands, the mean rates of necessary commands executed by the various designers were usually similar (~30 to ~80 commands per hour), with the plumbing designer presenting a more extended interquartile range (~10 to ~120 commands per hour). As was expected from the data in Table 2, the architectural designer had the highest rate of unnecessary commands, probably triggered by the aforementioned 'Cancels'. Again, experience in BIM modeling is reflected in the backwards commands chart, where the architectural designer had the lowest rate (~0 to ~20), while the plumbing specialist executed ~5 to ~60 backwards commands per hour. From this data and the balanced rate of forwards vs backwards commands, it can be inferred that the plumbing designer would need further training in BIM modeling. Therefore, it would be worth studying this designer's command patterns during the experiment in more detail in order to identify bad practices that need being addressed.

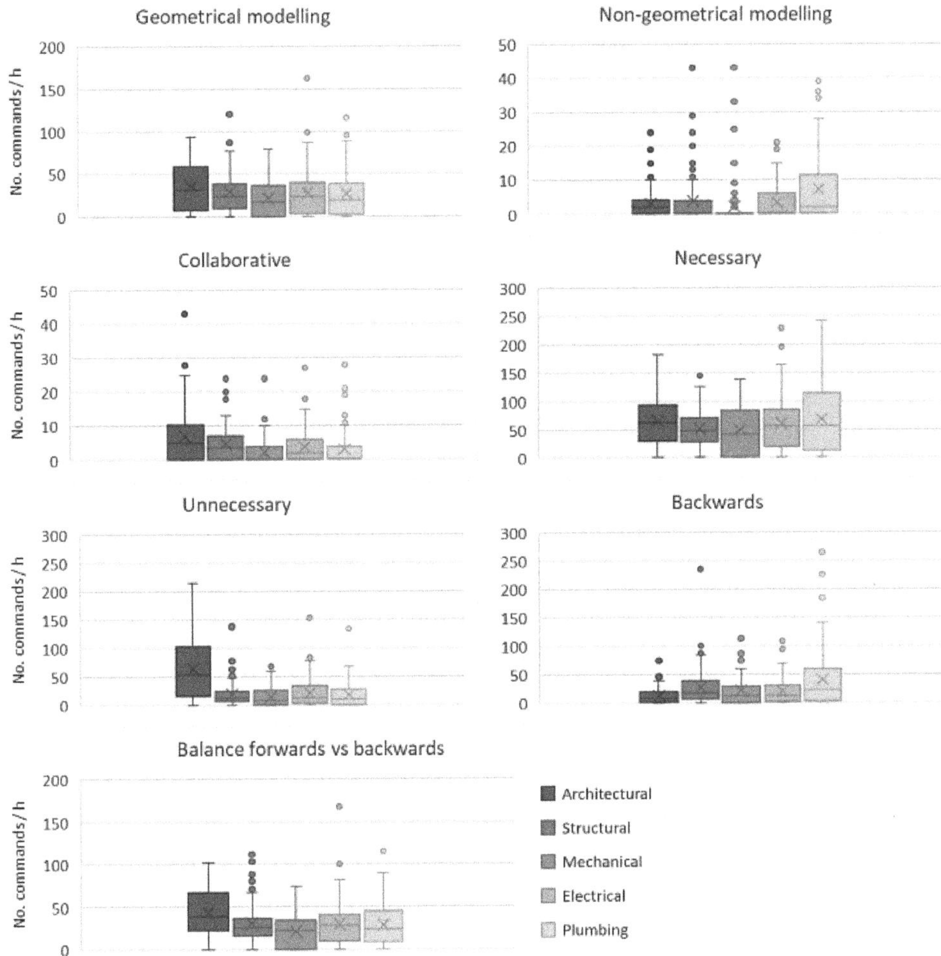

Figure 6. Distribution of the command execution rates for each sub-category by specialty during the experiment´s 90 time periods.

When these distributions are studied in detail for each of the experiment´s 90 time periods, the shares of each sub-category for the different designers provide new data worth analyzing. In Figure 7, the architectural designer shows a significant share of unnecessary commands throughout the experiment, with a maintained share of around 40% in the middle 30 time periods, while his share of backwards commands is the lowest of the five specialties (<10%). This confirms the aforementioned excessive use of the ESC key, as well as better skills in BIM modeling, hence requiring fewer corrections in the design process. His share of necessary commands stays nearly constant and balanced with contributory commands (~30% each). A higher share of collaborative commands can be detected during the initial time periods of the experiment (~10%), where this designer must provide basic information in the model to the other specialists, while constant interaction with the other designers can be identified throughout the rest of the process.

The structural designer shows a higher variability in the shares for each sub-category, with a high share of backwards commands in some time periods (~40–50%), which become negative contributions or massive corrections. This could indicate that a great number of mistakes had been made in previous time periods, or that the modeler was using auxiliary lines (detail lines) that were subsequently deleted. Nevertheless, both can be considered bad practices in BIM modeling that should be corrected. This designer's share of necessary commands is balanced with that of contributory commands, as well (~20–30%). The designer presents higher shares of non-geometrical modeling in the initial sessions (~20%), when families were being defined, and a higher share of geometrical modeling for the rest of the experiment (~30%), which are considered normal behaviors for this specialist. In addition, it was expected that the structural designer would demonstrate a more constant share of collaborative commands than he in fact did (~5%), since clashes are common between this and the other disciplines' models.

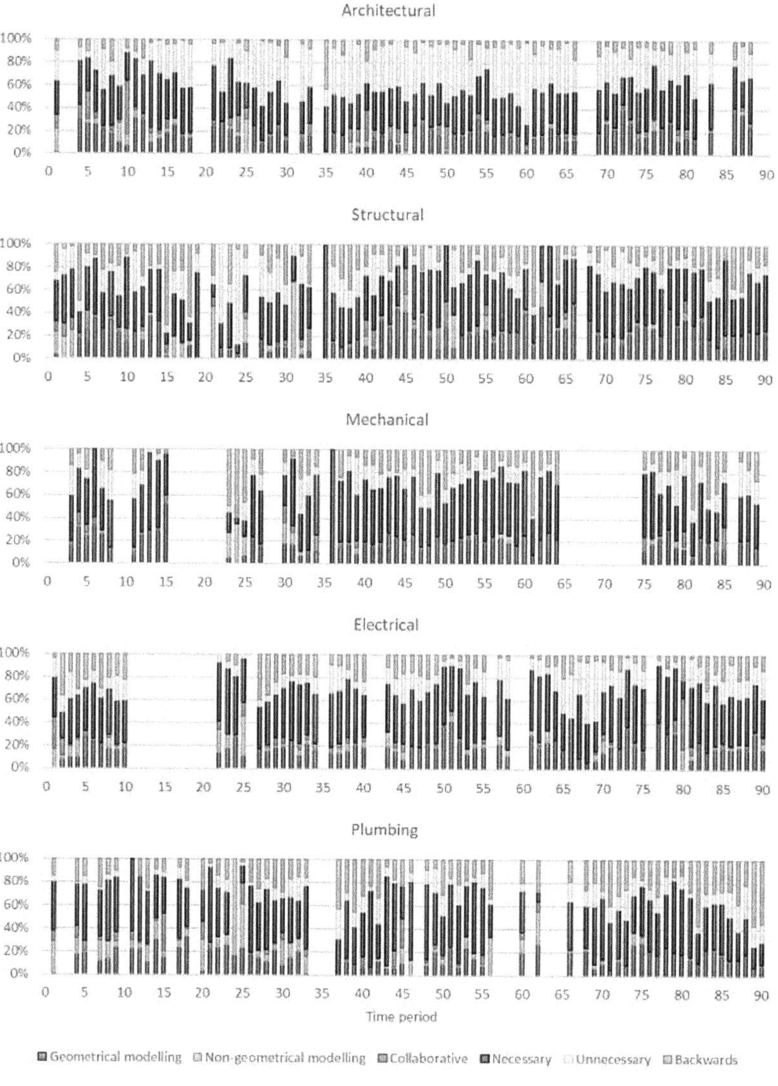

Figure 7. Share of commands by sub-category and time period for each designer.

The timeline for the mechanical designer shows at least three entire sessions without participation (periods 16–20 and 65–74), and several long pauses (periods 1–2, 9–10, 27–28, 35, 86, and 90). This non-contributory behavior is to be expected at the initial stages of the project since MEP specialists are waiting for the architect and structural engineer to provide the basic elements of the building in order to model the installations. However, the initial stages should be dedicated to defining the necessary families for the project, and in more advanced stages a better planning of workflows should provide the various specialists with tasks that ensure constant contribution to the project. It can also be observed that the necessary commands executed by the mechanical designer (~40%) approximately doubled the amount of contributory commands (~20%), which could indicate that this specialist used an excessive number of auxiliary commands, such as hiding elements or modifying visualization options. This kind of modeling practices require a more detailed analysis of the command patterns executed in order to increase the designer's performance by providing personalized training.

The electrical specialist did not participate in two entire sessions (periods 11–20) and took several long pauses throughout the experiment (periods 21, 26, 41–42, 56, 59–60, and 76). A significant share of backwards commands can be identified in the first half of the timeline (~20–40%), which decreases in the second half (~5–15%). This may indicate an improvement in the modeling skills of this designer. Again, his share of necessary commands (~40–50%) is virtually double that of contributory commands (~20–25%), while a nearly constant share of collaborative commands is evident (~3%), through which this specialist received updates from the rest of the designers.

Finally, the plumbing designer shows a large constant share of backwards commands (~20–40%), but the lowest share of unnecessary commands (~10%). Additionally, a high proportion of necessary commands with respect to contributory commands is apparent (~40–50% to ~20–25%), as well as a high provision of non-geometrical information to the model (~10%) and a variable share of collaborative commands (~5–30%).

According to this analysis, it would be advisable to train the architectural designer in order to reduce the use of unnecessary 'Cancels', which apparently this specialist does while thinking. The structural, mechanical and plumbing designers' training should focus on reducing the number of backwards commands, for which a deeper analysis of the specific causes should be performed previously. This analysis should be based on detecting bad practices through the study of command patterns. All the MEP designers should receive training in order to reduce the quantity of necessary commands that they execute to produce a more effective modeling process, since necessary commands do not really contribute to the model but are only auxiliary commands that sometimes can be avoided through good modeling practices. While the plumbing specialist's command intensity and distribution was more constant than that of the other MEP designers, the analysis indicates that this designer needs further training in BIM modeling in order to avoid using that significant share of backwards commands.

When working individually, where only information is required and not delivered to others, clashes between models are prone to occur. A greater emphasis on the delivery of information would imply a lower rate of errors and avoid the emergence of problems during the project. During this experiment, the flow of information from the MEP designers was not as regular with respect to other specialties, which is contradictory since these are usually the specialties that generate more conflicts during collaborative work. In general, this team of designers should receive training in a more comprehensive plan of interaction regarding the sharing of information through the 'Save to master' command to ensure better and more frequent synchronization of the local models. This would avoid unnecessary mistakes, and therefore some of the backwards commands executed, thus improving the individual performance of each designer.

5. Discussion

The increased use of BIM worldwide has created an opportunity to study the log files generated by the available software. The analysis of these files can be oriented towards analyzing the modelers' performance and behavior in order to improve productivity, but requires a preliminary data cleaning and organization process to be able to identify problems, causes, and possible solutions. In this research, a method for studying these log files was proposed in order to identify good and bad user practices during the design phase of a building project that could subsequently be useful for elaborating personalized training proposals and to increase both individual and collaborative performance of designers. By analyzing the events occurred during the experiment's timeline, specifically pauses in the specialists' work, it was also possible to detect behaviors directly related to their roles in the project development workflow.

BIM log mining has been used before to analyze the commands executed by a modeler. Zhang et al. [33] presented an experience where the most repeated commands could have been common to the five designers considered in the present study, including 'Cancel', 'Disallow join', 'Temporary hide', 'Delete', 'Align', 'Undo', and 'Reference plane'. In their analysis they noticed that some of the most used commands identified were somehow necessary or auxiliary. However, this method also allowed detecting bad modeling practices, such as an excessive use of 'Disallow join', which is not common and needs finding the cause for that behavior in order to correct it. The analysis applied in the present study also made it possible to detect bad practices with a general classification and overview of the executed commands. For example, it was noticed that the plumbing specialist used a significant amount of backwards commands, which made this designer's work less efficient. In addition, a positive evolution in this aspect was detected for the electrical designer, who started using more of these commands in the initial time periods, and then reduced them by the end of the experiment. Nevertheless, a combination of both analysis methods would certainly allow generating more personalized recommendations for further training.

In this study, the performance of BIM users participating in the experiment was measured through their command execution intensities, which usually were within the range of 200 to 400 commands per hour. Modeling performance has been measured before through BIM log mining. For example, Yarmohammadi et al. [26] obtained the time it took different modelers to execute a certain command pattern. Their method required finding for all the modelers a same pattern, which was obtained from the analysis of log files from different workstations in an architecture firm. As occurred in their study, finding command patterns common to the five specialties in the experiment here described would require these to be based on basic commands, such as hiding, copying, and moving, rotating, or aligning elements. However, most of the contributory work carried out by each specialist is based on commands exclusively related to their discipline, as was shown in Table 3 and Tables A1–A4. Nevertheless, the method used in the present study does not consider that some commands take more time to complete than others. Therefore, a more complex analysis might be necessary to measure modelers' performance, since current methods do not capture these differences between disciplines or commands.

Zhang et al. [23] also presented a command pattern discovery method based on BIM log mining, which they used to measure and analyze productivity. Their study was oriented identify command patterns associated to a certain task and to evaluate modelers' performance executing those patterns in order to better allocate resources to different design tasks in design projects. In the present study, the main focus was put on studying the use that the various specialists showed regarding a set of general categories of commands. Those categories were created according to the contribution of commands to the design process, classifying them into geometrical modeling, non-geometrical modeling, collaborative, necessary, unnecessary, and backwards. The study of the distributions of these categories for each designer during the experiment allowed detecting bad modeling practices and further training needs. By populating a timeline of the experiment with data specific for each time period, it was possible to study the evolution of users' behavior and performance regarding these categories, which proved useful for better planning information workflows in future projects.

As mentioned before, workflows in the development of BIM projects are hybrid and not solely BIM-based, making it necessary to use other platforms, such as sketches on paper or consulting documents online [32]. These activities cannot be detected by studying BIM log files and therefore need a more thorough observation of the designers throughout the design process. From that observation, qualitative data could arise that complements and validates the detection of certain events in the study of short and long pauses in the log files, thus allowing to review collective behavior.

The experience and method described here has clear applications in architecture firms developing BIM projects where coordinators and managers are willing to analyze behavioral patterns and efficiency of their modelers and to identify further training needs and opportunities to increase their performance. The collaborative work environment built for this experiment provided a singular means that strengthened teamwork and allowed a constant flow of information between the participants, who supported each other to find solutions to the issues that would arise during the design process. This method is useful not only to analyze users' performance in BIM modeling, but also to better understand their behavior and identify improvement strategies. This analysis was carried out using spreadsheets in a semi-manual process. However, given the large amount of data to be managed, it would be advisable to implement the entire process in software to increase the automation of the analysis.

Author Contributions: Conceptualization, E.F. and A.M.-R.; Data curation, A.M.-R.; Formal analysis, A.M.-R. and J.S.-M.; Funding acquisition, E.F., A.M.-R. and R.G.-A.; Investigation, E.F. and R.G.-A.; Methodology, A.M.-R. and A.N.-B.; Project administration, E.F.; Resources, E.F. and R.G.-A.; Software, J.S.-M. and A.N.-B.; Supervision, E.F.; Visualization, A.N.-B. and F.L.; Writing—original draft, E.F., J.S.-M. and F.L.; Writing—review & editing, A.M.-R. All authors have read and agreed to the published version of the manuscript.

Funding: This research was funded by the National Commission for Science and Technology Research in Chile (CONICYT), grant number 1171108 and the VI Own Research and Transfer Plan of the University of Seville (VI PPIT-US). The APC was partially funded by the Universidad Del Bío-Bío.

Conflicts of Interest: The authors declare no conflict of interest. The funders had no role in the design of the study; in the collection, analyses, or interpretation of data; in the writing of the manuscript, or in the decision to publish the results.

Appendix A. Most Used Commands by Each Designer

Table A1. Contributory commands most frequently used by the structural designer.

Command ID	Description	Absolute Frequency	Relative Frequency
ID_OBJECTS_BEAM	Create a beam	190	10.7%
ID_FILE_SAVE_TO_MASTER	Save the active project to the central model again	181	20.8%
ID_OBJECTS_CURVE_LINE	Create a line	132	28.2%
ID_FINISH_SKETCH	Finish sketch	125	35.2%
ID_EDIT_MOVE	Move selected objects or their copies	99	40.8%
ID_EDIT_MOVE_COPY	Move copies of selected objects	87	45.6%
ID_OBJECTS_PROJECT_CURVE	Create a straight line or an arc	84	50.3%
ID_VIEW_NEW_SCHEDULE	Create a schedule	69	54.2%
ID_OBJECTS_CURVE_RECT	Create a rectangle	52	57.1%
ID_TRUSS_WEB_CURVE	Create a truss web	44	59.6%
ID_OBJECTS_LEVEL	Create a level	42	61.9%
ID_FINISH_SWEEP	Finish sweep	40	64.2%
ID_OBJECTS_FOOTING_SLAB	Create a footing slab	37	66.3%
ID_END_INPLACE_FAMILY	Finish the family	34	68.2%
ID_REVIT_FILE_SAVE	Save the active project	33	70.0%
ID_LOAD_INTO_PROJECTS	Load document into open projects	33	71.9%
ID_OBJECTS_TRUSS	Create a truss beam	32	73.7%
ID_EDIT_MIRROR	Mirror selected objects	31	75.4%
ID_EDIT_ROTATE	Rotate selected objects	31	77.1%
ID_FINISH_SKETCH_PATH	Finish a sketch of sweep path	29	78.8%
ID_SKETCH_2D_PATH	Create or edit the path through a sketch in a plane	28	80.3%

Table A2. Contributory commands most frequently used by the mechanical designer.

Command ID	Description	Absolute Frequency	Relative Frequency
IDS_RBS_CREATE_PIPE	Create pipe	467	36.1%
ID_FILE_SAVE_TO_MASTER	Save the active project to the central model again	122	45.5%
ID_ALIGN	Align references	118	54.6%
IDS_RBS_CREATE_DUCT	Create duct	91	61.7%
ID_RBS_MECHANICAL_DIFFUSER	Insert an air diffuser	51	65.6%
ID_RBS_MECHANICAL_EQUIPMENT	Create mechanical equipment	50	69.5%
IDS_RBS_CREATE_FLEX	Create flexible duct	49	73.3%
ID_RBS_PIPE_PIPE	Create pipes	43	76.6%
ID_EDIT_ROTATE	Rotate selected elements	39	79.6%
ID_LOAD_INTO_PROJECTS	Load document into open projects	23	81.4%

Table A3. Contributory commands most frequently used by the electrical designer.

Command ID	Description	Absolute Frequency	Relative Frequency
ID_OBJECTS_DETAIL_CURVES	Create a detail line or arc	242	14.7%
ID_RBS_ELECTRICAL_DEVICE	Add electrical devices	190	26.3%
ID_RBS_LIGHTING_FIXTURE	Add lighting fixtures	170	36.6%
ID_FILE_SAVE_TO_MASTER	Save the active project to the central model again	163	46.5%
IDS_RBS_CREATE_CONDUIT	Create conduit	154	55.9%
ID_EDIT_ROTATE	Rotate selected elements	130	63.8%
ID_EDIT_PASTE_NO_EVENT	Paste element	106	70.2%
ID_RBS_ELECTRICAL_LIGHTING_DEVICE	Create a lighting device	46	73.0%
ID_VIEW_NEW_SCHEDULE	Create a schedule	34	75.1%
ID_FAMILY_LOAD	Load a family into the project	31	77.0%
ID_EDIT_COPY	Copy the selection and keep it in the clipboard	24	78.4%
ID_RBS_ELECTRICAL_EQUIPMENT	Add electrical equipment	22	79.8%
ID_RBS_PIPE_PIPE	Create pipe	21	81.0%

Table A4. Contributory commands most frequently used by the plumbing designer.

Command ID	Description	Absolute Frequency	Relative Frequency
ID_RBS_PIPE_PIPE	Create pipes	408	23.7%
ID_RBS_PLUMBING_FIXTURE	Insert a plumbing fixture	145	32.1%
IDS_RBS_CREATE_PIPE	Create a pipe	108	38.4%
ID_FAMILY_TYPE	Modify predefined types for this family	105	44.5%
ID_REVIT_FILE_SAVE	Save the active project	62	48.1%
ID_EDIT_ROTATE	Rotate selected elements	61	51.7%
ID_RBS_ADD_PIPE_CONNECTOR	Add pipe connector to the family	59	55.1%
ID_LOAD_INTO_PROJECTS	Load document into open projects	57	58.4%
ID_FILE_SAVE_TO_MASTER_SHORTCUT	Save the active project to the central model again	53	61.5%
ID_EDIT_MOVE	Move selected objects or their copies	52	64.5%
ID_VIEW_NEW_SCHEDULE	Create a schedule	48	67.3%
ID_FILE_SAVE_TO_MASTER	Save the active project to the central model again	47	70.0%
IDC_APPLY_MovePropsDialogBar	Apply to an object the same properties from another object	41	72.4%
ID_ANNOTATIONS_DIMENSION_ALIGNED	Create aligned annotation	40	74.7%
ID_OBJECTS_LEVEL	Create a level	36	76.8%
ID_OBJECTS_REFERENCE_CURVE	Create a reference line	35	78.8%
ID_EDIT_MOVE_COPY	Move copies of selected objects	32	80.7%

References

1. Fulford, R.; Standing, C. Construction industry productivity and the potential for collaborative practice. *Int. J. Proj. Manag.* **2014**, *32*, 315–326. [CrossRef]
2. Oh, M.; Lee, J.; Hong, S.W.; Jeong, Y. Integrated system for BIM-based collaborative design. *Autom. Constr.* **2015**, *58*, 196–206. [CrossRef]
3. Hannele, K.; Reijo, M.; Tarja, M.; Sami, P.; Jenni, K.; Teija, R. Expanding uses of building information modeling in life-cycle construction projects. *Work* **2012**, *41*, 114–119. [CrossRef]
4. Eadie, R.; Browne, M.; Odeyinka, H.; McKeown, C.; McNiff, S. BIM implementation throughout the UK construction project lifecycle: An analysis. *Autom. Constr.* **2013**, *36*, 145–151. [CrossRef]

5. Xu, X.; Ma, L.; Ding, L. A framework for BIM-enabled life-cycle information management of construction project. *Int. J. Adv. Robot. Syst.* **2014**, *11*, 1–13. [CrossRef]

6. Wong, J.K.W.; Zhou, J. Enhancing environmental sustainability over building life cycles through green BIM: A review. *Autom. Constr.* **2015**, *57*, 156–165. [CrossRef]

7. Elmualim, A.; Gilder, J. BIM: Innovation in design management, influence and challenges of implementation. *Archit. Eng. Des. Manag.* **2014**, *10*, 183–199. [CrossRef]

8. Sackey, E.; Tuuli, M.; Dainty, A. Sociotechnical systems approach to BIM implementation in a multidisciplinary construction context. *J. Manag. Eng.* **2014**, *31*, 1–12. [CrossRef]

9. Paavola, S.; Miettinen, R. Dynamics of Design Collaboration: BIM Models as Intermediary Digital Objects. *Comput. Support. Coop. Work CSCW An Int. J.* **2018**, *27*, 1113–1135. [CrossRef]

10. Bryde, D.; Broquetas, M.; Volm, J.M. The project benefits of building information modelling (BIM). *Int. J. Proj. Manag.* **2013**, *31*, 971–980. [CrossRef]

11. Rezgui, Y.; Beach, T.; Rana, O. A governance approach for BIM management across lifecycle and supply chains using mixed-modes of information delivery. *J. Civ. Eng. Manag.* **2013**, *19*, 239–258. [CrossRef]

12. Mutai, A. *Factors Influencing the Use of Building Information Modeling (BIM) within Leading Construction Firms in the United States of America*; Proquest, Umi Dissertation Publishing: Ann Arbor, MI, USA, 2011; ISBN 9781243709448.

13. Azhar, S. Building Information Modeling (BIM): Trends, Benefits, Risks, and Challenges for the AEC Industry. *Leadersh. Manag. Eng.* **2011**, *11*, 241–252. [CrossRef]

14. Navendren, D.; Manu, P.; Shelbourn, M.; Mahamadu, A.-M. Addressing Challenges to Building Information Modelling Implementation in Uk: Designers' Perspectives. In Proceedings of the 30th Annual ARCOM Conference, Cambridge, UK, 4–6 September 2017; pp. 733–742.

15. Succar, B. Building information modelling framework: A research and delivery foundation for industry stakeholders. *Autom. Constr.* **2009**, *18*, 357–375. [CrossRef]

16. del Caño, A.; de la Cruz, M.P.; Solano, L. Diseño, ingeniería, fabricación y ejecución asistidos por ordenador en la construcción: Evolución y desafíos a futuro. *Inf. la Construcción* **2007**, *59*, 53–71. [CrossRef]

17. Volk, R.; Stengel, J.; Schultmann, F. Building Information Modeling (BIM) for existing buildings—Literature review and future needs. *Autom. Constr.* **2014**, *38*, 109–127. [CrossRef]

18. Fruchter, R. M3R: Transformative Impacts of Mixed Media Mixed Reality Collaborative Environment in Support of AEC Global Teamwork. In *Transforming Engineering Education*; ASCE: Reston, PA, USA, 2018; pp. 229–257.

19. Muñoz-La Rivera, F.; Vielma, J.C.; Herrera, R.F.; Carvallo, J. Methodology for Building Information Modeling (BIM) Implementation in Structural Engineering Companies (SECs). *Adv. Civ. Eng.* **2019**, *2019*, 1–16. [CrossRef]

20. Bonchi, F.; Castillo, C.; Gionis, A.; Jaimes, A. Social Network Analysis and Mining for Business Applications. *ACM Trans. Intell. Syst. Technol.* **2011**, *2*, 1–37. [CrossRef]

21. Jara, C.; Alarcón, L.F.; Mourgues, C. Accelerating interactions in project design through extreme collaboration and commitment management—A case study. In Proceedings of the Annual Conference of the IGLC, Taipei, Taiwan, 15–17 July 2009; pp. 477–488.

22. Zhang, L.; Ashuri, B. BIM log mining: Discovering social networks. *Autom. Constr.* **2018**, *91*, 31–43. [CrossRef]

23. Zhang, L.; Wen, M.; Ashuri, B. BIM Log Mining: Measuring Design Productivity. *J. Comput. Civ. Eng.* **2018**, *32*, 04017071. [CrossRef]

24. Oraee, M.; Hosseini, M.R.; Edwards, D.J.; Li, H.; Papadonikolaki, E.; Cao, D. Collaboration barriers in BIM-based construction networks: A conceptual model. *Int. J. Proj. Manag.* **2019**, *37*, 839–854. [CrossRef]

25. Kouhestani, S.; Nik-Bakht, M. IFC-based process mining for design authoring. *Autom. Constr.* **2020**, *112*, 103069. [CrossRef]

26. Yarmohammadi, S.; Pourabolghasem, R.; Castro-lacouture, D. Automation in Construction Mining implicit 3D modeling patterns from unstructured temporal BIM log text data. *Autom. Constr.* **2017**, *81*, 17–24. [CrossRef]

27. van der Aalst, W.; Buijs, J.; van Dongen, B. Towards improving the representational bias of process mining. In Proceedings of the International Symposium on Data-Driven Process Discovery and Analysis, Campione d'Italia, Italy, 18–20 June 2012; Volume 116, pp. 39–54.

28. Rozinat, A.; van der Aalst, W.M.P. Conformance checking of processes based on monitoring real behavior. *Inf. Syst.* **2008**, *33*, 64–95. [CrossRef]
29. Pan, Y.; Zhang, L. BIM log mining: Exploring design productivity characteristics. *Autom. Constr.* **2020**, *109*, 102997. [CrossRef]
30. Das, R.; Turkoglu, I. Creating meaningful data from web logs for improving the impressiveness of a website by using path analysis method. *Expert Syst. Appl.* **2009**, *36*, 6635–6644. [CrossRef]
31. Grosfeld-Nir, A.; Ronen, B.; Kozlovsky, N. The Pareto managerial principle: When does it apply? *Int. J. Prod. Res.* **2007**, *45*, 2317–2325. [CrossRef]
32. Harty, C.; Whyte, J. Emerging hybrid practices in construction design work: Role of mixed media. *J. Constr. Eng. Manag.* **2010**, *136*, 468–476. [CrossRef]
33. Zhang, N.; Tian, Y.; Al-Hussein, M. A Case Study for 3D Modelling Process Analysis based on BIM Log File. In Proceedings of the MOC2019—Modular and Offsite Construction Summit, Banff, AB, Canada, 21–24 May 2019; pp. 309–313.

Article

Effects of BIM-Based Construction of Prefabricated Steel Framework from the Perspective of SMEs

Mooyoung Yoo, Jaejun Kim * and Changsik Choi

Department of Architectural Engineering, Hanyang University, 220 Wangsimni-ro, Seongdong-gu, Seoul 04763, Korea; yoomoos@gmail.com (M.Y.); ccs5530@hanyang.ac.kr (C.C.)
* Correspondence: jjkim@hanyang.ac.kr; Tel.: +82-2-2220-0307; Fax: +82-2-2296-1583

Received: 11 February 2019; Accepted: 24 April 2019; Published: 26 April 2019

Abstract: Small- and medium-sized enterprises (SMEs) are part of the building construction industry. Although many effect analyses of applying building information modeling (BIM) to projects have been conducted, analyses from the perspective of SMEs are lacking. We propose a BIM-based construction of prefabricated steel framework from the perspective of SMEs. We derive the essential functions of the system from the viewpoint of SMEs and verify the qualitative effect through a case analysis of prefabricated steel frame construction that is based on BIM. The following system functions and qualitative effects are analyzed according to project stages that are based on interviews of working groups participating in system development and case projects. (1) Preconstruction stage: extraction of fabrication drawing and review of shop drawing, (2) fabrication stage: prefabrication review, steel member removal, and field loading review, and (3) construction phase: integrated management of cost and schedule and quality management. The expected effects of applying the system are qualitatively and quantitatively analyzed through expert group interviews and surveys. For the quantitative analysis, an evaluation index is used for the end-user computing satisfaction survey. Further analysis of the finishing and installation work is required. Future research should also analyze the effect of system application on human resource management.

Keywords: SMEs; BIM; construction management system; steel frame construction

---

1. Introduction

Building information modeling (BIM) enables integrated design [1], and it can be applied to various fields (e.g., design verification, quantity calculation, and prefabrication) throughout the construction life cycle by utilizing the three-dimensional BIM model that is built for each construction type [2]. In the preconstruction phase, performing collision detection and design review using BIM can help to reduce the losses that occur in the construction phase by identifying the problems that can be solved beforehand at the design phase (e.g., design errors) [3,4]. Doing so improves the design quality by reducing design errors, design changes, and rework [5,6]. In the fabrication phase, the procurement process of construction frame members through BIM-based prefabrication can be simplified, and the productivity of workflow between the designers and constructors can be improved [7]. In addition, in terms of production productivity, the workers that are involved in the production of frame members can utilize the information inputted to BIM objects to support the manufacturing process; such information is parametric and it is not provided in the existing two-dimensional (2D)-based production work [8]. In the construction phase, a four-dimensional (4D) simulation can be implemented by linking the BIM model with the schedule [9]. A BIM-based 4D simulation combines information that is related to the process, equipment, and space to identify various uncertainties that arise during the process, and provides information that can be reviewed in advance (Moon et al., 2014). Thus, the effect of BIM on each phase of a construction project becomes clear.

However, previous studies on the effects of BIM on construction projects were conducted for large-scale projects [10]. Previous studies analyzed the return on investment (ROI) that was attained by general contractors for BIM-based construction projects [11–13]. Of course, many studies have analyzed the effects of BIM application from the viewpoint of small- and medium-sized enterprises (SMEs) [14–17]. However, studies on developing a BIM-based construction management system from the viewpoint of SMEs are lacking [18]. Due to the nature of SMEs, the sales and profits are not as large as those of large-scale enterprises, and many of them abandon the application of BIM, which involves high initial costs [19,20]. Therefore, we begin this study by analyzing the difficulties of managing a construction project from the viewpoint of SMEs. We propose a BIM-based construction management system framework to support SMEs. We also conduct a case study to verify the major functions that will be installed in the system for the potential SME users and their perceived performance.

2. Literature Review

2.1. Characteristics of SMEs in Construction Management

In 2016, 99.5% of the 5.5 million businesses in the United Kingdom (UK) were classified as SMEs, accounting for 60% of all private sector employment in the UK and a huge proportion of the 240,000 construction service providers that are operating within the UK [21]. The latest set of government business population estimates shows that the number of construction companies with less than 50 employees has increased to over 1,005,290, up from 972,475 in 2016. The data also indicate that these small businesses are responsible for an approximate total turnover of over £185 Bn a year, up from £172 Bn in 2016. The total turnover of all the businesses operating in the construction sector reached £296.8 Bn in 2017, up from £271.9 Bn in 2016 [22]. The Department for Business, Enterprise, and Regulatory Reform (BERR) [23] defines the size of a firm in terms of the number of employees (see Table 1). The BERR defines

- micro-sized companies as those having fewer than 10 employees,
- small-sized companies as those having fewer than 50 employees,
- medium-sized companies as those having between 50 and 249 employees, and
- large-sized companies as those having 250 employees or above.

A previous study outlined the characteristics of SMEs, as follows [24]. Many SMEs do not record procedures in a clear format, unlike large-scale companies [25]. Thus, SMEs' management style is flexible and informal. For example, one of the main methods for communicating information is through informal interviews between individuals. Therefore, a fundamental limitation in identifying human resources to be used in construction projects and undertaking construction management based on quantitative information exists. Moreover, SMEs may face difficulties in raising funds and maintaining adequate cash flow, which thus limits continued capital investment in employee education and technology development [26]. As such, SMEs that participate in construction projects have common limitations with regard to improving their business systems. However, owners/managers of SMEs play a key role in the decision-making process [27,28]. Thus, if the outcomes of business process reengineering (BPR) are clear, the SME may be more likely to change and in a more flexible manner, when compared to their large-scale counterparts. Therefore, this study proposes a BIM-based construction management system framework from the perspective of SMEs.

2.2. Limitations of BIM Adoption

Many previous studies have analyzed BIM application and its effects [1–9]. However, most of this research is limited to case studies, such as loss prevention due to design errors for a single project [3,4]. Moreover, the studies focused on improving design quality, which is difficult to quantify [5,6]. Some works analyzed the effect of BIM application on interference checking and conducted a 4D simulation for a specific task [9]. Ideally, BIM supports all of the project participants using data on the construction life cycle. However, it is difficult to improve the process and achieve the business integration effect using BIM in large organizations [7].

Table 1. Functions and effects of building information modeling (BIM).

BIM Function	Effect of BIM	Phase
3D BIM conversion design	- Create object information through 3D modeling - Improved drawing consistency	Design/Construction
Visualization	- Improved understanding of work scope and tasks through improved communication - Design suitability review and VE enhancement function	Design/Construction
Linking through object base	- Automation of design changes - Prevention of drawing errors and notation omissions	Design/Construction
Clash check	- Enabling advance production of members through accurate drawings - Reduced field work and construction period and increased productivity	Design/Construction
2D drawing creation	- Design, construction, and automatic extraction of tender drawings - Reduced field work and construction period and increased productivity	Design/Construction
Quantity calculation and estimation	- Quantity calculation and utilization depending on the part type, construction type, and phase - 4D + Cost = 5D (Estimate analysis)	Design/Construction
4D simulation	- Creation of schedule, material, and allocation plans for personnel - 3D + Time = 4D (Process analysis)	Design/Construction
Temporary work and construction management	- Transfer of equipment, material transfer and loading path, operator working path planning, and pre-work coordination with equipment operator(s)	Construction
Combination with various analyses	- Analysis of energy efficiency, structural analysis, and Leadership in Energy and Environmental Design (LEED) analysis	Design

Moreover, it is difficult to manage the inputted manpower by only using BIM data for the building. As a result, research on the application of the internet of things to manage workers by attaching wireless sensors to helmets is also being conducted [29]. In addition, BIM does not immediately reflect implementation via a virtual model. As a result, laser scan and image scan technologies have been introduced as tools for creating and integrating real models. Additionally, the concept of digital twin, which combines virtual models and real models to pre-simulate specific issues, may be used. Indeed, recent studies have tended to integrate not only BIM, but also various other technologies (e.g., three-dimensional (3D) laser scanning, augmented reality, mobile computing, wireless connection, quick response codes) with building construction management [30].

Existing BIM-based construction management systems (e.g., a project management information system; PMIS) should be accompanied by BPR, because the system is often constructed at the enterprise level [31]. In addition, the scope of the system construction is very wide, because the project characteristics of various business areas, rather than a single project, should be considered. Each division of the company should devise the system to address all of the diverse functions that are required. Some companies report issues that are caused by the presence of several project-related databases and the fact that security concerns limit system accessibility for the users. In this paper, we propose a BIM-based construction management system from the viewpoint of SMEs. In addition, we qualitatively verify the functioning of the system through a case analysis.

2.3. BIM for Prefabricated Steel Frame Construction

A general contractor manages the various specialty contractors that are involved in a construction project. A steel frame construction is implemented through a specialty contractor. Steel frame construction involves prefabricating materials at a factory that are based on a preconstruction blueprint [32]. Generally, in the preconstruction phase, the design drawings and specifications are reviewed, the construction companies are selected, and the construction plans and details are prepared [33]. The factory processing of the steel frame members is conducted in the order of full-size drawing, prototyping, deformation inspection, marking gauge, cutting and machining, drilling, joining, riveting and welding, metal cutting, rustproofing, and loading at the site. When the steel frame members are brought to the laydown area, they are then assembled by bolting or welding, depending on the installation drawing. Lastly, a crane is used to lift the steel frame members, and the workers complete the installation process [34,35].

However, if any design changes are made to the steel frame, the following problems may arise [33]: (1) Increase in construction costs and a delay in construction period due to the need to place new orders for materials and/or additional production, (2) surplus of preordered materials, and (3) duplicate production of steel frame members. These costs and risks due to the delay can impact the profits of the specialty contractors. It is necessary to introduce BIM and other recent technologies (e.g., laser scanning) to solve these problems.

Table 1 lists the main functions and effects of BIM. A 3D BIM model with object information is created from a 3D BIM conversion design based on the drawings that were generated at the design stage, and the 3D BIM model is visualized for collaboration among experts in different fields [36]. In addition, the clash check function helps to improve the quality of the 2D design drawings by examining the inter-member clash and reducing errors therein in terms of potential errors that may occur at the preconstruction site [13]. Parametric modeling—a popular BIM technique—is used to project and automatically change the information that is associated with each member when the design is changed. Thus, the design process is more quickly and accurately completed. Table 1 lists the effects of information utilization using BIM in the design and construction phases.

Table 2 lists the research highlights of previous studies that used BIM for modeling productivity, shop drawing calculation, and process management in steel frame construction.

Table 2. Research highlights of previous studies that studied BIM for steel frame construction.

Category	Researcher/s (Year)	Research Content
Design	Eom and Shin [37]	Development of an automation module for modeling steel frame joints that can be used in structural detail design and modeling stages
	Ko et al. [38]	Development of an automatic design system for steel connections based on set-based Design with structural building information modeling (S-BIM)
	Eom and Shin [39]	Development of an interface module that can exchange information between structural analysis software supporting structural design work and BIM software supporting detailed modeling and drawing work
	Li et al. [40]	Description of a modeling system for steel structure joints based of BIM at the design stage
	Oti and Tizani [41]	Introduction of a BIM-based structural sustainability appraisal system
Construction	Ryu and Kim [42]	Development of a 4D simulation system prototype through automatic-process production
	Yun et al. [43]	Development of a tracking method for lifting paths of a steel frame tower crane using global positioning system (GPS) in the BIM environment
	Shin and Yang [44]	Development of a smart creation process for shop drawings depending on drawing types
	Kim [45]	Application of real-time locating system (RTLS) technology for automating spray-applied fire-resistive covering work
	Xie et al. [46]	Using radio-frequency identification and real-time virtual reality simulation for optimization in steel construction
	Liu et al. [47]	Using BIM to improve the design and construction of bridge projects

Previous studies focused on improving the productivity of the unit tasks of steel frame construction (e.g., automatic modeling of joints and drawings) and managing the process. However, the SMEs need to derive the effects of BIM for tasks that are related to the design, fabrication, and construction, rather than applying the technology to the unit tasks. This study presents a case study on steel frame construction to verify the main BIM-based construction management function to be installed in the system for an SME.

2.4. Evaluation of Information System

Information systems (e.g., PMIS) that were used in existing enterprises possess different characteristics depending on their purpose and support tasks. Information systems are classified according to their purpose and functionalities that are based on the requirements of the company [48,49]. Other works have analyzed the characteristics of information systems according to their contribution to efficiency improvement and relationship improvement in the internal and external aspects of the company [50,51]. Research regarding the success of information systems is based on the work of Shannon and Weaver and Mason [52,53], and the results suggest that the level of information can be measured in terms of the technical, semantic, and effectiveness levels [54]. However, the usefulness of such evaluations is limited for the following reasons. First, it is difficult to quantify the measurement factors and objectively separate the effects of the information system from other aspects [55–57]. As a result, attempts have been made to evaluate the information systems in terms of their use as well as the end user's satisfaction (i.e., considering the user's cognitive aspect). However, Lucas [55] argued that, while indicators pertaining to an actual information system (such as measured time) are somewhat objective and easy to quantify, they are not a useful measure, because the use of the information system may be compulsory. Thus, in many studies, end-user satisfaction, rather than evaluation of use, is perceived as a key factor in measuring the success of an information system [58–62].

Several models have been proposed to measure the end-user satisfaction with information systems [63,64]. The model of Doll and Torkzadeh [64] presented 12 evaluation indicators, which consisted of five factors (content, accuracy, format, ease of use, and timeliness) that are based on existing results (Table 3). Other researchers have consistently verified this model [65–67]. In this study, we use these indicators to evaluate the system to be developed. In doing so, we intend to customize the system for SMEs.

Table 3. Evaluation index for measuring End-User Computing Satisfaction (EUCS).

Category	Evaluation Index	Contents of Evaluation Index
Content	C1	Does the system provide the precise information you need?
	C2	Does the information content meet your need?
	C3	Does the system provide reports that seem to be just about exactly what you need?
	C4	Does the system provide sufficient information?
Accuracy	A1	Is the system accurate?
	A2	Are you satisfied with the accuracy of the system?
Format	F1	Do you think the output is presented in a useful format?
	F2	Is the information clear?
Ease of Use	E1	Is the system user friendly?
	E2	Is the system easy to use?
Timeliness	T1	Do you get the information you need in time?
	T2	Does the system provide up-to-date information?

3. Research Method

3.1. BIM-Based Construction Management System Framework

Figure 1 shows the proposed BIM-based construction management system framework from the perspective of SMEs, based on the research problem and theoretical considerations.

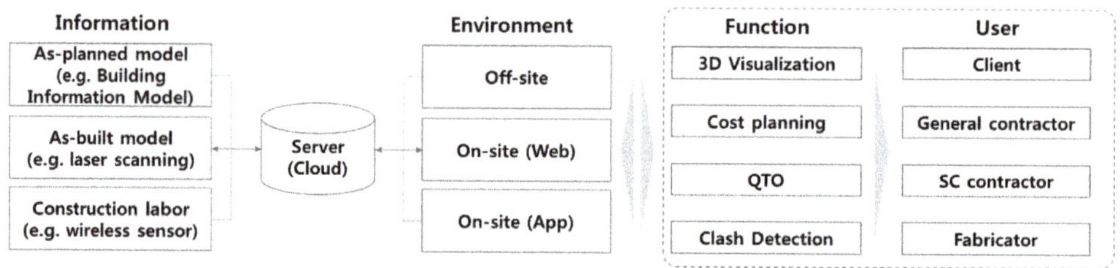

Figure 1. Conceptual diagram of BIM-based construction.

The system framework has the following features. First, it should be able to input and output various information (e.g., BIM model, wireless sensor data, laser scan, or image scan data). Even if the system only needs to manage the construction of a single process (e.g., steel frame construction), the as-planned model for the plan, the as-built model for the execution result, and the information about the construction labor can differ. Information on buildings can be obtained from the BIM Model and laser scan data, but that manpower needs to be collected using equipment, such as wireless sensors (e.g., safety helmets, safety vests, and safety shoes) [29]. Construction management and human resource management should be undertaken by utilizing this information. Second, from the viewpoint of SMEs, the system should only be equipped with the functions (e.g., 3D visualization, cost planning, quantity take-offs (QTOs), and clash detection) that are necessary for a BIM-based construction management system. It is necessary to supervise the construction progress while using this information and to establish accurate planning, input, and management of the workforce. Moreover, the system should be realistic, usable, and compatible in various system operating environments (e.g., Off-site and Web- and app-based on-site management environments). Third, the system must improve accessibility for all users in various formats (e.g., html, pdf, image, dwfx, etc.), which are typically used for the Web and apps. It should be possible to access the information that is necessary for construction management with smart device operation technology rather than the BIM software, which is difficult to access and operate.

3.2. Main Function Derivation for the System

We analyze a typical steel frame construction in order to derive the necessary functions for the system [35]. Table 4 summarizes the main work phases and tasks, the authorities that are responsible for each task, work location, and BIM usage in steel frame construction. For on-site and off-site steel frames constructions, BIM authoring creates the 3D model, visualizes the quality review, produces a shop drawing, conducts a 4D simulation for process management, and checks QTOs for cost management.

In the preconstruction phase, the drawings and specification reviews are used to check whether the design requirements are reflected and whether the object information that is related to the material and dimension of the steel frame is correctly entered. After reviewing the design information, a 2D steel frame shop drawing is extracted from the BIM model for factory production.

In the prefabrication phase, accurate quantity calculation data are used when ordering the steel material for frame member fabrication. In addition, the cutting work from the manufacturing process is numerically managed using the BIM model. As a result, it is possible to find not only the gross weight of the member, but also its net weight, excluding the cutting part. To minimize the loss of steel material, the BIM model can be used to efficiently cut the members. During this process, the installation of the steel frame is reviewed to reduce the work, and the site conditions, such as fabrication, transportation, lifting, and ease of installation, are considered. When the fabrication and inspection of the steel members are completed, the prefabricated steel members are procured on-site.

Table 4. Main tasks involved in steel frame construction and BIM use mapping.

Phase	Task	Authority	Location	BIM Use
Preconstruction	- Drawing and specification review	GC/SC *	On-site	-
	- Construction planning	GC/SC	On-site	Visualization, 4D simulation
	- Shop drawing	GC/SC	On-site	Creation of 2D shop drawings
	- Cross check	GC/SC	On-site	Visualization, clash check
Prefabrication	- Bringing steel framing members and reviewing the quantity	GC/SC	Off-site	Quantity calculation
	- Steel frame cutting	GC/SC	Off-site	Creation of 2D drawings (cutting plan)
	- Steel frame mounting	GC/SC	Off-site	-
	- Steel frame assembly and welding	GC/SC	Off-site	Quantity calculation
	- Painting steel frame members	GC/SC	Off-site	Quantity calculation
	- Marking	GC/SC	Off-site	3D BIM authoring
	- Precision inspection	GC/SC	Off-site	Visualization (e.g., laser scanning)
	- Carrying steel frame on and off the site	GC/SC	Off-site	Quantity calculation
Construction	- Review of quantity brought to a laydown area	GC/SC	On-site	Quantity calculation
	- Building and installing steel frame columns	GC/SC	On-site	Visualization, temporary works, and construction management
	- Steel girder, beam lifting, and installation	GC/SC	On-site	Temporary works and construction management
	- Vertical and horizontal inspection of the steel	GC/SC	On-site	Visualization (e.g., laser scanning)
	- Bolting and welding of the steel	GC/SC	On-site	Quantity calculation
	- Fireproof coating spray on steel	GC/SC	On-site	Quantity calculation
	- Steel frame installation finish (fastening)	GC/SC	On-site	Quantity calculation

*: GC and SC denote general contractor and specialty contractor, respectively

In the construction phase, the steel frame members that were procured from the factories are laid down when considering the order of installation. They are then temporarily assembled through steel bolting and welding. Thereafter, the construction is sequentially done using the shop drawing and 4D simulation. The installation procedure involves installing the column, large beam, small beam, and brace members.

3.3. Definition of System Database

3.3.1. Building Objects of Prefabricated Steel Frame

Figure 2 shows the objects that make up the BIM model of the prefabricated steel frame. The part that constitutes the prefabricated steel frame as per the BIM model is divided into the main part and the secondary part. Each part is automatically categorized with a number (double or int) or a string and stored as a BIM object. In addition, the bolts that are connected to the steel frame members and welded members are grouped, and the respective components are assembled. The number and dimensions of the holes for cutting and bolting the steel frame members are included [68].

3.3.2. Connecting Building Objects to Create a Prefabricated Steel Frame

Table 5 lists the numerical data regarding the BIM model of the steel frame joints for the case study project that is discussed in this paper. It shows the data structure of the connections of the steel frame parts, namely the main part, secondary part, and bolts, as well as welding lengths. The materials of the end plate and bolts and the hole types are stored as strings, whereas the size and dimension of the bolts, tolerance distance, and notch length are stored as the numerical values. The strings and numerical data can be utilized to numerically control the machines that are used to produce the steel frame members [70]. In addition, the integration of the components of each steel member results in the complete BIM model for the steel frame construction. It includes most of the building components that are used in the steel frame construction based on the attribute values of the steel members. The BIM model constructed using such an object-oriented modeling can be quickly and accurately modified if any changes to the design and/or information are made.

Figure 2. Building objects of the BIM model [69].

Table 5. Connecting building objects to create a prefabricated steel frame.

Figure	Input		Attribute	Value	Type
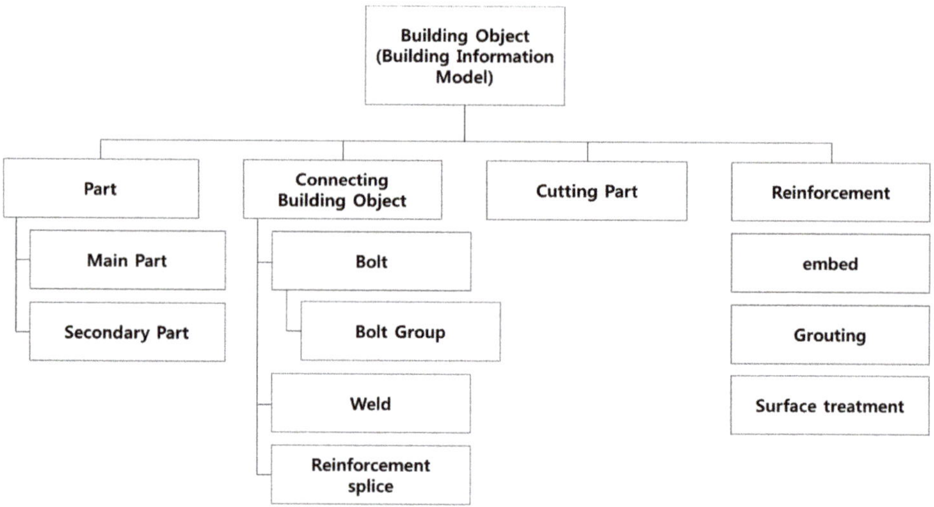 	End plate (a)	Material	mat	S275JR	string
		Thickness	tpl1	10	double
		Depth	hpl1	200	double
		Width	bpl1	180	double
	Bolt (c)	Diameter	diameter	20	double
		Grade	screw	7990	string
			lbd	60	string
			lwd	67.1	string
			lba	66	double
			nb	3	int
			nw	2	int
			rb1	40	double
			rb2	40	double
			rw1	40	double
			rw2	40	double
			lbtyp	1	int
	Weld (b)		w3_size	6	double
	Notch (d)		t_cut_length	82	double
			t_cope_length	26	double
			b_cut_length	82	double
			b_cope_length	26	double
			cope_fitting_type	3	int

4. Effects of BIM-Based Construction Management System for SMEs

4.1. Case Study

4.1.1. Project Description

To analyze the effect of BIM on the prefabricated steel frame construction from an SME perspective, the project, as summarized in Table 6, is chosen. The total construction cost was less than ₩1 Bn and the total load acting on the prefabricated steel frame was 91.78 t. The design phase of this case study project was based on a 2D Computer-Aided Design (CAD) drawing.

Table 6. Overview of the case study project.

Item	Description
Project name	Neighborhood residence facilities construction
Location	Gyeonggi-do Namyangju-si Sampae-dong 153-7
Main structure	Steel structure
Construction duration	04/12/2017–10/31/2017
Total Cost (₩)	761,406,020
Quantity of prefabricated steel frame	91.78 t
Building coverage	294.00 m^2

The project client majored in architectural engineering and has worked as a BIM manager. Thus, the client is aware of the usefulness of the BIM model. However, the client does not have firsthand experience of the effect of BIM on actual projects. After interviewing the client, we found that the accuracy of the information regarding the quantity of prefabricated steel frame, which accounted for a large portion of the total construction cost, significantly influenced the decision-making process. Table 7 summarizes the total construction cost (of the three bidding projects), the prefabricated steel frame construction, and whether BIM was applied.

Table 7. Comparative analysis for general contractor/ specialty contractor (GC/SC) who participated in the case study project bidding.

Item of Comparison	GC/Specialty Contractor 1 (Selected)	GC/Specialty Contractor 2	GC/Specialty Contractor 3
Total construction cost (₩)	761,406,020 (VAT included)	818,840,000 (VAT included)	1,017,514,000 (VAT included)
Prefabricated steel frame construction (₩)	161,406,202	155,700,000	253,658,750
Quantity of prefabricated steel frame (t)	91.78	- (No information on quantity)	105.13 (Lack of information on the quantity calculation)
Material, labor, and overhead costs	Constructed based on steel billets	Missing information	Missing information
Whether BIM was applied	Yes	No	No
BIM Software	TEKLA	-	-

The specialty contractor (SC) used TEKLA to create a BIM model of the steel frame based on the design drawings for the general contractor (GC). Accordingly, we calculated the quantity of the major components required for the steel frame construction. Using this information, the GC calculated the material, labor, and overhead costs required to estimate the total construction cost. As for the other companies, the total construction cost was higher, and no information on the cost of steel frame construction, including material, labor, and overhead costs, was provided to the client. Therefore, the client selected the contractor who presented the lowest total construction cost with the exact quantities for steel construction.

The selected GC has 25 employees, with a capital of less than ₩1 Bn and a revenue of ₩14.6 Bn as of 2016. The SC, who is responsible for the steel frame construction, is engaged in the construction of steel frames and related structures, with a capital of ₩10 Mn and a revenue of ₩1,810.33 Mn as of 2015. The GC and SC are both SMEs.

4.1.2. Description of the BIM Model of a Prefabricated Steel Frame

Figure 3 shows the main and secondary parts of the BIM model based on the contents of the case study that is shown in Figure 2. Figure 4 shows the BIM model, including the number of bolts, welding length, cutting parts, grouting, and surface treatment related to the steel joints. Based on these data, the standard, quantity, grade, area, weight, and unit weight of the steel members can be automatically calculated, as shown in Figure 5.

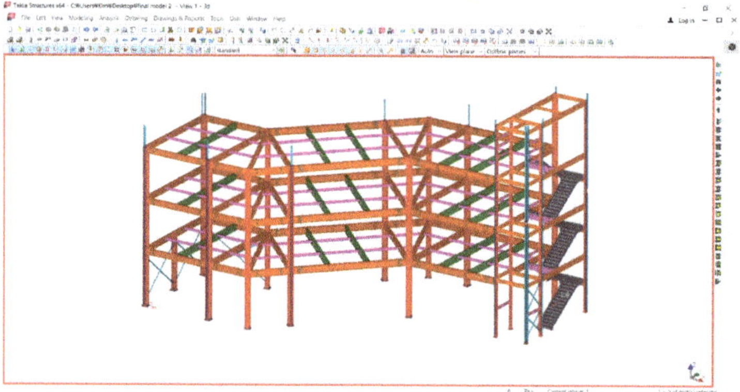

Figure 3. BIM model of a prefabricated steel frame.

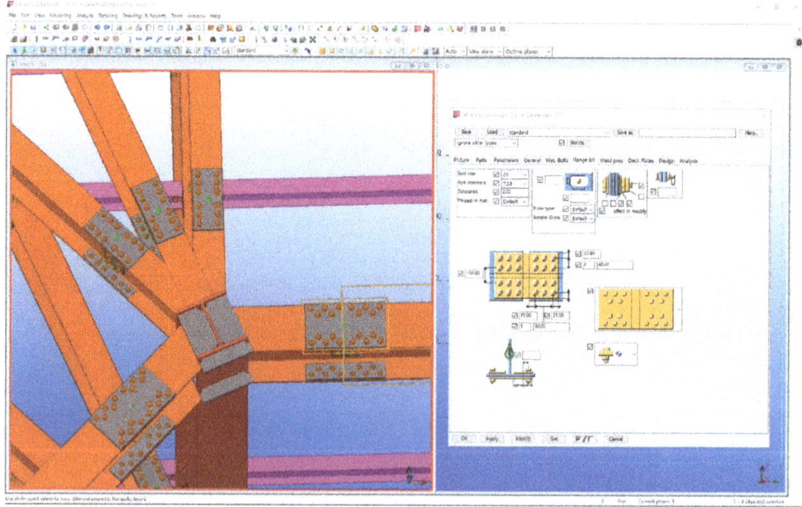

Figure 4. Detailed information regarding the joints of the steel frame members.

```
 List                                                                    —   □   ×
Report
 ----------------------------------------------------------------------------------
 VICTOR BUYCK JOB NO: 1                                          Page: 1
 Title: TEKLA KOREA                                              Date: 09.11.2017
 BUILDING CODE :
 ----------------------------------------------------------------------------------
 Assembly     Part         No.   Size              Grade         Length(mm) Area(m2) Weight(kg)
 ----------------------------------------------------------------------------------

 B0(?)                     4     H250X125X6X9      SS400              495     0.00       0.00
 ----------------------------------------------------------------------------------

 B128(?)                   1     H450X200X9X14     SS400              675    39.96    2499.97
 ----------------------------------------------------------------------------------
              MB140(?)     1     H450X200X9X14     SS400              675     1.11      51.27
              BP4(?)       1     PL25X450          SS400              450     0.45      39.74
              F6(?)        2     PL19X170          SS400              280     0.15      14.20
              MB140(?)     2     H450X200X9X14     SS400              675     2.23     102.54
              MB148(?)     3     H500X200X10X16    SS400              734     3.85     197.40
              MB152(?)     3     H500X200X10X16    SS400              722     3.79     194.21
              MC10(?)      1     H350X350X12X19    SS400            13195    26.92    1801.27
              R24(?)       12    PL20X169          SS400              312     1.46      99.34

 B129(?)                   1     H500X200X10X16    SS400              675    45.04    2651.00
 ----------------------------------------------------------------------------------
              MB147(?)     1     H500X200X10X16    SS400              675     1.18      60.48
              BP4(?)       1     PL25X450          SS400              450     0.45      39.74
              C-CHAN11(?   1     [100X50X5X7.5     SS400            15020     5.56     140.55
              F6(?)        2     PL19X170          SS400              280     0.15      14.20
              MB147(?)     2     H500X200X10X16    SS400              675     2.36     120.96
              MB184(?)     2     H488X300X11X18    SS400              576     2.43     147.92
              MB188(?)     1     H488X300X11X18    SS400              567     1.20      72.72
              MC10(?)      1     H350X350X12X19    SS400            13195    26.92    1801.27
              MG40(?)      3     H450X200X9X14     SS400              675     3.34     153.82
```

Figure 5. Information regarding the quantities of the steel frame members.

4.2. Effects of BIM-Based Construction Management System for SMEs in the Preconstruction Phase

4.2.1. Extraction of Fabrication Drawing

The BIM model of the prefabricated steel frame, which was developed before the construction, is used to create various shop drawings for factory construction. As listed in Table 7, the types of drawings are general arrangement (GA) drawings, which show information regarding the planes, elevations, and cross-sections; a single part drawing, which shows a single part without the welded parts, such as anchor bolts and plates; an assembly drawing, which contains information regarding the steel frame members; and, a cast unit drawing, which is used for the foundation and steel frame plinth. The design changes are automatically reflected in the shop drawings as the BIM model and shop drawings are created based on the objects, and they are linked by parameters. For this case study, a total of 107 detailed drawings were generated to construct the prefabricated steel frame from the BIM model (Table 8).

Table 8. Drawing list.

Drawing	Name	Quantity (Unit, EA)
General arrangement drawings	Block plan drawing, ground plan drawing, elevation drawing, cross-sectional drawing, and window and door drawing	24
Shop drawing	General arrangement drawing	13
	Single part drawing	20
	Assembly drawing	71
	Others (anchor detail and 3D)	3

4.2.2. Review of Shop Drawing

In the final phase, the BIM model was used to conduct a clash check before fabricating the members using the shop drawings. As a result, a total of 39 clashes occurred, which was similar to the ones shown in Figure 6. The clashes mainly occurred

- when steel frame members were crossed or penetrated,
- between the bolts and bolted plates, and
- between a steel part member and a steel plate member.

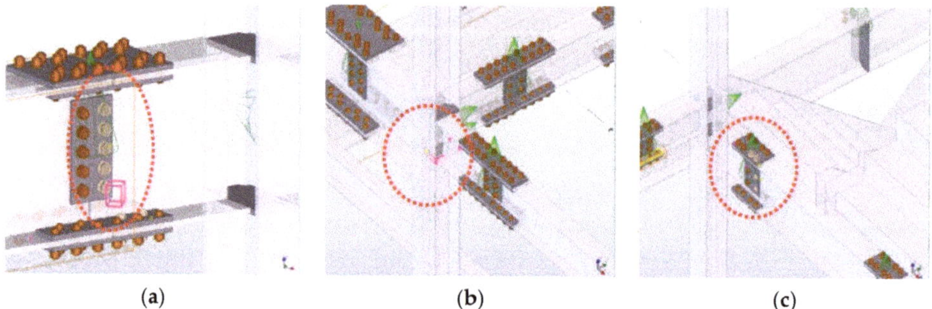

(a)   (b)   (c)

Figure 6. Clash check step applied to the BIM model: (**a**) Steel part members and bolts; (**b**) Steel part members and steel plate members; and, (**c**) Steel part members and stairs.

The members that were included in the clash check result can be classified into type, object ID, assembly ID, and object name. The members were modified prior to fabricating at the factory based on this information.

4.3. Effects of BIM-Based Construction Management System for SMEs in the Fabrication Phase

4.3.1. Prefabrication Review

In the fabrication phase, the necessary steel frame members should be accurately produced. Therefore, a fabrication review should be performed based on the information extracted from the BIM model. The following considerations are important in the manufacture of steel members:

- confirm whether it is easy to fabricate, transfer, lift, and field-install;
- confirm whether it contains additional steel frames for building the steel frames;
- check the welding position, welding method, and dimensions; and,
- check the size and shape of the steel frames.

The visual and quantity-related information regarding the unit steel frame members, including mark (Mark), specification (Description), quantity (Quantity), length (Length), material (Remark), and number of bolts, were extracted from the BIM model. Thus, the review could be performed smoothly. Table 9 lists the information required for steel frame fabrication.

4.3.2. Carrying Steel Frame Members from Off-Site to On-Site

The length, width, volume restriction of vehicles, and entrance road conditions around the site should be considered based on the surrounding traffic situation when carrying the steel frame members to the site from the factory in the order of construction. Furthermore, the order in which the steel frame members are to be unloaded and their exact arrival times should be known to the site supervisors. With the BIM model, it is possible to simulate the position of the laydown area in advance, as shown in Figure 7, and to estimate the quantity, depending on the construction progress.

Table 9. Prefabrication review of beams.

Girder Detail (2-Girders G24)

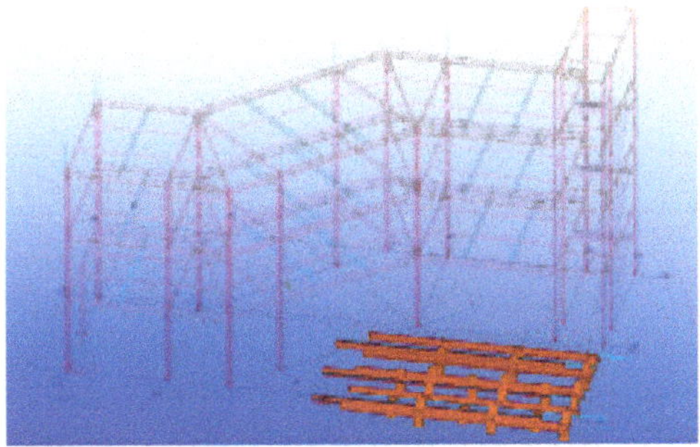

Bill of Material				
Mark	Description	Quantity	Length	Remark
MG19	H250 × 125 × 6 × 9	2	1580	SS400
250A2	PL16 × 121	8	530	SS400
250C2	PL6 × 170	8	200	SS400
GP8	PL9 × 149.5	2	232	SS400
SF15	PL9 × 59.5	2	232	SS400
Field Bolts		Grid Location		
24–M16 T.S.B × 45, F10T		G23	x4-x5/y3, EL+17.566	
128–M16 T.S.B × 50, F10T		G23	x4-x5/y3-y4, EL+17.566	

Figure 7. Preliminary review of building and member laydown area.

The BIM model helps construction companies and professional construction enterprises to efficiently lay the steel frame members by considering the type of construction. In addition, it is possible to minimize errors that may occur during the construction by accurately estimating the information regarding the steel frame members. Thus, the BIM model helps to save time and money, given that it optimizes the number of workers that are required in the field. In this project, the total weight of the steel frame members was approximately 92 t. It took two days (August 4th and 6th) to carry the steel frame members to the laydown area, as shown in Figure 8.

Figure 8. Transportation of the steel frame members to the laydown area.

4.4. Effects of BIM-Based Construction Management System for SMEs in the Construction Phase

4.4.1. Integrated Management of Cost and Schedule

After the steel members are carried to the laydown area, the installation work begins by referring to the shop drawings. To manage the cost and schedule associated with the prefabricated steel frames, these aspects of the BIM model are integrated, as seen in Table 10. Given that the project was small-scale, the construction that was carried out during the period of August 4th, 2017 to August 9th, 2017 was selected for the study. On August 4th, 2017, the steel frame members were carried to the laydown area, and the anchor bolts were installed. Based on the information that was provided in the BIM model and shop drawings, the materials were carried in, and the next phase of the work was prepared. Regarding the equipment, one crane was used for the member laydown, with five people working on-site. On August 5th, the main processes comprised building the steel columns and non-shrink grouting, and the total weight of the installed steel frame was 31.46 t. On August 6th, the welded plates and the steel frame beam members were carried to the area, and the total weight of the installed frame was 15.36 t. A total of 91.78 t of steel frame members were installed by August 9th.

4.4.2. Quality Management Using BIM and Laser Scanning

The construction errors in this case study project were considerably low, because the construction was small-scale, and the pre- and post-processes had little effect on the prefabricated steel frame construction. The construction errors can be measured by comparing the BIM model, which is an as-planned and already constructed model, with a laser scanning equipment, which is an as-built model that is created by a construction company and a professional construction enterprise. Figure 9 shows the construction errors that were identified in the case study project using the BIM model and laser scanning equipment. The errors ranged from 16 to 70 mm in the web or flange direction of the steel frame member.

Although not found in this case study project, a large risk may arise if there is a significant error in the pre-installed steel frame members. If the reference steel frame member, which is assembled with many members, is twisted or severely tilted toward a specific direction, the construction of the steel frame materials, which need to be installed through subsequent works, can be considerably affected. The damages can be significant due to the increase in the construction cost and any delay in the construction from new member orders or additional fabrications, disposing of pre-order members, and duplicated fabrication of steel frame members. Therefore, an as-built model for the pre-process should be created while using the laser scanning technology, the BIM model should be reviewed, and the prefabrication should be implemented by reflecting the actual conditions of the site.

Table 10. Four-/five-dimensional (4D/5D) model in terms of construction schedule.

Main progress	Building steel frame columns and non-shrink grouting			Main progress	Installation of plates on the first and second floors, and installation of steel beams on the first floor.		
Date	8/5/2017	Quantity	31.4 t	Date	8/6/2017	Daily construction quantity	15.36 t
Manpower	5 people	Rate of daily progress	34.19%	Manpower	5 people	Rate of daily progress	50.89%
Equipment	Crane 1 EA	Payment of daily progress	₩35,340,239	Equipment	Crane 1 EA	Payment of daily progress	₩22,999,953
Main progress	Carrying steel frame members to the laydown area, installation of steel beams on the first, second, and third floors, and installation of plates on the third floor			Main progress	Installation of steel beams on the second and third floors		
Date	8/7/2017	Daily construction quantity	19.48 t	Date	8/8/2017	Daily construction quantity	20.12 t
Manpower	5 people	Rate of daily progress	72.06%	Manpower	5 people	Rate of daily progress	93.93%
Equipment	Crane 1 EA	Payment of daily progress	₩24,196,437	Equipment	Crane 1 EA	Payment of daily progress	₩24,713,486

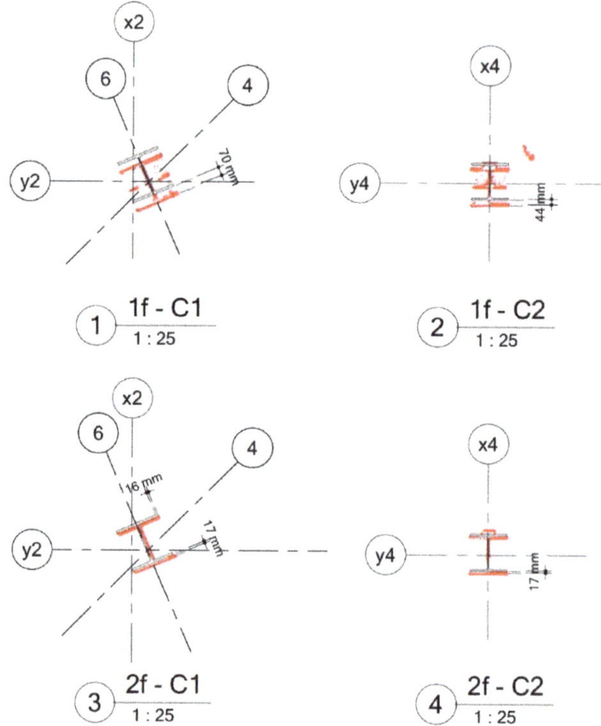

Figure 9. Column construction deviation.

5. Discussion

The features of the system framework described above are as follows: (1) capability to input and output various types of information, (2) the inclusion of only the necessary functions from the viewpoint of SMEs, and (3) improved accessibility, such that anyone may use the system online or with an app (Figure 10). Construction managers require access to detailed information on buildings, and clients need an environment that can utilize various types of reliable information to support decision-making.

Figure 10. Web and app-based operational samples using the proposed system.

In order to verify the effectiveness of the BIM-based construction management system framework that is proposed in this study, we interviewed three researchers (e.g., M-design, Hanyang University, Seoul National University of Science & Technology) who participated in the system development, and four participants (e.g., Client, GC, and SC) who implemented the case project. Table 11 shows the expected effects of applying the system according to the information that was provided during the interviews.

Table 11. Expected effects of application of the BIM-based construction management system.

Perspective	BIM Effects
Client	▪ Decision-making based on quantitative cost data ▪ Communication through visualization ▪ Execution of the built structures based on construction progress ▪ Confirmation of construction quality via surveying
GC (SME)	▪ Deciding construction orders by presenting quantitative construction estimate data ▪ Placing orders for steel frame members using shop drawings extracted from the BIM model ▪ Conducting simulations to carry the steel members to the site and confirm the location of the laydown area ▪ Executing volume management based on daily construction progress
SC (SME)	• Ensuring quality of shop drawings through clash detection review of members before preconstruction • Implementing fabrication and processing of steel frame members using shop drawings • Implementing construction using shop drawings and the BIM model

It is difficult to ensure the reliability of results that are only based on data from interviews with research groups and project groups participating in small-scale projects. Therefore, we also conducted an end-user computing satisfaction (EUCS) survey. The evaluation of the information system under development was based on five factors (Content, Accuracy, Format, Ease of Use, and Timeliness) and 12 indicators (Table 3). The survey covered the potential users of the system under development, namely 40 BIM experts, including architects, construction, construction, and the BIM consultant, all of whom have more than three years of experience in using BIM. The questionnaire was completed by 26 people. Table 12 shows the results of the survey.

Table 12. Results of the end-user computing satisfaction (EUCS) survey.

Category	Content				Accuracy		Format		Ease of Use		Timeliness	
Mean	3.81				3.79		3.58		3.00		3.37	
Index	C1	C2	C3	C4	A1	A2	F1	F2	E1	E2	T1	T2
Mean	3.69	3.77	3.92	3.85	3.81	3.77	3.73	3.42	2.65	3.35	3.38	3.35

The results of the analysis indicate that the satisfaction levels with regard to the content and accuracy categories were relatively high. This is because the system under development is based on the database (DB) of commercial BIM software (SW). Since the information that is needed for the task is extracted from the BIM model, the reliability of the information content and accuracy is ensured. On the other hand, the satisfaction levels for the ease of use and timeliness categories were relatively low. These results suggest that the pre-service training of the system to be developed is necessary from the perspective of SMEs. In addition, additional expert assistance is needed for tasks that involve a high degree of difficulty. We need to carefully consider user experience and user interface when considering the fact that the system should be usable on the Web and via an app. Finally, the survey results indicated that the format of the system should be improved in conjunction with ease of use and timeliness. The proposed BIM-based construction management system framework can then effectively contribute to improving the work productivity, as well as the cost effectiveness of the project unit from the viewpoint of SMEs. In order to avoid construction risks, such as rising labor costs in Korea, SMEs can utilize this system framework as a technology strategy to combat the rising trend of relatively low wages paid to overseas construction laborers (e.g., those from Vietnam).

6. Conclusions

In this paper, we analyzed the BIM-based construction effect from the perspective of SMEs. The main research results of the study are summarized, as follows. First, the BIM-based Construction Management System Framework was conceptually proposed from the viewpoint of SMEs to improve the reliability of analysis. Second, detailed work analysis of the steel frame construction was conducted to derive the main functions of the system. Through this, we derived the application areas of BIM that can be used step-by-step. Third, we analyzed the BIM-based construction effect through the case project of SMEs. We analyzed the effects of BIM on SMEs stakeholders and system development researchers. The case study analyzed the effects of BIM application on the preconstruction, fabrication, and construction phases. The fabrication documents were extracted from the BIM model, and a clash detection review process in the preconstruction phase reviews the quality of the fabrication documents. In the fabrication phase, the information in the fabrication documents was checked before the prefabrication. Furthermore, the process of carrying the steel frame members to the site was reviewed by considering the surrounding buildings and the available space, and the approximate location of the laydown area was confirmed. In the construction phase, the cost and schedule were linked based on the BIM model, such that the construction of the building could be requested by the building owner based on quantitative data depending on the construction progress. After the construction was completed, comparing the BIM model with the measurements that were performed with laser scanning equipment could control the quality. Most SMEs working on construction projects obtain qualitative results when the BIM is applied to a specific work process in the preconstruction, fabrication, and construction phases. Fourth, the end user's satisfaction with the system function was investigated to supplement the qualitative analysis result. In order to overcome the limitations of the single case analysis that was performed in this study, a EUCS survey was conducted among potential system users. The results of the analysis confirmed the direction for system development and improvement. In the future, we plan to analyze the effect of BIM application on structures other than steel frames (e.g., mechanical, electrical, and plumbing engineering) from the perspective of SMEs.

Author Contributions: Conceptualization, M.Y. and J.K.; methodology, M.Y. and J.K.; validation, M.Y., J.K. and C.C.; formal analysis, M.Y.; investigation, M.Y.; resources, M.Y.; data curation, M.Y.; writing—original draft preparation, M.Y. and J.K.; writing—review and editing, M.Y., J.K. and C.C.; visualization, M.Y.; supervision, M.Y.; funding acquisition, M.Y.

Funding: This work (Grant No. C0505520) was supported by the Business for Cooperative R&D between Industry, Academy, and Research Institute funded by the Korea Small and Medium Business Administration in 2019.

Conflicts of Interest: The authors declare that there are no conflicts of interest regarding the publication of this paper.

Data Availability: Data generated or analyzed during the study are available from the corresponding author upon request.

References

1. Cheng, J.C.P.; Won, J.; Das, M. Construction and demolition waste management using BIM technology. In Proceedings of the 23rd Annual Conference of the International Group for Lean Construction, Perth, Australia, 29–31 July 2015.
2. Sacks, R.; Eastman, C.; Lee, G.; Teicholz, P. *BIM Handbook: A Guide to Building Information Modeling for Owners, Designers, Engineers, Contractors, and Facility Managers*, 3rd ed.; Wiley: Hoboken, NJ, USA, 2018.

3.   Rajendran, P.; Gomez, C.P. Implementing BIM for waste minimisation in the construction industry: A literature review. In Proceedings of the 2nd International Conference on Management, Langkawi Kedah, Malaysia, 11–12 June 2012.

4.   Ahankoob, A.; Khoshnava, S.M.; Rostami, R.; Preece, C. BIM perspectives on construction waste reduction. In Proceedings of the Management in Construction Research Association (MiCRA), Langkawi Kedah, Malaysia, 5–6 December 2012.

5.   Anumba, C.; Dubler, C.; Goodman, S.; Kasprzak, C.; Kreider, R.; Messner, J.; Saluja, C.; Zikic, N. The BIM project execution planning guide and templates—Version 2.0. CIC Research Group, Department of Architectural Engineering, the Pennsylvania State University, University Park, PA, USA. 2010. Available online: https://www.bim.psu.edu/ (accessed on 12 January 2019).

6.   Williams, M. Building-information Modeling Improves Efficiency, Reduces Need for Changes. *Bus. J.* **2011**. Available online: http://www.bizjournals.com/louisville/print-edition/2011/07/08/building-information-modeling-improves.html (accessed on 17 March 2017).

7.   Ham, N.; Lee, S. Project benefits of digital fabrication in irregular-shaped buildings. *Adv. Civil Eng.* **2019**, *2019*, 1–14. [CrossRef]

8.   Hamid, M.; Tolba, O.; El Antably, A. BIM semantics for digital fabrication: A knowledge-based approach. *Autom. Constr.* **2018**, *91*, 62–82. [CrossRef]

9.   Zhang, J.P.; Hu, Z.Z. BIM- and 4D-based integrated solution of analysis and management for conflicts and structural safety problems during construction: 1. Principles and methodologies. *Autom. Constr.* **2011**, *20*, 155–166. [CrossRef]

10.   Bryde, D.; Broquetas, M.; Volm, J.M. The project benefits of building information modelling (BIM). *Int. J. Project Manag.* **2013**, *31*, 971–980. [CrossRef]

11.   Lee, G.; Park, K.H.; Won, J. D3 city project—Economic impact of BIM-assisted design validation. *Autom. Constr.* **2012**, *22*, 577–586. [CrossRef]

12.   Kim, S.; Chin, S.; Han, J.; Choi, C.H. Measurement of construction BIM value based on a case study of a large-scale building project. *J. Manag. Eng.* **2017**, *33*, 05017005. [CrossRef]

13.   Ham, N.; Moon, S.; Kim, J.H.; Kim, J.J. Economic analysis of design errors in BIM-based high-rise construction projects: Case study of Haeundae L Project. *J. Constr. Eng. Manag.* **2018**, *144*, 05018006. [CrossRef]

14.   Hong, Y.; Hammad, A.W.A.; Sepasgozar, S.; Akbarnezhad, A. BIM adoption model for small and medium construction organisations in Australia. *Eng. Constr. Archit. Manag.* **2018**, *26*, 154–183. [CrossRef]

15.   Hosseini, M.R.; Banihashemi, S.; Chileshe, N.; Namzadi, M.O.; Udeaja, C.E.; Rameezdeen, R.; McCuen, T. BIM adoption within Australian small and medium sized enterprises (SMEs): An innovation diffusion model. *Constr. Econ. Build.* **2016**, *16*, 71–86. [CrossRef]

16.   Dainty, A.R.J.; Leiringer, R.; Fernie, S.; Harty, C. BIM and the small construction firm: A critical perspective. *Build. Res. Inf.* **2017**, *45*, 696–709. [CrossRef]

17.   Lam, T.T.; Mahdjoubi, L.; Mason, J. A framework to assist in the analysis of risks and rewards of adopting BIM for SMEs in the UK. *J. Civ. Eng. Manag.* **2017**, *23*, 740–752. [CrossRef]

18.   Ghaffarianhoseini, A.; Doan, D.T.; Zhang, T.; Ur Rehman, A.; Amirhosein, G.; Nicola, N.; John, T. A BIM readiness & implementation strategy for SME construction companies in the UK. In Proceedings of the 33rd CIB W78 Conference 2016, Brisbane, Australia, 31 October–2 November 2016.

19.   Jernigan, F.E. *BIG BIM Little BIM: The Practical Approach to Building Information Modeling-Integrated Practice Done the Right Way!* 4Site Press: Salisbury, MD, USA, 2008.

20.   Erin, R.H. Achieving strategic ROI, measuring the value of BIM, in: Is it time for BIM? Achieving strategic ROI in your firm, Autodesk. 2016. Available online: https://damassets.autodesk.net/content/dam/autodesk/www/solutions/pdf/Is-it-Time-for-BIM-Achieving-Strategic-ROI-in-Your-Firm%20_ebook_BIM_final_200.pdf (accessed on 5 March 2017).

21.   Mark, R. SMEs: The key to the construction sector's future. Latest News, Scape Group. 2017. Available online: https://www.scapegroup.co.uk/news/2017/smes-the-key-to-the-construction-sectors-future (accessed on 2 September 2017).

22.   National Statistics. *Business Population Estimates 2017: Annual Business Population Estimates for the UK and Regions in 2017. Includes Revised Totals for 2013 to 2016, Taking Account of Updated Source Data*; Department for Business, Energy & Industrial Strategy: London, UK, 2017.

23. Department for Business, Enterprise and Regulatory Reform (BERR) (2006), SME STATISTICS 2006, BERR: Enterprise Directorate Analytical Unit. Available online: http://stats.berr.gov.uk/ed/sme/ (accessed on 22 January 2019).

24. Lu, S.L.; Sexton, M.G.; Abbot, C. Key characteristics of small construction firms: A United Kingdom perspective. In Proceedings of the CIB W065/055 Commissions: Transformation through Construction, Dubai, UAE, 15–17 November 2008; Available online: https://www.irbnet.de/daten/iconda/CIB17619.pdf (accessed on 17 February 2019).

25. Curran, J.; Blackburn, R.A. *Researching the Small Enterprise*, 1st ed.; Sage Publications: London, UK, 2001.

26. Robinson, R.B., Jr.; Pearce, J.A. Research thrusts in small firm strategic planning. *Acad. Manag. Rev.* **1984**, *9*, 128–137. [CrossRef]

27. Lu, S.; Sexton, M. Innovation in small construction knowledge-intensive professional service firms: A case study of an architectural practice. *Constr. Manag. Econ.* **2006**, *24*, 1269–1282. [CrossRef]

28. Sexton, M.; Barrett, P. Appropriate innovation in small construction firms. *J. Constr. Manag. Econ.* **2003**, *21*, 623–633. [CrossRef]

29. Yoo, M.; Choi, C.; Yoo, S.; Park, S. Communication performance of BLE-based IoT devices and routers for tracking indoor construction resources. *IJIBC* **2019**, *11*, 27–38. [CrossRef]

30. Alsafouri, S.; Ayer, S.K. Review of ICT implementations for facilitating information flow between virtual models and construction project sites. *Automat. Constr.* **2018**, *86*, 176–189. [CrossRef]

31. Singh, V.; Gu, N.; Wang, X. A theoretical framework of a BIM-based multi-disciplinary collaboration platform. *Automat. Constr.* **2011**, *20*, 134–144. [CrossRef]

32. Suh, S.W.; Yoon, Y.S.; Kim, S.K. The Development of Schedule Risk Management Tools at the Preconstruction Phase in Building Construction-Focused on the Steel Work. *J. Constr. Eng. Manag.* **2005**, *6*, 177–185.

33. Lim, C.H. Application of 3D information to reduce rework in a steel structure construction. Master's Thesis, Hanyang University, Seoul, Korea, 2017.

34. Jung, S.J. *Building Construction - Construction of Steel Structure*; Kimoondang: Seoul, Korea, 2013; ISBN 9788962255331.

35. Kim, Y.H.; Lee, J.S.; Oh, J.K.; Kim, J.J. A study for derivation of participant's information flow at framework construction based on BIM. *Korea J. Constr. Eng. Manag.* **2013**, *14*, 22–34. [CrossRef]

36. Park, C.S.; Park, H.T. Improving contractibility analysis tasks by applying BIM technology. *Korea J. Constr. Eng. Manag.* **2010**, *11*, 137–146. [CrossRef]

37. Eom, J.U.; Shin, T.S. A study on the automation of the connection modeling for steel structures based on BIM. *J. Korean Soc. Steel Constr.* **2010**, *22*, 99–108.

38. Ko, A.R.; Lee, S.I.; Cho, Y.S. Development of automatic design system for steel connection based on S-BIM. *Korea J. Constr. Eng. Manag.* **2013**, *29*, 21–28.

39. Eom, J.U.; Shin, T.S. A development of interface module between structural design and detail design based on BIM. *J. Korean Soc. Steel Constr.* **2011**, *23*, 113–124.

40. Li, K.; Gan, Y.; Ke, G.; Chen, Z. The Analysis and Application of BIM Technology in Design of Steel Structure Joints. In Proceedings of the 4th International Conference on Sensors, Measurement and Intelligent Materials (ICSMIM 2015), Shenzhen, China, 27–28 December 2015; pp. 1166–1171.

41. Oti, A.H.; Tizani, W. BIM extension for the sustainability appraisal of conceptual steel design. *Adv. Eng. Inform.* **2015**, *29*, 28–46. [CrossRef]

42. Ryu, J.S.; Kim, K.H. A study of 4D simulation system using automatic scheduling process-The focus on steel structural construction. *J. Architectural Inst. Korea* **2009**, *25*, 173–180.

43. Yun, S.H.; Park, C.W.; Lee, G.; Bongkeun, K. A study on a method for tracking lifting paths of a tower crane using GPS in the BIM environment. *J. Architectural Inst. Korea* **2008**, *24*, 163–170.

44. Shin, T.S.; Yang, J. A proposal for the automation process of creating shop drawings in steel constructions. *J. Architectural Inst. Korea* **2009**, *11*, 267–274.

45. Kim, K.T. A study on the application of RTLS technology for the automation of spray-applied fire resistive covering work. *J. Korea Inst. Build. Constr.* **2009**, *9*, 79–86. [CrossRef]

46. Xie, H.; Shi, W.; Issa, R.R.A. Using RFID and real-time virtual reality simulation for optimization in steel construction. *J. Inf. Technol. Constr.* **2011**, *16*, 291–308.

47. Liu, W.; Guo, H.; Li, H.; Li, Y. Using BIM to improve the design and construction of bridge projects: A case study of a long-span steel-box arch bridge project. *Int. J. Adv. Robot. Syst.* **2014**, *11*, 125. [CrossRef]

48. Ginzberg, M.J.; Zmud, R.W. Evolving Criteria for Information Systems Assessment. In Proceedings of the IFIP WG 8.2 Working Conference, San Francisco State University, San Francisco, CA, USA, 11–12 December 1987.

49. Will, P.; Olson, M.H. Managing investment in information technology: Mini case examples and implication. *MIS Q.* **1989**, *13*, 3–17. [CrossRef]

50. Benjamin, R.I.; Rockart, J.F.; Wyman, S.M.J. Information technology: A strategic opportunity. *Sloan Manag. Rev.* **1984**, *25*, 3–14.

51. Joan, S. IT Performance Measurement: A Discussion Paper on Measurement Practices. Available online: www.gsa.itpolicy.gov (accessed on 14 December 2018).

52. Shannon, C.E.; Weaver, W. *The Mathematical Theory of Communication*; University of Illinois Press: Urbana, IL, USA, 1949.

53. Mason, R.O. Measuring information output: A communication systems approach. *Inform. Manag.* **1978**, *1*, 219–234. [CrossRef]

54. Delone, W.H.; Mclean, E.R. Information systems success: The quest for the dependent variable. *Inf. Syst. Res.* **1992**, *3*, 66–95. [CrossRef]

55. Lucas, H.C., Jr.; Henry, C. Evolution of An Information System: From Keyman to Every Person. *Sloan Manag. Rev.* **1978**, *19*, 39–55.

56. Hamilton, S.; Chervany, N.L. Evaluating information system effectiveness—Part I: Comparing evaluation approaches. *MIS Q.* **1981**, *5*, 55–69. [CrossRef]

57. Saarineen, T. An expanded instrument for evaluating information system success. *Inform. Manag.* **1996**, *31*, 103–118. [CrossRef]

58. Bailey, J.; Pearson, S. Development of a tool for measuring and analyzing computer user satisfaction. *Manag. Sci.* **1983**, *29*, 530–545. [CrossRef]

59. Miller, J.; Doyle, B.A. Measuring the effectiveness of computer-based information systems in the financial services sector. *MIS Q.* **1987**, *11*, 107–124. [CrossRef]

60. Kim, J.S.; Cho, Y.B.; Kim, Y.I. A study on factor analysis for successful informatization of SMEs. *KMIS* **1994**, *1994*, 129–163.

61. Jung, K.E.; Lee, D.M. Reciprocal effect of the factors influencing the satisfaction of IS users. *Asia Pac. J. Inf. Syst.* **1995**, *5*, 199–226.

62. Suh, K.S. A study on the factors affecting the success of end-user computing. *APJIS* **1995**, *5*, 259–288.

63. Lves, B.; Olson, M.H.; Baroudi, J.J. The measurement of user information satisfaction. *Commun. ACM* **1983**, *26*, 785–793.

64. Doll, W.J.; Torkzadeh, G. The management of end-user computing satisfaction. *MIS Q.* **1988**, *12*, 259–274. [CrossRef]

65. Doll, W.J.; Xia, W. Confirmatory factor analysis of the end-user computing satisfaction instrument: A replication. *JUEC.* **1994**, *9*, 24–31. [CrossRef]

66. Kim, S.; McHaney, R. Validation of the end-user computing satisfaction instrument in case tool environments. *J. Comput. Inform. Syst.* **2000**, *41*, 49–55.

67. Hendrickson, A.R.; Glorfeld, K. On the repeated test-retest reliability of the end-user computing satisfaction instrument. *Decis. Sci.* **1994**, *25*, 655–667.

68. AEC Magazine. Tekla Structure 19. 2013. Available online: http://www.aecmag.com/software-mainmenu-32/563-tekla-structures-19 (accessed on 2 February 2019).

69. Kim, J.; Yoo, M.; Ham, N.; Kim, J.; Choi, C. Process of Using BIM for Small-Scale Construction Projects - Focusing on the Steel-frame Work. *KIBIM Mag.* **2018**, *8*, 41–50.

70. Robinson, C. Structural BIM: Discussion, case studies and latest developments. *Struct. Des. Tall Spec. Build.* **2007**, *16*, 519–533. [CrossRef]

applied sciences

Article

Reducing Critical Hindrances to Building Information Modeling Implementation: The Case of the Singapore Construction Industry

Longhui Liao [1,*], Evelyn Ai Lin Teo [2] and Ruidong Chang [3]

[1] Sino-Australia Joint Research Center in BIM and Smart Construction, Shenzhen University, Shenzhen 518060, China

[2] Department of Building, School of Design and Environment, National University of Singapore, Singapore 117566, Singapore; bdgteoal@nus.edu.sg

[3] Centre for Comparative Construction Research, Faculty of Society and Design, Bond University, Gold Coast 4226, Australia; rchang@bond.edu.au

* Correspondence: a0109736@u.nus.edu; Tel.: +86-755-8695-2433

Received: 29 June 2019; Accepted: 6 September 2019; Published: 12 September 2019

Abstract: The Singaporean government has made building information modeling (BIM) implementation mandatory in new building projects with gross floor areas over 5000 m^2, but the implementation is still plagued with hindrances such as lacking project-wide collaboration. The purposes of this study are to identify critical factors hindering BIM implementation in Singapore's construction industry, analyze their interrelationships, and identify strategies for reducing these hindrances. The results from a survey of 87 experts and five post-survey interviews in the Singaporean construction industry identified 21 critical hindrances, among which "need for all key stakeholders to be on board to exchange information" was ranked top. These hindrances were categorized into lack of collaboration and model integration (LCMI), lack of continuous involvement and capabilities (LCIC), and lack of executive vision and training (LEVT). LEVT and LCIC contributed to LCMI; LEVT caused LCIC. The proposed framework implying the key hindrances and their corresponding managerial strategies can help practitioners identify specific adjustments to their BIM implementation activities, which enables to efficiently achieve enhanced BIM implementation. The hindrances identified in this study facilitate overseas BIM implementers to customize their own lists of hindrances.

Keywords: building information modeling (BIM); building project; hindrance; factor analysis; structural equation modeling (SEM); managerial strategies; Singapore

---

1. Introduction

Building information modeling (BIM) refers to the integration of technological and organizational solutions. The solutions can not only enhance inter-organizational and multidisciplinary collaboration, but also improve the efficiency and quality of the planning, design, construction, and management of buildings [1–5]. Although the value of BIM is now widely recognized compared with the traditional drafting practices, it is not possible to reap the full benefits without awareness, commitment, and capabilities of implementing BIM as well as a realistic view of the adoption status. For example, Chelson [6] found that BIM operators may be typically young and lack enough field knowledge to incorporate new work processes into the project workflow. Khosrowshahi and Arayici [7] revealed that designers and contractors may psychologically contradict the new processes. Forsythe et al. [8] found that firms would start to implement BIM if policymakers already required, specified, or mandated them to do that. Otherwise, it would need many years before BIM is more often used. Juan et al. [9] reported that most firms would use BIM to maintain competitiveness in the market where other firms had already implemented BIM in an earlier time. Thus, both opportunities and risks appeal to exist in BIM implementation.

In the Singapore context, a top-down approach has been used in driving BIM implementation. The Building and Construction Authority (BCA) has been playing a dominant role and made much effort to promote BIM. Among which, the most important regulation was a five-year BIM adoption roadmap. Specifically, all new building projects (both private and public) that have gross floor areas (GFAs) of greater than 20,000 m^2 must submit their architectural plans in BIM format for regulatory approvals since July 2013 and submit their structural and mechanical, electrical, and plumbing (MEP) plans in BIM format since July 2014. Eventually, all new building projects with GFAs of 5000 m^2 and above are mandated to submit their building plans in BIM format, which came into force in July 2015 [10]. In the meantime, the local construction value chain has been encouraged to work collaboratively, with part of implementation cost being subsidized [11]. The local government has also drafted the BIM Particular Conditions to guide the local construction industry to address the procedures of digital data processing, roles and responsibilities, intellectual property rights, each party' extent of reliance on three-dimensional (3D) models, and contractual privity. Consequently, the overall BIM adoption rate had improved from 20% in 2009 to 65% in 2014; such implementation, however, tended to be fragmented BIM uses in individual parties, rather than based on project-wide collaboration [12]. In addition, the building contracts in Singapore are still developed on the basis of the traditional contractual framework that prohibits collective benefits and encourages individualism. When problems occur, such a contractual structure would easily thrust project participants into adversarial positions [13,14]. Overall, most practitioners are conservative to change.

The specific purposes of this study are to (1) identify critical hindrances to BIM implementation in the Singapore construction industry, (2) investigate interrelationships among the hindrances, and (3) identify managerial strategies for reducing these hindrances. Given that the BIM submissions policy in new building projects in Singapore is mandatory, the local practitioners have to be ready for moving towards full BIM implementation and gain an in-depth understanding of what really hinders their BIM implementation and how to reduce such hindrances. Although many studies have investigated BIM implementation in the global construction industry, no studies have been done so far to comprehensively study the hindrances to BIM implementation in Singapore as the present study does. Also, few studies have attempted to investigate the relationships among the critical hindrances and accordingly build a conceptual framework.

The Economic Strategies Committee (ESC) of Singapore has advocated that the local economy, especially labor-intensive industries, should improve work efficiency to maintain competitiveness [15]. Thus, the planning, design, construction, and management of building projects have a critical implication. The managerial strategies identified in this study can help the local industry players eliminate the critical hindrances' negative influence to enhance BIM implementation. Although this study focuses on building projects in Singapore, overseas practitioners may use the identified hindrances to prepare their own hindrances and follow the research method to formulate their strategies. Thus, this study may contribute to the existing literature related to BIM implementation.

2. Literature Review

Through the literature review analyzing 26 previous global studies on BIM implementation, this study has identified 47 hindrances, as shown in Table 1. It should be noted that in the Singapore context, the BCA drives the whole value chain to optimize building designs for off-site manufacture (OSM) which are encouraged in full BIM implementation [11], and intends to develop a BIM guide for Design for Manufacturing and Assembly [12]. Prefabricators can use the precision of geometric data contained in building information models to aid the manufacturing process and assembly of building components on site. Thus, in this study, hindrances related to the integration of BIM and OSM were also identified. Faced with these 47 hindrances, project teams may not implement BIM openly and collaboratively. Lam [12] reported that only 20% of building projects in Singapore had implemented BIM with a relatively high collaboration level. Thus, the local construction industry has been facing many issues: owners cannot see beyond initial cost; designers tend to over-emphasize the BIM e-submissions and lack time to perform design coordination; main contractors can rarely use the designers' models because in most cases such models were not developed in the same way the contractors intend to build the buildings; subcontractors, especially those small- and medium-sized enterprises (SMEs), lack investment for hardware, software, and training; and facility managers are rarely involved upfront and lack BIM uses [12]. Consequently, the contractors may need to re-build the models, taking much time, which in turn hinders the collaboration with the designers.

Table 1. Hindrances to building information modeling (BIM) implementation.

Code	Hindrances to BIM Implementation	1	2	3	4	5	6	7	8	9	10	11	12	13	14	15	16	17	18	19	20	21	22	23	24	25	26
H01	Executives failing to recognize the value of BIM-based processes and needing training						√			√					√				√			√					√
H02	Concerns over or uninterested in sharing liabilities and financial rewards							√	√	√			√	√						√							
H03	Construction lawyers and insurers lacking understanding of roles/responsibilities in new process				√								√							√							
H04	Lack of skilled employees and need for training them on BIM and off-site manufacture (OSM)			√	√			√	√	√	√	√	√		√	√	√		√		√	√					√
H05	Industry's conservativeness, fear of the unknown, and resistance to change comfortable routines				√			√				√						√	√	√	√						√
H06	Employees still being reluctant to use new technology after being pushed to training programs																		√	√							√
H07	Entrenchment in two-dimensional (2D) drafting and unfamiliarity to use BIM			√				√									√	√	√	√	√	√					√
H08	Financial benefits cannot outweigh implementation and maintenance costs			√						√								√	√	√							
H09	Lack of sufficient evidence to warrant BIM use																		√	√	√	√					√
H10	Liability of BIM such as the liability for common data for subcontractors												√						√	√	√						
H11	Resistance to changes in corporate culture and structure										√	√							√	√	√	√	√				
H12	Need for all key stakeholders to be on board to exchange information			√		√		√					√	√	√				√	√	√				√	√	
H13	Lack of trust/transparency/communication/partnership and collaboration skills	√	√		√			√				√	√						√	√	√						
H14	BIM operators lacking field knowledge		√									√									√						
H15	Field staff dislike BIM coordination meetings looking at a screen											√															
H16	Lack of consultants' feedbacks on subcontractors' model coordination											√															
H17	Few benefits from BIM go to designers while most to contractors and owners												√														
H18	Lack of legal support from authorities																	√	√		√						√
H19	Lack of owner request or initiative to adopt BIM		√																			√	√				√
H20	Decision-making depending on relationships between project stakeholders		√																	√							√
H21	Owners set minimal risk and minimum first cost as crucial selection criteria										√												√				
H22	Poor knowledge of using OSM and assessing its benefits											√															
H23	Requiring higher onsite skills to deal with low tolerance OSM interfaces											√															
H24	OSM relies on suppliers to train contractors to install correctly											√															
H25	Owners' desire for particular structures or finishes when considering OSM											√															
H26	Market protection from traditional suppliers/manufacturers and limited OSM expertise											√															
H27	Contractual relationships among stakeholders and need for new frameworks	√		√		√	√	√					√	√	√				√	√		√	√	√	√		
H28	Traditional contracts protect individualism rather than best-for-project thinking	√	√										√														
H29	Lack of effective data interoperability between project stakeholders	√		√			√	√			√	√	√			√				√					√	√	
H30	Owners cannot receive low-price bids if requiring three-dimensional models											√															
H31	Firms' unwillingness to invest in training due to initial cost and productivity loss								√			√						√	√								
H32	Assignment of responsibility/risk to constant updating for broadly accessible BIM information								√			√															
H33	Lack of standard contracts to deal with responsibility/risk assignment and BIM ownership							√			√	√							√						√		
H34	BIM model issues (such as ownership and management)				√			√				√													√		
H35	Poor understanding of OSM process and its associated costs											√															
H36	OSM requires design to be fixed early using BIM											√					√										
H37	Seeing design fees of OSM as more expensive than traditional process											√															

Table 1. *Cont.*

Code	Hindrances to BIM Implementation	1	2	3	4	5	6	7	8	9	10	11	12	13	14	15	16	17	18	19	20	21	22	23	24	25	26
																											References
H38	Difficulty in logistics and stock management of OSM										√														√		
H39	Unclear legislations and qualifications for precasters and inadequate codes for OSM varieties										√																
H40	Interpretations resulted from unclear contract documents																			√							
H41	Using monetary incentive for team collaboration results in blaming rather than resolving issues																			√							
H42	Costly investment in BIM hardware and software solutions			√	√					√	√								√		√	√	√				√
H43	Interoperability issues such as software selection and insufficient standards			√									√				√		√								√
H44	Need for increasingly specialized software for specialized functions												√						√		√						√
H45	Difficulty in multi-discipline and construction-level integration												√							√							
H46	Technical needs for multiuser model access in multi-discipline integration							√					√						√								
H47	Firms cannot make most use of Industrial Foundation Classes (IFC) and use proprietary formats												√									√					√

Note: 1 = American Institute of Architects and American Institute of Architects, California Council (AIA and AIACC) [16]; 2 = AIA and AIACC [17]; 3 = Aranda-Mena et al. [18]; 4 = Arayici et al. [19]; 5 = Autodesk [20]; 6 = Autodesk [21]; 7 = Azhar et al. [22]; 8 = Bernstein and Pittman [23]; 9 = Bernstein et al. [24]; 10 = Blismas and Wakefield [25]; 11 = Chelson [6]; 12 = Eastman et al. [26]; 13 = Fischer et al. [13]; 14 = Fischer [27]; 15 = Fox and Hietanen [28]; 16 = Gao and Fischer [29]; 17 = Gibb and Isack [30]; 18 = Juan et al. [9]; 19 = Kent and Becerik-Gerber [31]; 20 = Khosrowshahi and Arayici [7]; 21 = Kiani et al. [32]; 22 = Kunz and Fischer [33]; 23 = McFarlane and Stehle [34]; 24 = Ross et al. [35]; 25 = Sattineni and Mead [36]; 26 = Zahrizan et al. [37].

However, instead of studying all the 47 hindrances holistically, each of the 26 previous studies tended to only explore part of the hindrances. More importantly, little is known about how the hindrances may influence BIM implementation in the Singapore context. For example, Khosrowshahi and Arayici [7] identified the hindrances to BIM implementation at high maturity levels for the contractors in the United Kingdom, but failed to investigate the factors for other roles in the construction value chain. Zahrizan et al. [37] identified the factors hindering BIM diffusion in the Malaysian construction industry with respects to culture, people, technology, and government's recognition, but rarely studied BIM work processes and the key role of the local government regarding its active participation to specify BIM use and its financial support such as defraying a proportion of training and consultancy costs. Juan et al. [9] explored the hindrances affecting the Taiwan construction industry to be ready to adopt BIM, but was limited to the architectural firms. Although Oo [38] investigated some hindrances to move towards the new work processes in Singapore, its main focus was on the architectural discipline and the identification of driving factors rather than the hindrances. This present study would fill this gap by identifying the critical hindrances that significantly influenced BIM implementation in Singapore and analyzing the influence mechanisms among these hindrances, extending the relevant literature.

3. Method and Data Presentation

Figure 1 presents the research methodology. The literature search indicated that the use of questionnaire survey technique was appropriate in collecting professional views on critical factors in previous construction management studies [39,40]. Thus, a questionnaire survey was performed to investigate the 47 hindrances' influence on BIM implementation in the Singapore construction industry. The questionnaire was designed on the basis of the above literature review and refined according to the comments from five BIM experts who were interviewed face-to-face in a pilot study. All the experts, who were working for large firms and possessed at least three years' experience in implementing BIM in Singapore, were selected to pretest the questionnaire. The final questionnaire collected general information of respondents, and requested them to rate each hindrance's influence on BIM implementation. The ratings should be made regarding one of their ongoing or recently-completed building projects. A five-point Likert scale (1 = very insignificant; 2 = insignificant; 3 = neutral; 4 = significant; 5 = very significant) was used. Miller [41] found that a human usually can hold "seven plus or minus two" objects in working memory. In a one-dimensional absolute-judgment task, a person is presented with a number of stimuli and responds to each stimulus with a corresponding response. Performance is nearly perfect up to five or six stimuli but declines as the number is increased. Thus, to make it convenient for the respondents to judge, the five-point scale was adopted in this paper, which has been widely used in previous studies related to construction management [42–45].

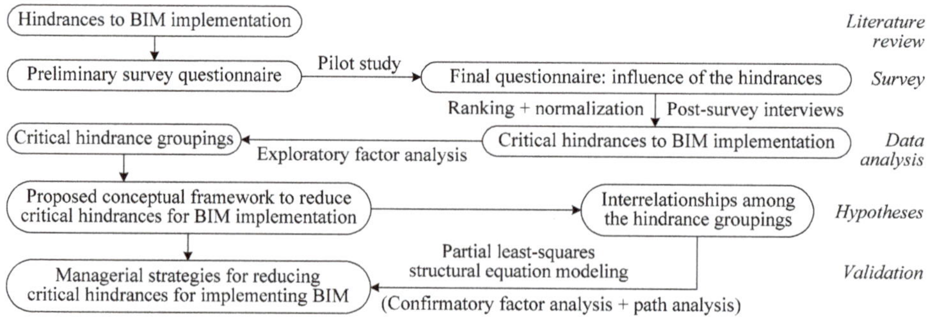

Figure 1. Research methodology.

The population was comprised of all the organizations that were operating in the Singapore construction industry. The sampling frame consisted of the BCA, the Urban Redevelopment Authority, the Housing and Development Board (HDB), the building developers registered with the Real Estate Developers' Association of Singapore, the architectural consultancy firms registered with the Singapore Institute of Architects, the structural and MEP consultancy firms registered with the Association of Consulting Engineers Singapore, the contractors registered with the BCA, and the facility management firms registered with the Association of Property and Facility Managers. Among the contractors, it was considered logical to select only the large ones because they tend to have adequate resources for BIM implementation. Since there was a sampling frame, a probability sample should be adopted. Simple random sampling was used in the data collection because each organization was as likely to be drawn as the others. Finally, the questionnaires were sent to 692 organizations via emails or handed to them personally. It was considered appropriate that 87 completed questionnaires were received based on willingness to participate in this study [46]. The response rate of 12.57% was acceptable because it fell within the general response rate of 10–15% for Singapore surveys [39]. The profile of the 87 respondents is shown in Table 2. The 14 organizations in the "others" category included the BCA, the HDB, developers, precasters, and other consultancy firms such as multidisciplinary consultancy firms and a BIM consultancy firm. Thus, the responding organizations could represent major BIM implementers in the local construction value chain. Moreover, because BIM implementation had been mandated in Singapore since July 2015, it was reasonable that under half (42.53%) had over three years' BIM implementation experience.

Table 2. Profile of respondents and their organizations.

Characteristics	Categorization	N	%	Characteristics	Categorization	N	%
	Respondents				*Organizations*		
	Government agent	2	2.3		Architectural firm	18	20.7
	Developer	5	5.7		Structural engineering firm	6	6.9
	Architect	21	24.1		MEP engineering firm	13	14.9
	Structural designer	9	10.3		General construction firm	30	34.5
Discipline	MEP designer	9	10.3	Main business	Trade construction firm	3	3.4
	General contractor	28	32.2		Facility management firm	3	3.4
	Subcontractor	6	6.9		Others	14	16.1
	Supplier/Manufacturer	2	2.3				
	Facility manager	5	5.7				
	5–10 years	39	44.8		0	9	10.3
Work experience	11–15 years	10	11.5	Years of BIM implementation	1–3	41	47.1
	16–20 years	8	9.2		4–5	22	25.3
	21–25 years	9	10.3		6–10	13	14.9
	>25 years	21	24.1		>10	2	2.3

In addition, after the survey was performed, five BIM experts who had participated in the survey and were experienced in BIM implementation in the Singapore construction industry were contacted for personal interviews. During which, they were presented with the results obtained from the survey. They commented that the findings were in agreement with what they expected. These professionals were also invited to explain the results which would be discussed in the following sections.

4. Results and Discussion

4.1. Ranking of Hindrances to BIM Implementation

The Cronbach's alpha coefficient was 0.974, much higher than the threshold of 0.70 [47]. Thus, the data collected in this study had high reliability. The ranking of the 47 hindrances to BIM implementation in the Singapore construction industry was shown in Table 3. A few methods were adopted to identify the critical factors when Likert scale data were collected in previous studies, but none of them was established as a standardized method. The threshold value was also not determined. Magal et al. [48] recognized all the factors that were deemed critical in previous studies. No cutoff point was established in this said study. Shen and Liu [49] set a cutoff value of 4 because it represented "significant" in the five-point scale in this study. Nitithamyong and Skibniewski [50] used the middle value of the Likert scale as the threshold. Xu et al. [51] and Zhao et al. [43] chose the factors with normalized values of 0.50 and above as the critical factors. Won et al. [52] selected the factors with mean values exceeding the total mean value of the data as the critical factors.

Table 3. Ranking of hindrances to BIM implementation.

Code	Mean	Rank	Normalization *	Code	Mean	Rank	Normalization *	Code	Mean	Rank	Normalization *
H01	3.644	8	0.782	H17	3.161	46	0.018	H33	3.540	13	0.618
H02	3.494	20	0.545	H18	3.184	45	0.055	H34	3.529	14	0.600
H03	3.241	41	0.145	H19	3.414	26	0.418	H35	3.506	19	0.564
H04	3.690	3	0.855	H20	3.264	40	0.182	H36	3.494	20	0.545
H05	3.678	5	0.836	H21	3.345	34	0.309	H37	3.448	22	0.473
H06	3.414	26	0.418	H22	3.333	35	0.291	H38	3.402	30	0.400
H07	3.690	3	0.855	H23	3.402	30	0.400	H39	3.241	41	0.145
H08	3.368	33	0.345	H24	3.241	41	0.145	H40	3.391	32	0.382
H09	3.529	14	0.600	H25	3.149	47	0.000	H41	3.218	44	0.109
H10	3.310	37	0.255	H26	3.414	26	0.418	H42	3.667	6	0.818
H11	3.414	26	0.418	H27	3.713	2	0.891	H43	3.517	17	0.582
H12	3.782	1	1.000	H28	3.655	7	0.800	H44	3.425	23	0.436
H13	3.310	37	0.255	H29	3.425	23	0.436	H45	3.529	14	0.600
H14	3.621	10	0.745	H30	3.299	39	0.236	H46	3.644	8	0.782
H15	3.425	23	0.436	H31	3.621	10	0.745	H47	3.563	12	0.655
H16	3.322	36	0.273	H32	3.517	17	0.582	–	–	–	–

Note: Total mean value = 3.451. * Normalized value = (mean − minimum mean)/(maximum mean − minimum mean). Hindrances with normalized values less than 0.50 would not be considered as critical hindrances.

In this study, the mean scores of the 47 hindrances to BIM implementation in Singapore were normalized, as shown in Table 3. The hindrances that obtained normalized values of 0.50 or above were recognized as critical ones. The results implied that out of the 47 factors, 21 were critical in hindering BIM implementation in building projects in Singapore. Besides, all the mean scores (3.494 and above) of the 21 critical hindrances exceeded 3.451 which was the total mean value of all the 47 hindrances. However, the 22nd hindrance (H37) obtained a mean score (3.448) below this total mean value. Thus, the results of applying the "comparing mean scores with the total mean value" method supported the normalization method which was adopted in this study. Among the critical hindrances, "need for all key stakeholders to be on board to exchange information" (H12) was ranked top. This result echoed El Asmar et al. [53], which found that early involvement of the contractors to share expertise and information upfront is key to creating optimal design models early in a building project. In the post-survey interviews, the experts mentioned that in Singapore downstream parties were generally not involved upfront, and also suggested that BIM implementation would be efficient if the entire team (the owner, the designers, the contractors, and the facility manager) could actively participate from the beginning of the design stage. Indeed, lots of details need to be developed by specialist contractors who, however, usually used the traditional approach in design detailing. Based on the ranking of the critical hindrances in Table 3, the practitioners would have a clear knowledge of the areas of activities of BIM implementation deserving more attention, and establish resources allocation priorities accordingly.

4.2. Underlying Hindrance Groupings

In this study, exploratory factor analysis (EFA) was conducted to obtain a manageable set of hindrance groupings. Such groupings should well represent the 21 critical hindrances to BIM implementation. The appropriateness of performing EFA with the data collected in this study was assessed. The Kaiser-Meyer-Olkin value was 0.911, suggesting that the critical hindrances had a high common variance. The value of the test statistic (chi-square) for Bartlett's sphericity was 1265.756, with a p-value of 0.000, indicating that sufficient correlations existed among the hindrances to proceed with EFA. Thus, the data were appropriate for EFA. The EFA process was terminated after meeting its widely-used threshold values: (1) 0.60 for cumulative percentage of variance (CPV), indicating that all the extracted hindrance groupings together should explain at least 60% of the variance of all the critical hindrances; (2) 0.50 for communalities of all the critical hindrances; and (3) ±0.40 for factor loadings of all the hindrances to be significant [54,55]. The communality of a critical hindrance presents the total amount of variance that this critical hindrance shares with the rest critical hindrances, and the factor loadings of a critical hindrance refer to the correlation between this hindrance and the hindrance groupings.

EFA was conducted using the software IBM SPSS Statistics 20 and the results indicated a well-defined three-factor structure, as shown in Table 4. The CPV explained from the three extracted groupings was 64.070%. Additionally, all the factor loadings of the hindrances were above 0.40 and the communalities above 0.50, indicating a robust EFA. The hindrance groupings were named as "lack of collaboration and model integration", "lack of continuous involvement and capabilities", and "lack of executive vision and training", respectively.

Table 4. Results of exploratory factor analysis (EFA) on critical hindrances to BIM implementation.

Code	Hindrances to BIM Implementation	Communality	Hindrance Grouping		
			1	2	3
Grouping 1: Lack of Collaboration and Model Integration (LCMI)					
H34	BIM model issues (such as ownership and management)	0.757	0.948	—	—
H09	Lack of sufficient evidence to warrant BIM use	0.515	0.707	—	—
H36	OSM requires design to be fixed early using BIM	0.701	0.707	—	—
H45	Difficulty in multi-discipline and construction-level integration	0.512	0.657	—	—
H35	Poor understanding of OSM process and its associated costs	0.712	0.646	—	—
H46	Technical needs for multiuser model access in multi-discipline integration	0.587	0.545	—	—
H33	Lack of standard contracts to deal with responsibility/risk assignment and BIM ownership	0.727	0.524	—	—
H27	Contractual relationships among stakeholders and need for new frameworks	0.649	0.516	—	—
H28	Traditional contracts protect individualism rather than best-for-project thinking	0.574	0.463	—	—
H43	Interoperability issues such as software selection and insufficient standards	0.594	0.431	—	—
Grouping 2: Lack of Continuous Involvement and Capabilities (LCIC)					
H12	Need for all key stakeholders to be on board to exchange information	0.673	—	0.780	—
H07	Entrenchment in 2D drafting and unfamiliarity to use BIM	0.712	—	0.760	—
H14	BIM operators lacking field knowledge	0.676	—	0.739	—
H42	Costly investment in BIM hardware and software solutions	0.630	—	0.652	—
H32	Assignment of responsibility/risk to constant updating for broadly accessible BIM information	0.568	—	0.600	—
H31	Firms' unwillingness to invest in training due to initial cost and productivity loss	0.555	—	0.557	—

Table 4. *Cont.*

Code	Hindrances to BIM Implementation	Communality	Hindrance Grouping		
			1	**2**	**3**
	Groping 3: Lack of Executive Vision and Training (LEVT)				
H01	Executives failing to recognize the value of BIM-based processes and needing training	0.758	—	—	0.814
H04	Lack of skilled employees and need for training them on BIM and OSM	0.725	—	—	0.770
H02	Concerns over or uninterested in sharing liabilities and financial rewards	0.537	—	—	0.654
H05	Industry's conservativeness, fear of the unknown, and resistance to change comfortable routines	0.699	—	—	0.620
H47	Firms cannot make most use of IFC and use proprietary formats	0.596	—	—	0.554
	Eigenvalue		11.142	1.256	1.057
	Variance (%)		53.059	5.979	5.033
	CPV (%)		53.059	59.038	64.070

4.2.1. Lack of Collaboration and Model Integration

This critical hindrance grouping accounted for about 53% of the total variance of the 21 critical hindrances and included 10 hindrances. All the 10 hindrances could be related to the collaboration and model integration of BIM implementation in the building project. The widely-accepted definition of BIM proposed by the National Institute of Building Sciences (NIBS) stated that "a basic premise of BIM is collaboration by different stakeholders at different phases of the life cycle of a facility to insert, extract, update or modify information in the BIM to support and reflect the roles of that stakeholder" [56]. The hindrance with the highest factor loading was "BIM model issues (such as ownership and management)" (H34). Even BIM technology has been used by the designers and possibly the contractors, the technological process has been suffering from physical and information fragmentation in different stages of the project. The creation, integration, and use of digital design models would potentially raise many liability issues because little collaboration was built within the typical project team [12], such as the anxiety about providing wrong information by the designers [26] and about offering advice by the contractors. The professionals involved in the post-survey interviews observed that due to potential liabilities, the team members do not fully exchange data. For instance, as the principal role in the Singapore context, the architect may change the design frequently without informing other designers and the contractors, hindering the creation of a composite design and construction model, whereby, all the parties can work on it. Such issues urged the roles and responsibilities in the model management process to be established in standard contracts. Even so, the liabilities may not be solved without trust-based collaboration and proper risk sharing among the key stakeholders. As mentioned earlier, the building contracts in Singapore are still developed on the basis of the adversarial contractual system, leading to individualism and isolated working environment. Thus, three critical hindrances (H27, H28, and H33) that described the fragmented contractual relationships were related to the trust-based collaboration needed for BIM implementation among the team.

"Lack of sufficient evidence to warrant BIM use" (H09) obtained the second highest loading. The post-survey interviewees highlighted that due to the lack of project-wide collaboration, BIM implementation appeared to remain in an early stage in Singapore; it is unrealistic to expect that in the short term, the project team can fully reap the benefits that BIM implementation brings. The lonely BIM adoption among the major stakeholders could also be attributed to the poor understanding of collaboration and data integration. Thus, "poor understanding of off-site manufacture (OSM) process and its associated costs" (H35) was also included in this grouping.

To implement BIM along with OSM, the digital models should be fixed early and fit for off-site fabrication and on-site assembly (H36), because any changes would be costly after the fabrication commences [25]. However, design environments were difficult for multi-disciplinary integration at the construction level (H45) [7] due to "interoperability issues such as software selection and insufficient standards" (H43) [37]. In addition, the multi-discipline model integration also requires "technical expertise, protocols, and infrastructure for multiuser model access" (H46) [9,22]. The post-survey interviewees also suggested that in Singapore different parties tended to use various software or software versions, creating a difficulty in integrating digital models across disciplines.

4.2.2. Lack of Continuous Involvement and Capabilities

This grouping explained about 6% of the total information of the 21 critical hindrances and included six of them. Among these, two significant hindrances (H12 and H32) were closely related to the continuous involvement of the major stakeholders. "Need for all key stakeholders to be on board to exchange information" (H12) obtained the highest factor loading. The early and continuous involvement of the owner and the key designers and contractors from early design through project completion has been advocated for BIM implementation [53]. This would pave the way for the project-wide collaboration in the subsequent stages of the project. Another high-loading hindrance was "assignment of responsibility/risk to constant updating for broadly accessible BIM information" (H32). In the collaborative project team, the processes of creating digital models in the design stage and using, updating, and managing the models in the construction and operations and maintenance stages would assign liabilities to different participants as the project proceeds.

Four significant hindrances (H07, H14, H31, and H42) were associated with relevant capabilities that are necessary for successful implementation of BIM. As recommended by the BIM Project Execution Planning Guide, the capabilities of each party can be defined as resources (personnel, tools and their training, and information technology support), competencies, and experience [57]. The second highest factor loading hindrance "entrenchment in two-dimensional (2D) drafting and unfamiliarity to use BIM" (H07) revealed that in Singapore many firms lacked competencies and experience to use BIM [32]. For example, the upfront BIM operators tend to lack enough field knowledge to know what they are modeling and its constraint in the actual construction (H14); consequently, the digital models may not be developed correctly. In addition, capital investment should also be included in the resources [42]. Thus, "firms' unwillingness to invest in training" (H31) and "costly investment in BIM hardware and software solutions" (H42) could also be related to the capabilities of the participants. Compared with the biggest firms in Singapore that can make most use of BIM, a huge number of SMEs face adoption challenges such as lacking the capital investment to build up BIM competencies [8]. The experts involved in the post-survey interviews also stressed the importance of the BIM implementers' financial capabilities, especially for the SMEs and foreign firms based in Singapore.

4.2.3. Lack of Executive Vision and Training

This hindrance grouping represented about 5% of the total variance and included five hindrances. "Executives failing to recognize the value of BIM-based processes and needing training" (H01), "concerns over or uninterested in sharing liabilities and financial rewards" (H02), and "industry's conservativeness, fear of the unknown, and resistance to change comfortable routines" (H05) were associated with the lack of executive vision among firms, resulting in negative mindsets and behaviors. In particular, H01 achieved the highest factor loading, which described that the executives may not commit on the new working method. In the Singapore context, they tended not to see BIM implementation as mainstream activity but as additional workload. This value proposition established the conservative and unsupportive culture of most firms (H05), hindering BIM implementation in Singapore. This finding was consistent with Khosrowshahi and Arayici [7] which found inadequate marginal utility to be realized by using BIM. Another high-loading hindrance was H02. It was unrealistic to expect the major stakeholders to be willing to share both liabilities and rewards, although such sharing would help build transparency, trust, and collaboration within the team.

In addition to the need for training the executives (H01), "lack of skilled employees and need for training them on BIM and OSM" (H04) was also related to training. Very often the employees tended to be reluctant to adopt the new technology and participate in the new workflow [37], while the post-survey interviewees highlighted the concern about the executives' willingness to train their employees in Singapore. One example of such reluctance is that many firms could not take advantage of the commonly-used data exchange format (Industrial Foundation Classes, IFC), and still use proprietary formats (H47) which would not enable smooth data exchange with other parties.

4.3. Conceptual Framework and Validation

The conceptual framework was constructed to depict the critical hindrances to BIM implementation and the hypothetical influence paths among the hindrances groupings (Figure 2). Three hypotheses were involved in the framework:

Hypothesis 1. *Lack of executive vision and training positively contributes to lack of continuous involvement and capabilities.*

Hypothesis 2. *Lack of continuous involvement and capabilities positively contributes to lack of collaboration and model integration; and*

Hypothesis 3. *Lack of executive vision and training positively contributes to lack of collaboration and model integration.*

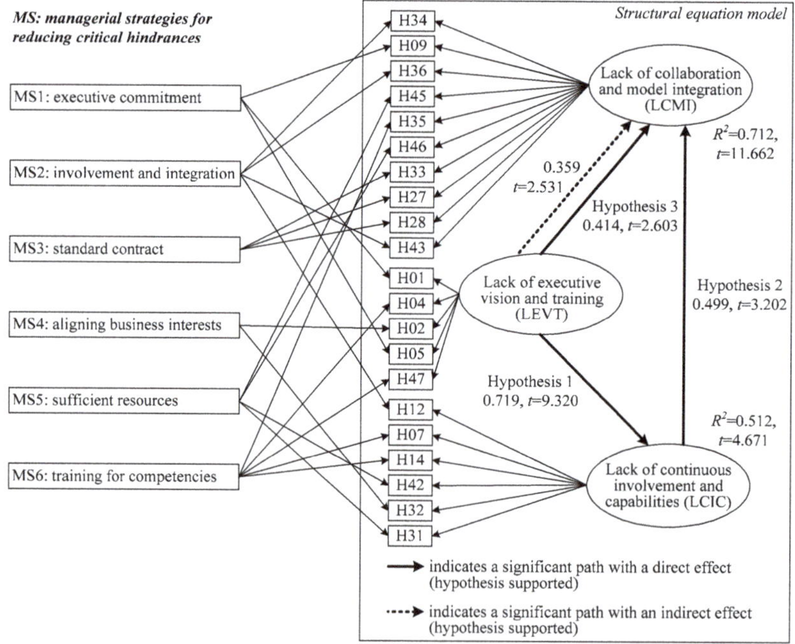

Figure 2. Conceptual framework to reduce critical hindrances for BIM implementation.

Structural equation modeling (SEM) has been recognized as a good technique in terms of relationship analysis among variables [42,58]. A SEM model includes structural and measurement models. The measurement model presents the relationships between measured variables (the critical hindrances) and latent variables or constructs (the hindrance groupings), and the structural one specifies the relationships among the three hindrance groupings. Two types of SEM are commonly-used: covariance-based SEM (CB-SEM) and partial least-squares structural equation modeling (PLS-SEM). Compared with CB-SEM, PLS-SEM: (1) ensures a good statistical power when the sample size is small [59]; (2) can deal with non-normal data sets [60,61]; and (3) can identify key driving constructs [62]. In this study, PLS-SEM was used as the sample size (87) tended to be inadequate for CB-SEM, and this study primarily aimed to investigate the intergroup relationships and driving hindrance groupings. In addition, this approach was widely used in project management research conducted both in Singapore [42,58] and globally [63,64].

SEM can efficiently fit the data and the SEM model. This was because confirmatory factor analysis (CFA) and path modeling analysis were simultaneously conducted. CFA was performed to test whether the 21 critical hindrances could represent well the three hindrance groupings. Furthermore, to check whether the path coefficients (related to hypotheses) were statistically significant, the bootstrapping technique [65] was applied in this study. Hair et al. [62] suggested that 5000 bootstrap subsamples should be used, which were randomly drawn from the 87 data sets.

The PLS-SEM analysis results required the following interpretations: (1) reliability and validity testing; and (2) relationships assessment according to the path coefficients obtained from the analysis. The reliability and validity of the measurement model should be assessed using the indicators and their thresholds: (1) 0.70 for the Cronbach's alpha coefficients [47]; (2) 0.55 for the factor loadings of the measured variables [66]; (3) 0.50 for the average variances extracted (AVEs) of the latent variables [67]; and (4) 0.70 for composite reliability (CR) scores [62]. In addition, discriminant validity should be assessed to test a construct's distinction from others. Two traditional methods were widely used in previous studies: (1) the Fornell-Larcker criterion that the square root of the AVE of each construct should exceed the correlation between this construct and any other constructs [67]; and (2) the cross-loading assessment that a measured variable's factor loading on its respective construct should be greater than its cross-loadings [68].

However, Henseler et al. [69] found that the two traditional assessment methods are not reliable in terms of detecting the lack of discriminant validity in PLS-SEM. This is because such methods would result in an unacceptably low sensitivity, especially in the situation that the AVE is low. Instead, this said study proposed that heterotrait-monotrait ratio of correlations (HTMT) should be a new approach. The HTMT is the average of the heterotrait-heteromethod correlations (the correlations of the measured variables across the constructs that measure different phenomena). The HTMT can be used in two ways: (1) being a criterion; or (2) being a statistical test. If the value of the HTMT is higher than a predefined cutoff point (0.85 or 0.90), this suggests that the SEM model lacks discriminant validity. Henseler et al. [69] suggested that 0.90 is more suitable (referred to $HTMT_{0.90}$) when the sample size is around 100 and the AVE values are not high. Alternatively, the HTMT can form the foundation of a statistical discriminant validity test. The bootstrapping procedure allows that confidence intervals for the HTMT can be constructed to test the null hypothesis (H0: $HTMT \geq 1.0$) against the alternative hypothesis (H1: $HTMT < 1.0$). If 1.0 (held in the H0) falls within a confidence interval, it can be concluded that the SEM model lacks sufficient discriminant validity; conversely, if 1.0 does not fall within the range of the interval, one may conclude that two constructs are empirically distinct.

Tables 5 and 6 show the results of CFA which was conducted using the software SmartPLS 3. As indicated in Table 5, all the factor loadings were greater than 0.55, which were significant at the 0.05 level (critical *t*-value = 1.96). Besides, all the values of the Cronbach's alpha, AVE, and CR exceeded their respective cutoff values. Meanwhile, it could be seen from Table 6 that the *p*-values for the HTMT tests among the three hindrance groupings were below 0.05, and that the HTMT values were smaller than 0.90. Therefore, both the HTMT test and the $HTMT_{0.90}$ criterion indicated that the SEM model met the requirements of the discriminant validity. Thus, the measurement model was reliable and valid for the structural path modeling. In addition, the absolute model fit statistics that were available in the PLS-SEM outputs were acceptable. Standardized root mean residual was 0.066, below 0.10 which, as recommended by Hair et al. [55], was the cutoff point, and the Chi-Square value (320.029) was large. Furthermore, the bootstrapping results in Figure 2 suggested that all the path coefficients (0.719, 0.449, and 0.414, respectively) for the three hypotheses were positive, with the p-values of 0.000. The coefficients of determination (R^2) for lack of collaboration and model integration and lack of continuous involvement and capabilities were 0.712 (with t-value at 11.662 and p-value at 0.000) and 0.512 (with t-value at 4.671 and p-value at 0.000), respectively. Hence, these statistics validated the conceptual framework (Figure 2). The subsequent sections will discuss this framework's rationale.

Table 5. Validity and reliability evaluation of measurement model.

Grouping	Hindrance Code	Factor Loading	p-Value	AVE	Cronbach's Alpha	CR
LCMI	H34	0.813	0.000 *			
	H09	0.680	0.000 *			
	H36	0.848	0.000 *			
	H45	0.706	0.000 *			
	H35	0.849	0.000 *	0.618	0.931	0.942
	H46	0.766	0.000 *			
	H33	0.850	0.000 *			
	H27	0.798	0.000 *			
	H28	0.770	0.000 *			
	H43	0.762	0.000 *			
LCIC	H12	0.798	0.000 *			
	H07	0.825	0.000 *			
	H14	0.833	0.000 *	0.622	0.878	0.908
	H42	0.803	0.000 *			
	H32	0.711	0.000 *			
	H31	0.757	0.000 *			
LEVT	H01	0.866	0.000 *			
	H04	0.847	0.000 *			
	H02	0.756	0.000 *	0.624	0.848	0.892
	H05	0.732	0.000 *			
	H47	0.739	0.000 *			

Note: LCMI = Lack of collaboration and model integration; LCIC = Lack of continuous involvement and capabilities; LEVT = Lack of executive vision and training. * The loading was significant at the 0.05 level (two-tailed).

Table 6. Discriminant validity of hindrance groupings.

HTMT	Original Sample	Sample Mean	t-Value	p-Value
LEVT→LCIC	0.829	0.831	9.797	0.000
LCIC→LCMI	0.879	0.880	10.937	0.000
LEVT→LCMI	0.866	0.866	19.653	0.000

Note: LCMI = Lack of collaboration and model integration; LCIC = Lack of continuous involvement and capabilities; LEVT = Lack of executive vision and training.

4.3.1. Hypothesis 1

This hypothesis that the lack of executive vision and training positively contributes to the lack of continuous involvement and capabilities was supported by the PLS-SEM analysis results. BIM implementation requires not only sufficient resources such as costly infrastructure, but also competent and experienced personnel that can be trained or engaged from the market [57]. Insufficiency of any kind of the capabilities would hinder BIM implementation. This tallies with the post-survey interviews that in practice the practitioners in Singapore should be provided with training and education programs as well as technical support for BIM adoption. This is because they may not be knowledgeable and experienced about a higher level of BIM implementation. In most cases, the executives can determine the allocation of the capital investment to purchase and upgrade the infrastructure, and the sponsorship of training programs [42]. However, the post-survey interviewees also reported that the management of many firms in Singapore tend to keep things under control as they previously did, because compared with the relatively straightforward implementation cost, the benefits of BIM implementation tend not to be so concrete. Thus, without the executive support (H01), training sessions could not be arranged (H04 and H47), and a higher resource allocation priority could not be obtained, leading to the lack of the capabilities among the major stakeholders (H07, H14, H31, and H42).

In addition, the continuous involvement of the major stakeholders is crucial to BIM implementation [53]. Nonetheless, the professionals involved in the post-survey interviews pointed out that in the Singapore context, the downstream parties are often vary of providing professional advice in the design modeling due to potential liabilities, and the upfront parties are cautious about providing design information in the later stages. The stakeholders are unaware of or unwilling to internally waive the liabilities in the project team (H02), and tend to be reluctant to change their customized ways of working and blaming (H05). Nothing is more critical to BIM implementation than a supportive project culture, which depends largely on the project leadership team. Thus, the lack of executive vision and mission would hinder the continuous involvement of the key stakeholders (H12) and the constant model updates (H32).

4.3.2. Hypothesis 2

The hypothesis that the lack of continuous involvement and capabilities positively contributes to the lack of collaboration and model integration was supported. Collaboration is the precondition to BIM implementation because the major stakeholders need to communicate and exchange information with others to complete their scopes of work. The early and active involvement of the downstream stakeholders is critical to building trust and sharing knowledge with others, which facilitates the project-wide collaboration needed for multidisciplinary model integration [53]. Thus, the lack of continuous stakeholder participation (H12) prevents the team from working collaboratively (H27). On the other hand, the collaboration between the design team and the construction team will not intrinsically lead to a blending of disciplines [16]. In the Singapore context, even in the construction phase, there was generally insufficient collaboration between the design consultants and the contractors. The communication between the designers and site managers was weak. Under the current contractual framework of Singapore, the relationships tend to be adversarial. The constant updating and management of broadly accessible digital information would potentially subject the designers and the contractors to increased liabilities as the project proceeds [26]. Therefore, the lack of proper assignment of responsibilities and allocation of risks (H32) would point to the need for collective risk management and model management (H28, H33, and H34). In addition, plenty of resources upfront are also needed to create and fix the model early (H36). Thus, the lack of capital resources (H31 and H42) creates a difficulty in the model integration (H43, H45, and H46). Moreover, the staff may unconsciously evaluate whether their knowledge, skills, and experience are good enough to enable them to be involved in the BIM-based work practices. If they regard themselves as less competent and experienced individuals in working with BIM (H07 and H14) or have not yet learnt about similar success stories, they would not actively use BIM (H09). Because of human nature, the negative mindset and passive behavior may in turn keep their inertia to continuously learn about relevant processes (H35). This would result in fragmented or discipline-specific BIM implementation, rather than project-wide collaboration and multidisciplinary design modeling and integration. In the post-survey interviews, the experts also emphasized the importance of building a high level of capabilities in the project team, as experienced and skilled personnel that can lead BIM modeling teams are still lacking in the current market in Singapore.

4.3.3. Hypothesis 3

This hypothesis that the lack of the executive vision and training positively contributes to the lack of collaboration and model integration was also supported. It is worth noting that when faced with change, people would possibly react in their accustomed and comfortable ways as well as be biased against the reality [70]. The executives may be unwilling to change (H01) when they have long been psychologically entrenched in the traditional drafting practices [7] and cannot see the value of BIM in increasing inter-organizational collaboration and improving work efficiency [1]. It has been verified in numerical studies and projects that BIM implementation can facilitate information integration across the lifecycle, and close collaboration among the participants [71,72]. People may still bias or ignore the reality (H09) and be reluctant to learn the new working method (H35). The post-survey interviewees also highlighted that although the executives of many firms in Singapore change to use 3D tools, the leadership style appeals to continue to keep a 2D mindset (H05). Additionally, without the pre-agreed liabilities and rewards sharing arrangements (H02), the downstream parties, if involved upfront, would work financially at risk (H28 and H33) and lack motivation and enthusiasm to work collaboratively with others (H27). Some stakeholders may take advantage of information asymmetry at the cost of other stakeholders. They think that the efforts spent in collaborating with others toward a "win–win situation" would discourage them from optimizing the benefits they could have obtained [8]. Consequently, the project-wide transparency and collaboration cannot be built; the construction and operations expertise cannot be incorporated into the upfront design modeling and multi-disciplinary coordination (H36), and potential liability issues would inevitably be raised in the dynamic model management in different phases (H34). Furthermore, if the IFC data exchange format is not used by all the parties (H47), the cross-enterprise design integration would be difficult (H43 and H45). Besides the interoperability standards and guidelines, local experimentation and continuous learning play a central role in BIM implementation [1]. Thus, the lack of training programs (H04) hinders the continuous improvement of the skill sets (H46) required by the model integration and management throughout the project.

4.4. Managerial Strategies

To reduce the 21 critical hindrances to BIM implementation, six managerial strategies (Figure 2) have been proposed based on the data analysis results. These strategies can not only help the leadership teams of building projects in Singapore to reduce possible hindrances in their BIM implementation practices, but also provide insights for the local government to refer to when rolling out new regulations related to BIM implementation.

Executive commitment (MS1). The planning, design, construction, and operations and maintenance processes have been increasingly relying on the information models [26]. The executives of the primary project participants should recognize the inevitable change to adapt to the information-oriented project delivery and continually commit on BIM diffusion which will probably be stunted if the participants cannot implement their part of BIM (H01 and H09). With the tone of changing at the top, the employees who carry out day-to-day work on the shop floor must change their passive mindsets and behaviors (H05). Moreover, the willingness of implementing BIM is also affected by the Singapore government's policies, competitor motivation, financial incentives, and technical support [9]. The post-survey interviewees recommended that in addition to the existing government mandate and support in Singapore, incentives such as additional GFA for the owner and a series of objective performance milestones for the designers and contractors need to be formulated.

Involvement and integration (MS2). The participants should be early involved and physically collocated to build trust and collaboration as the BIM adoption in Singapore tended to be firm-based. Specifically, the key contractors, manufacturers, and facility management team downstream can contribute their knowledge and experience in the digital design modeling upfront (H12), which enables to fix the design early and build constructability in the design (H36). Otherwise, problems may occur in the subsequent stages where design changes would be costly. The close collaboration and frequent communication upfront may help address the interoperability issues such as by using predetermined software or software versions and interoperability standards (H43). This can align fragmented BIM adoptions in different participants by integrating their discipline-specific models. Furthermore, these preparatory work would also serve as the basis of collaboratively managing the digital model in the later stages (H34).

Standard contract (MS3). An updated version of the BIM Particular Conditions in Singapore should be developed as the standard contract (H33) to incorporate the BIM work processes into the local contractual framework because the current version was partly completed [73]. As mentioned earlier, currently, the design processes may not be collaborative because the design team and the construction team in Singapore tend to individually create different models. The new multi-party contract should be set by the local government and incentivize the participants to openly share data, which helps improve the contractual relationships (H27) among these primary participants because they can act as a collaborative team (H28). Besides, the flow of information can also be ensured in the model management. In addition, the owner should set relevant contractual requirements on BIM when building the project team. Otherwise, the service providers' motivation and willingness of implementing BIM in practice would be affected [33].

Aligning business interests (MS4). Sharing business interests should be agreed on by all the key stakeholders (H02) in the contract. This will build the necessary trust and the continuous collaboration among the participants [53]. All the participants should have a clear understanding of the opportunities, risks, and responsibilities when incorporating BIM work practices into the project workflow in Singapore. Sharing risks (H32) forces them to be responsible for the project rather than shifting risks or blaming their partners, whereas sharing rewards drives the team to think and behave in a best-for-project manner. These avoid the downstream parties from working at risk upfront. Furthermore, the individual corporate goals are bound with the project outcomes. It is worth noting that individual parties can leverage BIM only when it is successfully implemented in the project.

Sufficient resources (MS5). The management should allocate sufficient resources which are not limited to the skilled personnel, tools (H45 and H46), and information technology support, but also the capital investment (H31 and H42). For example, BIM software have been constantly improving to enable the cross-discipline integration at the construction level. The post-survey interviewees, however, reported that the hardware in many firms in Singapore cannot support the advanced software applications efficiently. In addition to the local government's subsidies, the primary participants should also invest in improving their infrastructure. It should be noted that BIM implementation may span many years; those who achieve enhanced BIM implementation will gain a competitive advantage when bidding for building projects in future, which in turn helps convince them to guarantee the sufficiency of the resources [74].

Training for competencies (MS6). Apart from the awareness and willingness to implement BIM, technical knowledge and ability is also needed [8]. Training programs should be provided to build the staff's competencies (H04, H07, H14, H35, and H47) and thus reduce their anxieties about their qualifications for being involved in the new project workflow in Singapore. Once the project team is built, the owner may provide trainings to the service providers on how to use new software applications, reinvent workflow, assign responsibilities, and collaborate with others using interoperability tools like IFC. For instance, it is imperative in practice to maintain integrity across design models, because changes are made to the different models by their respective disciplines. Both manual updates using IFC and smart automated transactions in BIM servers require specialized expertise. In addition, the major stakeholders should also arrange constant in-house training and education programs to help their staff adapt to new policies and work rules and procedures.

5. Conclusions and Recommendations

This study has examined the hindrances to BIM implementation in building projects in Singapore and explored the interrelationships among these hindrances. The analysis results from the questionnaire survey and the post-survey interviews implied that 21 out of the 47 hindrances were deemed critical. If everything is important, nothing is manageable. The top-ranked hindrances represented the most important areas of activities of BIM implementation. Since resources are usually limited in a project, the project leadership team should allocate resources for such areas rather than all the key areas. Additionally, the 21 critical factors hindering BIM implementation were grouped into three categories:

lack of collaboration and model integration, lack of continuous involvement and capabilities, and lack of executive vision and training, which were confirmed by CFA. Furthermore, these hindrance groupings and the intergroup relationships formed a conceptual framework. This framework could depict the key hindrances of BIM implementation in the Singapore construction industry. According to the path modeling analysis results, the lack of executive vision and training and the lack of continuous involvement and capabilities contribute to the lack of collaboration and model integration, and that the lack of executive vision and training results in the lack of continuous involvement and capabilities. Thus, the key areas described by the hindrances in the driving hindrance grouping "lack of executive vision and training" should also have top management priority. According to the ranking of and the relationships among the critical hindrances, six managerial strategies are proposed for project teams to overcome these hindrances.

There are limitations to the conclusions. First, the critical factors hindering BIM implementation might not be exhaustive. These factors might not be continuously true as time passes. Second, the data analyzed in this study were collected from the BIM implementers in Singapore, which may restrict the interpretation and generalization of the analysis results.

Nonetheless, the managerial strategies proposed in this study are not only applicable in the project teams in Singapore, but also hold true in overseas building project teams. This is because: (1) overseas project teams may also use the hindrances to BIM implementation identified in this study and follow the methodology used in this study to customize their own lists of hindrances, with minor adjustments. Similar to Singapore, other countries are also encouraging, specifying, or mandating BIM uses in publicly-funded building and construction projects by issuing a variety of BIM policies and standards [75,76]. Other countries such as the United States, the United Kingdom, and Australia also only have gone a step further to include elements of relational contracting which stress on shared rewards and responsibilities to break out of the conservative industry culture; (2) the governments that have not made much effort to incentivize BIM implementation can refer to these managerial strategies to purposefully and efficiently formulate and roll out their plans and policies. They may conditionally mandate BIM uses in their construction and building projects, establish national data exchange standards for enhanced multidisciplinary model integration, and provide technical support such as by defraying a proportion of the capital investments in software purchase, subscription, updating, and training; and (3) the proposed conceptual framework indicating the groupings of the critical hindrances as well as the intergroup cause-effect relationships is novel and can help the BIM implementers identify specific adjustments to their BIM implementation activities for efficiently enhancing BIM implementation. Thus, the main findings of this study contribute to the scholarship in terms of BIM implementation.

Reducing the critical hindrances to BIM implementation requires the project team, including the executives and employees of the major stakeholders, to change the accustomed work practices. Thus, with support from the conceptual framework constituted and validated in this study, future publications would propose an extended change framework for the project organization to systematically guide the primary participants and their staff to change.

Author Contributions: Conceptualization, L.L.; methodology, L.L.; data curation, L.L. and E.A.L.T.; formal analysis, L.L. and R.C.; validation, L.L.; investigation, L.L.; writing—original draft preparation, L.L.; resources, L.L. and E.A.L.T.; writing—review and editing, E.A.L.T. and R.C.; supervision, E.A.L.T.; funding acquisition, L.L.

Funding: This research was supported by the full-time research scholarship of National University of Singapore and funded by the Natural Science Foundation of Shenzhen University (grant no. 2019090). The APC was funded by the Natural Science Foundation of Shenzhen University (grant no. 2019090).

Conflicts of Interest: The authors declare no conflict of interest.

References

1. Miettinen, R.; Paavola, S. Beyond the BIM utopia: Approaches to the development and implementation of building information modelling. *Autom. Constr.* **2014**, *43*, 84–91. [CrossRef]

2. Cha, H.S.; Lee, D.G. A case study of time/cost analysis for aged-housing renovation using a pre-made BIM database structure. *KSCE J. Civ. Eng.* **2015**, *19*, 841–852. [CrossRef]

3. Jiang, S.; Wang, N.; Wu, J. Combining BIM and Ontology to facilitate intelligent green building evaluation. *J. Comput. Civ. Eng.* **2018**, *32*, 04018039. [CrossRef]

4. Jiang, S.; Wu, Z.; Zhang, B.; Cha, H.S. Combined MvdXML and Semantic technologies for green construction code checking. *Appl. Sci.* **2019**, *9*, 1463. [CrossRef]

5. Ding, Z.; Liu, S.; Liao, L.; Zhang, L. A digital construction framework integrating building information modeling and reverse engineering technologies for renovation projects. *Autom. Constr.* **2019**, *102*, 45–58. [CrossRef]

6. Chelson, D.E. The Effects of Building Information Modeling on Construction Site Productivity. Ph.D. Thesis, University of Maryland, College Park, MD, USA, 2010.

7. Khosrowshahi, F.; Arayici, Y. Roadmap for implementation of BIM in the UK construction industry. *Eng. Constr. Archit. Manag.* **2012**, *19*, 610–635. [CrossRef]

8. Forsythe, P.; Sankaran, S.; Biesenthal, C. How far can BIM reduce information asymmetry in the Australian construction context? *Proj. Manag. J.* **2015**, *46*, 75–87. [CrossRef]

9. Juan, Y.K.; Lai, W.Y.; Shih, S.G. Building information modeling acceptance and readiness assessment in Taiwanese architectural firms. *J. Civ. Eng. Manag.* **2017**, *23*, 356–367. [CrossRef]

10. Cheng, J.C.; Lu, Q. A review of the efforts and roles of the public sector for BIM adoption worldwide. *J. Inf. Technol. Constr.* **2015**, *20*, 442–478.

11. BCA. *Reaching New Milestones with Design for Manufacturing and Assembly*; Build Smart, Building and Construction Authority: Singapore, 2016.

12. Lam, S.W. The Singapore BIM Roadmap. In Proceedings of the Government BIM Symposium 2014, Singapore, 13 October 2014. Available online: http://bimsg.org/wp-content/uploads/2014/10/BIM-SYMPOSIUM_MR-LAM-SIEW-WAH_Oct-13-v6.pdf (accessed on 15 June 2019).

13. Fischer, M.; Reed, D.; Khanzode, A.; Ashcraft, H. A Simple Framework for Integrated Project Delivery. In Proceedings of the 22nd Annual Conference of the International Group for Lean Construction, Oslo, Norway, 25–27 June 2014; pp. 1319–1330.

14. Liao, L.; Teo, E.A.L. Managing critical drivers for building information modelling implementation in the Singapore construction industry: An organizational change perspective. *Int. J. Constr. Manag.* **2019**, *19*, 240–256. [CrossRef]

15. ESC. *Report of the Economic Strategies Committee*; Economic Strategies Committee: Singapore, 2010.

16. AIA and AIACC. *Integrated Project Delivery: A Guide*; American Institute of Architects: Sacramento, CA, USA, 2007.

17. AIA and AIACC. *Experiences in Collaboration: On the Path to IPD*; American Institute of Architects: Sacramento, CA, USA, 2009.

18. Aranda-Mena, G.; Crawford, J.; Chevez, A.; Froese, T. Building information modelling demystified: Does it make business sense to adopt BIM? *Int. J. Manag. Proj. Bus.* **2009**, *2*, 419–434. [CrossRef]

19. Arayici, Y.; Coates, P.; Koskela, L.; Kagioglou, M.; Usher, C.; O'Reilly, K. BIM adoption and implementation for architectural practices. *Struct. Surv.* **2011**, *29*, 7–25. [CrossRef]

20. Autodesk. *Improving Building Industry Results Through Integrated Project Delivery and Building Information Modeling*; Autodesk Inc.: San Rafael, CA, USA, 2008.

21. Autodesk. *A Framework for Implementing a BIM Business Transformation*; Autodesk Inc.: San Rafael, CA, USA, 2012.

22. Azhar, N.; Kang, Y.; Ahmad, I.U. Factors influencing integrated project delivery in publicly owned construction projects: An information modelling perspective. *Procedia Eng.* **2014**, *77*, 213–221. [CrossRef]

23. Bernstein, P.G.; Pittman, J.H. *Barriers to the Adoption of Building Information Modelling in the Building Industry*; Autodesk, Inc.: San Rafael, CA, USA, 2004.

24. Bernstein, H.M.; Jones, S.A.; Russo, M.A. *The Business Value of BIM in North America: Multi-Year Trend Analysis and User Rating (2007–2012)*; McGraw-Hill Construction: Bedford, MA, USA, 2012.

25. Blismas, N.; Wakefield, R. Drivers, constraints and the future of offsite manufacture in Australia. *Constr. Innov.* **2009**, *9*, 72–83. [CrossRef]

26. Eastman, C.; Teicholz, P.; Sacks, R.; Liston, K. *BIM Handbook: A Guide to Building Information Modeling for Owners, Managers, Designers, Engineers and Contractors*, 2nd ed.; John Wiley & Sons: Hoboken, NJ, USA, 2011.

27. Fischer, M. Reshaping the life cycle process with virtual design and construction methods. In *Virtual Futures for Design, Construction and Procurement*; Brandon, P., Kocatürk, T., Eds.; Blackwell Publishing Ltd.: Malden, MA, USA, 2008; pp. 104–112.

28. Fox, S.; Hietanen, J. Interorganizational use of building information models: Potential for automational, informational and transformational effects. *Constr. Manag. Econ.* **2007**, *25*, 289–296. [CrossRef]
29. Gao, J.; Fischer, M. *Case Studies on the Implementation and Impacts of Virtual Design and Construction (VDC) in Finland*; Center for Integrated Facility Engineering, Stanford University: Stanford, CA, USA, 2006.
30. Gibb, A.; Isack, F. Re-engineering through pre-assembly: Client expectations and drivers. *Build. Res. Inf.* **2003**, *31*, 146–160. [CrossRef]
31. Kent, D.C.; Becerik-Gerber, B. Understanding construction industry experience and attitudes toward integrated project delivery. *J. Constr. Eng. Manag.* **2010**, *136*, 815–825. [CrossRef]
32. Kiani, I.; Sadeghifam, A.N.; Ghomi, S.K.; Marsono, A.K.B. Barriers to implementation of building information modeling in scheduling and planning phase in Iran. *Aust. J. Basic Appl. Sci.* **2015**, *9*, 91–97.
33. Kunz, J.; Fischer, M. *Virtual Design and Construction: Themes, Case Studies and Implementation Suggestions*; Center for Integrated Facility Engineering, Stanford University: Stanford, CA, USA, 2012.
34. McFarlane, A.; Stehle, J. DfMA: Engineering the Future. In Proceedings of the Council on Tall Buildings and Urban Habitat (CTBUH) 2014 Shanghai Conference, Shanghai, China, 16 September 2014; pp. 508–516.
35. Ross, K.; Cartwright, P.; Novakovic, O. *A Guide to Modern Methods of Construction*; IHS BRE Press: Bucks, UK, 2006.
36. Sattineni, A.; Mead, K. Coordination Guidelines for Virtual Design and Construction. In Proceedings of the 30th International Association for Automation and Robotics in Construction, Montreal, QC, Canada, 11–15 August 2013; pp. 1491–1499.
37. Zahrizan, Z.; Ali, N.M.; Haron, A.T.; Marshall-Ponting, A.; Hamid, Z.A. Exploring the adoption of Building Information Modelling (BIM) in the Malaysian construction industry: A qualitative approach. *Int. J. Res. Eng. Technol.* **2013**, *2*, 384–395.
38. Oo, T.Z. Critical Success Factors for Application of BIM for Singapore Architectural Firms. Master's Thesis, Heriot-Watt University, Edinburgh, UK, 2014.
39. Teo, E.A.L.; Chan, S.L.; Tan, P.H. Empirical investigation into factors affecting exporting construction services in SMEs in Singapore. *J. Constr. Eng. Manag.* **2007**, *133*, 582–591. [CrossRef]
40. Hwang, B.G.; Zhu, L.; Tan, J.S.H. Identifying critical success factors for green business parks: Case study of Singapore. *J. Manag. Eng.* **2017**, *33*, 04017023. [CrossRef]
41. Miller, G.A. The magical number seven, plus or minus two: Some limits on our capacity for processing information. *Psychol. Rev.* **1956**, *63*, 81–97. [CrossRef] [PubMed]
42. Zhao, X.; Hwang, B.G.; Low, S.P.; Wu, P. Reducing hindrances to enterprise risk management implementation in construction firms. *J. Constr. Eng. Manag.* **2014**, *141*, 04014083. [CrossRef]
43. Zhao, X.; Hwang, B.G.; Low, S.P. Enterprise risk management in international construction firms: Drivers and hindrances. *Eng. Constr. Archit. Manag.* **2015**, *22*, 347–366. [CrossRef]
44. Liao, L.; Teo, E.A.L. Critical success factors for enhancing the building information modelling implementation in building projects in Singapore. *J. Civ. Eng. Manag.* **2017**, *23*, 1029–1044. [CrossRef]
45. Liao, L.; Teo, E.A.L. Organizational change perspective on people management in BIM implementation in building projects. *J. Manag. Eng.* **2018**, *34*, 04018008. [CrossRef]
46. Wilkins, J.R. Construction workers' perceptions of health and safety training programmes. *Constr. Manag. Econ.* **2011**, *29*, 1017–1026. [CrossRef]
47. Nunnally, J.C. *Psychometric Theory*, 2nd ed.; McGraw-Hill: New York, NY, USA, 1978.
48. Magal, S.R.; Carr, H.H.; Watson, H.J. Critical success factors for information center managers. *MIS Q.* **1988**, *12*, 413–425. [CrossRef]
49. Shen, Q.; Liu, G. Critical success factors for value management studies in construction. *J. Constr. Eng. Manag.* **2003**, *129*, 485–491. [CrossRef]
50. Nitithamyong, P.; Skibniewski, M.J. Key success/failure factors and their impacts on system performance of web-based project management systems in construction. *J. Inf. Technol. Constr.* **2007**, *12*, 39–59.
51. Xu, Y.; Yeung, J.F.Y.; Chan, A.P.C.; Chan, D.W.M.; Wang, S.Q.; Ke, Y. Developing a risk assessment model for PPP projects in China—A fuzzy synthetic evaluation approach. *Autom. Constr.* **2010**, *19*, 929–943. [CrossRef]
52. Won, J.; Lee, G.; Dossick, C.; Messner, J. Where to focus for successful adoption of building information modeling within organization. *J. Constr. Eng. Manag.* **2013**, *139*, 04013014. [CrossRef]
53. El Asmar, M.; Hanna, A.S.; Loh, W.Y. Quantifying performance for the integrated project delivery system as compared to established delivery systems. *J. Constr. Eng. Manag.* **2013**, *139*, 04013012. [CrossRef]

54. Peterson, R.A. A meta-analysis of variance accounted for and factor loadings in exploratory factor analysis. *Mark. Lett.* **2000**, *11*, 261–275. [CrossRef]

55. Hair, J.F.; Black, W.C.; Babin, B.J.; Anderson, R.E. *Multivariate Data Analysis*, 7th ed.; Prentice Hall: Upper Saddle River, NJ, USA, 2009.

56. NIBS. *United States National Building Information Modeling Standard Version 1—Part 1: Overview, Principles, and Methodologies*; National Institute of Building Sciences: Washington, DC, USA, 2007.

57. Anumba, C.; Dubler, C.; Goodman, S.; Kasprzak, C.; Kreider, R.; Messner, J.; Saluja, C.; Zikic, N. *BIM Project Execution Planning Guide—Version 2.0*; Computer Integrated Construction Research Program, Pennsylvania State University: University Park, PA, USA, 2010.

58. Lim, B.T.H.; Ling, F.Y.Y.; Ibbs, C.W.; Raphael, B.; Ofori, G. Mathematical models for predicting organizational flexibility of construction firms in Singapore. *J. Constr. Eng. Manag.* **2012**, *138*, 361–375. [CrossRef]

59. Reinartz, W.; Haenlein, M.; Henseler, J. An empirical comparison of the efficacy of covariance-based and variance-based SEM. *Int. J. Res. Mark.* **2009**, *26*, 332–344. [CrossRef]

60. Fornell, C.; Bookstein, F.L. Two structural equation models: LISREL and PLS applied to consumer exit-voice theory. *J. Mark. Res.* **1982**, *19*, 440–452. [CrossRef]

61. Hair, J.F.; Sarstedt, M.; Pieper, T.M.; Ringle, C.M. The use of partial least squares structural equation modeling in strategic management research: A review of past practices and recommendations for future applications. *Long Range Plan.* **2012**, *45*, 320–340. [CrossRef]

62. Hair, J.F.; Ringle, C.M.; Sarstedt, M. PLS-SEM: Indeed a silver bullet. *J. Mark. Theory Pract.* **2011**, *19*, 139–151. [CrossRef]

63. Doloi, H. Rationalizing the implementation of web-based project management systems in construction projects using PLS-SEM. *J. Constr. Eng. Manag.* **2014**, *140*, 04014026. [CrossRef]

64. Le, Y.; Shan, M.; Chan, A.P.C.; Hu, Y. Investigating the causal relationships between causes of and vulnerabilities to corruption in the Chinese public construction sector. *J. Constr. Eng. Manag.* **2014**, *140*, 05014007. [CrossRef]

65. Davison, A.C.; Hinkley, D.V. *Bootstrap Methods and Their Application*; Cambridge University Press: New York, NY, USA, 1997.

66. Tabachnick, B.G.; Fidell, L.S. *Using Multivariate Statistics*, 5th ed.; Pearson/Allyn & Bacon: Boston, MA, USA, 2007.

67. Fornell, C.; Larcker, D.F. Evaluating structural equation models with unobservable variables and measurement error. *J. Mark. Res.* **1981**, *18*, 39–50. [CrossRef]

68. Barclay, D.W.; Higgins, C.A.; Thompson, R. The partial least squares approach to causal modeling: Personal computer adoption and use as illustration. *Technol. Stud.* **1995**, *2*, 285–309.

69. Henseler, J.; Ringle, C.M.; Sarstedt, M. A new criterion for assessing discriminant validity in variance-based structural equation modelling. *J. Acad. Mark. Sci.* **2015**, *43*, 115–135. [CrossRef]

70. Low, S.P. Managing total service quality: A systemic view. *Manag. Serv. Qual.* **1998**, *8*, 34–45.

71. Rezgui, Y.; Beach, T.; Rana, O. A governance approach for BIM management across lifecycle and supply chains using mixed-modes of information delivery. *J. Civ. Eng. Manag.* **2013**, *19*, 239–258. [CrossRef]

72. Ma, Z.; Ma, J. Formulating the application functional requirements of a BIM-based collaboration platform to support IPD projects. *KSCE J. Civ. Eng.* **2017**, *21*, 2011–2026. [CrossRef]

73. Wickersham, J. *Legal and Business Implications of Building Information Modeling (BIM) and Integrated Project Delivery (IPD), BIM-IPD Legal and Business Issues*; Rocket Press Publishing: Lafayette, LA, USA, 2009.

74. Teo, A.L.; Heng, P.S.N. Deployment Framework to Promote the Adoption of Automated Quantities Taking-Off System. In Proceedings of the CRIOCM2007 International Research Symposium on Advancement of Construction Management and Real Estate, Sydney, Australia, 8–13 August 2007; Zou, P.X.W., Newton, S., Wang, J., Eds.; Chinese Research Institute of Construction Management: Hong Kong, China, 2007.

75. Smith, P. BIM implementation—Global strategies. *Procedia Eng.* **2014**, *85*, 482–492. [CrossRef]

76. McAuley, B.; Hore, A.; West, R. *BICP Global BIM Study–Lessons for Ireland's BIM Programme*; Dublin Institute of Technology: Dublin, Ireland, 2017.

Article

Acceptance Model for Mobile Building Information Modeling (BIM)

Sim-Hee Hong [1], Seul-Ki Lee [2], In-Han Kim [3] and Jung-Ho Yu [1,*]

[1] Department of Architecture Engineering, Kwangwoon University, Seoul 01897, Korea
[2] ICEE, Seoul National University, Seoul 08826, Korea
[3] Department of Architecture, Kyung-Hee University, Seoul 17104, Korea
* Correspondence: myazure@kw.ac.kr; Tel.: +82-2940-5564

Received: 23 July 2019; Accepted: 23 August 2019; Published: 4 September 2019

Abstract: Mobile Building Information Modeling (BIM) is noted for tools that enable the systematic interchange of information and contribute to enhancing collaborative performance through BIM. BIM programs, which are continuously available in the mobile environment, have been developed. Moreover, in some sites, mobile BIM is applied to generate benefits in projects. Various efforts are being made to use mobile BIM; however, its utilization is low. Also, mobile BIM has lacked an analysis of the factors that affect actual users' acceptance of mobile BIM. Therefore, this study analyzes the factors that affect the acceptance of mobile BIM by construction practitioners and presents the association of factors as a model to activate mobile BIM use. To this end, this study analyzed a literature review for suggesting the factors that were expected to affect mobile BIM acceptance. The assessment items were decided based on the analysis result. Second, 111 copies were received by surveying the construction practitioners. Third, it identified factors that significantly affected the acceptance of mobile BIM and proposed models through factor analysis and structural equation models. Finally, based on the analysis, it presented the findings. This study expects to contribute to enhanced acceptance of mobile BIM technology by managing the significant factors properly. Also, it is expected that the result can be used to develop a variety of mobile BIM that is more easily acceptable to them. This study presented a model for accepting mobile BIM based on the survey results of Korean practitioners; therefore, it is necessary to explore ways to generalize the model in the future.

Keywords: BIM; open BIM; mobile BIM; mobile application; technology acceptance model (TAM)

---

1. Introduction

1.1. Research Background

Since the 2000s, building information management (BIM) has been used in construction projects for various purposes, and the devices used for BIM have become more diverse. Initially, BIM was implemented on an office environment; however, BIM in the office environment is difficult to share, so real-time communication with the construction site was challenging. As a result, various technologies that enable real-time information use were applied, such as web-based cloud applications [1,2], virtual reality, and augmented reality [3,4]. Further, handheld devices, such as mobile phones, have enabled real-time information use [5].

The use of mobile devices in the construction field for information utilization is changing. Further, the way the project stakeholders operate is also changing. Mobile phones allow portability, thus enabling real-time communication between the system and the project stakeholders [6]. Mobile-based BIM can enhance the collaborative performance of BIM, thus enabling the linked exchange of information [7]. Mobile programs in construction are used for viewing and managing drawings and documents, safety management, and communication. According to a survey by the Architectural Urban Research and Information Center (AURIC), approximately 40% of the respondents said they used mobile programs, and they used functions for managing drawings and documents, checking construction codes, checking drawings on the site, checking modifications in the construction situation, and so on [8].

With the development of various mobile programs for utilizing BIM, it is possible to quickly check and modify 3D information in real time. Since 2011, many companies have launched BIM programs that can be supported in mobile environments [8]. Most existing BIM tools were based on an office environment such as Revit and ArchiCAD; however, the use of mobile programs in the BIM field is increasing due to the development of tools based on mobile environments such as A360, BIMx, and Formit. Over time, it was observed that certain mobile BIM users reaped benefits in their projects, such as cost reduction and reduction of the rework rate [9]; therefore, there are efforts to improve mobile BIM usage even though the current utilization is low.

The reasons for this lack of mobile BIM usage in Korea were investigated, and it was noted that factors such as the age of the expected users [8] and inconsistencies between the actual working processes and BIM processes [10] were primarily responsible. However, another study reported that experience and age did not significantly affect the use of mobile BIM use [11]. Therefore, there is no overarching reason that can explain why mobile BIM as a technology has not been accepted by real users.

The technology acceptance model (TAM) defines external factors that affect the behaviors of users who accept a technology. Acceptance is defined as a psychological and technological state in which an individual or organization is prepared to use the technology smoothly. The TAM enables a definition of a significant factor that affects user accommodation of technology. TAMs are currently used in various fields, such as travel [12], sports [13], and shopping [14], among others, and are also used in the BIM field to analyze general BIM acceptance [15] and BIM acceptance in organizations [16,17]. If mobile BIM systems can also utilize the TAM, it is expected that an effective user-based technology acceptance analysis will be possible.

Therefore, this study aims to present significant findings on the activation of mobile BIM usage through a mobile BIM acceptance model through analyzing surveys of construction practitioners. First, we analyzed existing literature on mobile applications of the TAM and BIM TAM to find the external and internal variables that affect mobile BIM acceptance. Second, the hypothesis for the relationships between the variables was established based on the TAM2 and Information System (IS) models. Third, a survey was conducted based on the analysis results and hypothesis. Fourth, the results of the survey were analyzed using a structural equation, and the significant external variables for a mobile BIM TAM were suggested. Therefore, this study is expected to contribute toward basic research on the active adoption of mobile BIM technologies.

1.2. Research Methodology

As part of the efforts to improve the use of mobile BIM, this work analyzes the relationships among factors affecting acceptance of mobile BIM by construction practitioners and presents a mobile BIM TAM. The research processes followed are classified thusly.

(1)   Define the mobile BIM characteristics based on a literature review: The characteristics of mobile applications and BIM were defined through a literature review to suggest TAMs. The analysis results were assessed for external factors that affected mobile BIM acceptance.

(2)   Collect data to propose a mobile BIM TAM: The survey responses from construction practitioners (i.e., designers, contractors, Construction Managers (CMs), and BIM contractors) were collected. An overview of the data collection process is as follows.

The survey data were obtained from a sample of experienced BIM users or those experiencing BIM at work. The survey was conducted between 22 March and 29 April, and a total of 111 responses were received. Each question was measured on a 7-point Likert scale ranging from "strongly disagree" to "strongly agree."

Among the 111 respondents, 31 were designers, 37 were from construction management organizations, 11 were from construction organizations, 18 were engineers, and 14 were researchers. The respondents' average experience in construction was about 11.6 years and the average experience with BIM was about 3.5 years. Their average preference for using BIM tools was 5.84 points on the 7-point Likert scale. The respondents' characteristics are detailed in Table 1.

(3) Explore key factors for mobile BIM acceptance through factor analysis: The survey results were analyzed as follows. A principal component analysis (PCA) of the factors was conducted using the software package IBM Statistics SPSS 21.0. The results are shown using the Kaiser–Meyer–Olkin (KMO) measure and Bartlett's test. Through this analysis, variables that did not have a significant impact on mobile BIM acceptance were removed, and the assessment items were classified as significant factors.

(4) Validate the proposed model using a structural equation model (SEM): A hypothesis was established for the relationships between the factors from the factor analysis results. An analysis of the SEM was conducted using the software IBM Statistics AMOS 21.0 to validate the hypothesis. The result shows the ratio of χ^2 to the degrees of freedom (df), root-mean-square residual (RMR), parsimonious goodness of fit index (PGFI), Tucker–Lewis index (TLI), comparative fit index (CFI), and root-mean-square error of approximation (RMSEA). We eliminated the insignificant hypothesis of the set assumptions based on the path analysis results. Further, we checked the convergent validity and discriminant validity of the proposed model on the confirmatory factor analysis (CFA).

Table 1. Characteristics of the respondents ($n = 111$).

Measure		Frequency	%
the Respondent's Organization	Designer	31	27.9
	CM	37	33.3
	Contractor	11	9.9
	Engineer	18	16.3
	Researcher	14	12.6
Total			
Respondent's Average Experience	Construction Industry	Approx. 11.6 years	
	BIM	Approx. 3.5 years	
Preference for Using BIM Tools		5.84/7	

2. Literature Review

2.1. Mobile BIM in the Construction Industry

A mobile device is defined as a personal tool that is portable [18,19] and touchable [20,21]. Because mobile devices are portable, they enable checking and writing information regardless of the location. Devices referred to as being mobile include smartphones, smart pads, and service books [8], and mobile devices can use software applications that provide multipurpose convenience. Accordingly, mobile devices and applications allow users to add or delete functions according to their personal preferences while developing and using new applications as needed.

Mobile device use is increasing owing to characteristics such as the outdoor production of construction projects and the use of mobile devices by construction engineers to carry out field processes effectively [22]. Mobiles in construction are expected to contribute to reductions in construction time, capital costs, features, and components, while increasing the overall productivity [23].

BIM has been widely used in construction projects since the early 2000s. BIM is an integrated 3D object-based information storage system, which is used to pre-review tasks through model visualization, such as building design, clash detection, and quantity calculation. Currently, the scope of the original BIM work has expanded considerably. BIM is also used in specialized fields such as structural analysis, energy analysis, and regulation checks. It is used in the design and construction phases, as well as in building inspection during the maintenance phase. However, it is difficult to utilize BIM in construction sites for real-time communication because the information is thus far generated and shared only in the office environment. Mobile BIM appears to improve the difficulty in sharing information and better utilizes BIM.

Mobile BIM is defined as a mobile-based system that can be used on smartphones, smart pads, and surface books, and provides an environment in which BIM data can be utilized regardless of the location [8]. For example, mobile BIM is available in the applications of A360, BIM 360, BIM 360 Ops, BIM 360 Field, BIMx, Field 3D, FINAL CAD, Formit, etc.

Mobile BIM can address many issues that arise due to the difficulty in sharing information between teams in the collaboration process. On some sites, mobile BIM has significantly reduced costs. For example, FINAL CAD, a Mobile BIM, reduces construction time by an average of 11%, rework rate by an average of 25%, and defect rate by an average of 50% at construction sites worldwide [9]. Mobile BIM can reduce the safety, paperwork, and drawing management by three times. Moreover, it is expected that a mobile environment will enable providing desired functions according to personal preferences based on the characteristics of the mobile devices themselves.

Various attempts have been made to use mobile BIM in practical applications. Mobile BIM is used on mobile platforms in certain projects for safety management or quality management by focusing on the construction phase; however, it is still underutilized. The reasons for low utilization of mobile BIM could possibly be the low usage of BIM tools and the age and experience of the users. However, other studies have noted that age and experience do not significantly affect the use of mobile BIM [24]. Low mobile BIM usage can also be interpreted as a lack of analysis regarding the factors that affect actual user acceptance and utilization.

2.2. Theories Related Technology Acceptance

To promote the use of mobile BIM, this study analyzed the factors and relationships affecting the acceptance of mobile BIM by construction practitioners based on the TAM and information system (IS) success model. This study considered the TAM, IS success model, model of innovation resistance (Ram, 1987), and diffusion theory (Rogers, 2003), among others, and adopted the TAM and IS success models to establish the research hypothesis.

The TAM is the most common theory to evaluate technology usage. The TAM expresses the relationships between factors that affect the acceptance of the technology. The IS success model is also the basis for other similar theories since it was revised in 2003. The IS success model expresses the factors that affect the use of technology. It is still used to evaluate technology such as a digital library [25], online learning [26], a mobile catering app [27], and so on.

Mobile BIM is used to improve an individual's work and enhance the organization's work, such as BIM. As a result, the variables associated with an organization must be reviewed to evaluate mobile BIM. Accordingly, we adopted TAM2 to evaluate social factors and the IS success model to evaluate the success of the information system. The other theories were excluded because the model of innovation resistance diffusion theory contained many variables that have similar meaning with the TAM and IS success models.

The TAM originates from the theory of reasoned action (TRA). It uses the constructs "perceived usefulness" and "perceived ease of use," which are the individual hypotheses that are affected by external variables. The TAM is used to indicate the relationships between external variables that affect users' acceptance of technology and the factors that affect their actual behaviors. This model assumes the relationships among the external variables, perceived usefulness, and perceived ease of use (see Figure 1). However, because the TAM is founded on the individual hypothesis, it has a limitation in that the effects of social influence are ignored.

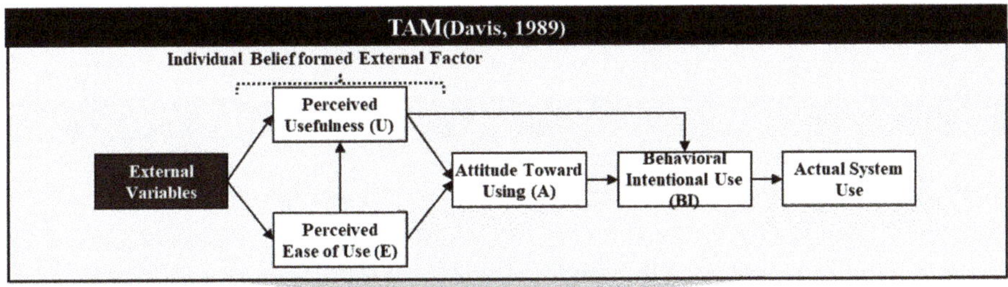

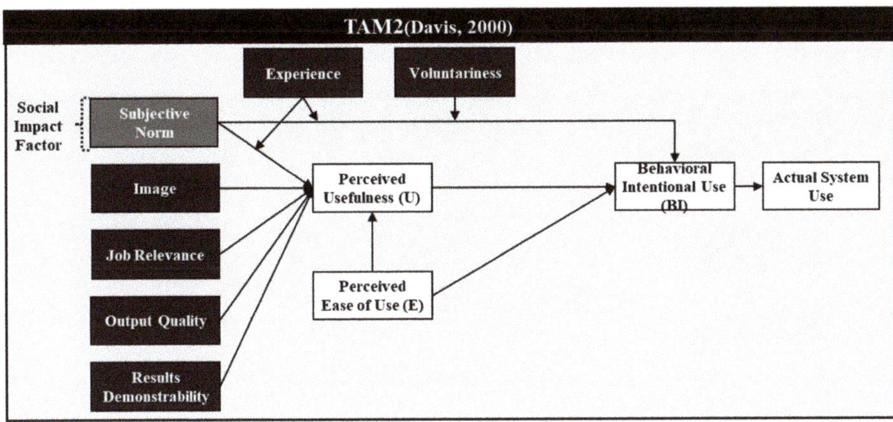

Figure 1. Davis' Technology Acceptance Model (TAM) (1998, 2000).

To improve this, Venkatesh and Davis [28] suggested the TAM2 (or ETAM) as an extension of the original TAM. TAM2 is included as external variables such as subjective norms, which are related to social influence [23]. The TAM2's external variables can be distinguished by both the social influence process and an identical tool process. The social influence process includes subjective norms, voluntariness, and image, whereas the identical tool process includes job relevance, output quality, and result demonstrability [29]. More recently, the TAM3 has been proposed, suggesting additional external variables that may affect perceived usefulness [30].

Also, DeLone and McLean proposed the IS success model [31,32] to assess the successful use of information systems. They proposed the original IS success model in 1992, which initially adopted system quality and information quality as the independent variables and expressed the impact as individual impact and organizational impact in four phases. Many researchers have expanded this model and the details; therefore, DeLone and McLean proposed an advanced IS success model in 2003 (see Figure 2).

In the advanced IS success model, service quality, which was perceived as a lower variable, was added as an independent variable, and the model was organized into three phases by integrating the individual and organizational impacts into net effects. The expanded IS success model is divided into six areas: system quality, information quality, service quality, use, user satisfaction, and net benefit.

Mobile BIM has the characteristics of BIM as an information system used for the benefit of an organization to enhance collaborative performance and to improve the sharing of information organically. Therefore, the research theory was established by referring to the expanded TAM2, including the social impact variables and the IS success model, which presented the success factors of the information management system. If essential factors that affect technology acceptance can be controlled, it is expected that the active adoption of such technology may be possible. Further, it is expected that these results will assist mobile BIM developments that can more easily accommodate users by reflecting factors that affect their acceptance of technology.

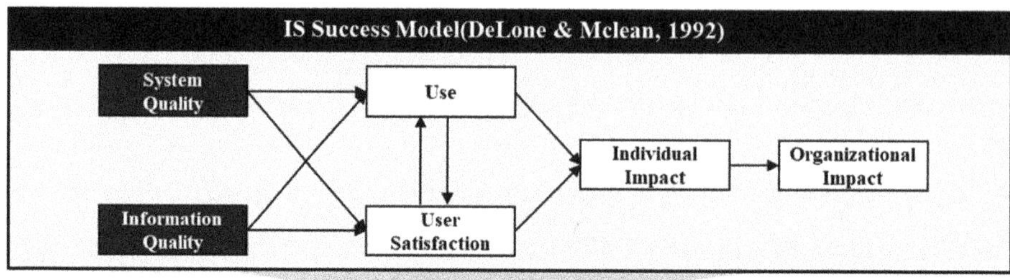

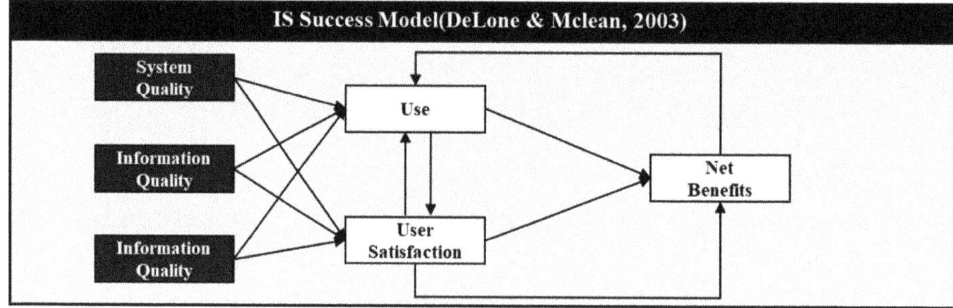

Figure 2. IS success model (1992, 2003).

3. Exploring the Key Factors for Mobile BIM Acceptance

3.1. Key Factors for Mobile BIM Acceptance

In this section, the exploratory factor analysis results are presented to establish a hypothesis for the mobile BIM acceptance model based on the survey results. The survey items comprise external variables that are expected to be significant to mobile BIM acceptance by referring to the TAM2 and IS success models, as well as previous studies related to the mobile TAM or BIM TAM. This study reviewed the relevant theories and literature reviews written in Korean and English related to the BIM TAM and mobile TAM that were published up until 2018 from around the world. Thus, the survey items are expected to be significant to construction practitioners worldwide. The representative literature is summarized in Table 2.

Literature reviews on the BIM TAM or mobile TAM aim to improve the users' skills and increase the amount of information that is targeted. First, mobile capability is based on constant internet access, ease of communication, and convenience of portability. The literature reviews related to mobile TAM adopted the quality and usability of application functions as external variables and verified the relationships between variables by adopting the perceived usefulness, usability, and use as the internal variables. Next, BIM improves individual convenience and organizational convenience using visual objects that contain information. The literature reviews related to the BIM TAM adopted external variables, not only on the quality of the system, but also on the organizational factors. Therefore, the external variables adopted the information technology characteristics in literature reviews.

Factors derived from literature reviews are summarized as follows: First, technology quality comprises key factors related to the performance of the tool that is being used, including system quality, information quality, service quality, and cost. Second, behavior control comprises key factors related to the environment in which the tool is expected to be used, including internal support, external support, internal pressure, and external pressure. Third, personal competency comprises key factors related to the individual's psychological state toward the tool that is to be used, including self-efficiency and personal innovativeness. Fourth, organization competency comprises key factors related to the group psychological state toward the tool that is to be used, including collective efficiency and organization innovativeness.

Table 2. Literature Reviews on key factor for Mobile-BIM Acceptance.

Category	Author	Details	Key Factors	Countries
Mobile	[20]	Analyze the impact of mobile application characteristics on user satisfaction.	User usability, information quality, stability of security	Korea
Mobile	[18]	Presents factors that affect mobile application usage. Analyze the effects of factors using statistical analysis.	Utility, usability contents, entertainment, cost	USA
Mobile	[21]	Suggest factors that affect mobile application acceptance. Analyze the impact relationship of factors.	Usability, efficacy, innovativeness, security	Korea
Mobile	[11]	Analysis of key factors that affect the mobile application acceptance and present the findings.	Performance expectancy, effort expectancy, social influence, entertainment motivation, social utility motivation, communication motivation	Korea
Mobile	[33]	Present the acceptance model for a mobile service.	Trust, innovativeness, relationship drivers, functionality	USA
BIM	[34]	Analyze the major factors of BIM software selection to improve the use of BIM.	Usability, functionality, business, experience	USA
BIM	[15]	Analyze factors affecting BIM acceptance and the impact relationship of factors.	Technology quality, behavior control, personal competency, organizational competency, cost	Korea
BIM	[35]	Analyze the key factors in BIM acceptance from the point of view of CMs. Present the strategy of using BIM.	Social impact, personal impact, business impact	USA
BIM	[36]	Analyze critical success factors that affect the acceptance of BIM technology and present the findings.	Human-related factors, industry-related factors, project-related factors, policy-related factors, pesource-related factors	Korea

3.2. Exploratory Factor Analysis of the Key Factors

The survey conducted as part of this study collected 111 sets of responses from stakeholders in the construction industry (i.e., designers, contractors, CMs, engineers, and BIM contractors). A factor analysis was performed to remove external variables that did not significantly affect the acceptance of mobile BIM and its separate assessment items that had a significant impact on similar factors. First, a PCA was performed for factor extraction, and the varimax method with Kaiser normalization was performed as a factor rotation method. As a result, 10 items were eliminated from the 39 items that impeded the validity of the analysis. These excluded items are related to screening composition of system quality, items related to cost, and items related to the education of internal support systems. Second, the remaining 29 items were analyzed using factor analysis (see Table 3).

First, the KMO measure provides the statistical figure to which relationship between variables is well explained by the other variables, and it indicates the fitness of the variables selected by the factor analysis. In general, a KMO measure of 0.90 or higher is considered highly appropriate, and a KMO measure of less than 0.50 is considered unacceptable. The KMO measure of this study was 0.870. This implies that the 29 selected variables were appropriately selected. Next, Bartlett's test of sphericity checks whether a factor analysis model is appropriate by determining the model's goodness of fit with the p-value. Bartlett's test of sphericity showed a p-value less than 0.05; thus, the factor analysis model was assumed to be appropriate for the study. Meanwhile, the cumulative dispersion value of the data was 77.746%, and the results of factor analysis had a high description.

Table 3. Results of the factor analysis.

Component	Items	Factor Loading	Eigenvalue	Cumulative %	Cronbach's α
Tool Quality	SVQ2	0.885	10.906	37.606	0.958
	IQ3	0.868			
	SVQ4	0.863			
	SVQ3	0.854			
	IQ2	0.844			
	SVQ5	0.819			
	IQ1	0.807			
	IQ4	0.785			
	SVQ1	0.772			
	SYSQ2	0.708			
Behavior Control	ES1	0.893	4.714	53.862	0.946
	EP1	0.871			
	EP2	0.861			
	EP3	0.857			
	IP1	0.808			
	IS3	0.805			
	IP2	0.796			
	ES2	0.742			
Personal Efficacy	PE2	0.922	3.218	64.957	0.897
	PE4	0.887			
	PE1	0.869			
	PE3	0.839			
	PI2	0.561			
Organization Innovativeness	OI3	0.905	2.654	74.109	0.907
	OI2	0.858			
	OI1	0.806			
Collective Efficacy	CE4	0.790	1.055	77.746	0.932
	CE3	0.764			
	CE1	0.696			
Kaiser–Meyer–Olkin Measure of Sampling Adequacy					0.870
Bartlett's Test of Sphericity			Approx. Chi-Square		3172.681
			df.		406
			Sig.		0.000

SYSQ: System Quality, IQ: Information Quality, SVQ: Service Quality, IS: Internal Support, IP: Internal Pressure, ES: External Support, EP: External Pressure, PI: Personal Innovativeness, PE: Personal//; 07uiii efficiency, OI: Organization Innovativeness, CE: Collective efficiency

The first factor included 10 items, the second factor included 8 items, the third factor included 5 items, and the fourth and fifth factors included 3 items each. The first factor was named technology quality, the second was named behavior control, the third was named personal efficacy, the fourth was named organization innovativeness, and the fifth was named collective efficacy. These factors were used as the external variables in the TAM model. The details of each factor are described in Section 4.2: Research Hypotheses.

4. Proposed Mobile BIM Acceptance Model

4.1. Overview of the Proposed Model

The relationships among the key factors presented in the literature review and factor analysis were established in the form of a hypothesis to present a model for mobile BIM acceptance. First, the external variables adopted were tool quality, personal efficacy, behavior control, organization innovativeness, and collective efficacy, as derived from the results of the factor analysis. Second, the internal variables adopted were the ease of use and usefulness and consensus on appropriation according to the IS success model and the characteristics of BIM. Finally, the intentions adopted were the individual intention and organization intention. Nine hypotheses on the relationships among the variables were established to suggest factors affecting mobile BIM acceptance (as shown in Table 4 and Figure 3).

Table 4. Research hypotheses.

Hypotheses		Definition
H1	a	Tool quality will positively effect ease of use.
	b	Tool quality will positively effect usefulness.
H2	a	Personal efficacy will positively effect ease of use.
	b	Personal efficacy will positively effect usefulness.
H3	a	Behavior control will positively effect ease of use.
	b	Behavior control will positively effect usefulness.
H4	a	Organization innovativeness will positively effect ease of use.
	b	Organization innovativeness will positively effect usefulness.
H5	a	Collective efficacy will positively effect ease of use.
	b	Collective efficacy will positively effect usefulness.
H6	a	Ease of use will positively effect usefulness.
	b	Usefulness will positively effect ease of use.
H7	a	Ease of use will positively effect consensus on appropriation.
	b	Ease of use will positively effect individual intention.
	c	Ease of use will positively effect organizational intention.
H8	a	Usefulness will positively effect consensus on appropriation.
	b	Usefulness will positively effect individual intention.
	c	Usefulness will positively effect organizational intention.
H9	a	Consensus on appropriation will positively effect individual intention.
	b	Consensus on appropriation will positively effect organizational intention.

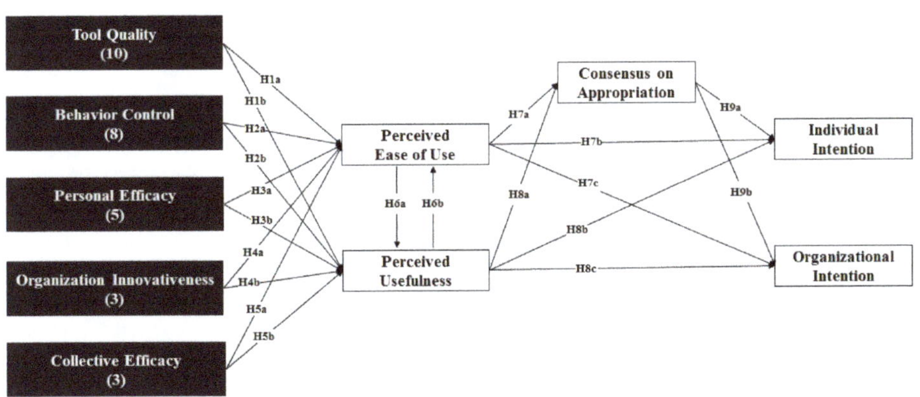

Figure 3. Proposed model for the mobile BIM acceptance model.

4.2. Research Hypotheses

4.2.1. External Variables for Mobile BIM Acceptance

The external variables presented in the literature review and factor analysis are as follows (shown in Table 5).

Table 5. Assessment items of external variables for Mobile-BIM acceptance.

Variables	Assessment Items
Tool Quality	It is easy to input and output data with a mobile BIM.
	Using a mobile BIM improves the accessibility of information.
	The information in a mobile BIM is accurate and detailed.
	A mobile BIM provides sufficient information for the task.
	The information in a mobile BIM is available throughout the life cycle.
	A mobile BIM is easy to learn how to use.
	A mobile BIM is quick to respond to questions about how to use it.
	It is easy to use a mobile BIM manual.
	A mobile BIM is fast at reflecting user requirements.
	A mobile BIM enables continuous updating and After Service(A/S).
Personal Efficacy	I do not have any resistance toward using a mobile BIM.
	I can easily get used to using a mobile BIM.
	I understand the benefits of using a mobile BIM.
	I am confident that I will learn (manual, training, etc.) how to use a mobile BIM.
	I have the technical ability to use new information technology.
Behavior Control	Our organization offers incentives toward using a mobile BIM.
	Our organization enforces the use of a mobile BIM as a policy.
	Our boss or colleague requires the use of a mobile BIM.
	A mobile BIM enables economic benefits from the industry or the government.
	I can get appropriate training for the use of a mobile BIM from industry or government.
	Our organization is required to use a mobile BIM as a project delivery or contract method.
	Our organization requires the use of a mobile BIM in your relationship with your partner.
	Our organization is required to use a mobile BIM to meet the requirements of the contract.
Organization Innovativeness	Our organization has no psychological resistance to using new information technology.
	Our organization has technical capabilities for the use of new information technology.
	Our organization is active in the use of new information technology.
Collective Efficacy	Our organization does not have any resistance to the use of a mobile BIM.
	Our organization understands the benefits of using a mobile BIM.
	Our organization is confident in learning (mechanics, training, etc.) about how to use a mobile BIM.

Tool quality is defined as the quality of the mobile BIM application. In this study, "technology" refers to the mobile BIM application, or the software running on mobile. Assessment items related to technology quality are system quality, information quality, and service quality.

System quality is an assessment item from the IS success model [30]. It has been presented as an external variable by many researchers [18,20,21,33,34,37]. The IS success model defines system quality as the quality of the measures of the information processing system itself [30], and the performance and convenience of the hardware and software [32]. Moreover, system quality has a similar meaning as job relevance in TAM2 and usability in the literature review.

Information quality is also an assessment item that is proposed in the IS success model [31]. It has been presented as an external variable by many researchers [15,35]. The IS success model defines information quality as measures of the information system output [31] and the quality of the information provided by the information system [30]. Accordingly, information quality is defined as the quality of information related to work and the quality of the system environment in which the user obtains that information.

Service quality is a variable that is added as an independent variable by DeLone & McLean in 2003. Service quality is defined as the quality of the service that can be obtained from the work provided by the system and the relevant departments. Service quality is also presented as an external variable by various researchers [18,20].

Personal efficacy is defined as the psychological state of the individual who utilizes the mobile BIM application for their work. The assessment items are personal efficiency and personal innovativeness.

Personal efficiency is defined as a belief that an individual will be able to achieve successful results through action. Personal efficiency was adopted as an external variable to evaluate the use of technology [15,35].

Personal innovativeness is defined as an individual's active psychological attitude or ability to introduce new information technology. Personal innovativeness was adopted as an external variable to evaluate the use of technology [15,35].

Behavior control is defined as an external variable related to the environment in which the tool is used. Behavior control factors are divided into internal assessment items and external assessment items. The internal assessment items refer to those inside the organization, whereas the external assessment items refer to those outside the organization. The external variables related to the behavior control include pressure and support. First, pressure is defined as compulsory factors in the environment. It is similar to the subjective norm in TAM2. Subjective norm is the awareness of how people who are important to themselves will feel about doing certain things. It is adopted as an external variable in various literature reviews [21,28,33,35,36,38,39]. Next, support is defined as factors that help users to use their skills more efficiently.

Internal support is defined as the assistance within an organization from the organization's policies, bosses, or colleagues in selecting an action. For example, providing software in an organization, providing free education, and providing incentives.

Internal pressure is defined as a pressure that the organization gives to an individual, such as the organization's policies, bosses, or colleagues, in selecting an action. For example, compulsory policies in an organization, pressure from superiors or colleagues, and so on.

External support is defined as assistance from outside an organization through policies, systems, etc. in selecting an action. For example, there are government economic benefits, free government education, etc.

External pressure is defined as the pressure from outside the organization from the framework, contract, and relationship in selecting an action. For example, there is forced pressure from the owner, compulsory selection under the contract, and so on.

Organization innovativeness is defined as the organization's active psychological attitude or capacity to introduce new information technology. Organization innovativeness adopts an external variable to assess the use of technology in an organization as a factor from the characteristics of the BIM and IS success model [15,31,32].

Collective efficacy is defined as the belief that an organization will be able to achieve successful results through action. Collective efficacy also adopts an external variable to assess the use of technology in an organization from the characteristics of the BIM and IS success model [15,31,32].

4.2.2. Internal Variables for Mobile BIM Acceptance

In the mobile TAM, consensus on appropriation, perceived ease of use, and perceived usefulness were adopted as internal variables (shown in Table 6).

Table 6. Assessment items of internal variables for mobile BIM acceptance.

Variables	Assessment Items
Consensus on Appropriation	The organization members show conformity on the tasks that apply a mobile BIM, which is set by the organization.
	The organization members show conformity regarding how to apply a mobile BIM, such as work guidelines and rules, which are set by the organization.
Perceived Ease of Use	It is easy to learn how to cooperate with a mobile-BIM.
	It is easy to exchange information with a mobile BIM.
	The guideline for collaboration with a mobile BIM is defined such that it can be followed quickly.
Perceived Usefulness	A mobile BIM improves to interoperability among stakeholders.
	A mobile-BIM allows for comprehensive management of life-cycle information.
	A mobile BIM reduces decision-making time.
	A mobile BIM can expand the utilization range of collaboration with other organizations.
	A mobile BIM reduces task-handling time.
	A mobile BIM improves task accuracy.
	A mobile BIM allows for easy collaboration with other organizations.

Consensus on appropriation is defined as a consensus among members of the organization on information technology. Consensus on appropriation was chosen as an internal variable in the TAM for evaluating the BIM acceptance [15]. Mobile BIM was also selected as an internal variable because it also has characteristics similar to the BIM.

Perceived ease of use defines the extent to which it is believed that using an information technology system does not require much effort [40]. Perceived ease of use is presented as an internal variable in various literature reviews [18,21,33,35,39,41–43].

Perceived usefulness is defined as believing that the use of information technology systems will improve one's performance [39]. Perceived usefulness is presented as an internal variable in various literature reviews [18,21,33,35,39,41–43].

4.2.3. Intention toward Mobile BIM Acceptance

Intention is defined the extent of intended or planned use of an information technology system [40]. Intention is divided into individual intention and organizational intention according to the target. It is adopted as an internal variable in various literature reviews [18,21,33,35,39,41–43]. A variable similar to intention, net benefit, is employed by the IS success model [15,31,39,42] (shown in Table 7).

Table 7. Assessment items of individual and organizational intention to accept a mobile BIM.

Variables	Assessment Items
Individual Intention	I have the intention to use a mobile BIM for my task.
	I have the intention to recommend a mobile BIM to others.
	I have the intention to take the time to learn how to use a mobile BIM.
Organizational Intention	My organization has the intention to encourage members toward using a mobile-BIM.
	My organization has the intention to be active in using a mobile BIM for the task.
	My organization has the intention to recommend a mobile BIM to other organizations that have a collaborative relationship with our organization.

Individual intention is defined as an individual's intention or use plan to use an information technology system. If an internal variable has a significant effect on the individual intention, the individual's acceptance of the technology is also deemed significant.

Organizational intention is defined as an organization's intention or plan to use an information technology system. If the internal variables have a significant effect on the organizational intention, the organization's acceptance of technology is also deemed significant.

5. Model Validation

5.1. Validation of the Proposed Model

In this chapter, the proposed model is validated as a structural equation model under key factors, which were presented as an exploratory factor analysis. To validate our measurement model, we analyzed the assessments of convergent and discriminant validity. This study used the following model fit measures to decide the model's goodness-of-fit: the ratio of χ^2 to df, RMR, PGFI, the TLI, CFI, and RMSEA. The model-fit indices of the proposed model and the acceptance level were compared.

The initial measurement model did not show a good fit compared with the recommended values of χ^2/df = 2.773, RMR = 0.312, PGFI = 0.365, TLI = 0.617, CFI = 0.635, and RMSEA = 0.131. Therefore, organization innovativeness and collective efficacy were removed. This was because the relationship between the latent variables in the initial measurement model was not significant, and the standardized factorial load was less than 0.5. In addition, we re-analyzed behavior control, which was smaller than the p-value and was excluded from some hypotheses, by dividing it into internal support, internal pressure, external support, and external pressure. The analysis results show that internal support and internal pressure were significant to the model. Further, consensus on appropriation and organizational intention were also excluded as factors that impeded the model's suitability. Therefore, the final measurement model had the values of χ^2/df = 2.08, RMR = 0.45, PGFI = 0.553, TLI = 0.854, CFI = 0.867, and RMSEA = 0.102 (shown in Table 8).

Table 8. Fit indices for research model.

Fit Indices	Recommended Value	Measurement Model	Structural Model
χ^2/df	≤3.0	2.08	2.02
RMR	≤0.1	0.145	0.145
PGFI	≥0.5	0.553	0.504
TLI	≥0.9	0.854	0.73
CFI	≥0.9	0.867	0.75
RMSEA	≤0.1	0.102	0.095

The SEM analysis showed that tool quality, personal efficacy, and internal control had a positive effect on the ease of use and usefulness. In other words, if the tool quality, personal efficacy, and internal control increased, the ease of use and usefulness also increased. Further, usefulness had a significant positive effect on individual intention; the individual intention increased with usefulness. The CFA results showed that the path from tool quality, personal efficacy, and behavior control among internal support to individual intention was significant (shown in Table 9).

The discriminant validity test was not satisfied between the factors of ease of use and technology quality (Table 10). However, technology quality was not removed because it has been referred to as a significant factor in literature reviews.

Table 9. Results of CFA.

Latent Constructs	Observed Indicators	Factor Loading	*t*-Value	Composite Reliability	AVE
Tool Quality (TQ)	TQ 1	0.837	-	0.906	0.493
	TQ 2	0.883	11.795		
	TQ 3	0.897	12.117		
	TQ 4	0.905	12.326		
	TQ 5	0.813	10.253		
	TQ 6	0.788	9.767		
	TQ 7	0.863	11.315		
	TQ 8	0.833	10.664		
	TQ 9	0.786	9.726		
Personal Efficacy (PE)	PE 1	0.723	8.593	0.865	0.571
	PE 2	0.472	-		
	PE 3	0.855	5.014		
	PE 4	0.83	4.962		
	PE 5	0.936	5.16		
	PE 6	0.904	5.108		
Behavior Control (BC)	BC 1	0.949	-	0.856	0.667
	BC 2	0.925	14.711		
	BC 3	0.750	9.994		
Perceived Ease of Use (EOU)	EOU 1	0.898	-	0.675	0.409
	EOU 2	0.887	13.068		
	EOU 3	0.885	13.004		
Perceived Usefulness (U)	U 1	0.919	-	0.904	0.535
	U 2	0.938	17.488		
	U 3	0.888	14.768		
	U 4	0.882	14.527		
	U 5	0.899	15.297		
	U 6	0.722	9.495		
	U 7	0.895	15.113		
Individual Intention to Accept Mobile-BIM (IIA)	IIA 1	0.971	-	0.800	0.572
	IIA 2	0.848	13.025		
	IIA 3	0.805	11.648		

* *t*-Value for these parameters were not available because they were fixed for scaling purposes.

Table 10. Results of discriminant validity test.

Observed Indicators		r2		AVE	Discriminant Validity
Tool Quality (TQ)	PE	0.094		0.571	Acceptable
	BC	0.089		0.667	Acceptable
	EOU	0.590	0.493	0.409	Unacceptable
	U	0.543		0.535	Unacceptable
	IIA	0.213		0.572	Acceptable
Personal Efficacy (PE)	BC	0.001		0.667	Acceptable
	EOU	0.200		0.409	Acceptable
	U	0.138	0.571	0.535	Acceptable
	IIA	0.348		0.572	Acceptable
Behavior Control (BC)	EOU	0.0847		0.409	Acceptable
	U	0.117	0.667	0.535	Acceptable
	IIA	0.024		0.572	Acceptable
Perceived Ease of Use (EOU)	U	0.605	0.409	0.535	Unacceptable
	IIA	0.272		0.572	Acceptable
Perceived Usefulness (U)	IIA	0.430	0.535	0.572	Acceptable

As expected, hypothesis H1a and H2a were supported ($\gamma = 0.666$ and $\gamma = 0.42$, respectively). Increased tool quality and personal efficacy were associated with increased perceived ease of use. Behavior control was not very supported ($\gamma = 0.123$). The tool quality and personal efficacy had an effect on perceived usefulness. H1b and H2b were supported ($\gamma = 0.424$ and $\gamma = 0.45$, respectively). Perceived ease of use appeared to be a significant determinant of perceived usefulness; H6a was supported ($\beta = 0.405$). In addition, perceived usefulness had a significant effect on individual intention in accepting a mobile BIM; H8b was supported ($\beta = 0.631$). Therefore, tool quality and personal efficacy exhibited a relatively strong influence on mobile BIM acceptance, and behavior control had relatively weak influence. H1a, H2a, and H3a explained 58.6% of the variance in the perceived ease of use. H1b, H2b, and H6a explained 71.6% of the variance in perceived usefulness. H8b explained 43.7% of the variance in individual intention to accept a mobile BIM. The final proposed model is shown in Figure 4, and the direct, indirect, and total effects of each construct are summarized in Table 11.

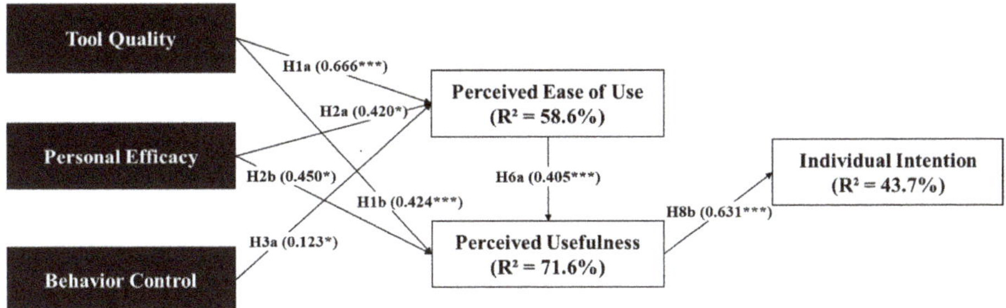

Figure 4. The validation results of the proposed model. $^{+}\,p < 0.1$, $^{*}\,p < 0.05$, $^{**}\,p < 0.01$, $^{***}\,p < 0.001$.

Table 11. Direct, indirect and total effects.

	Total Effects			Direct Effects			Indirect Effects		
	EOU	U	IIA	EOU	U	IIA	EOU	U	IIA
TQ	0.666	0.693	0.449	0.666	0.424	0	0	0.269	0.449
PE	0.42	0.619	0.398	0.42	0.45	0	0	0.17	0.398
BC	0.123	0.071	0.047	0.123	0.021	0	0	0.05	0.047
EOU	0	0.405	0.272	0	0.405	0.017	0	0	0.255
U	0	0	0.631	0	0	0.631	0	0	0
IIA	0	0	0	0	0	0	0	0	0

5.2. Findings and Discussion

As a result of the analysis, the external factors that significantly affected the mobile BIM acceptance were found to be technology quality, self-efficiency, and behavior control.

Tool Quality: Tool quality is the software performance of the mobile BIM application, suggesting that the quality of the tools that the users need for the mobile BIM applications must be ensured. As a result, it is expected that the application could contribute to the mobile BIM acceptance if the application reflects the assessment items of the technology quality analyzed in this study as being significant to the mobile BIM acceptance. First, the mobile BIM should facilitate the input and output of data and provide reasonable access to information. This is consistent with the mobile nature of having easy access to information regardless of the location. Moreover, a variety of ways to efficiently use the information on mobile hardware with small screens, such as a user-centered interface design, should be explored. Second, the information that can be obtained from the tool should be accurate and detailed. The information obtained should be available throughout the life cycle. This conforms to the characteristics of the BIM that detailed information generated during the project life cycle can be aggregated based on visual objects. Accordingly, various methods should be explored to effectively manage various information, such as BIM information, construction project information, and web cloud utilization. Third, it should be easy to understand the instructions and learn how to use. Replies to questions about how to use them should be answered quickly. Accordingly, there is a need to provide manuals or tutorials to help the users learn how to use the system. Moreover, there should be a window for exchanging questions and answers on how to use it. Finally, continuous updating of tools and A/S should be possible because users require speed. This suggests the need for a steady system performance improvement and suggests that a community should be provided for continuous feedback from the implementer on the system performance.

Personal Efficacy: Personal efficiency is the psychological state of an individual using the mobile BIM application in his or her work. It suggests that the mobile BIM induces a belief that the use of a mobile BIM will produce successful results. As a result, measures should be taken to appropriately promote various benefits that individuals can enjoy by using the mobile BIM.

Behavior Control: Behavior control is the environment in which tools are used. It suggests that mobile BIM applications require internal support and internal pressure, which are appropriate to the internal structure of the organization. First, the mobile BIM can be more easily used when receiving ancillary help from an organization's policies, members, etc. This suggests that appropriate policies should be established in the organization for the use of a mobile BIM, such as free educational opportunities and free software. Second, a mobile BIM can be more easily used when it is under appropriate pressure from the organization's policies, members, etc. This suggests that organizations, such as providing incentives, should have appropriate policies to encourage individuals to use tools, as well as provide assistance in using the mobile BIM.

Further, unlike the BIM TAM [15–17], it has been found that the factors associated with the organization do not affect the acceptance of the technology. A mobile BIM is a tool that organizations already use as part of their efforts to make BIM more comfortable to use. Accordingly, the acceptance of the mobile BIM technologies focuses on enhancing the ease of use for individuals, unlike the decision-making that determines the use of the BIM tools in groups without BIM. Thus, unlike BIM, areas related to the organization are interpreted as not affecting technology acceptance.

6. Conclusions

Mobile devices in construction help to operate an information system regardless of the location, and it is introduces many changes to information approach and to the working of project stakeholders. Besides, mobile BIM draws attention to tools that can contribute to enhancing collaborative performance through BIM by enabling the exchange of information organically. Accordingly, BIM programs supported by the mobile environment are continuously being released, and it is found that the mobile BIM applied to some sites generates actual benefits in carrying out projects.

Although various efforts are being made to use a mobile BIM, it is found that it is underutilized due to the lack of experience in using a mobile BIM, the high age of expected users, and inconsistency between the actual working processes and BIM processes. Moreover, some studies related to the use of mobile BIM have differing views on the same cause, which may be interpreted as a lack of analysis of factors that affect the actual users' acceptance and utilization of the mobile BIM.

Accordingly, this study analyzed the factors and their associations with the factors that affect the acceptance of a mobile BIM by construction practitioners as part of its efforts to activate the use of the mobile BIM. A questionnaire was organized to define the characteristics of the mobile and BIM through existing literature considerations and to analyze the external factors affecting the mobile BIM acceptance. The survey collected 111 copies from construction practitioners (designers, contractors, CMs, engineering, and BIM contractors). The results of the survey were broken down into factors and path analysis was undertaken using SPSS 21.0 and Amos 21.0, and the factors that significantly affected mobile BIM acceptance were expressed in models through the structured model by checking the concentration and determination feasibility.

As a result of the analysis, external factors that significantly affected the mobile BIM acceptance were found to be tool quality, personal efficacy, and behavior control. First, tool quality is the software performance of a mobile BIM application, suggesting that users need the quality of the tools they need for mobile BIM. Second, personal efficacy suggests that measures should be taken to appropriately promote the various benefits of using a mobile BIM. Third, behavior control suggests that there must be adequate support and pressure within the organization to use a mobile BIM.

It is expected that this study will contribute to the revitalization of the technology along with the acceptance of the technology, provided that significant factors affecting the acceptance of the technology are appropriately managed. Moreover, it is expected that a mobile BIM will be developed to allow users to develop a variety of applications that are more easily acceptable to them.

This study is limited in that it was difficult to generalize these results for all countries because our survey subject of this study is construction practitioners in Korea. If future studies conduct surveys on USA or Europe where the use level of mobile BIM is high, it will be able to present new findings due to comparing the results.

Author Contributions: Conceptualization, S.-H.H., S.-K.L., and J.-H.Y.; methodology, S.-H.H., S.-K.L., and J.-H.Y.; software, S.-K.L.; validation, S.-H.H., S.-K.L., I.-H.K. and J.-H.Y.; formal analysis, S.-H.H. and S.-K.L.; investigation, S.-H.H. and S.-K.L.; resources, S.-H.H., I.-H.K. and J.-H.Y.; data curation, S.-H.H. and S.-K.L.; writing—original draft preparation, S.-H.H.; writing—review and editing, S.-K.L., and J.-H.Y.; visualization, S.-H.H. and S.-K.L.; supervision, J.-H.Y.; project administration, J.-H.Y.; funding acquisition, I.-H.K. and J.-H.Y.

Funding: This research was supported by a grant (19AUDP-B127891-03) from the Architecture & Urban Development Research Program funded by the Ministry of Land, Infrastructure and Transport of the Korean government.

Conflicts of Interest: The authors declare no conflict of interest.

References

1. Alan, R.; Alan, H.; Mustafa, A.; Roger, W. Exploring how information exchanges can be enhanced through Cloud BIM. *Autom. Constr.* **2012**, *24*, 175–183.
2. Kim, B.K. A Study on the Developing Method of the BIM (Building Information Modeling) Software Based on Cloud Computing Environment, World Academy of Science. *Eng. Technol. Int. J. Struct. Constr. Eng.* **2012**, *6*, 1131–1135.
3. Sebastjan, M.; Žiga, T.; Matevž, D. Component based engineering of a mobile BIM-based augmented reality system. *Autom. Constr.* **2014**, *42*, 1–12.
4. Yeh, K.C.; Tsai, M.H.; Kang, S.C. On-Site Building Information Retrieval by Using Projection-Based Augmented Reality. *J. Comput. Civ. Eng.* **2012**, *26*, 342–355. [CrossRef]
5. Park, J.W.; Cho, Y.K. Use of a Mobile BIM Application Integrated with Asset Tracking Technology over a Cloud, Invited Speech. In Proceedings of the CRIOCM 21st International Conference on "Advancement of Construction Management and Real Estate", Hong Kong, China, 14–17 December 2016.
6. Brad, H.; Dave, M. *BIM and Construction Management: Proven Tools, Methods, and Workflows*, 2nd ed.; Wiley: Hoboken, NJ, USA, 2015.
7. Rebekka, V.; Julian, S.; Frank, S. Building Information Modeling (BIM) for existing Building-Literature review and future needs. *Autom. Constr.* **2014**, *38*, 109–127.
8. Hong, S.H.; Lee, S.K.; Yu, J.H. A Study on the Direction of Developing the Mobile-based BIM Tool in Design Phase. *J. Korea Facil. Manag. Assoc.* **2018**, *13*, 5–13.
9. Ministry of Land News. Available online: http://www.ikld.kr/news/articleView.html?idxno=70708, (accessed on 16 March 2017).
10. Song, K.M.; John, Y.S. A Comparative Analysis on BIM usage between Korean & USA Architectural Firms regarding Effective Design Process. *Korea Digit. Des. Soc.* **2012**, *12*, 524–533.
11. Seok, K. Factors influencing intention of mobile application use. *Int. J. Mob. Commun.* **2014**, *12*, 360–379.
12. Park, I.W. Utilization and Circulation of Smart Phone Mobile Application of Travel Agencies through Expansion of the Technology Acceptance Model (TAM). Ph.D. Thesis, Kyung-Hee University, Seoul, Korea, 2012.

13. Cho, J.H. *Extended Technology Acceptance Model (ETAM) Through Acceptance of Sports Applications*; Master course; Chung-Ang University: Seoul, Korea, 2015.

14. Shin, M.A. The Effect of Mobile Shopping Application Characteristics on Continuous Use Intention: Focused on Technology Acceptance Model. Master's Thesis, Inha University, Inshenon, Korea, 2016.

15. Lee, S.K.; Yu, J.H. Key Factors Affecting BIM Acceptance in Construction. *J. Archit. Inst. KOREA Plan. Des.* **2013**, *29*, 79–86.

16. Lee, S.K.; Yu, J.H.; David, J. BIM Acceptance Model in Construction Organizations. *J. Manag. Eng.* **2015**, *31*, 04014048. [CrossRef]

17. Lee, S.K. BIM Acceptance Readiness Evaluation Model Considering Organization Culture. Ph.D. Thesis, Kwangwoon University, Seoul, Korea, 2014.

18. Kim, S.S. An Empirical Study on Users' Intention to Use Mobile Applications. Ph.D. Thesis, Soongsil University, Seoul, Korea, 2012.

19. Seo, I.R. A Study on the Causes of Mobile Application Acceptance (Focusing on the Education Category). Master's Thesis, Hong-Ik University, Seoul, Korea, 2011.

20. Chen, Z.Y. A Study of the Effect of Mobile Application Characteristics on User Satisfaction. Master's Thesis, Shilla University, Busan, Korea, 2015.

21. Kim, J.K. A Study on the Usage Intention of Category Types in the Mobile Application Based on the Technology Readiness and Acceptance Model. Ph.D. Thesis, Kong-Ju National University, Gongju, Korea, 2013.

22. Kim, S.J.; Kim, T.H.; Ok, H. A Study on Developing the System for Supporting Mobile-based Work Process in Construction Site. *Smart Media J.* **2017**, *6*, 50–57.

23. Sarah, B.; Alex, D.; Tony, T.; Chimay, A. Mobile ICT support for construction process improvement. *Autom. Constr.* **2006**, *15*, 664–676.

24. Hong, S.H.; Lee, S.K.; Yu, J.H. An Analysis of Factors Affecting the Use of Mobile-based BIM Tool. In Proceedings of the General Assembly & Spring Annual Conference of AIK, Seoul, Korea, 11–12 July 2019.

25. Ahmed, I.A.; Imran, M.; Ramayah, T.; Osama, A.; Nasser, A. Modeling digital library success using the Delone and McLean information system success model. *J. Librariansh. Inf. Sci.* **2019**, *51*, 291–306.

26. Osama, I.; Adnan, A.; Zaini, A.; Ramayah, T. Online learning usage within Yemeni higher education: The role of compatibility and task-technology fit as mediating variables in the is success model. *Comput. Educ.* **2019**, *136*, 113–129.

27. Yi-shun, W.; Timmy, H.T.; Wei-Tsong, W.; Ying-Wei, S.; Ping-Yu, C. Developing and validating a mobile catering app success model. *Int. J. Hosp. Manag.* **2019**, *77*, 19–30.

28. Venkatesh, V.; Davis, F.D. A theoretical extension of the technology acceptance model: Four longitudinal field studies. *Manag. Sci.* **2000**, *46*, 186–204. [CrossRef]

29. Song, Y.H.; Hur, W.M. Exploratory Study of Adoption and Diffusion of Premium Digital Convergence Product: Moderating Effecting of Social Value. *Korea Knowl. Manag. Soc.* **2011**, *12*, 53–76. [CrossRef]

30. Venkatesh, V.; Bala, H. TAM 3: Advancing the technology acceptance model with a focus on interventions. *Decis. Sci.* **2008**, *39*, 273–315. [CrossRef]

31. DeLone, W.H.; McLean, E.R. Information system success: The quest for the dependent variable. *Inf. Syst. Res.* **1992**, *3*, 60–95. [CrossRef]

32. DeLone, W.H.; McLean, E.R. The DeLone and McLean model of information system success: A ten-year update. *J. Manag. Inf. Syst.* **2003**, *19*, 9–30.

33. Theodora, Z.; Vaggelis, S.; Angelos, M.; Maro, V. Modeling user's acceptance of mobile services. *Electron. Commer. Res.* **2012**, *12*, 225–248.

34. Lee, C.J.; Lee, G.; Won, J.S. An Analysis of the B1M Software Selection Factor. *J. Archit. Inst. KOREA Plan. Des.* **2009**, *25*, 153–163.

35. Chung, Y.C. Factors Affecting the BIM Acceptance of Construction Managers. Master's Thesis, Sungkyunkwan University, Seoul, Korea, 2015.

36. Beliz, O.; Ugur, K. Critical Success Factors of Building Information Modeling Implementation. *J. Manag. Eng.* **2017**, *33*, 04016054.

37. Arayici, Y.; Coates, M.; Koskela, L.; Kagioglou, M.; Usher, C.; Reilly, K.O. Technology adoption in the BIM Implementation for lea architectural practice. *Autom. Constr.* **2011**, *20*, 189–195. [CrossRef]

38. Hsu, C.L.; Lin, J.C.C. What drives purchase intention for paid mobile apps? An expectation confirmation model with perceived value. *Electron. Commer. Res. Appl.* **2015**, *14*, 46–57. [CrossRef]

39. Nam, Y.W. The Determinants Affecting Net Benefit of Smartphone App Usage. Master's Thesis, Pukyoung National University, Busan, Korea, 2014.

40. Davis, F.D. Perceived usefulness, perceived ease of use, and user acceptance of information technologies. *J. MIS Q.* **1989**, *13*, 319–340. [CrossRef]

41. Ju, S.G.; Son, Y.K.; Jun, S.G. A Study on the Acceptance Decision Factors of Mobile Services: Focusing on Expanding Technology Acceptance Model by Intrinsic Motivation. *Manag. Inf. Syst. J.* **2011**, *30*, 117–146.

42. Lee, J.W.; Seo, D.Y.; Park, H.G.; Sohn, Y.K. A Study on the Acceptance Decision Factor of the Mobile WCDMA service. *Korean J. Advert.* **2009**, *20*, 231–250.

43. Park, C.H. Assessment of BIM Acceptance Degree of Korean AEC Participants. Master's Thesis, Sungkyunkwan University, Seoul, Korea, 2014.

MDPI

St. Alban-Anlage 66

4052 Basel

Switzerland

Tel. +41 61 683 77 34

Fax +41 61 302 89 18

www.mdpi.com

Applied Sciences Editorial Office

E-mail: applsci@mdpi.com

www.mdpi.com/journal/applsci